Lecture Notes in Computer Science 5067

Commenced Publication in 1973
Founding and Former Series Editors:
Gerhard Goos, Juris Hartmanis, and Jan van Leeuwen

W0192983

Sotiris E. Nikoletseas   Bogdan S. Chlebus
David B. Johnson   Bhaskar Krishnamachari (Eds.)

Distributed Computing in Sensor Systems

4th IEEE International Conference, DCOSS 2008
Santorini Island, Greece, June 11-14, 2008
Proceedings

 Springer

Volume Editors

Sotiris E. Nikoletseas
CTI and University of Patras
Computer Engineering and Informatics Department
Patras, Greece
E-mail: nikole@cti.gr

Bogdan S. Chlebus
University of Colorado at Denver
Department of Computer Science and Engineering
Denver CO 80217, USA
E-mail: Bogdan.Chlebus@cudenver.edu

David B. Johnson
Rice University
Department of Computer Science
Houston, TX 77005-1892, USA
E-mail: dbj@cs.rice.edu

Bhaskar Krishnamachari
University of Southern California
Department of Electrical Engineering
Los Angeles, CA 90089, USA
E-mail: bkrishna@usc.edu

Library of Congress Control Number: 2008928150

CR Subject Classification (1998): C.2.4, C.2, D.4.4, E.1, F.2.2, G.2.2, H.4

LNCS Sublibrary: SL 5 – Computer Communication Networks
and Telecommunications

ISSN   0302-9743
ISBN-10   3-540-69169-3 Springer Berlin Heidelberg New York
ISBN-13   978-3-540-69169-3 Springer Berlin Heidelberg New York

Springer is a part of Springer Science+Business Media

springer.com

© Springer-Verlag Berlin Heidelberg 2008

Typesetting: Camera-ready by author, data conversion by Scientific Publishing Services, Chennai, India
Printed on acid-free paper   SPIN: 12275786   06/3180   5 4 3 2 1 0

Message from the General Co-chairs

We are pleased to welcome you to Santorini Island, Greece, for DCOSS 2008, the IEEE International Conference on Distributed Computing in Sensor Systems, the fourth event in this series of annual conferences. The DCOSS meetings cover the key aspects of distributed computing in sensor systems, such as high-level abstractions, computational models, systematic design methodologies, algorithms, tools, and applications. This meeting would not be possible without the tireless efforts of many volunteers. We are indebted to the DCOSS 2008 Program Chair, Sotiris Nikoletseas, for overseeing the review process, composing the technical program, and making the local arrangements. We appreciate his leadership in putting together a strong and diverse Program Committee, whose members cover the various aspects of this multidisciplinary research area. We would like to thank the Program Committee Vice Chairs, Bogdan Chlebus, Bhaskar Krishnamachari, and David B. Johnson, as well as the members of the Program Committee, the external referees consulted by the PC, and all of the authors who submitted their work to DCOSS 2008. We also wish to thank the keynote speakers for their participation in the meeting.

Several volunteers contributed significantly to the realization of the meeting. We wish to thank the organizers of the workshops collocated with DCOSS 2008 as well as the DCOSS Workshop Chair, Koen Langendoen, for coordinating workshop activities. We would like to thank Yang Yu and Thiemo Voigt for their efforts in organizing the poster session and demo session, respectively. Special thanks goes to Kay Romer for organizing the Work-in-Progress session and to Luis Almeida for the DCOSS competition event. Special thanks also goes to Tian He and Cristina Pinotti for handling conference publicity, and to Zachary Baker for his assistance in putting together this proceedings volume. Many thanks also go to Animesh Pathak for maintaining the conference webpage and Germaine Gusthiot for handling the conference finances.

We would like to especially thank Jose Rolim, DCOSS Steering Committee Chair, for inviting us to be the General Chairs. His invaluable input in shaping this conference series and his timely intervention in resolving meeting-related issues are gratefully acknowledged.

Finally, we would like to acknowledge the sponsors of DCOSS 2008. Their contributions are always a key enabler of a successful conference. The research area of sensor networks is rapidly evolving, influenced by fascinating advances in supporting technologies. We sincerely hope that this conference series will continue to serve as a forum for researchers working in different, complementary areas of this multidisciplinary field to exchange ideas and interact, cross-fertilizing research on the algorithmic and foundational side, as well as of high-level approaches and the more applied and technological issues related to the tools and applications of wireless sensor networks.

We hope you enjoy the meeting.

June 2008   Tarek Abdelzaher
                                                          Viktor K. Prasanna

Message from the Program Chair

This proceedings volume contains the accepted papers of the Fourth International Conference on Distributed Computing in Sensor Systems. DCOSS 2008 received a record of 116 submissions to its three tracks covering the areas of Algorithms, Systems, and Applications. During the review procedure at least two reviews for all papers and three (or more) reviews for most papers were solicited. After a fruitful exchange of opinions and comments during the final stage, 29 papers (25% acceptance ratio) were accepted as regular papers. Also, 12 papers were accepted as short papers.

The research contributions in these proceedings span diverse important aspects of sensor networking, including energy management, communication, coverage and tracking, time synchronization and scheduling, key establishment and authentication, compression, medium access control, code update, and mobility. A multitude of novel algorithmic design and analysis techniques, systematic approaches, and application development methodologies are proposed for distributed sensor networking, a research area in which complementarity and cross-fertilization are of vital importance.

I would like to thank the three Program Vice Chairs, Bogdan Chlebus (Algorithms), David B. Johnson (Systems), Bhaskar Krishnamachari (Applications) for agreeing to lead the review process in their Track and for an efficient and smooth cooperation; also, the members of the strong and broad DCOSS 2008 Program Committee, as well as the external reviewers who worked with them. I wish to thank the Steering Committee Chair, Jose Rolim, and the DCOSS 2008 General Chairs, Tarek Abdelzaher and Viktor Prasanna, for their trust and their valuable contribution to the organization of the conference, as well as the Proceedings Chair, Zachary Baker, for his tireless efforts in preparing these conference proceedings.

June 2008   Sotiris Nikoletseas

Organization

General Chair

Tarek Abdelzaher   Univ. of Illinois, Urbana Champaign, USA

Vice General Chair

Viktor K. Prasanna   University of Southern California, USA

Program Chair

Sotiris Nikoletseas   University of Patras and CTI, Greece

Program Vice Chairs

Algorithms
Bogdan Chlebus   Univ. of Colorado at Denver, USA

Applications
Bhaskar Krishnamachari   Univ. of Southern California, USA

Systems
David B. Johnson   Rice University, USA

Steering Committee Chair

Jose Rolim   University of Geneva, Switzerland

Steering Committee

Sajal Das	University of Texas at Arlington, USA
Josep Diaz	UPC Barcelona, Spain
Deborah Estrin	University of California, Los Angeles, USA
Phillip B. Gibbons	Intel Research, Pittsburgh, USA
Sotiris Nikoletseas	University of Patras and CTI, Greece
Christos Papadimitriou	University of California, Berkeley, USA
Kris Pister	University of California, Berkeley, and Dust, Inc., USA
Viktor Prasanna	University of Southern California, Los Angeles, USA

Poster Chair

Yang Yu   Motorola Labs, USA

Workshops Chair

Koen Langendoen   Delft University of Technology,
                                 The Netherlands

Proceedings Chair

Zachary Baker   Los Alamos National Lab, USA

Publicity Co-chairs

Tian He   University of Minnesota, USA
Cristina Pinotti   University of Perugia, Italy

Web Publicity Chair

Animesh Pathak   Univ. of Southern California, USA

Finance Chair

Germaine Gusthiot   University of Geneva, Switzerland

Work-in-Progress Chair

Kay Römer   ETH, Zurich, Switzerland

Demo Chair

Thiemo Voigt   Swedish Institute of Computer Science, Sweden

Competition Chair

Luís Almeida   Universidade de Aveiro, Portugal

Sponsoring Organizations

IEEE Computer Society Technical Committee on Parallel Processing (TCPP)
IEEE Computer Society Technical Committee on Distributed Processing
  (TCDP)

INTRALOT (www.intralot.com)
University of Patras (www.upatras.gr)
Greek Ministry of National Education and Religious Affairs (www.ypepth.gr)
University of Geneva (www.unige.ch)
TCS-Sensor Lab (Theoretical Computer Science and Sensor Nets) at the
    University of Geneva (http://tcs.unige.ch/)

Support from

Computer Engineering and Informatics Department of U. of Patras
    (www.ceid.upatras.gr)
Research Academic Computer Technology Institute (CTI, www.cti.gr)
SensorsLab at CTI/Research Unit 1 (ru1sensorslab.cti.gr)
EU R&D Project AEOLUS (Algorithmic Principles for Building Efficient
    Overlay Computers, aeolus.ceid.upatras.gr)
EU R&D Project FRONTS (Foundations of Adaptive Networked Societies
    of Tiny Artefacts, fronts.cti.gr)
EU R&D Project ProSense (Promote, Mobilize, Reinforce and Integrate
    Wireless Sensor Networking Research and Researchers: Towards Pervasive
    Networking of WBC and the EU)
EU R&D Project WISEBED (Wireless Sensor Network Testbeds)

Held in Co-operation with

ACM Special Interest Group on Computer Architecture (SIGARCH)
ACM Special Interest Group on Embedded Systems (SIGBED)
European Association for Theoretical Computer Science (EATCS)
IFIP WG 10.3

Program Committee

Algorithms

Matthew Andrews	Bell Labs, USA
James Aspnes	Yale, USA
Costas Busch	Louisiana State University, USA
Bogdan Chlebus (Chair)	University of Colorado Denver, USA
Andrea Clementi	University of Rome 'Tor Vergata', Italy
Eric Fleury	ENS Lyon/INRIA, France
Rachid Guerraoui	EPF Lausanne, Switzerland
Evangelos Kranakis	Carleton University, Canada
Shay Kutten	Technion, Israel
Miroslaw Kutylowski	Wroclaw Technical University, Poland
Andrew McGregor	University of California San Diego, USA

Referees

Joon Ahn
Novella Bartolini
Tiziana Calamoneri
Alessio Carosi
Jerry Chaing
Jihyuk Choi
Anshuman Dasgupta
Miriam Di Ianni
Shane Eisenman
Vissarion Ferentinos
Emanuele Fusco
Sachin Ganu
Maciej Gebala
Amitabha Ghosh
Luciano Guala
Min Guo
Jason Haas
Elyes Ben Hamida
Bijit Hore
Kévin Huguenin
Hojjat Jafarpour

Ravi Jammalamadaka
Vikas Kawadia
Marcin Kik
Mirosła Korzeniowski
Michał Koza
Prashant Krishnamurthy
Rajesh Krishnan
Rajnish Kumar
Nic Lane
Chih-Kuang Lin
Francesco Lo Presti
Mihai Marin-Perianu
Daniel Massaguer
Michele Mastrogiovanni
Jonathan McCune
Alberto Medina
Ghita Mezzour
Emiliano Miluzzo
Gianpiero Monaco
Mirco Musolesi
Michele Nati

Sundeep Pattem
Michał Ren
Niky Riga
Gianluca Rossi
Amit Saha
Stefan Schmid
Jens Schmitt
Divyasheel Sharma
Simone Silvestri
Avinash Sridharan
Mario Strasser
Ahren Studer
Ronen Vaisenberg
Paola Vocca
Yi Wang
Matthias Woehrle
Bo Xing
Xingbo Yu
Marcin Zawada

Table of Contents

Short Papers

Performance of a Propagation Delay Tolerant ALOHA Protocol for Underwater Wireless Networks

Joon Ahn and Bhaskar Krishnamachari

Ming Hsieh Department of Electrical Engineering
Viterbi School of Engineering
University of Southern California, Los Angeles, CA 90089
joonahn@usc.edu, bkrishna@usc.edu

Abstract. Underwater wireless networks have recently gained a great deal of attention as a topic of research. Although there have been several recent studies on the performance of medium access control (MAC) protocols for such networks, they are mainly based on simulations, which can be insufficient for understanding the fundamental behavior of systems. In this work we show a way to analyze mathematically the performance of an underwater MAC protocol. We particularly analyze a propagation delay tolerant ALOHA (PDT-ALOHA) protocol proposed recently [1]. In this scheme, guard bands are introduced at each time slot to reduce collisions between senders with different distances to the receiver, which have a great impact on acoustic transmissions. We prove several interesting properties concerning the performance of this protocol. We also show that the maximum throughput decreases as the maximum propagation delay increases, identifying which protocol parameter values realize the maximum throughput approximately. Although it turns out that exact expression for the maximum throughput of PDT-ALOHA can be quite complicated, we propose a useful simple expression which is shown numerically to be a very good approximation. Our result can be interpreted to mean that the throughput of PDT-ALOHA protocol can be 17-100% higher than the conventional slotted ALOHA, with proper protocol parameter values

1 Introduction

Underwater wireless networks have recently gained a great deal of attention as a topic of research, with a wide range of sensing and security applications just starting to be explored [2,3,4]. Since radio-frequency (RF) electromagnetic waves, except those of very low frequencies, decay very rapidly in water, underwater networks need to adopt acoustic transmissions instead in most cases. The long propagation delays associated with acoustic transmission, however, can significantly affect the performance of traditional network protocols, requiring the design and analysis of new approaches for different layers, including medium access.

The problem of designing a simple medium access protocol appropriate for underwater networks was addressed recently in [1]. The authors of this work show that the performance of classical slotted ALOHA deteriorates in an underwater setting where transmissions from one slot can overlap with future slots. They propose the introduction of guard bands in each time slot to address this problem; we refer to their scheme as

S. Nikoletseas et al. (Eds.): DCOSS 2008, LNCS 5067, pp. 1–16, 2008.

the propagation delay tolerant ALOHA (PDT-ALOHA) protocol. Although they have demonstrated that the throughput can be increased using guard bands through simulations, [1] have failed to investigate rigorously the performance of the protocol, which can be done through the theoretical analysis.

We analyze mathematically the performance of the PDT-ALOHA protocol in this work. We investigate different metrics of performance – expected number of successful packet receptions in a time slot, throughput, and maximum throughput. We obtain exact expressions for the number of receptions and throughput. Although the exact expressions are not closed-form, they can be easily calculated numerically with given parameters. Further, we obtain very simple expressions for the maximum throughput and its maximizers which are shown to be very good approximations.

We also prove a number of interesting and useful properties concerning the performance of the PDT-ALOHA protocol. We prove that the expected number of successful packet receptions is independent of the propagation speed; and that its maximum is non-decreasing as the size of guard band increases; and derive a bound on the network load that offers the maximum throughput. We also show that the maximum throughput decreases as the maximum propagation delay increases.

As per the original study where this protocol is proposed [1], we focus the analysis on two-dimensional underwater wireless networks. Although underwater applications can naturally allow for 3D placements, it is also possible to envision practical applications involving the deployment of sensor nodes in a two-dimensional plane just below the water surface, or close to the bottom of the water body. In any case, we should note that the methodology for analyzing performance in three-dimensions would be essentially similar to the analysis that we present in this work.

Related Works

There are several literatures which have studied on the throughput of ALOHA protocols in the underwater networks. [5] has investigated the impact of the large propagation delay on the throughput of selected classical MAC protocols and their variants through simulations. [6] has compared the performance of ALOHA and CSMA with RTS/CTS protocols in underwater wireless networks. And [1] has studied on the throughput of PDT-ALOHA through simulations producing rough idea of the performance. While these works are mainly based on simulations we approach the problem from the theoretical view point.

There have been other works which take the theoretical approach as we do. [7] has analyzed slotted ALOHA without the guard band and concluded that slotted ALOHA degrades to unslotted ALOHA under high propagation delay. [8] have analyzed the performance of ALOHA in a linear multi-hop topology. However, these works do not consider the guard band to relieve the negative effect of the large propagation delay. [9,10] have taken consideration of the guard band for the slotted ALOHA protocols in their analysis. However, they have assumed satellite networks where the imperfection or sloppiness of each node's implementation causes variable propagation delay, and they have focused on how to deal with the sloppiness using the guard bands. The main difference from their problem is that nodes are located on the ground approximately same distance away from the satellite in their problem so that the propagation delay is more or less same for each node. But, the distance to the receiver can vary greatly from node to node in underwater wireless sensor networks.

2   PDT-ALOHA Protocol and Assumptions

The PDT-ALOHA is essentially a simple enhancement of the slotted ALOHA protocol; the difference is that it has a guard band at the end of each time slot. Specifically, all transmitters in the network maintain synchronized time slots for communication in the PDT-ALOHA. A transmitter sends a packet at the start of a time slot when it has one to send, then waits until the start of the next time slot although it has another packet in its queue (See the time diagram of transmission of sender A in Fig. 1). We refer to the time duration between the end of transmission and the start of the next time slot as a guard band.

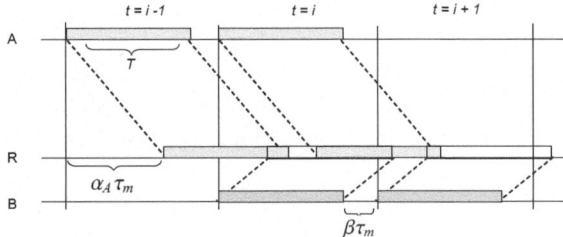

Fig. 1. Time diagram of packet transmission; A and B are transmitters and R is the receiver. B locates closer to the receiver than A.

The guard band in the PDT-ALOHA is designed in order to relieve collisions between packets of different senders transmitted in consecutive time slots. For example, if the guard band is at least as much as the maximum propagation time among all pairs of a transmitter and a receiver in the network, there would be no collision between packets transmitted in different time slots. Likewise, increasing the guard band reduces collisions, but it also increases the length of each time slot potentially reducing the throughput. Thus the selection of the appropriate guard-band length to get the maximum throughput can be formulated as an optimization problem.

We made following assumptions to analyze the performance of the PDT-ALOHA protocol unless stated otherwise.

- The network has one receiver and n transmitters, which are deployed in the two-dimensional disk area.
- The receiver locates at the center of the disk area, and the transmitters are deployed uniformly at random in the area.
- The propagation speed of communication is positive finite constant regardless of the location in the network, so that the maximum propagation time from the receiver to the farthest transmitter is a positive finite constant τ.
- The transmission rate is constant for every transmitter.

– The packet size is constant so that, along with the constant transmission rate, the transmission time for a packet is constant, which we assume is one; this doesn't incur loss of generality in our analysis. Only a proper scaling is needed for some parameters, particularly τ, in order to cope with the general transmission time.
– We assume packet arrivals to the network follow the Poisson distribution. Since the Poisson arrival in a slotted time system can be well approximated through the Binomial distribution, we assume the packet departure per node at a given time slot is I.I.D. Bernoulli. Specifically, a transmitter sends a packet to the receiver with probability p in each time slot.
– If the receiver receives more than one packet simultaneously at any time in a time slot, all the packets involved fail to get delivered successfully causing a collision.
– The links over which transmissions take place are lossless (e.g., using blacklisting).
– A transmitter always transmits a packet at the start of the time slot if the transmitter wants to send the packet.
– All the nodes have the globally synchronized time slots.
– The transmission time is no less than the maximum propagation time so that $\tau \leq 1$.

The assumption that $\tau \leq 1$ is to make sure that the collision between a time slot and another is confined to the consecutive time slots. So, with this assumption, there is no possibility that a packet sent in i-th time slot collides with another in j-th time slot, where $j \notin \{i-1, i, i+1\}$.

3   Throughput Analysis

In this section we analyze the throughput of the PDT-ALOHA protocol. Let us first look at the time slot. Each time slot consists of a transmission time and a guard band following the former. Since the guard band of the size of maximum propagation time τ would eliminate all the collision between different time slots, it does not make sense to have the guard band whose size is more than τ only decreasing the throughput without any further gain. Hence, we use the normalized factor β s.t. $0 \leq \beta \leq 1$ in expressing the size of guard band so that the time slot size is $1 + \beta \cdot \tau$.

3.1   Expected Number of Successful Packet Receptions

In order to analyze the throughput we first derive the expected number of successful packet receptions in a time slot. We use the linearity of expectations and conditional probabilities to calculate the expected number. Let the indicator variable I_i denote whether or not the receiver receives the packet from i-th transmitter successfully in the time slot.

$$I_i = \begin{cases} 1, & \textit{if successful reception} \\ 0, & \textit{otherwise} \end{cases} \tag{1}$$

Let N denote the random variable of the number of successful reception. Then, $N = \sum_i I_i$. Hence, the expected number is, by the linearity of expectations and conditional probability, as follows;

$$E[N] = \sum_{i=1}^{n} E[I_i] = \sum_{i=1}^{n} \Pr\{I_i = 1\}$$

$$= \sum_{i=1}^{n} \Pr\{no\ collision \mid i\text{-}th\ sender\ sends\} \cdot \Pr\{i\text{-}th\ sender\ sends\}$$

$$= n \cdot p \cdot \Pr\{NC|n_i\} \tag{2}$$

where $NC|n_i$ denotes the event that no collision occurs given that i-th sender transmits. The last equality of the above equations holds since the collision probability is symmetrical among all the senders.

Therefore, in order to calculate the expected number, we need to find out the probability of no collision for the transmitted packet from the i-th sender whose location is uniform at random over the network area.

3.2 Probability of No Collision

Let us consider some situations where collision can occur in order to get some intuition. Suppose a simple network with two senders, A and B, and one receiver R. B locates right next to R while A is very far from R, and the size of guard band is small enough. Then, if A transmits in the i-th time slot, R would receive last part of the packet in the beginning of the $(i + 1)$-th slot, which would produce collision with the packet transmitted in the $(i + 1)$-th slot by B although the two packets are sent in different time slots. The time diagram in Fig. 1 visually shows this situation, where α_i is the normalized propagation time distance of $i \in \{A, B\}$ from R defined by Definition 1, the normalized guard band size β is less than 1, and $\alpha_A > \alpha_B + \beta$. However, if $\alpha_A = \alpha_B$ there is no collision between packets in different time slots. Therefore, we can see that the collision depends on nodes' locations and two packets transmitted in different time slots can experience collision between each other.

Definition 1. *The **normalized (propagation) time distance** of sender X from the receiver is the propagation time from the receiver to X divided by the maximum propagation time τ in the network.*

After all, it is not hard to identify collision regions $R_p(\alpha)$, $R_c(\alpha)$, $R_n(\alpha)$ for the interested transmitter I which has the normalized time distance of α, where $R_p(\alpha)$ denotes the region such that a packet sent from I collides with a packet sent in the previous consecutive time slot by a node in $R_p(\alpha)$; $R_c(\alpha)$ in the same time slot; and $R_n(\alpha)$ in the next consecutive time slot. Equation (3), (4), (5) specify the regions in terms of the normalized time distance and Fig. 2 visually presents the regions.

$$R_p(\alpha) = \{\tau_p | \alpha + \beta \leq \tau_p \leq 1\} \tag{3}$$
$$R_c(\alpha) = \{\tau_c | 0 \leq \tau_c \leq 1\} \tag{4}$$
$$R_n(\alpha) = \{\tau_n | 0 \leq \tau_n \leq \alpha - \beta\} \tag{5}$$

The probability of no collision given a packet sent by an arbitrary i-th sender n_i is then as follows conditioning on the n_i's normalized time distance α;

$$\Pr\{NC|n_i\} = \int_0^1 \Pr\{NC|n_i, n_i \text{ at } \alpha \text{ away}\} \cdot pdf\{\alpha \text{ away} \mid n_i \text{ trans.}\}d\alpha$$

$$= \int_0^1 2\alpha \Pr\{NC|\alpha\}d\alpha \tag{6}$$

The last equation holds because the location of a node is independent of the packet transmission in our assumption.

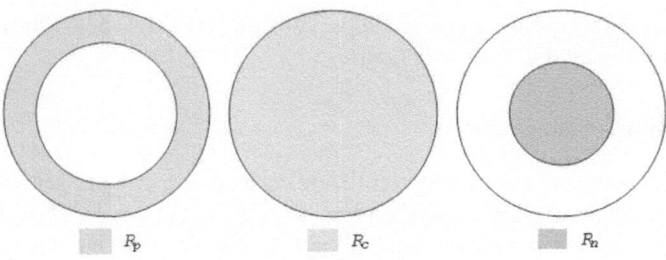

R_p R_c R_n

Fig. 2. Collision regions

Meanwhile, the probability of (no) collision of a specific packet does depend on the location of its sender because it defines the three collision regions, $R_n(\alpha)$, $R_c(\alpha)$, and $R_p(\alpha)$; and the regions' areas affect the probability. Hence, further conditioning on the numbers of transmitters in those three collision regions, we can get the following equation:

$$\Pr\{NC|n_i, n_i \text{ at } \alpha \text{ away}\} = \Pr\{NC| \text{ sent at } \alpha \text{ away}\}$$

$$= \sum_{x+y+z=n-1} \begin{array}{l} \Pr\{NC \mid \alpha, N_n = x, N_c = y, N_p = z\} \\ \times P\{N_n = x, N_c = y, N_p = z|\alpha\} \end{array} \tag{7}$$

where N_n, N_c, and N_p denote the number of other transmitters in R_n, R_c, and R_p respectively. Note that there are $(n-1)$ other transmitters (or interferers) because we focus on one specific transmitter's success.

Note also that the event of $N_n = x, N_c = y, N_p = z|\alpha$ has the multinomial distribution with parameters $n-1$, $p_n(\alpha)$, $p_c(\alpha)$, and $p_p(\alpha)$, where $p_i(\alpha)$, $i \in \{n, c, p\}$ denotes the probability that a transmitter lies in $R_i(\alpha)$. And each of these probabilities is the ratio of its area to the entire area of the network;

$$p_n(\alpha) = \begin{cases} (\alpha - \beta)^2 & , if\ 0 < \alpha - \beta < 1 \\ 0 & , otherwise \end{cases}$$

$$p_c(\alpha) = 1$$

$$p_p(\alpha) = \begin{cases} 1 - (\alpha + \beta)^2 & , if\ 0 < \alpha + \beta < 1 \\ 0 & , otherwise \end{cases}$$

Now we have two cases, each of which has three sub-cases. In the first case A ($0 \leq \beta \leq 0.5$), we have three sub-cases; (i) $0 \leq \alpha \leq \beta$, where $R_n(\alpha) = \emptyset$ so that $p_n(\alpha) = 0$; (ii) $\beta \leq \alpha \leq 1 - \beta$, where all three collision regions can have areas larger than zero; and (iii) $1 - \beta \leq \alpha \leq 1$, where $R_p(\alpha) = \emptyset$ so that $p_p(\alpha) = 0$. In the other case B ($0.5 \leq \beta \leq 1$), we have another three sub-cases; (i) $0 \leq \alpha \leq 1 - \beta$, where $R_n(\alpha) = \emptyset$; (ii) $1 - \beta \leq \alpha \leq \beta$, where $R_p(\alpha) = R_n(\alpha) = \emptyset$; and (iii) $\beta \leq \alpha \leq 1$, where $R_p(\alpha) = \emptyset$.

Let us consider Case A.(i) first. In this case, the conditional probability of no collision turns out to involve the binomial series as follows;

$$
\begin{aligned}
\Pr\{NC|\alpha\} &= \sum_{z=0}^{n-1} \Pr\{NC|\alpha, N_p = z, N_c = n - 1, N_n = 0\} \cdot \Pr\{N_p = z|\alpha\} \\
&= \sum_{z=0}^{n-1} \binom{n-1}{z} (1-p)^z (1-p)^{n-1} \cdot (1 - (\alpha + \beta)^2)^z ((\alpha + \beta)^2)^{n-1-z} \\
&= (1-p)^{n-1} (1 - p + p(\alpha + \beta)^2)^{n-1}
\end{aligned}
\tag{8}
$$

Equation (9) and (10) are the summary after calculating the other cases in the similar way; most of them involve the binomial series although Case A.(ii) involves the multinomial series.

In Case A,

$$
\Pr\{NC|\alpha\} = \begin{cases}
(1-p)^{n-1} (1 - p + p(\alpha + \beta)^2)^{n-1}, & \text{if } 0 \leq \alpha \leq \beta \\
(1-p)^{n-1} (1 - p + 4p\alpha\beta)^{n-1}, & \text{if } \beta \leq \alpha \leq 1 - \beta \\
(1-p)^{n-1} (1 - p(\alpha - \beta)^2)^{n-1}, & \text{if } 1 - \beta \leq \alpha \leq 1
\end{cases}
\tag{9}
$$

In Case B,

$$
\Pr\{NC|\alpha\} = \begin{cases}
(1-p)^{n-1} (1 - p + p(\alpha + \beta)^2)^{n-1}, & \text{if } 0 \leq \alpha \leq 1 - \beta \\
(1-p)^{n-1}, & \text{if } 1 - \beta \leq \alpha \leq \beta \\
(1-p)^{n-1} (1 - p(\alpha - \beta)^2)^{n-1}, & \text{if } \beta \leq \alpha \leq 1
\end{cases}
\tag{10}
$$

Substituting (9) or (10) into (6) we can obtain the expression for the probability of no collision which can be evaluated easily with the numerical method.

Note that the expression for probability of no collision does not involve the maximum propagation delay τ implying the probability is independent of τ so that the expected number of successful reception is also independent of τ. It turns out from Theorem 1 that the expected number is independent of τ even after relaxing the assumption of 2D unit disk of the network and the identical distribution of packet transmission for each node.

Theorem 1. *Given a network of nodes with fixed spatial locations of nodes, a fixed transmission probability p_i in a time slot for each node i, and a transmission time T for a packet, the expected number of successful packet reception f in a time slot is independent of the maximum propagation time τ in the network as long as $0 < \tau \leq T$. In other words, it is independent of the propagation speed v_p.*

Proof. Since the spatial locations of nodes are fixed, the spatial distance r_m from the receiver to the farthest node is constant; $r_m = \tau \cdot v_p = const.$

The spatial distance r_i of an arbitrary i-th transmitter is also fixed, and so the normalized propagation time delay α_i of the node is constant regardless of r_m as long as $r_m > 0$ or $0 < v_p < \infty$ because of the following:

$$r_i = \alpha_i \cdot \tau \cdot v_p = \alpha_i \cdot r_m \quad \Rightarrow \quad \alpha_i = \frac{r_i}{r_m} = const.$$

Let $r(R_i)$ denote the spatial region associate with the collision region R_i. Then, the spatial region of R_n, R_c, and R_p are all fixed regardless of τ because

$$r(R_n) = \{r : 0 \le r \le (\alpha_i - \beta)\tau v_p = (\alpha_i - \beta)r_m\}$$
$$r(R_c) = \{r : 0 \le r \le \tau v_p = r_m\}$$
$$r(R_p) = \{r : (\alpha_i + \beta)r_m \le r \le r_m\}$$

and α_i, β, and r_m are all constants.

Hence, the number of nodes in each of R_n, R_c, and R_p is constant regardless of the speed of propagation, and so the probability of no collision of the i-th transmitter is constant. Therefore,

$$f = \sum_i p_i \Pr\{NC|n_i\} = const. \text{ with respect to } \tau \qquad \square$$

3.3   Throughput for Finite Number of Nodes

In this paper we consider the throughput S in packets per transmission time. Because the expected number of successful packet receptions $f(n, \beta, p)$ in a time slot is independent of the propagation time as long as it is positive finite (Theorem 1), S can be expressed as follows:

$$S(n, \beta, p, \tau) = \frac{f(n, \beta, p)}{1 + \beta\tau} = \frac{np \Pr\{NC|n_i\}}{1 + \beta\tau} \qquad (11)$$

where we know the probability of no collision from the previous section.

Using the numerical evaluation of (11), Fig. 3 shows the characteristics of the throughput depending on the size of guard band β; in (a) the maximum propagation delay τ is fixed, but the number of nodes n is varying. In (b) n is fixed but τ is varying. These plots show how the throughput responds to the variables; the optimizer β values are similar for one case, but different in the other case. And the throughput converges rapidly as n increases.

3.4   Throughput for Infinite Number of Nodes

In this section, we investigate the throughput of PDT-ALOHA protocol with an infinite number of nodes with the traffic load λ over the network, i.e, $n \to \infty$ while $p = \lambda/n$. Hence, the throughput in this case is given by

$$S' = \lim_{n \to \infty} S\big|_{p=\frac{\lambda}{n}} = \frac{\lambda}{1 + \beta\tau} \lim_{n \to \infty} \Pr\{NC|n_i\} \qquad (12)$$

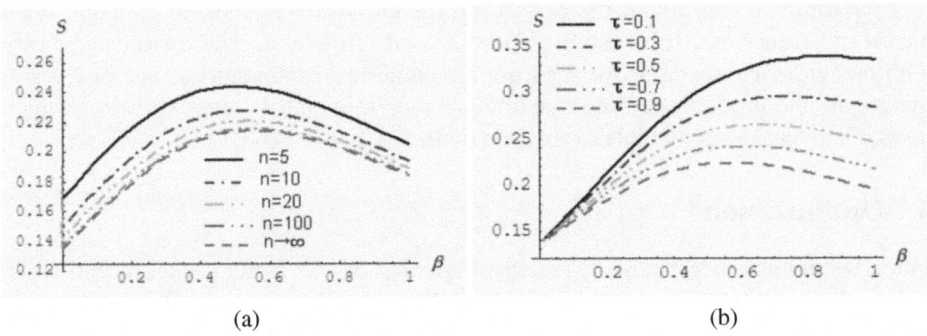

(a)   (b)

Fig. 3. Throughput of PDT-ALOHA vs. β; (a) when $\tau = 1$ and n is variable, (b) when $n = 100$ and τ is variable

Because the integrand of (6) converges uniformly over $[0, 1]$ (see Appendix of [11]), we can exchange integral and limitation by Theorem 7.16 of [12]. Hence, with the equalities in Table 1, we can achieve the conditional probability of no collision in this limiting case as follows:

If $0 \leq \beta \leq 0.5$;

$$\Pr\{NC|\alpha\} = \begin{cases} e^{-2\lambda + \lambda(\alpha+\beta)^2}, & 0 \leq \alpha \leq \beta \\ e^{-2\lambda + 4\lambda\alpha\beta}, & \beta \leq \alpha \leq 1 - \beta \\ e^{-\lambda - \lambda(\alpha-\beta)^2}, & 1 - \beta \leq \alpha \leq 1 \end{cases} \quad (13)$$

If $0.5 < \beta \leq 1$;

$$\Pr\{NC|\alpha\} = \begin{cases} e^{-2\lambda + \lambda(\alpha+\beta)^2}, & 0 \leq \alpha \leq 1 - \beta \\ e^{-\lambda}, & 1 - \beta \leq \alpha \leq \beta \\ e^{-\lambda - \lambda(\alpha-\beta)^2}, & \beta \leq \alpha \leq 1 \end{cases} \quad (14)$$

Table 1. Equalities to use as building blocks to calculate the throughput in the limiting case

$$\lim_{n\to\infty} \left(1 - \frac{\lambda}{n}\right)^{n-1} = e^{-\lambda}$$

$$\lim_{n\to\infty} \left(1 - \frac{\lambda}{n} + \frac{\lambda}{n}(\alpha+\beta)^2\right)^{n-1} = e^{-\lambda + \lambda(\alpha+\beta)^2}$$

$$\lim_{n\to\infty} \left(1 - \frac{\lambda}{n} + 4\frac{\lambda}{n}\alpha\beta\right)^{n-1} = e^{-\lambda + 4\lambda\alpha\beta}$$

$$\lim_{n\to\infty} \left(1 - \frac{\lambda}{n}(\alpha-\beta)^2\right)^{n-1} = e^{-\lambda(\alpha-\beta)^2}$$

Table 2. Constants for Approximation Models

a	0.2464
b	-2.9312
c	-0.9887
p	0.0784
q	0.2638
r	0.9173
p_1	0.1805
q_1	0.6543
r_1	0.8898
p_2	0.2257
q_2	0.6959
r_2	0.9049

Therefore, we can obtain the expression for the probability of no collision for a packet of a transmitter substituting (13) or (14) into (6) which can be evaluated easily with the numerical method. We shall use the numerical evaluations as one of tools to investigate the properties of the maximum throughput and its approximation, which turns out to have a very simple expression in Sect. 4.2.

4   Optimization

In this section we investigate the *maximum* number of successful packet receptions in a time slot and the *maximum* throughput of PDT-ALOHA protocol. We also have an interest in the protocol parameters, particularly the size of guard band and the traffic load, which realize the maximum throughput.

We start with special cases, i.e. $\beta = 0$, or $\beta = 1$, which can be analyzed analytically. Then, we examine general cases given a network size in terms of the maximum propagation delay. Because it is very hard to obtain the closed form expression (if possible), we resort to use the numerical method to analyze the optimum behavior of the system. Based on the result of the numerical analysis, we propose simple approximations for the optimum behavior and its protocol parameters.

In this section we also consider the traffic load per transmission time λ_r because it is useful to compare traffic loads between systems of different size of time slot and it turns out it gives simpler approximation for the optimum values. The traffic load per transmission time has the relationship as $\lambda_r = \lambda/(1+\beta\tau)$ with the traffic load per time slot which we have dealt with so far.

Although we assume in this section the limiting case where the number of nodes in the network is infinite, it is fairly straightforward to adapt the method we used here for the finite number of nodes.

4.1   Special Cases

When there is no guard band (i.e $\beta = 0$) or the guard band is full so that there is no collision between packets from different time slots (i.e. $\beta = 1$) we have a closed-form expression for throughput which is simple enough to analyze analytically the maximum throughput. When $\beta = 0$ it is easy to see from (6), (12), and (13) that the throughput is as follows:

$$S_0(\lambda) \doteq S(\beta = 0, \lambda, \tau) = \lambda e^{-2\lambda} \tag{15}$$

There is no guard band in this case which makes the maximum propagation time irrelevant to the throughput, which (15) confirms. Note that the throughput in this case become the throughput of the classical unslotted ALOHA protocol [13,14] as pointed by [7,1].

The maximum throughput can be obtained simply using the derivative since S_0 is convex. The maximum is achieved at $\lambda = 0.5$ (i.e. $\lambda_r = 0.5$) as follows:

$$S_0^* = e^{-1}/2 \tag{16}$$

When $\beta = 1$ the throughput is as follows from (6), (12), and (14):

$$S_1(\lambda, \tau) \doteq S(\beta = 1, \lambda, \tau) = \frac{\lambda e^{-\lambda}}{1 + \tau} \tag{17}$$

Because S_1 is convex regarding λ at any $\tau \in (0, 1]$, we can obtain its maximum given τ using the partial derivative as follows:

$$S_1^*(\tau) = \frac{e^{-1}}{1 + \tau} \tag{18}$$

where the maximizer is $\lambda = 1$ while the corresponding traffic load per transmission time is $\lambda_r = 1/(1 + \tau)$.

Figure 4.(a) shows S_0^* and S_1^* as well as the maximum throughput S^* which is discussed in Sect. 4.2; it can be seen that both of them are suboptimal although S_1^* approaches the maximum as τ goes to 0. When $\tau = 1$ for which the maximum propagation delay is equal to the transmission time of a packet, the throughput becomes same whether PDT-ALOHA has the full guard band or no guard band at all.

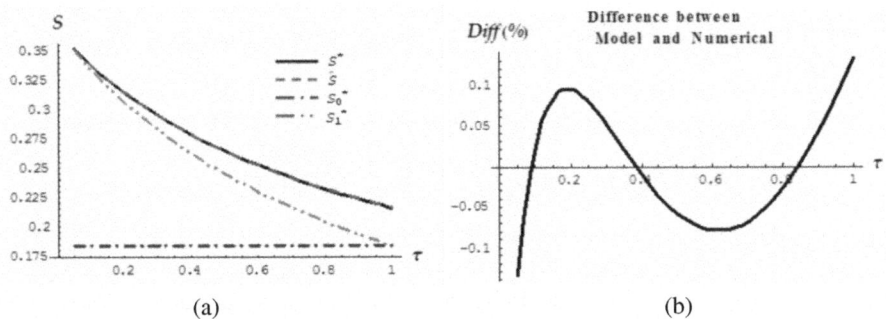

(a)   (b)

Fig. 4. The maximum throughput: (a) numerically calculated values and its approximation, (b) the difference between numerical values and its approximation

4.2   Maximum Throughput

Now we investigate the maximum throughput S^* over all possible non-negative guard band β and network load per time slot λ given the network size in terms of the maximum propagation delay. Note that it is sufficient to look into only $\beta \in [0, 1]$ and $\lambda \in [0, 1]$ because $S(\beta, \lambda, \tau) \leq S(1, 1, \tau), \forall \beta \geq 1, \forall \lambda \geq 1$ due to Theorem 2 (for a finite number of nodes), Theorem 3 (for an infinite number of nodes), and Theorem 4.

Theorem 2. *Suppose a network of n number of nodes is assumed as that of Sect. 3.3 with $p = \lambda/n$. Then, the throughput S_n of the PDT-ALOHA protocol with $\lambda \geq 1$ for the network is no higher than when $\lambda = 1$. That is,*

$$\lambda \geq 1 \Rightarrow S_n(\beta, \lambda, \tau) \leq S_n(\beta, 1, \tau), \forall \beta \in [0, 1]$$

Proof. The expected number of successful receptions $f_n(\beta, \lambda)$ in a time slot can be expressed as follows using (2), (6), (9), and (10):

$$f_n(\beta, \lambda) = \lambda \int_0^1 2\alpha \Pr\{NC|\alpha\} d\alpha = \lambda \left(1 - \frac{\lambda}{n}\right)^{n-1} \int_0^1 g_n(\beta, \lambda) d\alpha$$

where $g_n(\beta, \lambda)$ is a proper function after extracting $\left(1 - \frac{\lambda}{n}\right)^{n-1}$.

Suppose $\lambda \geq 1$. Since $0 < \lambda_1 \leq \lambda_2 < n$ implies $g_n(\beta, \lambda_1) \geq g_n(\beta, \lambda_2)$ for all $\beta \in [0, 1]$,

$$f_n(\beta, \lambda) = \lambda \left(1 - \frac{\lambda}{n}\right)^{n-1} \int_0^1 g_n(\beta, \lambda) d\alpha$$

$$\leq \lambda \left(1 - \frac{\lambda}{n}\right)^{n-1} \int_0^1 g_n(\beta, 1) d\alpha \leq \left(1 - \frac{1}{n}\right)^{n-1} \int_0^1 g_n(\beta, 1) d\alpha = f_n(\beta, 1)$$

where the last inequality holds since $x(1 - x/n)^{n-1} \leq (1 - 1/n)^{n-1}$ for $\forall x \geq 1$ and $\forall n \geq 2$.

Therefore,

$$S_n(\beta, \lambda, \tau) = \frac{f_n(\beta, \lambda)}{1 + \beta\tau} \leq \frac{f_n(\beta, 1)}{1 + \beta\tau} = S_n(\beta, 1, \tau) \qquad \square$$

Theorem 3. *Theorem 2 holds for the infinite number of nodes as long as the throughput limit exists.*

Proof. Since $S_n(\beta, \lambda, \tau) \leq S_n(\beta, 1, \tau)$ for $\forall \lambda \geq 1$ and $\forall n \geq 2$ from Theorem 2,

$$S(\beta, \lambda, \tau) = \lim_{n \to \infty} S_n(\beta, \lambda, \tau) \leq \lim_{n \to \infty} S_n(\beta, 1, \tau) = S(\beta, 1, \tau)$$

as long as the limits exist. $\qquad \square$

Theorem 4. *The throughput S with the normalized guard band size $\beta \geq 1$ of an arbitrary network is no higher than that of $\beta = 1$. That is,*

$$\beta \geq 1 \Rightarrow S(\beta, \boldsymbol{p}, \tau) \leq S(1, \boldsymbol{p}, \tau)$$

Proof. If $\beta \geq 1$, there is no longer collision of packets between different time slots and β does not have any effect on packets sent in the same time slot. Hence, the expected number of successful packet receptions in a time slot is same for $\beta \geq 1$ as that of $\beta = 1$. However, increasing β makes the size of time slot increases. Therefore, the claim follows. $\qquad \square$

We evaluate the maximum throughput for 20 values of τ starting from 0.05 to 1 incrementing 0.05 using the numerical method. After examining the behavior of S^*, we propose the following simple expression as an approximation of S^*:

$$\hat{S}(\tau) = p + \frac{q}{\tau + r} \tag{19}$$

where $p, q,$ and r are constants.

The black solid line of Fig. 4.(a) shows the interpolation of 20 data points of the maximum throughput found in aforementioned way. The red dash line is of our approximation $\hat{S}$ with constants achieved through numerical curve fitting (Table 2). And Fig. 4.(b) visually shows the accuracy of $\hat{S}$. As can be seen, our approximation has reasonably good accuracy. From the figure we can also see that the PDT-ALOHA with proper parameter values can achieve about 17% (when $\tau = 1$) to 100% (when $\tau \to 0$) improvement on throughput over the traditional ALOHA protocol.

The optimum values of protocol parameters which realize the optimum throughput are also of interest. In particular, we are interested in the optimum size of the guard band β^* and the optimum traffic load given the network size in terms of τ.

Through the numerical analysis we are able to propose a simple approximation model for the optimizer β^* as follows:

$$\hat{\beta}(\tau) = p_1 + \frac{q_1}{\tau + r_1} \tag{20}$$

where p_1, q_1, and r_1 are constants; their proper values are given in Table 2 through curve fitting.

As for the optimum traffic load per transmission time, we propose the following approximation model:
$$\hat{\lambda}_r(\tau) = p_2 + \frac{q_2}{\tau + r_2} \tag{21}$$

where p_2, q_2, and r_2 are constants (Table 2).

The black solid lines of Fig. 5.(a) and (c) show the interpolation of 20 data points of β^* and λ_r^*, respectively, which are found numerically. The associated red dash lines are of our approximation $\hat{\beta}$ and $\hat{\lambda}_r$ with constants in Table 2. And Fig. 5.(b) and (d) visually show the accuracy of corresponding optimizers. As you can see, both of our approximations have reasonably good accuracy.

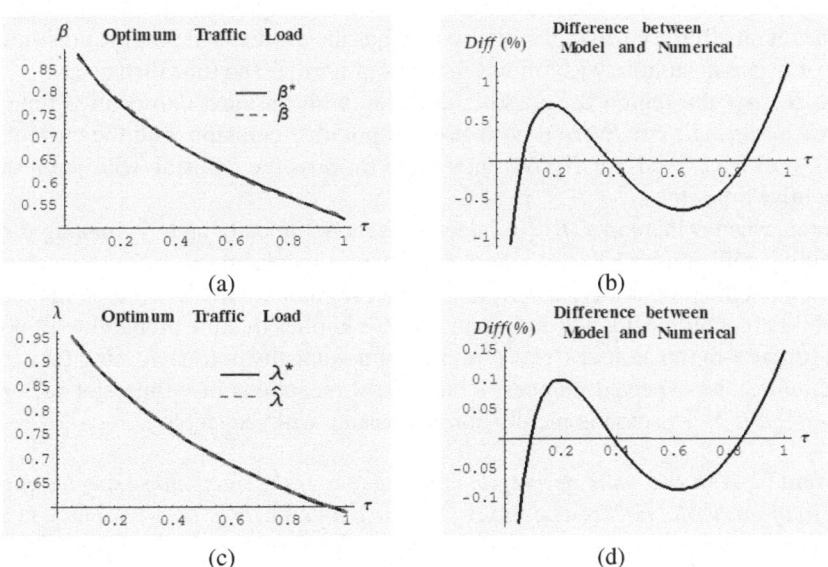

Fig. 5. The optimizers: (a) numerically calculated β^* values and its approximation, (b) the difference between numerical β^* values and its approximation, (c) λ_r^* values and its approximation, (d) the difference between numerical λ_r^* and its approximation

4.3   Maximum Expected Number of Successful Packet Receptions

In this subsection we consider the maximum expected number of successful receptions. We first present the analytic findings about the properties of the maximum successful number. The findings are more general than what we assume previously. We prove through Theorem 5 that the maximum expected number of receptions is monotonically non-decreasing with respect to the guard band β even when the network area is no longer 2D disk and the sending probability is not identical for each node as long as the maximum propagation delay is less than the transmission time of a packet.

Lemma 1. *Given the arbitrary location distribution of n transmitters and the probability p_i that i-th transmitter transmits a packet in a time slot, the expected number of successful reception in a time slot, $f(\beta, \boldsymbol{p})$, is monotonically non-decreasing as the normalized guard band β increases when the maximum propagation delay τ in the network is less than the transmission time T.*

In other words,

$$0 \le \beta_1 \le \beta_2 \le 1 \quad \Rightarrow \quad f(\beta_1, \boldsymbol{p}) \le f(\beta_2, \boldsymbol{p}),$$
$$\text{for all } \boldsymbol{p} = (p_1, \ldots, p_n) \text{ s.t } 0 \le p_i \le 1, \forall i \in \{1, \ldots, n\}$$

Proof. Because $\tau \le T$, a transmission can interfere only with the transmission of immediate previous, current, and/or immediate next time slot. Hence, there are at most three collision regions given a transmitter as we investigated in Sect. 3.2. The three regions for an arbitrary i-th transmitter which has the normalized propagation time distance of α_i are in summary as follows in terms of normalized time distance;

For $R_n(\alpha_i)$ the region for possible collision with the next consecutive time slot: $0 \le \tau_n \le \alpha_i - \beta$. For $R_c(\alpha_i)$ the region for possible collision with the current time slot: $0 \le \tau_c \le 1$. And, for $R_p(\alpha_i)$ the region for possible collision with the previous consecutive time slot: $\alpha_i + \beta \le \tau_p \le 1$

Hence, when β increases, $R_n(\alpha_i)$ decreases monotonically up to $\emptyset$, making the corresponding collision probability monotonically non-increasing; $R_c(\alpha_i)$ stays constant, not changing the probability; and $R_p(\alpha_i)$ decreases monotonically up to $\emptyset$, making the probability monotonically non-increasing. These implies that the probability of no collision for the i-th transmitter $\Pr\{NC|n_i\}$ is monotonically non-decreasing for each i.

Therefore, the expected number of successful receptions in a time slot $f(\beta, \boldsymbol{p}) = \sum_i p_i \Pr\{NC|n_i\}$ is monotonically non-decreasing with respect to β.   □

Theorem 5. *With the same assumptions of Lemma 1, the maximum expected number (over $\boldsymbol{p}$) of successful packet receptions $f^*(\beta)$ in a time slot (i.e. $f^*(\beta) = \max_{\boldsymbol{p}} f(\beta, \boldsymbol{p})$) is monotonically non-decreasing with respect to the normalized guard band size β. In other words,*

$$0 \le \beta_1 \le \beta_2 \le 1 \quad \Rightarrow \quad f^*(\beta_1) \le f^*(\beta_2)$$

Proof. From the definition of f^* and Lemma 1,

$$f^*(\beta_2) \ge f(\beta_2, \boldsymbol{p}) \ge f(\beta_1, \boldsymbol{p}), \quad \forall \boldsymbol{p}$$

Therefore, $f^*(\beta_2)$ is an upper bound of $f(\beta_1, p)$ for all p, which implies the following:

$$f^*(\beta_2) \geq \max_p f(\beta_1, p) = f^*(\beta_1) \qquad \square$$

As same as the previous section, we use the numerical method to evaluate f^*. The black solid line of Fig. 6.(a) shows the interpolation of 21 data points of f^* found numerically.

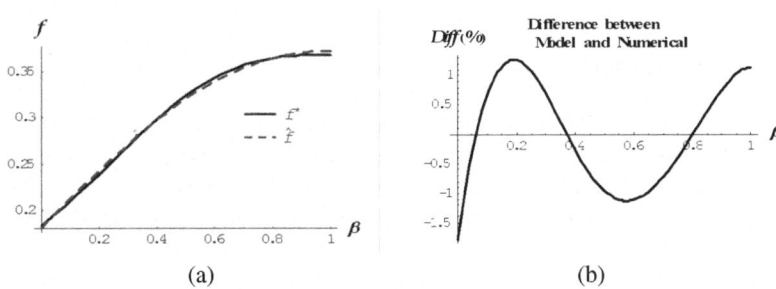

Fig. 6. The maximum number of successful receptions in a time slot: (a) numerically calculated values and its approximation, (b) the difference between numerical values and its approximation

Although it is hard to obtain the exact expression of f^*, we know from Theorem 5 that the maximized function $f^*(\beta) = \max_\lambda f(\beta, \lambda)$ is monotonically non-decreasing. From this fact and the observation that the log-scale plot of the numerically evaluated $f^*(\beta)$ is approximately of cubic function, we are able to propose the following approximation model for $f^*(\beta)$:

$$\hat{f}(\beta) = e^{a(\beta-1)^2(\beta+b)+c} \qquad (22)$$

where a, b, and c are constants and the constraint that $b < -1$ makes sure that the function is monotonically increasing.

The red dash line of Fig. 6.(a) is plotted by this approximation with proper constants suggested in Table 2 acquired through the numerical curve fitting. And Fig. 6.(b) shows its accuracy.

5   Conclusion

We have presented theoretical analysis of an underwater MAC protocol in this work. Specifically, we have analyzed mathematically the performance of the PDT-ALOHA protocol. We have investigated different metrics of performances – expected number of successful packet receptions in a time slot, throughput, and maximum throughput. We have obtained exact expressions for the number of receptions and throughput. Although it is very hard to obtain meaningful closed-form expressions from the exact expressions, it is fairly fast to numerically calculate them with given parameters. Further, we have obtained simple expressions for the maximum throughput and optimum protocol parameter values which are shown to be very good approximations. From a practical point of view, our result shows that the throughput of optimized PDT-ALOHA protocol is 17-100% better than that of conventional slotted ALOHA.

We have also proven a number of interesting and useful properties concerning the performance of the PDT-ALOHA protocol. We have proven that the expected number of successful packet receptions is independent of the propagation speed; that its maximum is non-decreasing as the size of guard band increases; and derive a bound on the network load that offers the maximum throughput.

In the future, we would like to extend these results for 3-dimensional underwater networks.

Acknowledgement

We would like to thank Affan A. Syed, John Heidemann, and Wei Ye from USC/ISI, who developed and discussed with us the original PDT-ALOHA algorithm. Affan A. Syed also helped greatly in discussing the original problem formulation.

This work was funded in part by NSF through the CAREER grant CNS-0347621.

References

1. Syed, A.A., Ye, W., Heidemann, J., Krishnamachari, B.: Understanding spatio-temporal uncertainty in medium access with ALOHA protocols. In: WuWNet 2007 (2007)
2. Akyildiz, I.F., Pompili, D., Melodia, T.: Underwater Acoustic Sensor Networks: Research Challenges. Ad Hoc Networks Journal, 257–279 (March 2005)
3. Partan, J., Kurose, J., Levine, B.N.: A Survey of Practical Issues in Underwater Networks. In: WUWNet 2006 (2006)
4. Heidemann, J., Li, Y., Syed, A., Wills, J., Ye, W.: Underwater Sensor Networking: Research Challenges and Potential Applications. In: WCNC 2006 (April 2006)
5. Guo, X., Frater, M., Ryan, M.: A Propagation-delay-tolerant Collision Avoidance Protocol for Underwater Acoustic Sensor Networks. In: Proc. IEEE Oceans Conference (2006)
6. Xie, P., Cui, J.H.: Exploring Random Access and Handshaking Techniques in Large-Scale Underwater Wireless Acoustic Sensor Networks. In: OCEANS 2006 (September 2006)
7. Vieira, L.F.M., Kong, J., Lee, U., Gerla, M.: Analysis of ALOHA protocols for underwater acoustic sensor networks. In: WUWNet (extended abstract, 2006)
8. Gibson, J.H., Xie, G.G., Xiao, Y., Chen, H.: Analyzing the Performance of Multi-Hop Underwater Acoustic Sensor Networks. In: OCEANS 2007 (June 2007)
9. Crozier, S., Webster, P.: Performance of 2-dimensional sloppy-slotted ALOHA random accesssignaling. In: Wireless Communications, Conference Proceedings, IEEE International Conference on Selected Topics, pp. 383–386 (1992)
10. Crozier, S.: Sloppy-slotted ALOHA. In: Proceedings of the 2nd International Mobile Satellite Conference, JPL, California Inst. of Tech., pp. 357–362 (1990)
11. Ahn, J., Krishnamachari, B.: Performance of Propagation Delay Tolerant ALOHA Protocol for Underwater Wireless Networks. Technical Report CENG-2007-13, USC (November 2007)
12. Rudin, W.: Principles of Mathematical Analysis, 3rd edn. McGraw-Hill, New York (1976)
13. Rosner, R.: Distributed Telecommunications Networks via Satellite and Packet Switching. Wadsworth, Inc. (1982)
14. Stallings, W.: Data and Computer Communications. Macmillan Publishing Company, New York (1985)

Time Synchronization in Heterogeneous Sensor Networks

Isaac Amundson[1], Branislav Kusy[2], Peter Volgyesi[1], Xenofon Koutsoukos[1], and Akos Ledeczi[1]

[1] Institute for Software Integrated Systems (ISIS)
Department of Electrical Engineering and Computer Science
Vanderbilt University
Nashville, TN 37235, USA
isaac.amundson@vanderbilt.edu
[2] Department of Computer Science
Stanford University
Stanford, CA 94305, USA

Abstract. Time synchronization is a critical component in many wireless sensor network applications. Although several synchronization protocols have recently been developed, they tend to break down when implemented on networks of heterogeneous devices consisting of different hardware components and operating systems, and communicate over different network media. In this paper, we present a methodology for time synchronization in heterogeneous sensor networks (HSNs). This includes synchronization between mote and PC networks, a communication pathway that is often used in sensor networks, but has received little attention with respect to time synchronization. In addition, we evaluate clock skew compensation methods including linear regression, exponential averaging, and phase-locked loops. Our HSN synchronization methodology has been implemented as a network service and tested on an experimental testbed. We show that a 6-hop heterogeneous sensor network can be synchronized with an accuracy on the order of microseconds.

1 Introduction

Wireless sensor networks (WSNs) consist of large numbers of cooperating sensor nodes that can be deployed in practically any environment, and have already demonstrated their utility in a wide range of applications including environmental monitoring, transportation, and healthcare. These types of applications typically involve the observation of some physical phenomenon through periodic sampling of the environment. In order to make sense of the individually collected samples, nodes usually pass their sensor data through the network to a centralized *sensor-fusion* node where it can be combined and analyzed. We call this process *reactive data fusion*.

One important aspect of reactive data fusion is the need for a common notion of time among participating nodes. For example, acoustic localization requires

S. Nikoletseas et al. (Eds.): DCOSS 2008, LNCS 5067, pp. 17–31, 2008.

the cooperation of several nodes in estimating the position of the sound source based on time-of-arrival data of the wave front [1], [2]. This may require up to 100-microsecond accurate synchronization across the network. Another example is acoustic emissions (AE), the stress waves produced by the sudden internal stress redistribution of materials caused by crack initiation and growth [3]. As the speed of sound is typically an order of magnitude higher in metals than in the air, the synchronization accuracy required for AE source localization can be in the tens of microseconds.

Accurately synchronizing the clocks of all sensor nodes within a heterogeneous network is not a trivial task. Existing WSN time synchronization protocols (e.g. [4], [5], [6], [7], [8]) perform well on the devices for which they were designed. However, these protocols tend to break down when applied to a network of *heterogeneous* devices. For example, the Routing-Integrated Time Synchronization (RITS) protocol [8] was designed to run on the Berkeley motes, and assumes the operating system is tightly integrated with the radio stack. Attempting to run RITS on an 802.11 network can introduce an unacceptable amount of synchronization error because it requires low-level interaction with the hardware, which is difficult to attain in PCs. Reference Broadcast Synchronization (RBS) [5], although portable when it comes to operating system and computing platform, is accurate only when all nodes have access to a common network medium. A combined network of Berkeley motes and PDAs with 802.11b wireless network cards, for example, would be difficult to synchronize using RBS alone because they communicate over different wireless channels. In addition, the communication pathway between a mote and PC is realized using a serial connection. Synchronization across this interface is essential when attempting to combine time-dependent sensor data from each device. Furthermore, it may be desirable to have several mote-PC gateways, since often it is not always possible to extract data fast enough through a single base station. In large networks, this also enables data packets to reach the base station in a fewer number of hops, thus minimizing delay and conserving energy. Although time synchronization across this interface has previously been explored (see Section 2), it has not been implemented in software.

Our work focuses on achieving microsecond-accuracy synchronization in heterogeneous sensor networks (HSNs). HSNs are a promising direction for developing large sensor networks for a diverse set of applications [9], [10], [11]. We consider a multi-hop network consisting of Berkeley motes and Linux PCs, a dominant configuration for reactive data fusion applications. Mote networks consist of resource-constrained devices capable of monitoring environmental phenomena. PCs can support higher-bandwidth sensors such as cameras, and can run additional processing algorithms on the collected data before routing it to the sensor-fusion node. In this sense, we model both motes and PCs as sensor nodes. Time synchronization in both mote and PC networks have been studied independently, however, a sub-millisecond software method to synchronize these two networks has not yet been developed to the best of our knowledge.

In this paper, we present a methodology for HSN time synchronization that utilizes a combination of existing synchronization protocols. Our methodology supports reactive data fusion, incurs little overhead of network resources, and has synchronization error on the order of microseconds. In addition, we have implemented a time synchronization service for reactive data fusion applications, which allows the application developer to focus on aspects of sensor fusion, and not the underlying aggregation mechanism. To achieve accurate cross-platform synchronization, we have developed a technique for synchronization between a mote and PC that is implemented completely in software, and remains effective when multiple mote-PC connections exist within the network.

The rest of this paper is organized as follows. Section 2 describes existing synchronization protocols for WSNs. We present the problem of HSN time synchronization in Section 3. We discuss the sources of synchronization error in Section 4, and clock skew compensation in Section 5. In Section 6, we present our methodology and implementation for HSN time synchronization, and in Section 7 our evaluation results. Section 8 concludes.

2   Related Work

Synchronization protocols can be classified as *sender-receiver*, in which one node synchronizes with another, or *receiver-receiver*, in which multiple nodes synchronize to a common event. Both have their advantages, and each can provide synchronization accuracy on the order of microseconds using certain configurations [12]. An in-depth survey on time synchronization in WSNs can be found in [12].

Several sender-receiver synchronization protocols have been developed for the Berkeley motes and similar small-scale devices that provide microsecond accuracy. Elapsed Time on Arrival (ETA) [13] provides a set of application programming interfaces for an abstract time synchronization service. In ETA, sender-side synchronization error is essentially eliminated by taking the timestamp and inserting it into the message *after* the message has already begun transmission. On the receiver side, a timestamp is taken upon message reception, and the difference between these two timestamps estimate the clock offset between the two nodes. RITS [8] is an extension of ETA over multiple hops. It incorporates a set of routing services to efficiently pass sensor data to a network sink node for data fusion.

In [14], mote-PC synchronization was achieved by connecting the GPIO ports of a mote and IPAQ PDA. The PDA timestamped the output of a signal, which was captured and timestamped by the mote. The mote then sent the timestamp back to the PC, which was able to calculate the clock offset between the two. Although using this technique can achieve microsecond-accurate synchronization, it was implemented as a hardware modification rather than in software.

Reference Broadcast Synchronization (RBS) [5] is a receiver-receiver protocol that minimizes error by taking advantage of the broadcast channel found in most networks. Messages broadcast at the physical layer will arrive at a set of receivers within a tight time bound due to the almost negligible propagation time of sending an electromagnetic signal through air. Nodes then synchronize their clocks to the arrival time of the broadcast message.

3   Problem Statement

Our goal is to provide accurate time synchronization to reactive data fusion applications in HSNs. We refer to the combination of these components as a *configuration*. Our testbed configuration consists of Mica2 motes and Linux PCs. There are two physical networks, the B-MAC network [15] formed by the motes and the 802.11 network containing the PCs. The link between the two is achieved by connecting a mote to a PC using a serial connection. This mote-PC configuration is chosen because it is representative of HSNs containing resource-constrained sensor nodes for monitoring the environment and resource-intensive PCs used for high-bandwidth sensing and computation.

Ideally, a single synchronization methodology would suffice for the entire HSN. However, no protocol has been developed that can achieve this. Instead, we turn to the underlying methodologies found in existing protocols, and use them in conjunction. Individually, the mote and PC networks have been studied extensively. However, the interaction between the two has not been sufficiently investigated. To understand how synchronization affects these connections, we first adopt a system model for time synchronization.

System Model Each sensor node in a WSN is a computing device that maintains its own local clock. Internally, the clock is a piece of circuitry that counts oscillations of a quartz crystal, energized at a specific frequency. When a certain number of these oscillations occur, a *clock-tick* counter is incremented. This counter is accessible from the operating system and its accuracy (with respect to atomic time) depends on the quality of the crystal, as well as various operating conditions such as temperature, pressure, humidity, and supply voltage. When a sensor node registers an event of interest, it will access the clock-tick counter and record a *timestamp* reflecting the time at which the event occurred.

Some protocols synchronize nodes to atomic time, often referred to as *real-time* or *Coordinated Universal Time* (UTC). Irrespective of whether a given protocol synchronizes to UTC, it is often convenient to represent the occurrence of an event according to some universal time standard. We use the notation t to represent an arbitrary UTC time, and the notation t_e to represent the UTC time at which an arbitrary event e occurred. Because each node records a timestamp according to its own clock, we specify the local time on node N_i at which event e was detected by the timestamp $N_i(t_e)$.

Although the two timestamps $N_i(t_e)$ and $N_j(t_e)$ correspond to the same real-time instant t_e, this does not imply that $N_i(t_e) = N_j(t_e)$; the clock-tick counter on node N_i may be offset from N_j. Therefore, from the perspective of node N_i, we define the *clock offset* with N_j at real-time t as $\phi^i_j(t) = N_i(t) - N_j(t)$. It may be the case that the offset changes over time. In other words, the *clock rate* of node N_i, $\frac{dN_i(t)}{dt}$, may not equal the ideal rate ($\frac{dN_i(t)}{dt} = 1$). We define the ratio of clock rates of two nodes as the *relative rate*, $rr^i_j = \frac{dN_i(t)}{dN_j(t)}$. The relative rate is a quantity directly related to the *clock skew*, defined as the difference between clock rates, and is used in our clock skew compensation methods. We refer to clock offset and clock rate characteristics as a node's *timescale*.

Practically all synchronization protocols can be implemented using *timescale transformation*. Rather than setting one clock to another, clocks advance uninhibited, and instead a reference table is maintained that keeps track of the clock offsets and clock drift between a node and its neighbors. The reference table is used to transform a clock value from one timescale to another, providing each node with a common notion of time. Clock adjustment is disadvantageous in WSNs because it leads to increased overhead and possible loss of monotonicity [16]. Therefore, in subsequent sections, we discuss synchronization solely from the perspective of timescale transformation.

4   Sources of Synchronization Error

Synchronization requires passing timestamped messages between nodes. However, this communication has associated message delay, which has both deterministic and nondeterministic components that introduce error into the timescale transformation. We call the sequence of steps involved in communication between a pair of nodes the *critical path*. Figure 1 illustrates the critical path in a wireless connection. The critical path is not identical for all configurations, however, it can typically be characterized by the steps outlined in the figure (for more details, see for example [5], [6], [7]).

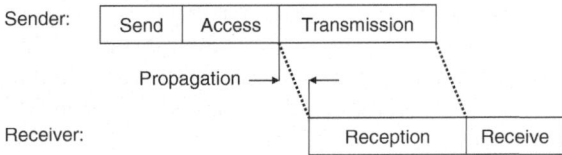

Fig. 1. Critical path

In both the mote and PC networks, the Send and Receive times are the delays incurred as the message is passed between the application and the MAC layer. These segments are mostly nondeterministic due to system call overhead and kernel processing. The Access time is the least deterministic segment in the critical path. It is the delay that occurs in the MAC layer while waiting for access to the communication channel. The Transmission and Reception times are the delays incurred from transmitting and receiving a message over the physical layer, bit by bit. They are mostly deterministic and depend on bit rate and message length. The Propagation time is the time the message takes to travel between the sender and receiver. Propagation delay is highly deterministic.

For the mote-PC serial pathway, the Send and Receive times are similar to wireless communication, however, the Access time is nonexistent. Because the mote-PC connection does not have flow control, pending messages are immediately transmitted without having to wait for an available channel. The Transmission time starts when the data on the sender is moved in 16-byte chunks to the buffer on the UART, and ends after the data is transmitted bit-by-bit across the serial port to the receiver. Similar to wireless networks, the Propagation time is minimal. The UART on the receiver places the received bits into its own 16-byte buffer. When the buffer is almost full, or a timeout occurs, it sends an interrupt to the CPU, which notifies the serial driver, and the data is transferred to main memory where it is packaged and forwarded to the user-space application.

In addition to the error from message delay nondeterminism, synchronization accuracy is also affected by clock skew when a pair of nodes operate for extended periods of time without correcting their offset. For example, suppose at real-time t, nodes N_1 and N_2 exchange their current clock values at local times $N_1(t)$ and $N_2(t)$, respectively. At some later time, an event e occurs that is detected and timestamped by N_1, which sends its timestamp $N_1(t_e)$ to N_2. If the clock rates on each node were equal, N_2 would simply be able to take the previously calculated offset and use it to transform the event timestamp $N_1(t_e)$ to the corresponding local time $N_2(t_e) = N_1(t_e) + \phi_2^1(t)$. However, if the relative rate $rr_2^1(t)$ is not equal to 1, but $1 + 20$ ppm[1], for example, an attempt to convert $N_1(t_e)$ to the local timescale would result in an error of $20 * 10^{-6} * (N_2(t_e) - N_2(t))\mu s$. If the interval between the last synchronization and the event was one minute, the resulting error due to clock skew alone would amount to 1.2 milliseconds!

For accurate synchronization, it is therefore necessary to minimize the nondeterministic sources of error in the critical path, and account for the deterministic sources by appropriately adjusting the timescale transformation. Clock skew compensation is necessary for minimizing synchronization error when nodes run for long periods of time without updating their offset. With an estimation of clock offset and relative rate, a complete timescale transformation, which converts an event timestamp $N_j(t_e)$ from the timescale of node N_j to the timescale of N_i, can be defined as

$$N_i(t_e) = N_i(t_s) + rr_j^i(t_s)[(N_j(t_e) - N_j(t_s)]$$

where $N_i(t_s)$ and $N_j(t_s)$ are the respective local times at which nodes N_i and N_j exchanged their most recent synchronization message, s.

5   Clock Skew Compensation

Independent of synchronization protocol, there are several options for clock skew compensation. The simplest is to do nothing. In some applications, event detection and synchronization always occur so close together in time that clock skew compensation is unnecessary. However, when it does become necessary, nodes must exchange timestamps periodically to ensure their clocks do not drift too far apart. In resource-constrained WSNs, this may be undesirable because it can result in high message overhead. To keep message overhead at a minimum, nodes can exchange synchronization messages less frequently and instead maintain a

---

[1] Parts per million (10^{-6}). A relative rate of 1 ppm means that one clock ticks $1\mu s/s$ faster than the other.

history of their neighbors' timestamps. Statistical techniques can then be used to produce an accurate estimate of clock offset at any time instant.

A linear regression fits a line to a set of data points such that the square of the error between the line and each data point is minimized overall. By maintaining a history of n local-remote timestamp pairs, node N_i can derive a linear relation $N_i(t) = \alpha + \beta N_j(t)$ and solve for the coefficients α and β. Here, β represents an estimation of $rr_j^i(t)$. A problem arises when attempting to improve the quality of the regression by increasing the number of data points. This can result in high memory overhead, especially in dense networks. However, it has been shown that sub-microsecond clock skew error can be achieved with as few as six timestamps in mote networks [7].

Exponential averaging solves the problem of high memory overhead by keeping track of only the current relative rate and the most recent neighbor-local synchronization timestamp pair. When a new timestamp pair is available, the relative rate is adjusted. Because the relative rate estimate is partially derived from its previous value, there will be a longer convergence time before an accurate estimate is reached. This can be reduced by providing the algorithm with an initial relative rate, determined experimentally.

The phase-locked loop (PLL) is a mechanism for clock skew compensation used in NTP [17]. The PLL compares the ratio of a current local-remote timestamp pair with the current estimate of relative rate. The PLL then adjusts the estimate by the sum of a value proportional to the difference and a value proportional to the integral of the difference. PLLs generally have a longer convergence time than linear regression and exponential averaging, but have low memory overhead. A diagram of a PLL implementation is illustrated in Figure 2. The Phase Detector calculates the relative rate between two nodes and compares this with the output of the PLL, which is the previous estimate of the relative rate. The difference between these two values is the phase error, which is passed to the second-order Digital Loop Filter. Because we expect there to be some amount of phase error, we choose a filter with an integrator, which allows the PLL to eliminate steady-state phase error. To implement this behavior in software, a digital accumulator is used, and is represented by $y(t) = (K_1 + K_2)u(t) - 10K_2K_1u(t-1) + 10K_2y(t-1)$. The resulting static

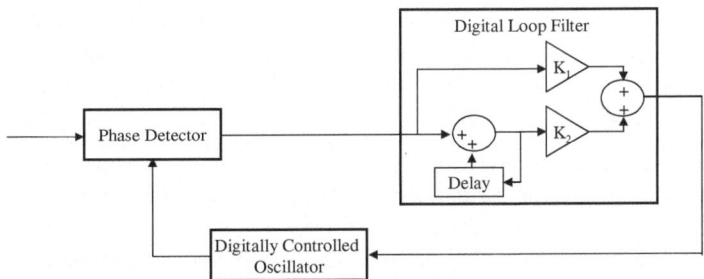

Fig. 2. Phase-locked loop

phase error is passed to the Digitally Controlled Oscillator (DCO). The DCO sums the previous phase error with the previous output, which produces the current estimate of relative clock rate, and is fed back into the PLL. Techniques for selecting the gains are presented in [17].

6   Time Synchronization in HSNs

In this section, we present our HSN synchronization methodology and the architecture of our synchronization service.

Synchronization Methodology. The accuracy of a receiver-receiver synchronization protocol (such as RBS) is comparable to sender-receiver synchronization (such as RITS) in mote networks. However, receiver-receiver synchronization has greater associated communication overhead, which can shorten the lifetime of the network. Therefore, we selected RITS to synchronize the mote network in our HSN.

Synchronization of PC networks has been studied extensively over the past four decades, however, popular sender-receiver protocols such as the Network Time Protocol (NTP) [18] only provide millisecond accuracy. This is acceptable because PC users typically do not require greater synchronization precision for their applications. For microsecond-level synchronization accuracy in PC networks, a receiver-receiver protocol such as RBS outperforms sender-receiver protocols because it has less associated message delay nondeterminism. We therefore use RBS to synchronize our PC network.

To synchronize a mote with a PC in software, we adopted the underlying methodology of ETA and applied it to serial communication. On the mote, a timestamp is taken upon transfer of a synchronization byte and inserted into the outgoing message. On the PC, a timestamp is taken immediately after the UART issues the interrupt, and the PC regards the difference between these two timestamps as the PC-mote offset, ϕ_{mote}^{pc}. Serial communication bit rate between the mote and PC is 57600 baud, which approximately amounts to a transfer time of 139 microseconds per byte. However, the UART will not issue an interrupt to the CPU until its 16-byte buffer nears capacity or a timeout occurs. Because the synchronization message is six bytes, reception time in this case will consist of the transfer time of the entire message in addition to the timeout time and the time it takes to transfer the date from the UART buffer into main memory by the CPU. This time is compensated for by the receiver, and the clock offset between the two devices is determined as the difference between the PC receive time and the mote transmit time.

Architecture. We have developed a PC-based time synchronization service for reactive data fusion applications in HSNs[2]. Figure 3 illustrates the interaction of each component within the service. The service collects sensor data from applications that run on the local PC, as well as from other service instances running

---

[2] Our synchronization service implementation is available as open source, and can be found at http://www.isis.vanderbilt.edu/Projects/NEST/HSNTimeSync.html

on remote PCs. It accepts event messages on a specific port, converts the embedded timestamps to the local timescale, and forwards the messages toward the sensor-fusion node. To maintain synchronization with the rest of the PC network, the service uses RBS. The arrival times of the reference broadcasts are stored in a reference table and accessed for timescale transformation. In addition, the service accepts mote-based event messages, and converts the embedded timestamps using the ETA serial timestamp synchronization method outlined above. The messages are then forwarded toward the sensor-fusion node. The service instance that resides on the sensor-fusion node transforms incoming timestamps into its local timescale before passing the event messages up to the sensor-fusion application.

Kernel modifications in the serial and wireless drivers were required in order to take accurate timestamps. Upon receipt of a designated synchronization byte, the time is recorded and passed up to the synchronization service in the user-space. The mote implementation uses the TimeStamping interface, provided with the TinyOS distribution [19]. A modification was made to the UART interface to insert a transmission timestamp into the event message as it is being transmitted between the mote and PC. The timestamp is taken immediately before a synchronization byte is transmitted, then inserted at the end of the message.

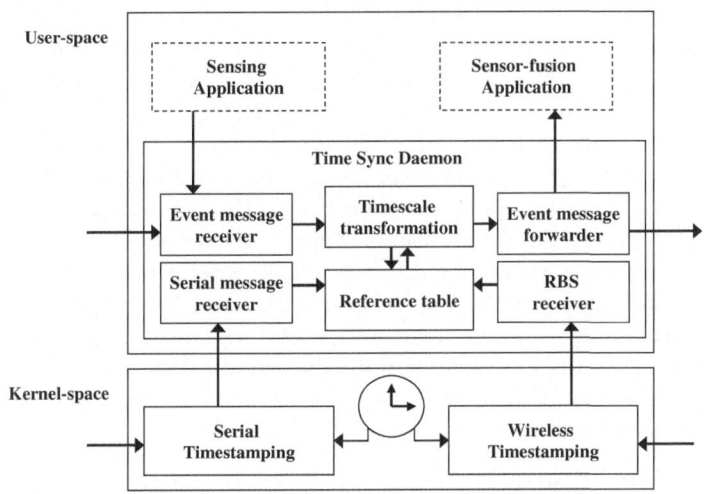

Fig. 3. PC-based time synchronization service

7   Evaluation

Experimental Setup. Our HSN testbed consists of seven Crossbow Mica2 motes and four stationary ActivMedia Pioneer robots with embedded Redhat Linux 2.4 PCs, as illustrated in Figure 4. In addition we employ a Linux PC to transmit RBS beacons. We chose this testbed because the issues that arise here

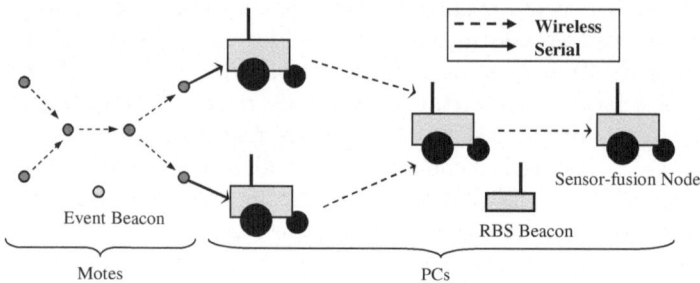

Fig. 4. Our sensor network testbed. Arrows indicate communication flow.

are representative of practical HSN configurations such as hierarchical clustering and networks with multiple sinks. In addition, routing sensor data from a mote network to a PC base station is a dominant communication pathway in sensor network architectures.

The reference broadcast node transmits a reference beacon containing a sequence number once every ten seconds. The arrival of these messages are timestamped in the kernel and stored in a reference table. Simultaneously, a designated mote broadcasts event beacons, once every $4000 \pm \epsilon$ milliseconds, where ϵ is a random number in the range (0,1000). Six hundred event beacons are broadcast per experiment. The motes timestamp the arrival of the event beacon, and place the timestamp into a message. The message is routed over three hops in the mote network to the mote-PC gateways, using RITS to convert the timestamp to the local timescale with each hop. The message is next transferred from the mote network to the PC network over the mote-PC serial connections, and the event timestamp is converted to the timescale of the gateway PCs. The gateway PCs forward the message two additional hops to the sensor-fusion node. The experiment was repeated using the different clock skew compensation techniques in the PC network, as described in Section 5. Because RITS synchronizes a single sender with a single receiver at the time of data transfer, and because event data is forwarded to the base station immediately after the event is detected or a data message is received, clock skew compensation in the mote network provides negligible improvement.

Subsystem Synchronization Results. We performed a series of experiments to quantify synchronization error on the individual critical paths in the HSN. The results allow us to justify our selection of synchronization protocols for the entire HSN. Note that no clock skew compensation was performed for these initial tests. To determine synchronization error, we used the *pairwise difference* evaluation method. Two nodes, N_1 and N_2, simultaneously timestamp the occurrence of an event, such as a reference beacon. These timestamps are then transformed to the timescale of node N_3, and the absolute value of their difference represents the error in the timescale transformation.

Experimental results in the literature (e.g. [8], [13]) indicate that RITS works well for synchronizing the mote network. We confirmed this on a 3-hop network

of Mica2 motes. A beacon was broadcast to two outlying motes, which timestamped its arrival and forwarded the timestamp to a network sink node 3 hops
away. At each intermediate node, the timestamps were converted to the local
timescale. Synchronization error was calculated as timestamps arrived at the
network sink node and, over 100 synchronizations, the average error was $7.27\mu s$,
with a maximum of $37\mu s$.

Based on the implementation described in [5], we synchronized our PC network using RBS. We used a separate transmitter to broadcast a reference beacon
every ten seconds (randomized) for 100 runs. Two PCs received reference broadcast r at local times $PC_1(t_r)$ and $PC_2(t_r)$, respectively. Synchronization error
was $8.10\mu s$ on average, and $43\mu s$ maximum. Results are displayed in Figure 5a.

Figure 5b plots the synchronization error between two PCs using RITS. Every two seconds, PC_1 sent two synchronization messages to PC_2. Immediately
before the command to send the first message was issued to the network interface controller on the sender, a transmission timestamp $PC_1(t_{tx})$ was taken
in the kernel. This was found to be the latest possible time for the sender to
take the timestamp. However, by the time the timestamp had been acquired,
the message had already been transmitted, so a second message was needed to
deliver the timestamp to the receiver. PC_2 recorded the timestamp $PC_2(t_{rx})$ in
the kernel interrupt function upon receipt of the first message, and obtained the
sender timestamp in the second message shortly after. The results show that we
cannot expect consistent synchronization accuracy using RITS with the 802.11
networked Linux PCs. This is partly due to sender-side message delay error in
RITS, which is nonexistent in RBS. In addition, the PC-based operating system
is not tightly integrated with the network interface, and therefore the timestamping precision of the transmission and reception of sync bytes is degraded.

To synchronize the mote with the PC, we used the synchronization methodology described in Section 6. To evaluate synchronization accuracy, GPIO pins
on the mote and PC were connected to an oscilloscope, and set high upon timestamping. The resulting output signals were captured and measured. The test
was performed over 100 synchronizations, and resulting error was $7.32\mu s$ on average. The results are displayed in Figure 6. The majority of the error is due to
nondeterministic message delay resulting from jitter, both in the UART and the
CPU. A technique to compensate for such jitter on the motes is presented in [7],
however, we did not attempt it on the PCs.

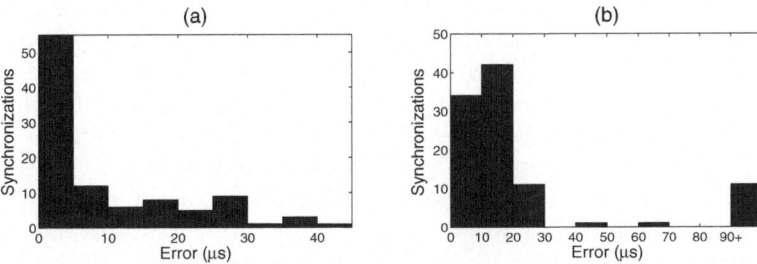

Fig. 5. (a) RBS and (b) RITS synchronization error between two PCs

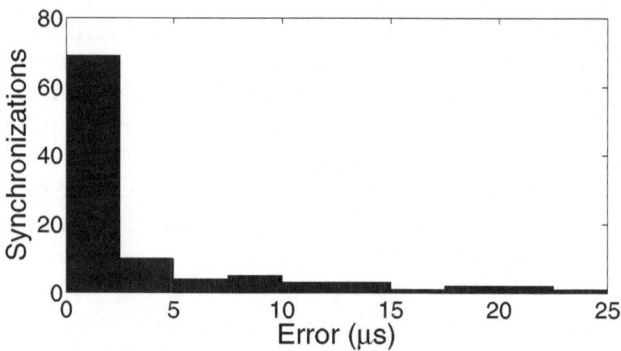

Fig. 6. Mote-PC error using ETA

HSN Synchronization Results. Figure 7 summarizes the synchronization error for each type of clock skew compensation technique under normal operating conditions, high network congestion, and high I/O load. To simulate a high network load, an extra mote and PC (not pictured in Figure 4) were introduced, each broadcasting messages of random size every 100 milliseconds. We first examined synchronization without clock skew compensation. The sensor-fusion node reported the average difference between the source timestamps as $12.05\mu s$, with a maximum of $270\mu s$. Next, we synchronized the PC network using linear regression as our clock skew compensation technique. Linear regression was implemented with a history size of 8 local-remote timestamp pairs for each neighbor, and the relative rate was initialized to 1. As expected, there was a notable improvement in synchronization accuracy, with an average of $7.62\mu s$ error, and a maximum of $84\mu s$. Repeating the experiment using exponential averaging gives errors similar to linear regression. For exponential averaging, we chose a value of 0.10 for α, and initialized the average relative rate to 1. These values were determined experimentally for rapid synchronization convergence. The average error recorded was $8.82\mu s$, with a maximum of $112\mu s$. The average synchronization error using phase-locked loops was $7.85\mu s$, with a maximum of $196\mu s$. For the digital loop filter, we used gains of $K_1 = 0.1$ and $K_2 = 0.09$, determined experimentally.

Memory overhead is minimal for each clock skew compensation technique. Nodes require 8 bytes for the current relative rate estimate for each neighbor within single-hop range. In the case of linear regression, a small history buffer for each neighbor is also required. Our implementation has no message overhead (except for the beacons transmitted by the RBS server). In fact, the only modification to the data message is the addition of a four-byte timestamp. Because these synchronization timestamps piggyback on data messages, no additional messages are required. Our methodology is therefore energy efficient, because message overhead directly impacts energy consumption. Convergence time depends on input parameters to the clock skew compensation algorithm. We found that, on average, it took the network 80 seconds to synchronize with linear

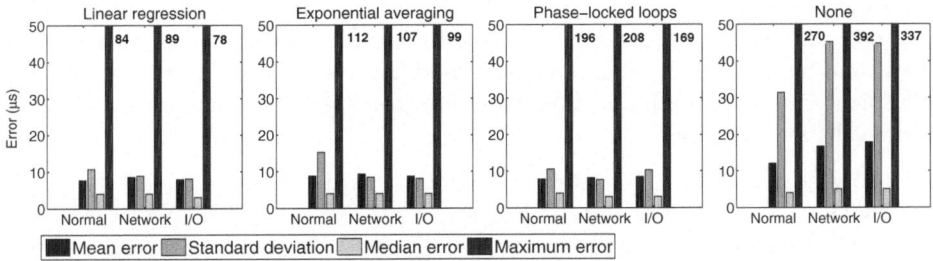

Fig. 7. HSN synchronization error with different types of clock skew compensation under normal operating conditions (Normal), high network congestion (Network), and high I/O load (I/O)

regression, 200 seconds with exponential averaging, and 1300 seconds with phase-locked loops. Note that these convergence times reflect an inter-beacon delay of four seconds.

Discussion. These results show that we are able to achieve microsecond-accurate synchronization in the HSN. Because error accrues with each timescale transformation, achieving this level of synchronization accuracy over a 6-hop network of heterogeneous devices is significant. Although the maximum synchronization error extends to tens of microseconds, it was principally caused by a small number of synchronization attempts in which prolonged operating system interrupt operations occurred. Although these were difficult to avoid, they did not occur frequently, and the worst-case synchronization error for each type of clock skew compensation technique was acceptable for most kinds of HSN data fusion applications. Furthermore, time synchronization was not affected by the most common types of nondeterministic behavior, such as network congestion and I/O operations.

The accuracy of the mote-PC synchronization is quite good. Although we did see error in the tens of microseconds, the maximum did not exceed $50\mu s$, and the average was below $10\mu s$. This is significant, because it demonstrates that microsecond accuracy synchronization can be achieved between mote and PC networks, enabling the development of HSN applications that require precision time synchronization. In addition, because this technique uses UART communication, it can easily be adapted for synchronization with UART-supported peripheral sensing devices.

8   Conclusion

Time synchronization is an important and necessary component in most wireless sensor network applications. However, as we describe in this paper, its implementation is non-trivial, especially when high-precision synchronization is required. In networks of heterogeneous devices, the problem is compounded by issues of system integration, and therefore alternative techniques must be employed to reduce synchronization error. We have shown that with certain configurations, microsecond accuracy can be achieved with careful selection of hardware, software, and network components. Our methodology is generally portable to other platforms, provided mechanisms exist within the target configuration that enable low-level timestamping.

Acknowledgements. This work was supported in part by a Vanderbilt University Discovery Grant, ARO MURI grant W911NF-06-1-0076, NSF CAREER award CNS-0347440, and NSF grant CNS-0721604. The authors would also like to thank Manish Kushwaha and Janos Sallai for their help with this project.

References

1. Williams, S.M., Frampton, K.D., Amundson, I., Schmidt, P.L.: Decentralized acoustic source localization in a distributed sensor network. Applied Acoustics 67 (2006)
2. Ledeczi, A., Nadas, A., Volgyesi, P., Balogh, G., Kusy, B., Sallai, J., Pap, G., Dora, S., Molnar, K., Maroti, M., Simon, G.: Countersniper system for urban warfare. ACM Transactions on Sensor Networks 1(2) (2005)
3. Huang, M., Jiang, L., Liaw, P.K., Brooks, C.R., Seeley, R., Klarstrom, D.L.: Using acoustic emission in fatigue and fracture materials research. JOM 50(11) (1998)
4. Romer, K.: Time synchronization in ad hoc networks. In: ACM MobiHoc. (2001)
5. Elson, J., Girod, L., Estrin, D.: Fine-grained network time synchronization using reference broadcasts. In: OSDI (2002)
6. Ganeriwal, S., Kumar, R., Srivastava, M.B.: Timing-sync protocol for sensor networks. In: ACM SenSys. (2003)
7. Maroti, M., Kusy, B., Simon, G., Ledeczi, A.: The flooding time synchronization protocol. In: ACM SenSys. (2004)
8. Sallai, J., Kusy, B., Ledeczi, A., Dutta, P.: On the scalability of routing integrated time synchronization. In: EWSN (2006)
9. Yarvis, M., Kushalnagar, N., Singh, H., Rangarajan, A., Liu, Y., Singh, S.: Exploiting heterogeneity in sensor networks. In: IEEE Infocom (2005)
10. Duarte-Melo, E., Liu, M.: Analysis of energy consumption and lifetime of heterogeneous wireless sensor networks. In: IEEE Globecom (2002)
11. Lazos, L., Poovendran, R., Ritcey, J.A.: Probabilistic detection of mobile targets in heterogeneous sensor networks. In: IPSN (2007)
12. Romer, K., Blum, P., Meier, L.: Time synchronization and calibration in wireless sensor networks. In: Stojmenovic, I. (ed.) Wireless Sensor Networks, Wiley and Sons, Chichester (2005)
13. Kusy, B., Dutta, P., Levis, P., Maroti, M., Ledeczi, A., Culler, D.: Elapsed time on arrival: A simple and versatile primitive for time synchronization services. International Journal of Ad hoc and Ubiquitous Computing 2(1) (2006)
14. Girod, L., Bychkovsky, V., Elson, J., Estrin, D.: Locating tiny sensors in time and space: A case study. In: ICCD: VLSI in Computers and Processors (2002)
15. Polastre, J., Hill, J., Culler, D.: Versatile low power media access for wireless sensor networks. In: ACM SenSys. (2004)
16. Elson, J., Romer, K.: Wireless sensor networks: A new regime for time synchronization. In: HotNets-I (2002)

17. Mills, D.L.: Modelling and analysis of computer network clocks. Technical Report 92-5-2, Electrical Engineering Department, University of Delaware (1992)
18. Mills, D.L.: Internet time synchronization: The network time protocol. IEEE Transactions on Communications 39(10) (1991)
19. Levis, P., Madden, S., Gay, D., Polastre, J., Szewczyk, R., Woo, A., Brewer, E., Culler, D.: The emergence of networking abstractions and techniques in TinyOS. In: NSDI (2004)

Stochastic Counting in Sensor Networks, or: Noise Is Good*

Y.M. Baryshnikov[1], E.G. Coffman[2], K.J. Kwak[2], and Bill Moran[3]

[1] Bell Labs
600 Mountain Ave.
Murray Hill, NJ 07974
ymb@research.bell-labs.com
[2] Electrical Engineering Dept.
Columbia University
New York, NY 10027
{egc,kjkwak}@ee.columbia.edu
[3] Electrical Engineering Dept.
University of Melbourne
Australia
b.moran@ee.unimelb.edu.au

Abstract. We propose a novel algorithm of counting indistinguishable objects by a collection of sensors. The information on multiple counting is recovered from the stochastic correlation patterns.

1 Introduction

One of the tasks fundamental to sensor applications is *counting*; that is, receiving and processing often noisy data, and returning the *number of objects* of interest located within a given sensor field. Such problems arise, for example in the context of ecological or agricultural monitoring where, for example, a sensor might report the number of animals of a certain kind that it observes. In such situations, discriminating between different animals is almost certainly infeasible. Accordingly, we adopt the *minimality paradigm* (cf. [2]), and postulate that the sensors have the simplest functionality required to perform their task. In our case, while sensors can count the objects within their sensing ranges, they cannot identify them on the basis of location. in particular, locations relative to the domains of other sensors are unknown. One immediate hurdle to achieving the goal is then the over-counting of the objects: if one has no means to identify the objects, naïve summation of the counts reported by different sensors inflates the total count. Is it possible to correct this over-count without resorting to explicit identification of the counted objects? At face value, this goal seems to be completely unattainable, so it may come as something of a surprise that, by a process of *stochastic data fusion*, in this case fusing data with noise in the sensing process, we can design efficient counting procedures.

---

* Research supported by DARPA DSO # HR0011-07-1-0002 via the project SToMP.

S. Nikoletseas et al. (Eds.): DCOSS 2008, LNCS 5067, pp. 32–45, 2008.

2   Problem Formulation

We assume that the ultimate task of the sensor network is to recover the total number of objects, henceforth referred to as *targets*. Targets τ potentially in the sensor field comprise a set $\mathcal{T}$. The set of targets actually in the sensor field is given by $\{\tau \in \mathcal{T} | X_\tau = 1\}$, where the indicator X_τ is 1 or 0 according as τ is or is not in the sensor field. Let $N := \sum_\tau X_\tau$ be the number of targets in the sensor field, which are to be observed by a collection $\mathcal{S}$ of sensors, σ. Typically, the sensing domain of a sensor is a disk of given radius. The sensor field is usually modeled as a rectangular subset of the union of the sensor domains, an assumption being that every point of the rectangular field is covered by at least one sensor. While perhaps useful to keep in mind as describing a canonical model, these assumptions are not needed in what follows.

We formalize target location relative to sensor domains by introducing an *incidence coefficient* $Z_{\sigma\tau}$ for each $\sigma \in \mathcal{S}$, $\tau \in \mathcal{T}$, which is 1 or 0 according as τ is or is not in the sensing area of σ. Thus, τ is in the intersection of $\sum_\sigma Z_{\sigma\tau}$ sensing domains. We can now define the target count C_σ by sensor σ as $C_\sigma = \sum_{\tau \in \mathcal{T}} Z_{\sigma\tau} X_\tau$, so the task of the sensor network we intend to design and analyze becomes the following:

> Given the counts C_σ of the targets registered by each sensor σ, estimate the total number N of targets in the area covered by the sensors within the sensor field.

It is immediately clear that the problem as stated does not have a well-defined solution; indeed, the total count depends on the number of targets counted several times by different sensors. We have only the rough bounds $\max_{\sigma \in \mathcal{S}} C_\sigma \leq N \leq \sum_{\sigma \in \mathcal{S}} C_\sigma$. More precisely, as the well-known inclusion-exclusion formula indicates,

$$N = \sum_\sigma C_\sigma - \sum_{\sigma_1 < \sigma_2} C_{\sigma_1 \sigma_2} + \ldots + (-1)^\ell \sum_{\sigma_1 < \cdots < \sigma_\ell} C_{\sigma_1 \sigma_2 \cdots \sigma_\ell} + \ldots \qquad (1)$$

where for an ordered subset $\sigma_1 < \sigma_2 < \ldots < \sigma_\ell$, $C_{\sigma_1 \sigma_2 \cdots \sigma_\ell}$ denotes the number of targets detected simultaneously by the sensors $\sigma_1, \sigma_2, \ldots, \sigma_\ell$. (Recall that the summation terms can be interpreted as alternately compensating for the total under and over counts of the partial sums to their left.)

2.1   Model Specifications

In the model above, we abstracted to the extreme the properties of sensor counting systems. It is instructive to map the model to some concrete realizations, to indicate the situations in which the techniques developed in this paper might be applied.

- *Domain coverage.* Many areas of application of sensor networks involve a domain, planar or spatial, and a finite collection of subdomains (sensing areas), labeled by the sensors, and consisting of the points where a target can be registered by the labeling sensor. An example of such applications would be an environmental monitoring system (counting the number of animals), or area surveillance systems, aiming at keeping track of the total number of persons or items of equipment in a facility. In this case each count (reading) returns instantaneous data on the number of targets in the sensed domain. Realistically, the domains are bound to overlap (to ensure complete coverage) making the estimation of the total count highly nontrivial.

- *Linear coverage.* In applications to transportation, the sensors can be placed along a transportation link, and register the number of targets passing by. In this case, the number of targets registered by a sensor would encompass the routes crossing the location of the sensor. Notice that in this setting, the readings need not, and in fact cannot be instantaneous, as the targets require time to move between the locations of the sensors.
- *Communication networks.* Application to communication networks requires an extension of the previous case to more general graphs. In this context targets represent the unique packets (or communication sessions) between some source-destination pairs, and a sensor corresponds to a counter of such packets or sessions traversing a router.

In each of these contexts, the problem is nontrivial, because of the overlap issue described earlier, together with the anonymous nature of the targets.

2.2 Stochastic Overlap Recovery

The total lack of information describing the overlap of the sensing areas precludes any rational hypothesis building when it comes to eliminating target over-counts. On the other hand, one can reason that *if the target visibility is not deterministic, but stochastic, then the contributions to the counts by different sensors coming from overlapping sensing areas might be correlated in some useful way.* One is led to speculate on whether the extra information given by correlations can in turn lead to a solution to our problem.

This paper shows indeed that exploitation of this simple idea parlays into a solution to our counting problem. In a sense, we acquire the ability to detect overlaps of the sensing areas at a sacrifice in the deterministic nature of the observations. "Noise is good" in that, by measuring randomly perturbed signals, we gather additional information, as compared to the situation when the measurements are noiseless; in the latter case, all measurements by a sensor would be identical and contribute nothing to our knowledge of the target counts.

The basic quantitative measure of correlation between two random variables is the covariance. We will see that the covariance between the counts by different sensors reflects the nature of the intersection between the sensing areas. To capture intersections of higher orders, however, one needs correlations of higher order. There are many statistical tools suitable for exploring the correlation structure of the data. This paper applies the notion of *cumulants*, which generalizes that of covariance coefficients, and which carries all the information of stochastic interdependency of the random variables being observed.

3   Related Work

The counting problem has been discussed in many papers in the system engineering context, see e.g. [10]. Most of the papers, however, assume that there is either explicit or implicit identification of the targets, or incremental counting, where the targets are counted one-by-one.

Earlier work by Marzullo [8] proposed an interval-based algorithm to detect faults and recover from errors. Assume at most f out of n sensors can be faulty and each sensor records the interval in which an object is detected. Marzullo calculates the smallest interval that contains all of the intersections of the $n - f$ intervals guaranteed to contain the true value that the event occurred. Based on Marzullo's work, Iyengar et al [7] propose another interval-based algorithm that produces a smaller interval under the assumption that the intervals of false sensing are sufficiently close.

In [5], the author addresses the question of counting anonymous targets in the context of robotic exploration. This poses challenges quite different from those arising in the context of this paper.

Brooks pointed out in [3] that integration (or fusion) of sensor readings is crucial to automating sensor networks, and that such distributed computing reinforces integrity of the system. To increase the precision of data, Brooks proposes sensor fusion and Byzantine agreement. To overcome faults in sensor readings, Clouqueur el al [4] propose two collaborative signal processing algorithms: value fusion and decision fusion. Sensors exchange a parameter (this can be consumed energy, required communication bandwidth, etc) to reach agreement as to whether a decision was correct or not, so recovery from the fault will be possible. The concept of *spatial sensor mining* (derivation of conclusions from distributed information collected over time), is introduced in [6]. They propose a change of focus from many sensors on a single target to many sensors on many targets.

Natheta [9] addresses the interesting but rather narrow problem of counting the number of nodes in the sensor network itself.

From our literature survey, it appears that no previous research has considered how to eliminate redundancy in data, to recover from errors, and hence to retrieve correct total target counts.

4   Preliminaries

This section introduces the necessary probabilistic tools (see e.g. [1] for greater detail). For a random variable X define the *cumulant generating function* as

$$\kappa(t) = \sum_n c_n(X) \frac{t^n}{n!} = \ln \mathbb{E} e^{tX}, \tag{2}$$

where we make the assumption that the Laplace transform of X is defined in some neighborhood of 0. In particular, the first two *factorial coefficients* are the mean and variance

$$c_1(X) = \mathbb{E} X \qquad c_2(X) = \mathbb{V} X \tag{3}$$

Given a collection of random variables $\mathbf{X} = (X_1, X_2, \ldots, X_n)$, not necessarily independent, define the cumulants $\kappa_n(X_1, \ldots, X_n)$ as

$$\kappa_n(X_1, \ldots, X_n) = \frac{\partial^n}{\partial t_1 \cdots \partial t_n} g(\mathbf{t})|_{\mathbf{t}=0},$$

where $\mathbf{t} = (t_1, \ldots, t_n)$ and $g(\mathbf{t}) = ln(\mathbb{E}(e^{\mathbf{t} \cdot \mathbf{X}}))$ is the logarithm of the Laplace transform of the vector-valued random variable $\mathbf{X}$. For example, the cumulant of the first order, $\kappa_1(X_1)$ is just the expected value of X_1; the joint cumulant of two random variables equals their covariance,

$$\kappa_2(X_1, X_2) = \mathbb{E}X_1X_2 - \mathbb{E}X_1\mathbb{E}X_2, \tag{4}$$

and the cumulant of three random variables X_1, X_2, and X_3 is given by

$$\begin{aligned}
\kappa_3(X_1, X_2, X_3) = {}& \mathbb{E}(X_1X_2X_3) - \mathbb{E}(X_1X_2)\mathbb{E}(X_3) \\
& - \mathbb{E}(X_1X_3)\mathbb{E}(X_2) - \mathbb{E}(X_2X_3)\mathbb{E}(X_1) + 2\mathbb{E}(X_1)\mathbb{E}(X_2)\mathbb{E}(X_3)
\end{aligned} \tag{5}$$

In general, the cumulant $\kappa_n(X_1, \ldots, X_n)$ of order n is a polynomial in joint moments of the random variables and can be recovered recursively from the general formula

$$\mathbb{E}\prod_{i=1}^{n} X_i = \sum_{\Pi = P_1 \amalg P_2 \ldots \amalg P_k} \prod_{j=1}^{k} \kappa(P_j), \tag{6}$$

where Π is a partition of the index set $\{1, 2, \ldots, n\}$ into blocks P_j, and $c(P_j)$ is the cumulant of order $|P_j|$ of the random variables with indices in P_j. Thus, $\mathbb{E}X_1X_2 = \kappa_2(X_1, X_2) + \kappa_1(X_1)\kappa_1(X_2)$ and

$$\begin{aligned}
\mathbb{E}X_1X_2X_3 = {}& \kappa_3(X_1, X_2, X_3) + \kappa_2(X_1, X_2)\kappa_1(X_3) + \kappa_2(X_2, X_3)\kappa_1(X_3) \\
& + \kappa_2(X_1, X_3)\kappa_1(X_2) + \kappa_1(X_1)\kappa_1(X_2)\kappa_1(X_3).
\end{aligned}$$

We will be using the following important facts concerning cumulant functions.

PROPERTY 1: The cumulants are symmetric functions of their arguments;
PROPERTY 2: The cumulants are multilinear:

$$\kappa_n(X_1' + X_1'', X_2, \ldots, X_n) = \kappa_n(X_1', X_2, \ldots, X_n) + \kappa_n(X_1'', X_2, \ldots, X_n);$$

PROPERTY 3: If the random variables $X_1, \ldots, X_n$ can be split into two groups, say $\mathbf{X}' = (X_1, \ldots, X_k)$ and $\mathbf{X}'' = (X_{k+1}, \ldots, X_n)$, so that $\mathbf{X}'$ and $\mathbf{X}''$ are independent, then

$$\kappa_n(X_1, \ldots, X_n) = 0.$$

PROPERTY 4: If the X_i are different notations for the same random variable X, then $\kappa_n(X, \ldots, X) = c_n(X)$, the n-th factorial coefficient of the cumulant generating function of X.

5   Sensing Area Intersections and Cumulants

Let us return to the original problem of estimating the number of targets detected in the sensing domains of a collection of sensors.

5.1   Bernoulli Visibility Variables

We introduce the often justifiable assumption that the targets are visible to the sensors not continuously, but intermittently, in some time-dependent fashion. *Further, we assume that the visibility of the targets is random and independent across the targets.* Situations where such assumptions are valid range from environmental deployment of sensor networks, where the targets might be animals not covered by foliage, to perimeter protection applications, where moving targets are detected and counted. We do not go into further details here, for our interest is restricted to the algorithms underlying the solutions to the counting problem complicated by overlapping sensing areas and imperfect visibility.

Up to this point, X_τ has indicated (deterministically) whether τ was located somewhere in the sensor field. It is convenient now to consider that all targets are in the sensor field and let the X_τ model the random *visibility* of the targets. Formally,

$$X_\tau = \begin{cases} 1 \text{ if the target is } visible; \\ 0 \text{ otherwise} \end{cases}$$

are *independent* $\{0,1\}$-*valued random variables*. The sensor counts $C_\sigma = \sum_\tau Z_{\sigma\tau} X_\tau$, then become random variables. Sensing noise is defined precisely by the following visibility assumption.

The i.i.d. random variables X_τ are Bernoulli(p) *distributed, that is, they independently take the value 1 with probability p, and 0 with probability $1 - p$.*

We infer that the count C_σ has the Binomial(p, N_σ) distribution, where $N_\sigma = \sum_\tau Z_{\sigma\tau}$ is the number of targets in σ's sensing area.

To appreciate the basic ideas behind our counting algorithm, we need only consider a system of two sensors. For this case, if A_1 and A_2 denote the two sensing areas, then the sensor-field partition of interest consists of the disjoint subsets $A_1 \setminus A_2$, $A_2 \setminus A_1$, and $A_1 \cap A_2$, which we denote by $A_{1\backslash 2}, A_{2\backslash 1}, A_{12}$. A similar notation applies to the number of targets in the respective areas, e.g., $N_{1\backslash 2}, N_{12}, \dots$ and to the counts $C_{1\backslash 2}, C_{12}, \dots$. To verify the fact that the cumulant for the two sensor counts C_1 and C_2 is simply their covariance, write

$$\kappa_2(C_1, C_2) = \kappa_2(C_{1\backslash 2} + C_{12}, C_2) = \kappa_2(C_{12}, C_2)$$
$$= \kappa_2(C_{12}, C_{2\backslash 1} + C_{12}) = \kappa_2(C_{12}, C_{12})$$
$$= N_{12} c_2(\texttt{Bernoulli}(p)) = p(1 - p)N_{12}$$

where Properties 2 and 3 have been applied in the first two lines and Property 4 in the last line.

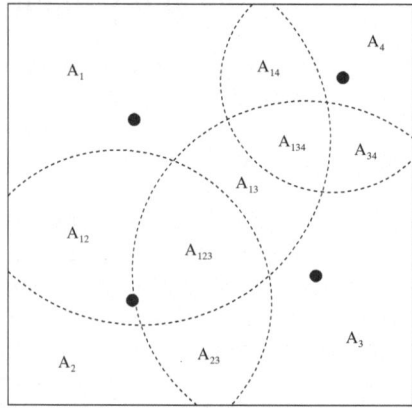

Fig. 1. Sensor Coverage Area

This simple formula carries two useful messages. Firstly, assuming p is known, knowledge of the covariance gives information about the size of the intersections of the sensing areas A_1 and A_2. Secondly, this information can be recovered only when target visibility is *truly stochastic*; that is, when the probability of visibility is neither 0, in which case the sensor system would not function, nor 1, in which case we would be left with our original intractable problem.

To recover the counts in intersections of more than two sensing areas (see Figure 1, for example), one takes higher cumulants. More precisely, using again the standard properties of cumulants, and the simpler notation $c_n(p) \equiv c_n(\texttt{Bernoulli(p)})$, we have

Lemma 1. *For sensors $\sigma_1, \sigma_2, \ldots, \sigma_n$ (repetitions allowed), let $C_k = \sum_\tau Z_{\sigma_k \tau} X_\tau$, $1 \le k \le n$, be the corresponding counts. To account for repetitions, suppose that the sensors can be divided into two groups, say $\{\sigma_1, \ldots, \sigma_i\}$ and $\{\sigma_{i+1}, \ldots, \sigma_n\}$, such that each sensor in the latter group is also a sensor in the former. Then the n-th order cumulant is given by*

$$\kappa_n(C_1, C_2, \ldots, C_n) = N_{12\ldots i} \cdot c_n(p).$$

where, extending earlier notation, $N_{12\ldots i}$ denotes the number of targets in the intersection of the sensing areas of the first i (i.e., the distinct) sensors.

Thus, to recover the count in the intersection of n sensing areas, $A_1, A_2, \ldots, A_n$ (again assuming that the visibility probability p is known), it is enough to estimate the cumulant $\kappa_n(C_1, C_2, \ldots, C_n)$ given that the n-th factorial coefficient of the Bernoulli random variable satisfies $c_n(p) \neq 0$.

But this last condition presents us with a potential problem. The cumulant $c_n(p)$ as a function of p is a polynomial of degree n with zeros at 0 and 1 and

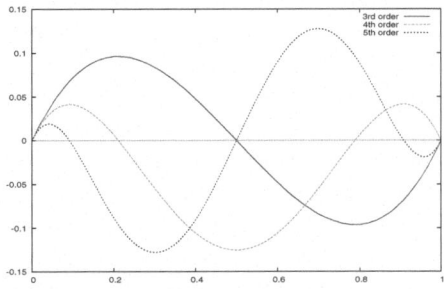

Fig. 2. Cumulants for Bernoulli variables as functions of p: $c_3(1/2) = 0$; in fact, more generally, cumulants of odd orders vanish at $1/2$

a further $(n - 2)$ *interior* zeros in the interval $(0, 1)$; a fact that can be easily derived from the well-known formula

$$c_{n+1} = p \cdot (1 - p) \cdot \frac{dc_n}{dp} \qquad (7)$$

which in turn follows from the explicit expression for the cumulant generating function of Bernoulli random variables,

$$\ln \mathbb{E} e^{s \texttt{Bernoulli}(p)} = \ln[pe^s + (1 - p)].$$

For example, plots of the third and fourth cumulant polynomials in p are shown in Figure 2. However, it is easy to verify that the interior zeros of consecutive cumulant polynomials interlace, and hence, if $c_n(p) = 0$, then $c_{n+1}(p) \neq 0$. Therefore, if the cumulant of order n vanishes for a given probability of visibility p, the next cumulant $c_{n+1}(p)$ will be non-vanishing at p, and the size of the intersection of the sensing areas $A_1, A_2, \ldots, A_n$ can be reconstructed, by Lemma 1, from the cumulant

$$\kappa_{n+1}(C_1, C_2, \ldots, C_n, C_n) = N_{12\ldots n} c_{n+1}(p). \qquad (8)$$

5.2   Poisson Counts

The preceding theory addressed the situation where the number of targets in each sensing domain is fixed and finite and only the visibility patterns are intermittent. Another realistic scenario would be that of *infinitely* many potential targets in each sensing area, with the *measure of a sensing area* being quite general. For simplicity, this measure will be taken here as the classical Lebesgue measure, in which case the measure is just the area, a_σ. Correspondingly, in lieu of the binomial random variables describing the observed counts visible by each sensor, one is driven to the assumption that, for each sensor σ, the count of visible targets is *Poisson distributed*, with the parameter of the Poisson law being proportional to a_σ. Of course, by extension, the number of targets visible in the intersections of the sensing areas is Poisson distributed as well, again with a parameter proportional to the areas of the intersections.

More formally, we assume that, as before, a finite number of sensors $\sigma \in \mathcal{S}$ are given, and that corresponding sensing domains A_σ have areas A_σ. For any collection of sensors $\mathcal{S}' \subseteq \mathcal{S}$, we denote the intersection of their sensing areas by $A_{\mathcal{S}'}$, and the area of this intersection by $A_{\mathcal{S}'}$. The visible objects in A_σ are counted by C_σ and are assumed to be a sample of a Poisson point process with intensity λa_σ; equivalently, we have an intensity-λ Poisson pattern of points in the plane, with those points of the pattern falling in A_σ being stochastically the same as a sample of a rate λa_σ Poisson process. Note particularly that, according to this definition, the counts of the visible targets in non-intersecting areas are independent. The overall goal is to recover the total number, N, of objects, which is Poisson distributed with mean $\mathbb{E}N = \lambda |\cup_{\sigma \in \mathcal{S}} A_\sigma|$ from the individual counts C_σ, which requires estimates of the counts for all intersections, as before.

As in the case of Bernoulli random variables, the cumulants are ideally suited to our problem:

Lemma 2. *The n-th order cumulant is given by*

$$\kappa_n(C_1, C_2, \ldots, C_n) = \lambda a_{12\ldots n}.$$

The proof of this lemma again follows immediately from the standard properties of cumulants and the fact that the cumulant generating function for the Poisson random variable of parameter λa is $e^{\lambda a} - 1$. In other words, for all cumulants $c_n(\texttt{Poisson}(\lambda a)) = \lambda a$.

6   Target Count Recovery

The results presented above indicate how a procedure might be designed for target-count recovery using statistical estimates of the cumulants. We concentrate on the more challenging case of Bernoulli visibility variables and start by giving the algorithm in broad outline; we then flesh it out with details that depend on specific circumstances.

TARGET COUNT RECOVERY ALGORITHM
Assume that the counts are sampled at unit rate from the same population of targets over a time period of duration, T, a given parameter.
1. For our Bernoulli case, the moments in (3) can be expressed as $\mathbb{E}C_\sigma = pN_\sigma$, $\mathbb{V}C_\sigma = p(1-p)N_\sigma$, which imply

$$p := 1 - \mathbb{V}C_\sigma / \mathbb{E}C_\sigma.$$

Computation the first two empirical moments and then substitution into the above gives an estimate for the visibility probability, p.

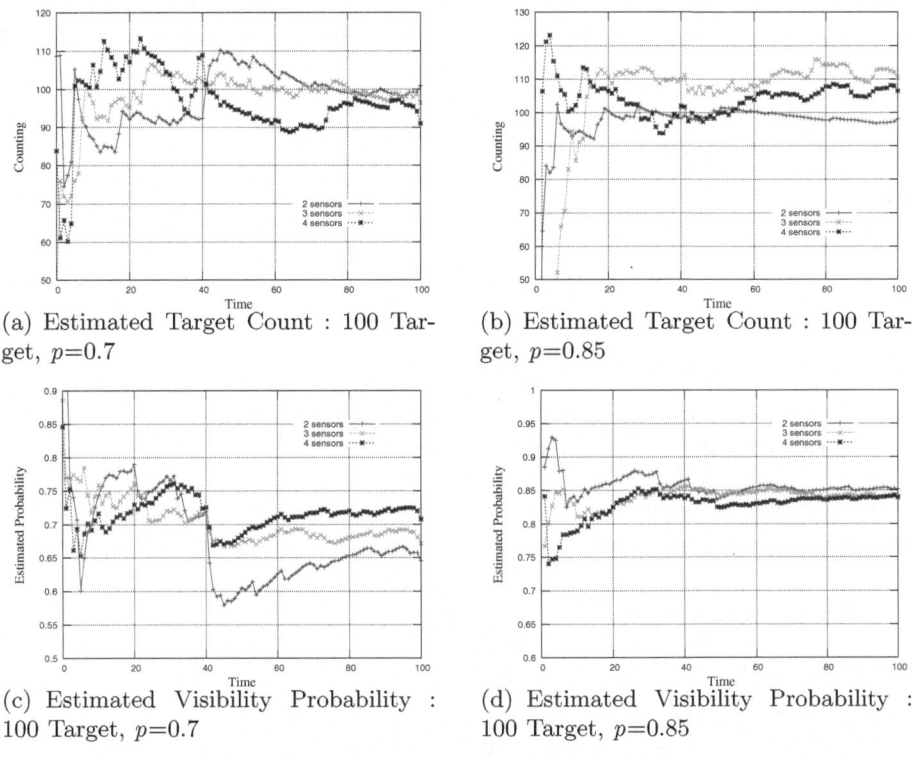

(a) Estimated Target Count : 100 Target, $p=0.7$

(b) Estimated Target Count : 100 Target, $p=0.85$

(c) Estimated Visibility Probability : 100 Target, $p=0.7$

(d) Estimated Visibility Probability : 100 Target, $p=0.85$

Fig. 3. Time trace of estimated target count and estimated visibility probability with 2, 3, and 4 sensors

2. From (7) compute the numbers $c_n(p)$ to one greater, say $s+1$, than the maximum depth (order) of intersections in the sensor field.
3. Compute the remaining cumulants $\kappa_3, \ldots, \kappa_s$ using formulas derived from (6) (recall that the first two are given by the coefficients computed in step 1).
4. Compute the estimates of the $N_{12\ldots n}$ from (2) unless $c(p)$ is too close to 0, in which case use (8) instead.
5. Compute the estimate for N from the data in step 4 substituted into the inclusion-exclusion formula (1).

Remarks: The estimate for p in step 1 can be averaged over several sensors, to improve its quality.

As the order of cumulants grows, the convergence of sampled values to the limit becomes slower. Thus, it makes sense to try to limit the order of cumulants needed for estimation by stopping the procedure when the count precision reaches a desired degree. To this end, the monotone character of the partial sums in the inclusion-exclusion formula allows one to halt the computations when the upper and lower bounds are close enough.

Indeed, as

$$\sum_{\sigma} N_{\sigma} \geq \sum_{\sigma} N_{\sigma} - \sum_{\sigma_1 < \sigma_2} N_{\sigma_1 \sigma_2} + \sum_{\sigma_1 < \sigma_2 < \sigma_3} N_{\sigma_1 \sigma_2 \sigma_3} \geq \ldots \geq N \geq$$

$$\geq \sum_{\sigma} N_{\sigma} - \sum_{\sigma_1 < \sigma_2} N_{\sigma_1 \sigma_2} + \sum_{\sigma_1 < \sigma_2 < \sigma_3} N_{\sigma_1 \sigma_2 \sigma_3} - \sum_{\sigma_1 < \sigma_2 < \sigma_3 < \sigma_4} N_{\sigma_1 \sigma_2 \sigma_3 \sigma_4} \geq \sum_{\sigma} N_{\sigma} - \sum_{\sigma_1 < \sigma_2} N_{\sigma_1 \sigma_2},$$

one can iteratively compute the upper bounds (with the deepest intersections involved of odd order) and lower bounds (with the deepest intersections of even order) until the required precision is achieved.   ■

7   Experimental Results

In this section, we simulate stochastic count recovery in the two-dimensional (planar) and one-dimensional sensoria.

7.1   Planar Sensorium

Here is our setup:

- 2, 3, and 4 sensors are located in a 50×50 area of interest.
- The sensing radius is equal to 35 for all three cases, so that the sensing regions provide at most a 2-cover of the area of interest.
- 100 targets are distributed uniformly at random over the sensing area and sensors detect each target with probability p. Each sensor counts the number of detected targets within its sensing radius.

From the sampled data, we calculate the sample mean and variance to estimate the visibility probability and the sample cumulants to estimate and recover the total number of targets in sensing area. Experiments are performed with different p: 0.7 and 0.85 respectively. Over time, the estimated target counting and detection probability converge to the actual value as in Figure 3. This demonstrates the statements in the preceding sections. Note that as the visibility probability increases, the estimated target count and the estimated visibility probability converge faster with smaller errors.

Next we extend our experiment to a more general case. We assume 24 sensors are distributed uniformly at random over 100×100 sensing area. Each sensor has sensing radius of 15. (This ensures that any point in the area can be covered by at most 3 sensors.) We assume the visibility probability p is equal to 0.85.

Experiments are performed with 100, 200, 300, and 400 targets respectively. We have run extensive simulations to get a mean estimated target count and estimated visibility probability. The standard deviation is provided as a measure of error. As in the previous experiments, estimated target count, estimated visibility probability and cumulants are calculated from the collected data of 24

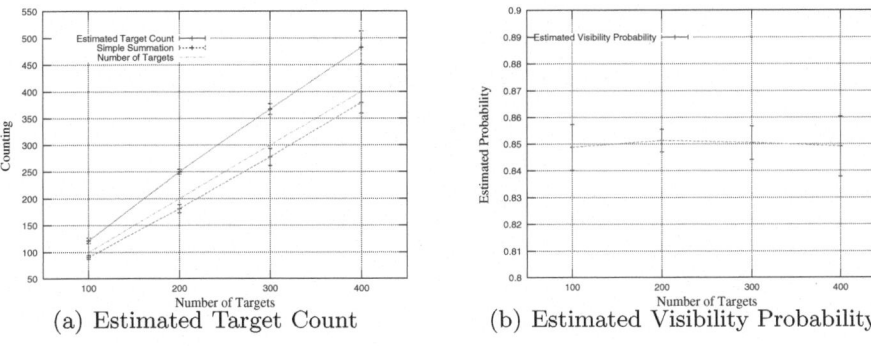

(a) Estimated Target Count   (b) Estimated Visibility Probability

Fig. 4. Estimated target count and estimated visibility probability. 24 sensors with 100, 200, 300, and 400 targets (p=0.85).

sensors. For comparison, to help understand how well the estimation scheme performs, we provide the simple summation of all counting from 24 sensors. As the number of target increases, so do the errors, but the error of our proposed scheme is fairly small compared to that of simple summation. The estimated visibility probability is reasonably accurate and provides up to double digit precision as in Figure 4.

7.2   Linear Sensorium

Here the substrate carrying the targets is one-dimensional, with the point-wise targets. The sensors are identified with certain intervals on the serorium, counting the number of visible targets falling within the corresponding interval.

Unlike the previous setup, in this set of experiments we randomized also over the sensor positions. We consider random placements of $S = 5$ sensors on an interval so that the total length of the sensors (not necessarily the total covered length) is equal to the interval length, see Figure below. The intervals are chosen independently, with uniformly distributed endpoints in the unit interval; the targets are uniformly distributed.

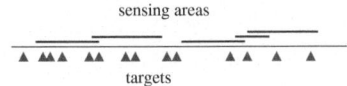

We again apply our stochastic count recovery algorithm taking into account only the cumulants of the second order. We can see that second order cumulants (via imputed second order overlaps) already give quite reasonable approximation of the actual number of targets in the covered area.

The results are shown on Figure 5.

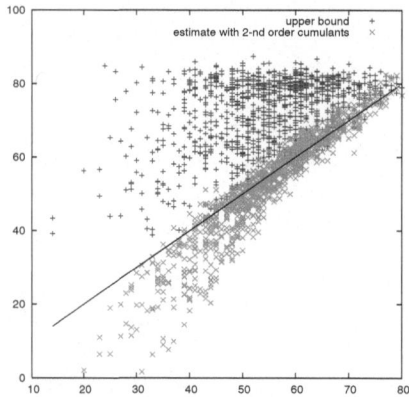

Fig. 5. The plot shows the scatter plots of the actual number of targets in the linear sensorium and upper bound (the sum total of the counts by each sensor, in red) and the estimates based on the inferred overlaps (in green)

8   Conclusion

We have described an algorithm for target count recovery from measurements by a sensor network, where each sensor is capable of counting the number of targets of interest in its sensing region. Overlaps of these regions preclude the possibility of using simple summation of counts as an effective estimate of the total number of targets.

We show that, by exploiting the stochastic nature of the measurements, and stochastic independence between sensors and between targets, it is possible over multiple measurements to arrive at an estimate of the total number of targets present. To do this we employ a method based on cumulants that quantifies non-independence between measurements and thereby permits estimation of the number of targets in the (multiple) overlaps of sensing regions of different sensors. Together with the inclusion-exclusion these numbers provide the estimated total number of targets. The method is described for both Bernoulli and Poisson detection models, and simulations are given to demonstrate the effectiveness of the method.

In a future work we extend the described methodology to the problem of Internet traffic measurements and monitoring.

References

1. Kenney, J.F., Keeping, E.S.: Mathematics of Statistics, pp. 77–82. Princeton, NJ: Van Nostrand (1951)
2. Baryshnikov, Y., Ghrist, R.: Target enumeration via integration over planar sensor networks. Technical report, UIUC (2007)
3. Brooks, R., Iyengar, S.S.: Robust distributed computing and sensing algorithm. Computer 29(6), 53–60 (1996)

4. Clouqueur, T., Ramanathan, P., Saluja, K.K., Wang, K.: Value-fusion versus decision-fusion for fault-tolerance in collaborative target detection in sensor networks. In: Proceedings of the Conference on Information Fusion, pages TuC2/25–TuC2/30 (2001)
5. Gfeller, B., Mihalák, M., Suri, S., Vicari, E., Widmayer, P.: Counting targets with mobile sensors in an unknown environment (preprint)
6. Heidemann, J., Bulusu, N.: Using geospatial information in sensor networks. In: Proceedings of the Workshop on Intersection of Geospatial Information and Information Technology (2001)
7. Iyengar, S.S., Prasad, L.: A general computational framework for distributed sensing andfault-tolerant sensor integration. IEEE Transactions on System, Man and Cybernetics 25(4), 643–650 (1995)
8. Marzullo, K.: Tolerating failures of continuous-valued sensors. ACM Transactions on Computer Systems 8, 284–304 (1990)
9. Nath, S., Gibbons, P., Seshan, S., Anderson, Z.: Synopsis diffusion for robust aggregation in sensor networks. In: Proceedings of SenSys 2004, pp. 250–262 (2004)
10. Son, B.R., Shin, S.C., Kim, J.: Implementation of the real-time people counting system using wireless sensor networks. International Journal of Multimedia and Uniquitous Engineering 2(2), 63–80 (2007)

On the Deterministic Tracking of Moving Objects with a Binary Sensor Network

Yann Busnel[1], Leonardo Querzoni[2], Roberto Baldoni[2], Marin Bertier[3],
and Anne-Marie Kermarrec[4]

[1] IRISA / University of Rennes 1 – France
[2] MIDLAB / University of Rome – "La Sapienza" – Italy
[3] IRISA / INSA Rennes – France
[4] INRIA Rennes – Bretagne Atlantique – France

Abstract. This paper studies the problem of associating deterministically a track revealed by a binary sensor network with the trajectory of a unique moving anonymous object, namely the *Multiple Object Tracking and Identification* (MOTI) problem. In our model, the network is represented by a sparse connected graph where each vertex represents a binary sensor and there is an edge between two sensors if an object can pass from a sensed region to another without activating any other remaining sensor. The difficulty of MOTI lies in the fact that trajectories of two or more objects can be so close (track merging) that the corresponding tracks on the sensor network can no longer be distinguished, thus confusing the deterministic association between an object trajectory and a track.

The paper presents several results. We first show that MOTI cannot be solved on a general graph of ideal binary sensors even by an omniscient external observer if all the objects can freely move on the graph. Then, we describe some restrictions that can be imposed *a priori* either on the graph, on the object movements or both, to make MOTI problem always solvable. We also discuss the consequences of our results and present some related open problems.

1 Introduction

Context. *"Tracking the movements of anonymous objects and associating each object trajectory with an unique identifier"* is a basic problem in many applicative contexts such as surveillance [8], rescue, traffic monitoring [3], pursuit evasion games, *etc.* Tracking objects with a sensor network is a challenging task. Potential inaccuracy of sensors (*e.g.* [1]) and the complexity of the localization subsystem computability (*e.g.* [2]) significantly complicate tracking initiation, maintenance and termination of object trajectories leading to false detection and missing observations.

Even without considering false detections and missing observations, the problem of associating an unique identifier to a track corresponding to the trajectory of one object is difficult due to the potential merging of tracks. This may confuse this association if sensors do not have adequate capabilities [7]. In other words, two tracks may be so close to each other at a certain time that they become indistinguishable, and then impossible to identify after splitting. Deciding, after a track merging, what is a relevant association among multiple hypothesis is usually achieved by looking at the behaviors of tracks

S. Nikoletseas et al. (Eds.): DCOSS 2008, LNCS 5067, pp. 46–59, 2008.

before the merging happened [7]. In the following, we refer to this one-to-one mapping between tracks and object trajectories as the *Multiple Object Tracking and Identification* (MOTI) problem.

We investigate MOTI by using a binary sensor network, which means that each sensor reports only a binary value indicating if an object is in its sensing region or not. The interest in focusing on such minimalist and simple devices is motivated by the fact that we want to study the essence of MOTI's solvability without looking at the powerfulness of sensors capabilities. As in [8], in our model, the sensor network is represented by a sparse connected graph, namely the *passage connectivity graph* (PCG), where each vertex represents a binary sensor and there is an edge between two sensors u and v only if an object can pass from the sensing region of u to the one of v without activating any other remaining sensor.

Contribution. In this paper, we show that it is impossible to solve the MOTI problem in a generic graph. To make the impossibility as strong as possible, this last is proved considering an omniscient observer that has complete knowledge of the state of the graph; moreover, we assume ideal conditions for object tracking such as perfect binary sensors (*i.e.* no false detections or missing observations), ideal coverage of the sensing areas (*i.e.* disjoint sensing regions), so that at each instant of time one object activates only one sensor and each sensor is activated by at most one object. We prove that the impossibility of solving MOTI is a structural problem that depends on the topology of the underlying passage connectivity graph. Then, we describe some restrictions that can be imposed *a priori* either on the graph, on the object movements, or on both, to make MOTI always solvable. More specifically, we show that if the passage connectivity graph is acyclic, MOTI can always be solved. Also, if the graph contains cycles with a length greater than ℓ, MOTI can be solved only if the maximum number of objects that can move concurrently is less than $\left\lceil \frac{\ell}{2} \right\rceil$.

Roadmap. In Section 2, we discuss some related works. Section 3 introduces the system model. Section 4 defines the Multiple Object Tracking and Identification (MOTI) problem and shows that this problem is impossible to solve in a general setting. Section 5 gives two different characterizations of MOTI solvability, which are used in Section 6 to introduce some interesting classes of systems where the problem can always be solved. Finally, Section 7 concludes the paper.

2 Related Work

Tracking mobile objects through sensors is a problem with a large spectrum of applications [6]. It is treated by the literature from various perspectives, but most of the works are concerned with the problem of correctly tracking one [1] or more [9] objects in a network of binary sensors characterized by noise and false detections under various constraints (*e.g.* noise levels, power consumption [5], limited computational or network resources [7], *etc.*). In our work, we take a more theoretical approach to the problem considering a setting characterized by a network of *ideal* sensors (no noise, no false detection, no limitation on communication) and show that, in the general case, it is impossible to correctly associate sensed tracks to moving objects. This result is a consequence of the possibility of track merging and splitting during the observation period [7]: once two tracks have merged due to the excessive proximity of two objects, it is impossible to deterministically maintain the identity of the two objects once their tracks split again; To the best of our knowledge, this is the first work to thoroughly examine and discusse this issue as well as prove its strong impact on the general problem solvability.

Issues related to track merging and splitting are common to every setting where two or more objects can move freely. However, in order to provide the reader with proofs of our statements, in this paper, we limit the problem analysis to a specific environment where object movements are partially constrained, and that can be abstracted as a *passage connectivity graph* [8]. Note that, despite the obvious limitations of this model, it still perfectly maps a lot of indoor applications like tracking the movements of people inside a building.

3   System Model

We consider a system composed of a set of generic objects moving in an environment where sensors can detect their presence. Such an environment can be modeled as a passage connectivity graph (PCG) $G(V, E)$ where the set of vertices V represents binary sensors; it exists an edge $e_{i,j} \in E$ linking two vertices $v_i, v_j \in V$ if and only if (*iff*) it is possible for an object to move from the position where it is detected by the sensor v_i to the one where it is detected by the sensor v_j without activating, during the movement, any other sensor. We assume that, in the considered environment, movements are always possible in both directions. Therefore, the graph G is considered undirected. Moreover, we assume that there is always one single way to move from one position to another: only an edge can connect two distinct vertices in the graph. For each vertex v_i, a special edge $e_{i,i}$ exists in E representing the possibility for an object to remain in the same position.

The set of moving objects $\{o_1, \ldots, o_x\}$ is denoted by $\mathbb{O}$. Time is represented as a discrete and infinite sequence $T = [t_0, t_1, \ldots]$ of instants starting at t_0. At every time instant, each object o occupies a position represented by a vertex $v_i \in V$. The position of an object o at a specific time instant t is given by the function $loc : T \times \mathbb{O} \to V$ that returns the corresponding vertex v_i. When an object, positioned on a vertex v_i, decides to move to a new position, it can choose as its destination any vertex v_j such that $e_{i,j} \in E$; the set of possible destinations is returned by the function $adj : \mathbb{G} \times V \to \mathcal{P}(V)$. The movement of an object o at a specific time instant t is given by the function $mov : T \times \mathbb{O} \to E$ that returns the edge used for the movement $e_{i,j}$ where $v_i = loc_t(o)$ and $v_j \in adj_G(v_i)$.

We assume that, regardless of a node's position and movement, the two following conditions always hold in our system: $\forall t \in T, \forall o_i, o_j \in \mathbb{O}$, (*i*) $loc_t(o_i) \neq loc_t(o_j)$ (*i.e.* two objects cannot be located on the same vertex at the same time) and (*ii*) $mov_t(o_i) \neq mov_t(o_j)$ (*i.e.* two objects cannot move on the same edge at the same time).

Each object $o \in \mathbb{O}$ moving in the system describes a trajectory. The *trajectory* described by object o between $t_i, t_j \in T$ (with $j > i$) is defined as $P_{t_i, t_j, o} = [v^{t_i}, \ldots, v^{t_j}]$ where $v^{t_k} = loc_{t_k}(o), k \in [i, j]$. Given t_i and t_j, we define the *global trajectory* as

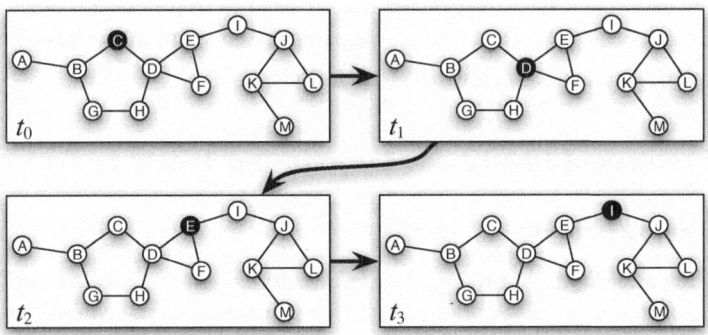

Fig. 1. Example of a trajectory of an object o in the interval $[t_0, t_3]$: $P_{t_0,t_3,o} = [C, D, E, I]$

the set containing all object trajectories. The global trajectory is denoted as $P_{t_i,t_j} = \left\{ P_{t_i,t_j,o} \right\}_{o \in \mathbb{O}}$, or simply P if the time interval is precisely defined by the context. A specific global trajectory P_{t_i,t_j} is an element of $\mathbb{P}_{t_i,t_j}$, the set containing all possible global trajectories described by the objects in $\mathbb{O}$ on G during the period $[t_i, t_j]$.

3.1   System State

The state of the system at time t is described by the state of each sensor (object detected or not) at that time. This state is represented as a vector of boolean values, one for each vertex in G. In the following, we denote as $\mathbb{S}$ the set of all possible state vectors with respect to x objects in a PCG G. We obviously have $\binom{|V|}{x}$ different state vectors.

Definition 1 (State Vector). *A state vector at time t, denoted S_t, is a vector of size $|V|$ where:*

$$\forall v \in V, S_t[v] = \begin{cases} 1 & \text{if } \exists\, o \in \mathbb{O} : loc_t(o) = v \\ 0 & \text{otherwise} \end{cases}$$

3.2   The Observer

The system has an observer that is able to read, at any time t, the state vector S_t. The aim of the observer is to identify objects and trace their trajectories over time. Given a state vector S_t, the observer can use a function tag to assign a unique identifier $\bar{o} \in \overline{\mathbb{O}}$ to each vertex v such that $S_t[v] = 1$ (and a predefined value $\bot$ to all the other vertices), with the only constraint that no two vertices can share a same identifier (with the exception of $\bot$). For instance, if $S_t = [1, 0, 0, 1, 0, 1, 1, 0, 0]$ and $\overline{\mathbb{O}} \subset \mathbb{N}$, we can have $tag(S_t) = [3, \bot, \bot, 2, \bot, 1, 4, \bot, \bot]$. Each identifier represents an object identified by the observer. For the sake of convenience, we introduce the function $\overline{loc} : T \times \overline{\mathbb{O}} \to V$ that returns the vertex v that was tagged by $\bar{o} \in \overline{\mathbb{O}}$ at time $t \in T$.

Given an object identified by the observer, and its corresponding tag $\bar{o}$, we can define the *observed trajectory* between two time instants t_i, t_j as $\overline{P}_{t_i,t_j,\bar{o}} = [v^{t_i}, \ldots, v^{t_j}]$

where $v^{t_k} = \overline{loc}_{t_k}(\overline{o}), k \in [i,j]$. Effectively, we have to introduce the *observed global trajectory* $\overline{P}_{t_i,t_j} = \{\overline{P}_{t_i,t_j,\overline{o}}\}_{\overline{o} \in \overline{\mathbb{O}}}$, corresponding to the set of all observed trajectories perceived by the observer. Obviously, $\overline{P}_{t_i,t_j}$ belongs to $\mathbb{P}_{t_i,t_j}$.

4 The Problem of Identifying Objects and Tracking Trajectories

4.1 The MOTI Problem

Let us consider an interval of time $[t_i, t_j]$. The *Multiple Object Tracking and Identification* (MOTI) problem is the problem of defining a function *tag* such that the following condition holds: $\forall o \in \mathbb{O}, \exists \overline{o} \in \overline{\mathbb{O}} : P_{t_i,t_j,o} = \overline{P}_{t_i,t_j,\overline{o}}$. In a global view, we can define MOTI as the following condition:

$$P_{t_i,t_j} = \overline{P}_{t_i,t_j} \tag{1}$$

That means that the set of observed trajectories is exactly the same than the real ones. Given that trajectories (both real and observed) are a consequence of object locations, at a finer level of granularity we have:

MOTI is solved iff $\forall o \in \mathbb{O}, \exists \overline{o} \in \overline{\mathbb{O}}, \forall t \in [t_i, t_j] : \overline{loc}_t(\overline{o}) = loc_t(o)$

The difficulty of this problem comes from the fact that there could be situations in which the observer can confuse trajectories of two or more objects.

4.2 An Impossibility Result

MOTI is not solvable in the cases which are precisely defined by the following theorem:

Theorem 1 (MOTI Unsolvability). *Given an interval of time $[t_i, t_j]$, a PCG $G(V, E)$, a set $\mathbb{O}$ of x objects and a tag function, MOTI cannot be solved iff*

$$\exists P, P' \in \mathbb{P}_{t_i,t_j} : P \neq P' \land \overline{P} = \overline{P'}.$$

where $\overline{P}$ and $\overline{P'}$ are obtained by the observer using the function tag *from real global trajectories P and P'.*

Proof. Let us consider an interval of time $[t_i, t_j]$, a PCG $G(V, E)$ with x objects.

- Assume that (1) $\exists P, P' \in \mathbb{P}_{t_i,t_j} : P \neq P' \land \overline{P} = \overline{P'}$ and (2) MOTI can be solved. Given that MOTI can be solved we have $P = \overline{P}$ and that $P' = \overline{P'}$ (*cf.* MOTI definition in section 4). But, we have $P \neq P'$ and $\overline{P} = \overline{P'}$. Therefore, there is a contradiction.
- Let prove now that if MOTI cannot be solved $\Longrightarrow \exists P, P' \in \mathbb{P}_{t_i,t_j} : P \neq P' \land \overline{P} = \overline{P'}$. Let prove the contrapositive (i.e. *modus tollens*) of this proposition.
  Consider $\forall P, P' \in \mathbb{P}_{t_i,t_j} : P \neq P' \Longrightarrow \overline{P} \neq \overline{P'}$ ($P = P' \lor \overline{P} \neq \overline{P'}$).
  Let call $map : \mathbb{P}_{t_i,t_j} \to \mathbb{P}_{t_i,t_j}$ the function which associate the real trajectory to the observed one, according to the given *tag* function. Each bijective *map* function

verifies the following expression: $\forall P, P' \in \mathbb{P}_{t_i,t_j} \; : \; P \neq P' \implies map(P) \neq map(P') \implies \overline{P} \neq \overline{P'}$ as $map(\cdot)$ corresponds to the observed trajectories. Let map be the identity function. We have: $\forall P \in \mathbb{P}_{t_i,t_j} \; : \; P = map(P) = \overline{P}$. Then, by definition of Section 4, MOTI can be solved. Thus, we have:

MOTI cannot be solved $\implies \exists P, P' \in \mathbb{P}_{t_i,t_j} \; : \; P \neq P' \wedge \overline{P} = \overline{P'}$.

We then obtain the equivalence, *i.e.* a characterisation of the MOTI unsolvability. □

Corollary 2 (MOTI Solvability). *Given an interval of time* $[t_i, t_j]$, *a PCG* $G(V, E)$, *a set* $\mathbb{O}$ *of* x *objects and a tag function. MOTI can be solved if, and only if,*

$$\forall P, P' \in \mathbb{P}_{t_i,t_j} \; : \; P \neq P' \implies \overline{P} \neq \overline{P'}.$$

Consequently, there is an *impossibility to solve MOTI* if, and only if, there exist at least two global trajectories which respect the condition of Theorem 1. Therefore, we can state that:

Theorem 3 (MOTI Impossibility). *Given the system model presented in Section 3, MOTI is impossible to solve.*

Proof. Consider a 4 vertex PCG G with a 4 edge loop around its vertices, as shown in Figure 2.*a*. Consider two objects moving in this graph G and a time interval constituted by two consecutive time instants $[t_0, t_1]$ with $t_1 = t_0 + 1$. Consider the two global trajectories $P, P' \in \mathbb{P}_{t_0,t_1}$, presented in Figure 2.*b* and 2.*c*. They are defined as:

$$\begin{cases} P = \left\{ \begin{matrix} [A \, , \, B] \\ [D \, , \, C] \end{matrix} \right\} \\ P' = \left\{ \begin{matrix} [A \, , \, C] \\ [D \, , \, B] \end{matrix} \right\} \end{cases}$$

Obviously, we have $P \neq P'$.

According to the system model and the observer capabilities, the information known by the observer for both case P and P', is presented in Figure 2.*d*. We assume that t_0 is the initial time. Then, the observer does not have any other information available about the system than the two following state vectors (which are the same for both P and P'):

$$S_{t_0} = \begin{bmatrix} 1, 0, 0, 1 \end{bmatrix} \text{ and } S_{t_1} = \begin{bmatrix} 0, 1, 1, 0 \end{bmatrix}$$

As the relation tag is a function (a single output for a given input element), it exists a unique tagging for a state, according to the corresponding state vector and previous state available. Thus, assume that:

$$tag(S_{t_0}) = \begin{bmatrix} 1, \perp, \perp, 2 \end{bmatrix}.$$

According to $tag(S_{t_0})$ and S_{t_1}, only two values are possible for $tag(S_{t_1})$:

$$(1) \; tag(S_{t_1}) = \begin{bmatrix} \perp, 1, 2, \perp \end{bmatrix} \text{ or } (2) \; tag(S_{t_1}) = \begin{bmatrix} \perp, 2, 1, \perp \end{bmatrix}.$$

For each of these cases, $tag(S_{t_1})$ is the same for the observation of P and P'. So, given a function tag, $\overline{P} = \overline{P'}$. Then, due to Theorem 1, MOTI cannot be solved. □

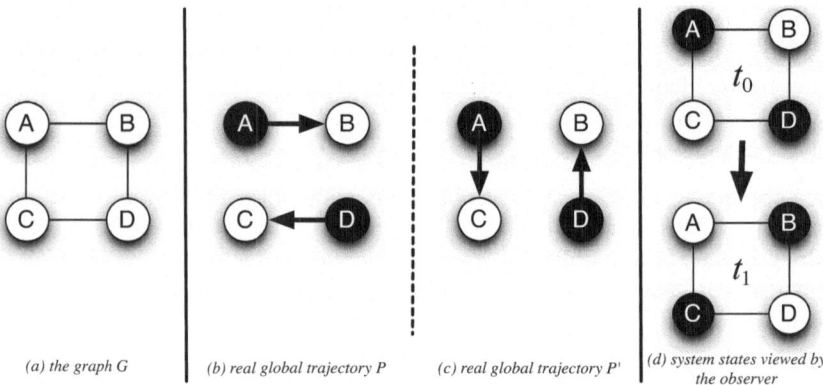

(a) the graph G | (b) real global trajectory P | (c) real global trajectory P' | (d) system states viewed by the observer

Fig. 2. A simple example in which the observer cannot track precisely the trajectories of objects: For a PCG G presented in (a) with two objects, two global trajectories P and P', presented in (b) and (c) respectively, can occur and the observer is not able to distinguish them with only the information presented on (d) if objects have moved following trajectory P or P'

For better understanding of this result, let's take a closer look at each case.

Case (1). In this case, the observed global trajectory computed by the observer is:

$$\overline{P} = \overline{P'} = \left\{ \begin{array}{c} [A\ ,\ B] \\ [D\ ,\ C] \end{array} \right\}.$$

Then, we have $P = \overline{P} = \overline{P'} \neq P'$ and MOTI is not solvable.

Case (2). By symmetry, in this case, the observed global trajectory computed by the observer is:

$$\overline{P} = \overline{P'} = \left\{ \begin{array}{c} [A\ ,\ C] \\ [D\ ,\ B] \end{array} \right\}.$$

Then, we have $P' = \overline{P'} = \overline{P} \neq P$ and MOTI is not solvable.

So, it exists at least one case in which MOTI cannot be solved. Then, Theorem 3 is proved.

5   MOTI Solvability

Before delving into the details about how the system model can be constrained to make MOTI solvable, we need to introduce some other notations.

5.1   Safe and Unsafe Characteristics

Given an object o and the trajectory it describes on the graph as time passes by, we can identify its single movements.

Definition 2 (Movement). *Let $o \in \mathbb{O}$ be an object moving in the system represented by $G(V, E)$. For each time instant $t \in T$, we define its movement as $m_{t,o} = P_{t,t+1,o}$.*

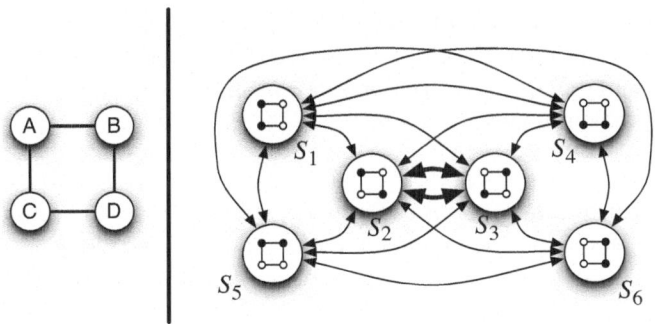

Fig. 3. Example of a state graph for a 4 vertex PCG G (shown on the left) including 2 objects

Definition 3 (Movement Set). *Consider a system represented by $G(V, E)$ where objects belonging to $\mathbb{O}$ can move. For each time instant $t \in T$, we define the movement set as $M_t = P_{t,t+1}$.*

The movement set is defined, for each time instant t, as the set of all movements done by objects; therefore, it represents how the system "evolves" just after time t.

If we now consider the system state at a specific time t, we can identify all the possible movements that objects are able to do. Each possible combination of these movements corresponds to a different movement set. All these movement sets are defined on the basis of the position of objects on the graph, *i.e.* given a state vector, we can define all the possible movements. From this point of view, movement sets are not necessarily tied to time, as they can be considered as the sets of *possible* movements that objects can do if, at a certain time, they are located on a specific subset of vertices. This idea leads us to the definition of the *State Graph*, which is a graph representing possible system states (in terms of state vectors) and possible movement sets linking them.

Definition 4 (State Graph). *Let $G(V, E)$ be the PCG representing the environment where x objects can move. The corresponding state graph is defined as $SG(\mathbb{S}, \mathbb{M})$, where $\mathbb{S}$ is the set of all the possible state vectors and $\mathbb{M}$ is the set of all the possible movement sets.*

Figure 3 shows an instance of a state graph, depicted on the right side, when considering two moving objects and the 4 vertex graph introduced above, depicted on the left side. Now we can define which edges and which vertices of the state graph should be considered as *unsafe* with respect to the solvability of MOTI.

Definition 5 (Unsafe Movement). *Consider a system represented by $G(V, E)$ where objects belonging to $\mathbb{O}$ can move, and the corresponding state graph $SG(\mathbb{S}, \mathbb{M})$. Consider two states $S, S' \in \mathbb{S}$ such that it exists a movement set $M \in \mathbb{M}$ that links these two states ($S \xrightarrow{M} S'$).*
M is unsafe iff $\exists M' \in \mathbb{M}$ such that $S \xrightarrow{M'} S'$ and $M \neq M'$.

We consider unsafe all movement sets linking two system states that are yet linked by some other movement sets. The idea behind this definition is that an observer can not distinguish which movement set has really occurred between all the unsafe movement sets linking the same system states (because the trajectories it observes are calculated only considering system states). The presence of unsafe movements in a state graph make MOTI problem impossible to solve. In the same way, we can define what an unsafe state is.

Definition 6 (Unsafe state). *Consider a system represented by $G(V, E)$ where objects belonging to $\mathbb{O}$ can move, and the corresponding state graph $SG(\mathbb{S}, \mathbb{M})$.*

A state $S \in \mathbb{S}$ is unsafe iff $\exists M \in \mathbb{M}, \exists S' \in \mathbb{S}$ such that $S \xrightarrow{M} S'$ and M is unsafe.

Considering again the state graph depicted in Figure 3, states S_2 and S_3 are both unsafe because there are two distinct edges (two unsafe movement sets) linking them (bold arrows). All the other states (and so, all the other movement sets) are safe.

5.2   Characterizing MOTI Solvability

On the basis of these definitions, we revise Theorem 2 and propose two different definitions of MOTI solvability. In the first case, we assume that the global trajectory is known (*e.g.* when analyzing the behaviour of the system *a posteriori*).

Theorem 4 (P-solvability). *Let $SG(\mathbb{S}, \mathbb{M})$ be the state graph defined over a PCG $G(V, E)$ representing the system where objects in $\mathbb{O}$ can move. Consider a specific trajectory $P \in \mathbb{P}_{t_i, t_j}$: MOTI can be solved iff $\forall t \in [t_i, t_{j-1}]$, M_t is safe.*

Proof.   Consider a specific trajectory $P \in \mathbb{P}_{t_i, t_j}$.

- Assume that $\exists t \in [t_i, t_{j-1}]$ such that M_t is *unsafe*. Let call $P^* = P_{t,t+1}$ the sub-trajectory of the considered P_{t_i, t_j} such that $M_t = P^*$. Due to Definition 5, it exists at least one different unsafe movement M' in $\mathbb{M}$ between the two same system states S, S' such that $M' = P'^*$. Obviously, $P^* \neq P'^*$ but they share the same initial and final state vectors S_t and S_{t+1}. Given that the relation tag available to the observer is a function, it exists a unique tagging for these states. Then, we have:

$$\exists P^*, P'^* \in \mathbb{P}_{t,t+1} : P^* \neq P'^* \wedge \overline{P^*} = \overline{P'^*}.$$

  Therefore, due to Theorem 1, MOTI is unsolvable for P^* and P'^*. Therefore, given that P^* is a sub-trajectory of P_{t_i, t_j}, MOTI is unsolvable for P_{t_i, t_j}.
- Assume here that $\forall t \in [t_i, t_{j-1}]$, M_t is *safe*. Then, we have: $\forall t \in [t_i, t_{j-1}]$, $\nexists M \neq M_t$ such that $S_t \xrightarrow{M} S_{t+1}$. Then, $\forall t \in [t_i, t_{j-1}], \exists! P \in \mathbb{P}_{t,t+1}$ from S_t to S_{t+1} and, as it is unique, this P is the sub-trajectory between time t and $t+1$ of the considered P_{t_i, t_j}. So, $\forall P' \in \mathbb{P}_{t,t+1}$ such that $P \neq P'$, the initial (or respectively the final) system state of P' is not equal to the initial (or respectively the final) system state of P. Then, consider a bijection function map as introduced in the proof of Theorem 1. We have: $\forall P' \in \mathbb{P}_{t,t+1} : P \neq P' \implies map(P) \neq map(P')$. If map is the identity function, we have: $\forall P' \in \mathbb{P}_{t,t+1} : P \neq P' \implies \overline{P} \neq \overline{P'}$. Therefore, due to Corollary 2, MOTI can be solved.   □

Even though characterizing MOTI solvability with respect to a specific global trajectory is useful for all those systems where we want to decide at any point of time if MOTI is solvable or not, it is also possible to define a set of cases where MOTI is always solvable. This new characterization generalizes Theorem 4 with all possible trajectories that can occur in the system:

Theorem 5 (ℙ-solvability). *Let $SG(\mathbb{S}, \mathbb{M})$ be the state graph defined over a PCG $G(V, E)$ representing the system where objects in $\mathbb{O}$ can move.*
$\forall P \in \mathbb{P}_{t_i, t_j}$: *MOTI can be solved iff* $\forall S \in \mathbb{S}$, S *is safe.*

Proof. Assume that $\forall P \in \mathbb{P}_{t_i, t_j}$ MOTI can be solved. Then, $\forall P \in \mathbb{P}_{t_i, t_j}, \forall t \in [t_i, t_{j-1}]$: M_t is safe (Theorem 4) and $\forall P \in \mathbb{P}_{t_i, t_j}, \forall t \in [t_i, t_{j-1}]$: S_t is safe (Definition 6). Given that each system state can occur as an initial state of a trajectory, we have:

$$\forall S \in \mathbb{S}, \exists P \in \mathbb{P}_{t_i, t_j} : S_{t_i} = S.$$

Then, given that all system states are safe for all possible trajectories, we have that $\forall S \in \mathbb{S}$, S is safe.

Assume now that $\forall S \in \mathbb{S}$, S is safe. Due to Definition 6, we have that $\forall M \in \mathbb{M}, M$ is safe. It follows that $\forall P \in \mathbb{P}_{t_i, t_j}, \forall t \in [t_i, t_{j-1}], M$ is safe and, due to Theorem 4 speaks that $\forall P \in \mathbb{P}_{t_i, t_j}$, MOTI can be solved. □

6   A Sufficient Condition for Making MOTI ℙ-solvable

In this section, we show how MOTI can be solved by constraining some characteristics of the system model. More specifically, we show how the problem becomes easily solvable if we assume that object movements are limited in some way.

As we previously explained, the state graph associated to a system may present one or more unsafe states that make MOTI unsolvable. One way to avoid the presence of unsafe states in the state graph, is to remove some of the unsafe movement sets such that the remaining movement sets are all safe. From a practical point of view, this means limiting object movements in the environment, modifying the environment itself or constraining their actions. Let think about a building where people can move from one room to another through doors, this can be realized simply by locking some doors. Therefore, that means deleting the corresponding edges between two vertices of PCG G and modifying accordingly the state graph.

Basically, the only cause of the MOTI non solvability in a system is the presence of cycles in the graph: when two or more objects move inside a cycle, it might be impossible to distinguish their trajectories based on the sole observation of the state vectors. This problem can be avoided by limiting the number of objects in $\mathbb{O}$, which can move concurrently, to a specific k that strictly depends from some characteristics of the PCG G. In this case, we can guarantee that MOTI is solvable for all the possible trajectories P as long as at most k objects move concurrently at each time unit. The set of all these trajectories is a subset of $\mathbb{P}$ and will be denoted as $\mathbb{P}^k$.

Theorem 6. *Let $k \leq |\mathbb{O}|$ be the maximum number of objects that can move concurrently, $\ell > 1$ be an integer and $G(V, E)$ be a PCG which does not contain cycles of length $1 < l < \ell$. $\forall P \in \mathbb{P}^k$: MOTI is P-solvable iff $\lceil \frac{\ell}{2} \rceil > k$.*

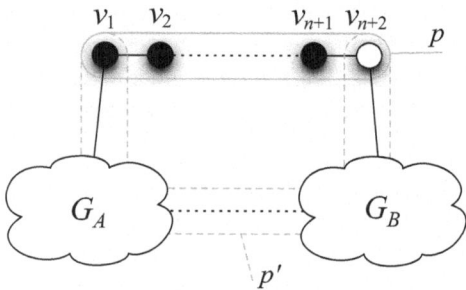

Fig. 4. Representation of the PCG G which illustrates the possibility of exiting paths between v_1 and v_{n+2}

Proof. We first prove that if $k < \lceil \frac{\ell}{2} \rceil \Rightarrow \forall P \in \mathbb{P}^k$, P does not contain any unsafe movement. The proof is done by induction on the number $n < k$ of objects moving concurrently in the system at each step. In the following, given S and S' two state vectors, we refer to $diff(S, S')$ to denote the number of vertices in G whose state changes between S and S'. Note that $diff(S, S')$ is always an even number because for each object that moves, two vertices change their state in the state vector.

Base step on n. Consider the State Graph construction algorithm reported in Appendix A. At the first iterative step of that algorithm, edges labelled with the movement of a single object are added to the edge-free state graph, linking all the possible couples of state vectors S, S' such that $diff(S, S') = 2$ and it exists an edge in G linking the two vertices that changed their state. For each of these couples (S, S'), a single edge is added as only one object in the system can do the movement associated to the two vertices whose state changes between S and S'. Therefore, for a system where only one object at a time can move (*i.e.* $n = 1$), the resulting state graph does not contain any unsafe movements.

Induction hypothesis. Assume that if $n < k$ objects move concurrently, none of the possible trajectories $P \in \mathbb{P}^n$ contains an unsafe movement.

Induction step on n. Now, consider the case where $n + 1$ objects move concurrently. We want to show that, beside this change, no unsafe movement is added to SG. More specifically, we want to prove that $\forall S, S' \in SG$ such that $2 \leq diff(S, S') \leq 2(n + 1)$, if we can add an edge in SG between S and S' labelled with $n + 1$ concurrent movements, then there cannot exists another edge between them labelled with $n + 1$ or less movements.

Let us first consider the case where $diff(S, S') = 2$. Assume that the two vertices changing their state between S and S' are labelled v_1 and v_{n+2}. An edge labelled with $n + 1$ movements can be added in SG between S and S' only if it the path $p = v_1, v_2, \ldots, v_{n+1}, v_{n+2}$ exists in G and $\forall v \in \{v_1, \ldots, v_{n+1}\}, S[v] = 1$. Now, we show that the only possible movement bringing the system from S to S' is the one where each object located in v_i, with $i \in [1, n + 1]$, moves to v_{i+1}. Assume, without loss of generality, that there is another possible movement that does not involve objects located on vertices $v_2, \ldots, v_{n+1}$. Figure 4 represents this case where the object located on v_1 must move to a different subgraph G_A of G and an object located on subgraph G_B must move to node v_{n+2}. If the only path connecting vertices in G_A with vertices in G_B is p (it does not exist a link p' as in Figure 4), then $|\{v \in G_A : S'[v] = 1\}| > |\{v \in G_A : S[v] = 1\}|$ (and respectively, $|\{v \in G_B : S'[v] = 1\}| < |\{v \in G_B : S[v] = 1\}|$). This imples that $\exists v \in G_A \cup G_B : S[v] \neq S'[v] \Rightarrow diff(S, S') > 2$, which is impossible due to our initial hypothesis. On the contrary, if there is a path $p' \neq p$ connecting G_A to G_B, then p is part of a cycle $c \subset G$. The length of c is, by assumption, at least ℓ. In order to have $diff(S, S') = 2$, in S, there must be an object located on all vertices in c but v_{n+2} and all the objects located on these vertices, with the exception of those located on $v_2, \ldots, v_{n+1}$, must be moved (otherwise $diff(S, S') > 2$). But this means that $m \ (\geq \ell - n)$ objects will move concurrently. Given that $\lceil \frac{\ell}{2} \rceil > k \geq n + 1$, we have $m > 2(n + 1) - n > n + 1$, *i.e.* every other edges connecting S to S' in SG must be labelled with more than $n + 1$ movements.

Now, consider the case where $diff(S, S') = 2(x + 1)$ with $x \leq n$. In this case, there are $n + 1$ distinct objects moving on $x + 1$ distinct paths, each characterized by the presence of an object on each vertex but the last one (as p in Figure 4). The same reasoning shown for the previous case can be applied to every single path, considering that each of these paths contains strictly less than $n + 1$ objects.

Now, we prove by contradiction that if $\forall P \in \mathbb{P}^k$, MOTI is P-solvable $\Rightarrow \lceil \frac{\ell}{2} \rceil > k$. Assume for the moment that $k \geq \lceil \frac{\ell}{2} \rceil$. Consider the smallest cycle c in G constituted by vertices $v_1, \ldots, v_\ell$. Now consider, without loss of generality, a state vector S where $\forall i \in \{2 \cdot k - 1 | k \in [1, \lceil \frac{\ell}{2} \rceil]\} : S[v_i] = 1$ and $\forall i \in \{2 \cdot k | k \in [1, \lfloor \frac{\ell}{2} \rfloor]\} : S[v_i] = 0$. Consider also the state vector S' that is identical to S but where $\forall i \in \{2 \cdot k - 1 | k \in [1, \lfloor \frac{\ell}{2} \rfloor]\} : S[v_i] = 0$ and $\forall i \in \{2 \cdot k | k \in [1, \lfloor \frac{\ell}{2} \rfloor]\} : S[v_i] = 1$ (Note that, if ℓ is odd then $S'[v_\ell]$ remains unchanged and equals to 1). Then, the states of all vertices – but the last in case of ℓ odd – indexed from 1 to ℓ are inverted. Such two state vectors are certainly in SG because we are assuming that $k \geq \lceil \frac{\ell}{2} \rceil$. Now consider the following movements $M = \{[v_i, v_{i+1} \mod \ell]\}_{i \in \{2 \cdot k - 1 | k \in [1, \lfloor \frac{\ell}{2} \rfloor]\}}$ and $M' = \{[v_i, v_{i-1} \mod \ell]\}_{i \in \{2 \cdot k - 1 | k \in [1, \lceil \frac{\ell}{2} \rceil]\}}$. Both movements link S to S' in SG and are labeled with lower or equals concurrent moves than k. Therefore, it exists unsafe movements. This is in contradiction with the initial assumption that $\forall P \in \mathbb{P}^k$, MOTI is P-solvable. $\square$

In a practical setting, there are two possible methods to guarantee that the conditions at the basis of Theorem 6 always hold. The first method requires to choose, as the PCG, a topology characterized by cycles of length strictly larger than $2 \cdot x$. As a consequence, MOTI is $\mathbb{P}$-solvable in any system characterized by an acyclic graph as presented in the following Theorem 7. The second method requires to limit the number of objects that can move concurrently in the system.

Theorem 7. *MOTI is $\mathbb{P}$-solvable in any system characterized by an acyclic graph $G(V, E)$.*

Proof. We can consider G as a graph with a cycle of infinite length. Due to Theorem 6, MOTI is $\mathbb{P}$-solvable as long as no more than an infinite number of objects move concurrently in the system. Given that $\mathbb{O}$ is finite, the $\mathbb{P}$-solvability is always guaranteed. $\square$

7 Conclusion

In this paper, we have considered the possibility of assigning trajectories to moving objects on a generic sparse passage connectivity graph. We have shown that assigning a unique identifier to a track revealed by a binary sensor network, namely the MOTI problem, is impossible. To make the result as strong as possible, it has been proved under strong assumptions i.e., perfect binary sensors, perfect coverage and omniscient observer of the state of the graph. We proved that this impossibility depends primarily of the topology of the graph and secondly of the number of moving objects.

Following this impossibility result, we have considered restricting the model to make such assignments possible. More specifically, we have shown that MOTI can be solved either if the graph is acyclic or the length of the smallest cycle in the graph is strictly greater than the double of the number of objects that can concurrently move in the system. This leaves the opportunity to modify the topology of the passage connectivity graph like for example in an indoor scenario. Once the maximum number of moving objects is known, our results mainly impact on the deployment phase of a sensor network.

Lot of open questions have still to be considered. For instance, some characterizations can be reformulated in order to reduce the computation of the observer. In a distributed way, identifying distributed algorithms able to impose the aforementioned restrictions to the system model, without relying on a global knowledge of the system is not trivial. This leaves the space for an exciting research agenda.

References

1. Aslam, J., Butler, Z., Constantin, F., Crespi, V., Cybenko, G., Rus, D.: Tracking a moving object with a binary sensor network. In: Proc. of SenSys 2003 (2003)
2. Aspnes, J., Goldengerg, D., Yang, Y.R.: On the computational complexity of sensor network localization. In: Proc. of 1st Intl Workshop on Algorithmic Aspects of WSN (2004)
3. Bloisi, D., Iocchi, L., Leone, G.R., Pigliacampo, R., Tombolini, L., Novelli, L.: A distributed vision system for boat traffic monitoring in the venice grand canal. In: VISAPP 2007 (2007)
4. Floyd, R.W.: Algorithm 97: Shortest path. Comm. of the ACM (1962)
5. Gui, C., Mohapatra, P.: Power conservation and quality of surveillance in target tracking sensor networks. In: Proc. of MobiCom 2004 (2004)
6. Han, M., Xu, W., Tao, H., Gong, Y.: An algorithm for multiple object trajectory tracking. In: Proc. of CVPR 2004 (2004)
7. Liu, J., Liu, J., Reich, J., Cheung, P., Zhao, F.: Distributed Group Management for Track Initiation and Maintenance in Target Localization Applications. In: Zhao, F., Guibas, L.J. (eds.) IPSN 2003. LNCS, vol. 2634, pp. 113–128. Springer, Heidelberg (2003)
8. Oh, S., Sastry, S.: Tracking on a graph. In: Proc. of IPSN 2005 (2005)
9. Singh, J., Madhow, U., Kumar, R., Suri, S., Cagley, R.: Tracking multiple targets using binary proximity sensors. In: Proc. of IPSN 2007 (2007)
10. Warshall, S.: A theorem on boolean matrices. Journal of the ACM (1962)

A   An Algorithm for Computing SG

In this appendix, we propose a state graph construction algorithm, based on the conditional transitive closure of a one-movement only state graph. This condition ensures that no inconsistent movement will be included in the state graph generated by the algorithm.

This protocol is composed of two main parts. The first part creates an empty edge set and a complete vertices set. Then, starting from line 3 to line 6, the algorithm puts in the state graph all one-object-movement edges. From line 7 to line 13, following the same mechanism as the one used in the Floyd-Warshall algorithm for transitive closure computation, three nested loops compute iteratively all possible movements with any number of concurrent movements. Finally, the last lines from 14 to 15, merge symmetric movement edges in order to return an undirected state graph.

The termination of this protocol is trivial as it is composed only of nested loops of finite length. Moreover, the complexity of this algorithm is obviously $O(|\mathbb{S}|^3)$ as it contains three nested loops on $\mathbb{S}$.

The correctness of this algorithm is based on the correctness of the Floyd-Warshall algorithm [4,10]. The condition of line 12 only ensures that no inconsistent movement can be included in the state graph, and given that new edges in SG are generated from consistent movements, no existing movement can be ignored.

Algorithm 1. State graph construction

Data: a graph G, the number of moving object x
Result: the associated state graph $SG(\mathbb{S}, \mathbb{M})$

1 $\mathbb{S} \leftarrow \{S \in \{0,1\}^{|V|} \mid \sum_{v \in V} S[v] = x\}$;
2 $\mathbb{M} \leftarrow \emptyset$;
3 **foreach** $S \in \mathbb{S}$ **do**
4   **foreach** $S' \in \mathbb{S} \backslash \{S\}$ **do**
5   **if** $\exists!(v,v') \in V2$ *such that* $S[v] = S'[v'] = 1$ *and* $S[v'] = S'[v] = 0$ **then**
6              $\mathbb{M} \leftarrow \mathbb{M} \cup \{S \xrightarrow{\{[v-v']\}} S'\}$;

7 **foreach** $S \in \mathbb{S}$ **do**
8   **foreach** $S' \in \mathbb{S} \backslash \{S\}$ **do**
9   **foreach** $S'' \in \mathbb{S} \backslash \{S\}$ **do**
10   **if** $(S' \rightarrow S \in \mathbb{M}) \wedge (S \rightarrow S'' \in \mathbb{M})$
11   **and** $(\forall S' \rightarrow S'' \in \mathbb{M}, \; e_{S',S''} \neq e_{S',S} \cup e_{S,S''})$
12   **and** $\forall [v',v_1] \in e_{S',S}, \; \forall [v_2,v''] \in e_{S,S''}, \; v_1 \neq v_2$ **then**
13                 $\mathbb{M} \leftarrow \mathbb{M} \cup \{S' \xrightarrow{e_{S',S} \cup e_{S,S''}} S''\}$;

14 **if** G *is undirected* **then**
15   merge symmetric edges in $\mathbb{M}$;
16 **return** $(\mathbb{S}, \mathbb{M})$;

An Adaptive and Autonomous Sensor Sampling Frequency Control Scheme for Energy-Efficient Data Acquisition in Wireless Sensor Networks

Supriyo Chatterjea and Paul Havinga

Pervasive Systems Group, Faculty of EEMCS, University of Twente,
P.O. Box 217 7500AE, Enschede, The Netherlands
{supriyo,havinga}@cs.utwente.nl

Abstract. Wireless sensor networks are increasingly being used in environmental monitoring applications. Collecting raw data from these networks can lead to excessive energy consumption. This is especially true when the application requires specialized sensors that have very high energy consumption, e.g. hydrological sensors for monitoring marine environments. We describe an adaptive sensor sampling scheme where nodes change their sampling frequencies autonomously based on the variability of the measured parameters. The sampling scheme also meets the user's sensing coverage requirements by using information provided by the underlying MAC protocol. This allows the scheme to automatically adapt to topology changes. Our results based on real and synthetic data sets, indicate a reduction in sensor sampling by up to 93%, reduction in message transmissions by up to 99% and overall energy savings of up to 87%. We also show that generally more than 90% of the collected readings fall within the user-defined error threshold.

1 Introduction

Large-scale, dense sensor networks are increasingly being used in a variety of environmental monitoring applications where the user requires data to be collected from every sensor in the network at regular intervals. This requires a large amount of data to be transmitted to the sink node. However, extracting large amounts of data leads to excessive energy consumption. The limited bandwidth of sensor nodes may also cause data quality to deteriorate due to dropped packets caused by buffer overflows.

The primary source of energy consumption in a sensor node is generally attributed to the operation of the radio transceiver. However, certain applications require specialized sensors that have very high power consumption. As an example, we are currently working together with the Australian Institute of Marine (AIMS) Science to deploy a sensor network on the Great Barrier Reef (GBR) [6]. Apart from using basic temperature sensors, scientists at AIMS require more sophisticated sensors which are capable of monitoring parameters such as salinity or the level of dissolved oxygen [4]. Such sensors can have power consumption

S. Nikoletseas et al. (Eds.): DCOSS 2008, LNCS 5067, pp. 60–78, 2008.
© Springer-Verlag Berlin Heidelberg 2008

levels ranging from 105mW to 720mW. Thus these specialized sensors could typically consume more than 30 times more power than a standard sensor node transceiver (e.g RFM TR1001 consumes $21mW$ in transmit mode [10]). Note also that these sensors commonly have very long start-up and sampling times (e.g. around 2 seconds). Thus in such applications, since sensor sampling operations could have a drastic impact on network lifetime, it is essential to not only reduce usage of the transceiver but also to reduce the number of sensor sampling operations.

In this paper we take a two-pronged approach to reducing energy consumption by not only reducing sensor sampling operations but also reducing the number of messages transmitted. This is carried out by exploiting the temporal correlations that may exist between successive sensor readings. The basic idea is to use *time-series forecasting* to try and predict future sensor readings. When the trend of a particular sensor reading is fairly constant and thus predictable, we reduce the sensor sampling frequency and message transmission rate. However, when the trend changes more frequently, both the sampling rate and message transmission rates are increased.

However, the lower sampling frequency used in such a sampling scheme may result in certain important events being missed out. In order to minimize the chance of this from occurring, our sampling scheme tries to ensure an acceptable level of coverage by allowing nodes in the network to adjust various parameters autonomously rather than using fixed values that are predefined prior to deployment. This allows the nodes to operate in a more energy-efficient manner as they are able to automatically adapt their operation to not only the variations in the environment but also to the user-specified coverage requirements. Another novel feature of our scheme is that nodes make use of topology information provided by the underlying MAC protocol (LMAC [10]) in order to adjust the sampling frequency. We are not aware of any existing work where cross-layer optimization is performed between the MAC and application layers. Usually optimizations are restricted to adjacent layers of the OSI model. Our cross-layered approach allows nodes to automatically re-adjust their sampling frequencies if topology changes are reported by the MAC layer.

We envision our sampling scheme to be used in non-critical applications, e.g. monitoring various parameters in the waters in the GBR, in a coffee beans storage warehouse, etc. This scheme is not meant for critical applications where lives may be at risk if a particular event is missed, e.g. a nuclear plant. In order to evaluate our scheme, we use both real-life and synthetically generated data sets. The real-life data sets are based on temperature readings obtained from outdoor (GBR [5]) and indoor environments (Intel Berkeley Laboratory [2]).

Our contributions are stated as follows:

1. We present a technique to generate synthetic data sets using altitude information and then classify their behavior according to a Variability Index. This helps us evaluate the performance of our sampling scheme under different conditions.

2. We present a localized sensor sampling frequency control scheme that helps reduce sampling operations by up to 93% and transmission operations by up to 99%.
3. We illustrate how a node can adjust certain parameters autonomously in order to ensure coverage is kept at an acceptable level across different data sets with different characteristics.
4. We show overall energy savings due to our adaptive sampling scheme can be up to 87% while around 90% of the collected data falls within the the user-specified threshold.

The following section provides a brief overview of time-series forecasting. This is followed by a description of how we use elevation data to generate synthetic data sets. We then present details of our adaptive sensor sampling scheme and provide the simulation results. Section 5 mentions the related work and finally the paper is concluded in Section 6.

2   Preliminaries of Time-Series Forecasting

Time-series forecasting is a technique that has been used in a wide variety of disciplines such as engineering, economics, and the natural and social sciences to predict the outcome of a particular parameter based on a set of historical values. These historical values, often referred to as a "time series", are spaced equally over time and can represent anything from monthly sales data to temperature readings acquired periodically by sensor nodes.

The general approach to time-series forecasting can be described in four main steps:

1. Analyze the data and identify the existence of a trend or a seasonal component.
2. Remove the trend and seasonal components to get *stationary* (defined below) residuals. This may be carried out by applying a transformation to the data.
3. Choose a suitable model to fit the residuals.
4. Predict the outcome by forecasting the residuals and then inverting the transformations described above to arrive at forecasts of the original series.

Before describing details of how we perform each of the above steps in our data aggregation framework, we first present some basic definitions.

Definition 1. *Let X_t be a time series where $t = 1, 2, 3, \ldots$ We define the mean of X_t as,*

$$\mu_t = E(X_t) \tag{1}$$

Definition 2. Covariance *is a measure of to what extent two variables vary together. Thus the covariance function between X_{t_1} and X_{t_2} is defined as,*

$$\gamma(t_1, t_2) = Cov(X_{t_1}, X_{t_2}) = E[(X_{t_1} - \mu_{t_1})(X_{t_2} - \mu_{t_2})] \tag{2}$$

Definition 3. *We define the autocovariance at lag* h *of* X_t *for* $h = 0,$
$1, 2, ..., T$ *as*

$$\gamma(h) = \frac{\sum_{t=h+1}^{T}(X_t - \bar{X})(X_{t-h} - \bar{X})}{T} \tag{3}$$

where $\bar{X} = T^{-1}\sum_{t=1}^{T} X_t$ *is the mean of the time series* X_t*. Note that* $\gamma(0)$ *is simply the variance of* X_t*.*

Definition 4. *The* autocorrelation function *(ACF),* ρ_h*, which indicates the correlation between* X_t *and* X_{t+h}*, is*

$$\rho(h) = \frac{\gamma(h)}{\gamma(0)} \tag{4}$$

Definition 5. *We consider the time series* X_t *to be* stationary *if the following two conditions are met:*

$$E(X_t) = \mu_t = \mu_{t+\tau} \forall \tau \in \mathbb{R} \tag{5}$$

$$\gamma(t + h, t) = \gamma(t + h + \tau, t + \tau) \forall \tau \in \mathbb{R} \tag{6}$$

Equation 5 and Equation 6 imply that the mean and covariance remain constant over time respectively. In the case of Equation 6, the covariance remains constant for a given lag h*.*

Definition 6. *A process is called a* white noise *process if it is a sequence of uncorrelated random variables with zero mean and variance,* σ^2*. We refer to white noise using the notation* $WN(0, \sigma^2)$*. By definition, it immediately follows that a white noise process is stationary with the autocovariance function,*

$$\gamma(t + h, t) = \begin{cases} \sigma^2 & \text{if } h = 0, \\ 0 & \text{if } h \neq 0 \end{cases} \tag{7}$$

2.1   Analysis of Data and Identification of Trend

As mentioned earlier, the first step is to either identify the trend or seasonal component. However, as we make predictions using a small number of sensor readings taken over a relatively short period of time (e.g. 20 mins), we make the assumption that the readings do not contain any seasonal component. Instead, given that t represents time, we model the sensor readings, R_t using a slowly changing function known as the *trend component*, m_t and an additional stochastic component, X_t that has zero mean. Thus we use the following model: $R_t = m_t + X_t$.

The main idea is to eliminate the trend component, m_t, from R_t so that the behavior of X_t can be studied. There are various ways of estimating the trend for a given data set, e.g. using polynomial fitting, moving averages, differencing, double exponential smoothing, etc. Due to the highly limited computation and

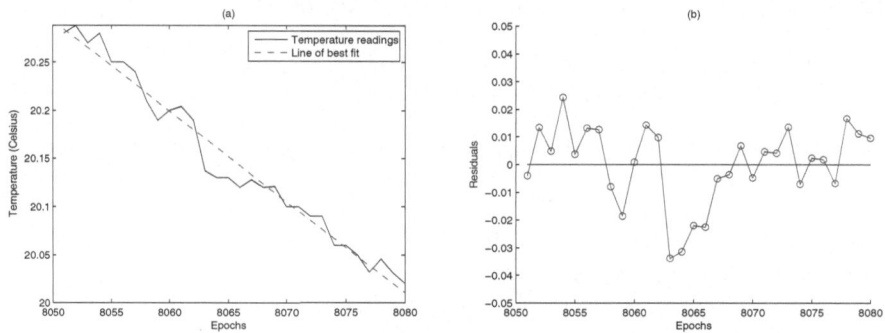

Fig. 1. Temperature sensor readings and the corresponding residuals

memory resources of sensor nodes, we make use of a first degree polynomial, i.e. $m_t = a_0 + a_1 t$.

The coefficients a_0, and a_1 can be computed by minimizing the sum of squares, $Q = \sum_{t=1}^{T}(R_t - m_t)^2$. In order to find the values of a_0 and a_1 that minimize Q, we need to solve the following equations:

$$\frac{\partial Q}{\partial a_0} = -2 \sum_{t=1}^{T}(R_t - a_0 - a_1 t) = 0 \tag{8}$$

$$\frac{\partial Q}{\partial a_1} = -2 \sum_{t=1}^{T}(R_t - a_0 - a_1 t)t = 0 \tag{9}$$

Solving equations 8 and 9 leads to:

$$a_0 = \frac{\sum_{t=1}^{T}(t - \bar{t})(R_t - \bar{R})}{\sum_{t=1}^{T}(t - \bar{t})^2} \tag{10}$$

$$a_1 = \bar{R} - a_0 \bar{t} \tag{11}$$

Eliminating the trend component from the sensor readings results in the residuals shown in Figure 1(b). The residuals display two distinct characteristics. Firstly, there is no noticeable trend and secondly there are particular long stretches of residuals that have the same sign. This smoothness naturally indicates a certain level of dependence between neighboring readings. Our aim is to study this dependence characteristic which in turn would help understand the behavior of the residuals so that predictions can be made.

Now that a stationary time series has been obtained, the next step is to choose an appropriate model that can adequately represent the behavior of the time series.

Stationary processes can be modelled using *autoregressive moving average* (ARMA) models. The ARMA model is a tool for understanding and subsequently predicting future values of a stationary series. The model consists of an autoregressive part, AR and a moving average part, MA. It is generally referred to as the ARMA(p, q) model where p is the order of autoregressive part and q is the order of the moving average part. The AR(p) model is essentially a linear regression of the current value of the series against p prior values of the series, $X_{t-1}, X_{t-2}, ..., X_{t-p}$. The MR model on the other hand is a linear regression of the current value of the series against the white noise of one or more prior values of the series, $Z_{t-1}, Z_{t-2}, ..., Z_{t-p}$. The complete ARMA$(p, q)$ model is defined as follows,

$$X_t = \phi_1 X_{t-1} + ... + \phi_p X_{t-p} + Z_t + \theta_1 Z_{t-1} + ... + \theta_q Z_{t-q} \qquad (12)$$

where $Z_t \sim \text{WN}(0, \sigma^2)$ and ϕ_i, $i = 1, 2, ..., p$ and θ_i, $i = 1, 2, ..., q$ are constants.

However, due to the limited computation and memory resources on a sensor node, we use an AR(1) model instead of the full ARMA model (i.e. $q = 0$) to predict the value R_t, i.e. $X_t = \phi_1 X_{t-1} + Z_t$. The constant ϕ_1 can now be estimated using the Yule-Walker estimator, i.e. $\hat{\phi}_1 = \hat{\rho}_1 = \frac{\gamma(1)}{\gamma(0)}$. We can then state that the general form of the minimum mean square error m-step forecast equation is

$$\hat{X}_{t+m} = \mu + \phi^m (X_t - \mu), m \geq 1 \qquad (13)$$

3   Data Set Generation and Classification

The algorithms presented in this paper are designed specifically for sensor networks that are deployed for environmental monitoring purposes. In order to extensively evaluate the performance of the various algorithms, it is essential to test them using several data sets that display different characteristics, e.g. while the sensor readings in one data set may be relatively constant, the readings in another data set may change rapidly. Unfortunately, due to the lack of large scale sensor network deployments, large, real-life data sets are still not readily available. While there are a few small real-life data sets available [2], they do not display the wide range of characteristics that is required for having a complete evaluation of the performance of the algorithms.

In order to circumvent this problem of the unavailability of certain types of data sets, we now present a technique to not only generate different data sets but also describe how the different data sets can be classified based on their variability. The technique allows the the generation of data sets that are correlated both spatially and temporally.

We generate synthetic sensor readings using elevation data obtained from the Seamless Data Distribution System provided by the U.S. Geological Survey (USGS) [1]. The site provides elevation data of the entire United States up to a resolution of 10 meters. The first step is to choose an appropriate area on the map. The area chosen depends on the required variability of the synthetic data set. For example, if the synthetic data set needs to have high variability, one may opt to choose the western part of the United States which is generally more mountainous, i.e. the *variation* of elevation changes significantly within a small

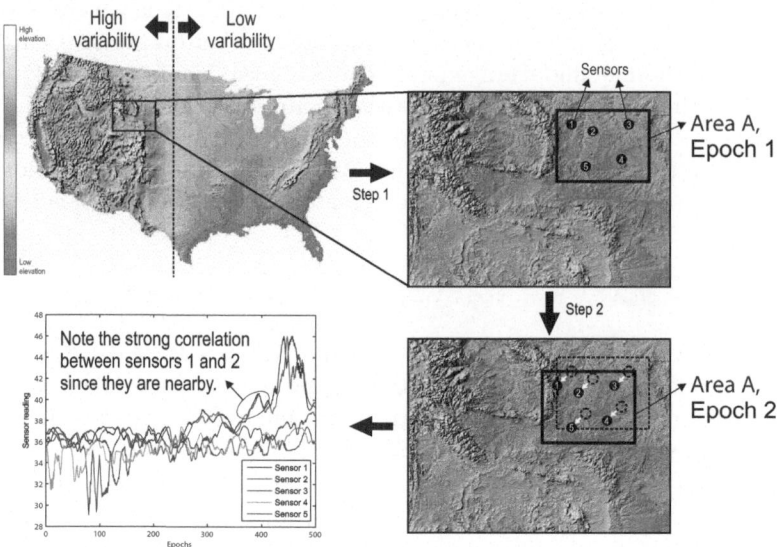

Fig. 2. Overview of how a synthetic data set is generated

distance. Conversely, if the user requires sensor readings where the changes are fairly small, one may choose an area that lies in the central part of the map. This is illustrated in Step 1 in Figure 2. Next an area is chosen within the area selected in the previous step. This area is labeled as A in Figure 2. Area A indicates the size of the deployment area of the sensor network as specified in the simulations. Nodes are then positioned randomly within this area and elevation readings are obtained at the location of every node. Thus if there are 100 randomly deployed nodes, a total of 100 elevation readings will be obtained and these are taken to be the sensor readings for the first epoch. Note that readings between nodes situated close to one another will be spatially correlated. Next, area A is shifted one unit in a particular direction together with all the nodes within it. New elevation readings are once again obtained at the location of every node. These readings are assumed to be the sensor readings for the second epoch. Thus readings of a particular node will be temporally correlated over the various epochs. This process is repeated until readings for the required number of epochs is obtained. The graphs in Figure 2 also show how the generated synthetic data set has both spatial and temporal correlations.

Another inherent property of data sets generated using this technique is the fact that different areas of the deployment area may have different levels of variation over a fixed duration. This not only mimics the characteristics of real-life data sets but is also very useful for testing the localized behavior of nodes, e.g. while some nodes in the network may experience high variability in the sensed readings, other nodes may experience low variability. Thus in order to operate in an optimized manner it is essential for the algorithms running on the nodes to adapt to the nature of the data that is being sensed. Figure 3 illustrates

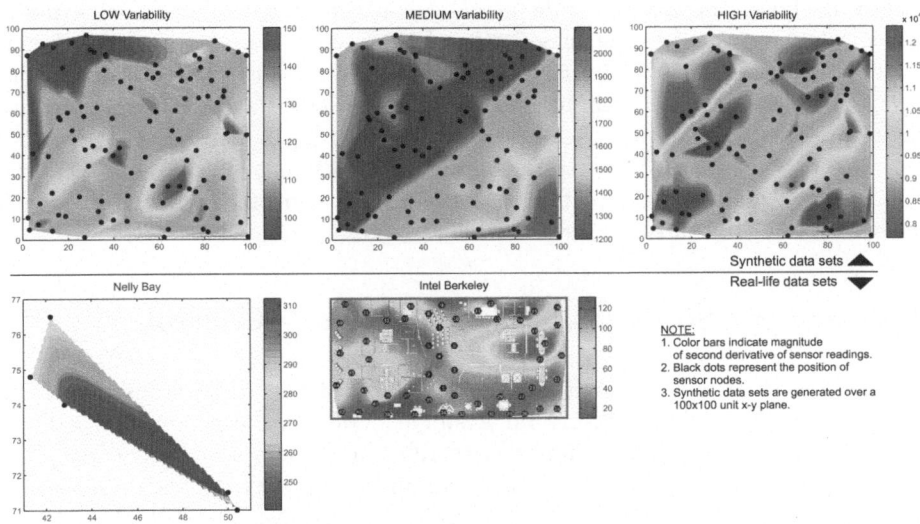

Fig. 3. Variability of the various real and synthetic data sets

how both a real life data set (obtained from Intel Berkeley [2])and synthetic data sets display different levels of variability in sensor readings in different regions. Note that because the performance of the algorithms described in this paper depend on the *rate of change of the trend* of a sensor reading and not the sensor reading itself, the contour maps in Figure 3 indicate *how often the trend* of the sensor readings changes. In other words, the contour maps are drawn using the magnitude of the second derivative of the sensor readings.

In order to test our algorithms extensively under different conditions, we shall use three different data sets which have varying levels of variability. The graph in Figure 4(a) shows how we classify data based on a *Variability Index* (V.I.).

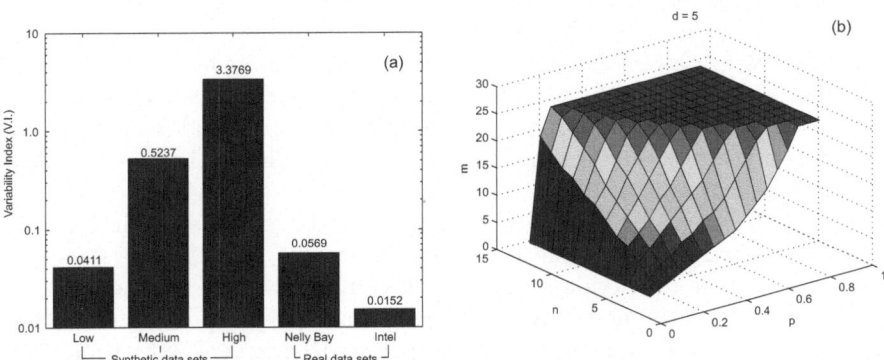

Fig. 4. (a)Variability Indices of different synthetic and real-life data sets, (b) Computation of MSSL given p (user-specified) and n (topology dependent)

The V.I. shows the *average rate of change of the trend of a sensor reading per epoch per node* (i.e. the magnitude of the second derivative of the sensor readings). For the sake of comparison, we have also included data obtained from two real life data sets: (i) temperature data from Nelly Bay in the GBR [5] and (ii) temperature data from the Intel Berkeley Lab [2]. It can be seen that the real-life data set clearly falls within the *low variability* category. We have chosen this metric to classify different data sets as the performance of our approach depends on how often the trend of a sensor reading changes.

4   Localized Sensor Sampling Frequency Control

As mentioned earlier, apart from the operation of the radio transceiver, the sampling of the various sensors on a single node may also consume a large amount of energy. In this section we describe a local adaptive sensor sampling frequency mechanism where the sensor sampling frequency depends not just on the predictability of the physical parameter being measured but also on the user-specified error threshold.

In general terms, when the reading of a particular sensor on a node can be predicted based on the recent past, we reduce the frequency of sampling the sensors by skipping a larger number of sensor sampling operations and performing predictions instead. However, the moment the prediction differs from the actual sampled reading by an amount specified by the user, the sampling frequency is increased. This local prediction mechanism also helps reduce the number of sensor readings that need to be transmitted to the sink node. We now describe the precise steps in greater detail.

After all nodes in the network have initialized (i.e. acquired LMAC [10] slots), a query is injected into the network specifying the required time interval (e.g. t time units) between successive sensor samples. The query also describes the error levels($\pm\delta$ units) that may be tolerated by the user. Upon receiving a query, all nodes initially acquire the first r readings at intervals of t time units and store these readings together with their corresponding sample times in the buffer. Note that the maximum number of (time, sensor reading) tuples that can be stored in the buffer is r. Once the first r readings have been acquired, the reading for epoch $(r + 1)$ is not only acquired but also predicted. We refer to the nth reading *acquired* after the r readings in the buffer as $R_{A(r+n)}$ and the nth reading *predicted* after the r readings in the buffer as $R_{P(r+n)}$. The method used to carry out the predictions is described in Section 4.1.

If $|R_{P(r+1)} - R_{A(r+1)}| \leq \delta$, it means that the prediction falls within the user-specified error margin. Since the prediction is accurate enough, we make the assumption that the prediction accuracy will continue to hold for the following epoch $(r + 2)$. Thus the node skips acquiring a sample in epoch $(r + 2)$. This is done by setting the variable CurrentSkipSamplesLimit (CSSL) to 1. CSSL indicates the number of samples that need to be skipped before the next sample is taken. The variable SkipSamples (SS) is then set to the value held by CSSL. For every sample skipped, the value of SS is decremented by 1. Every time SS reaches a value of 0, the node samples the sensor to acquire a reading and a prediction is also computed. Thus in this case, SS reduces to 0 in epoch $(r + 3)$ and a sensor reading, $R_{A(r+3)}$ is acquired and $R_{P(r+3)}$ is calculated. This time, if $|R_{P(r+3)} - R_{A(r+3)}| \leq \delta$, CSSL is incremented by one (i.e. it is set to 2) and SS is then set to the value held by CSSL. In other words, every time an accurate prediction is made, CSSL is incremented by 1 and SS decreases to 0 from an initial starting value of CSSL. If however, at any time $|R_{P(r+3)} - R_{A(r+3)}| > \delta$, i.e. the prediction made is inaccurate, both CSSL and SS are set to zero so that no samples are skipped. Since the node samples its sensor and compares it with the prediction based on the past r readings stored in its buffer every time CSSL and SS are set to 0, a node can once again start skipping samples the moment it detects that the sensor readings can be predicted accurately.

However, if a node continues to make a long line of correct predictions, there is the possibility of CSSL increasing infinitely. This of course is undesirable as it would mean that eventually, when a change in the measured parameter does take place, it would be impossible to detect it. Thus we set a maximum limit to the value of CSSL known as the MaximumSkipSamplesLimit (MSSL). Once this maximum limit is reached, the variable CSSL continues to remain at that limit even if further correct predictions are made. Figure 5 summarizes the method presented above with an example illustrating how the values of CSSL and SS vary. While we initially use static values of the various variables to illustrate the benefits of this technique, in Section 4.1 we describe how a node may autonomously adjust the values of the variables as the conditions in the environment and network change.

4.1   Prediction of Sensor Readings

As mentioned earlier, we make use of time series forecasting to predict sensor readings. In basic time series forecasting, the historical values based on which predictions are supposed to be made, *must* be equally spaced. However, in the approach presented above it is evident that sensor readings are not acquired at regular intervals, e.g. Figure 5 clearly shows that an increasing number of samples are skipped when several correct predictions are made consecutively. While there are ways to estimate missing observations, such strategies would not only involve additional computation but may also not be very effective. This is because in our sampling scheme, in many instances, the majority of sensor readings may be missing (e.g. a buffer which is capable of storing 20 readings, may contain 18 missing readings and only 2 sampled readings).

As shown in Figure 6(a), every prediction is made based on the most recent r readings (representing the previous r epochs) stored in the buffer. Note that *all* the r readings are used regardless of whether they are sampled or predicted readings. Since the initial prediction is made for the epoch that *immediately* follows the epoch in the rth position of the buffer, the prediction is made using Equation 13 and setting m to 1 (i.e. we carry out a 1-step forecast), i.e.

$$\hat{X}_{t+1} = \mu + \phi(X_t - \mu) \tag{14}$$

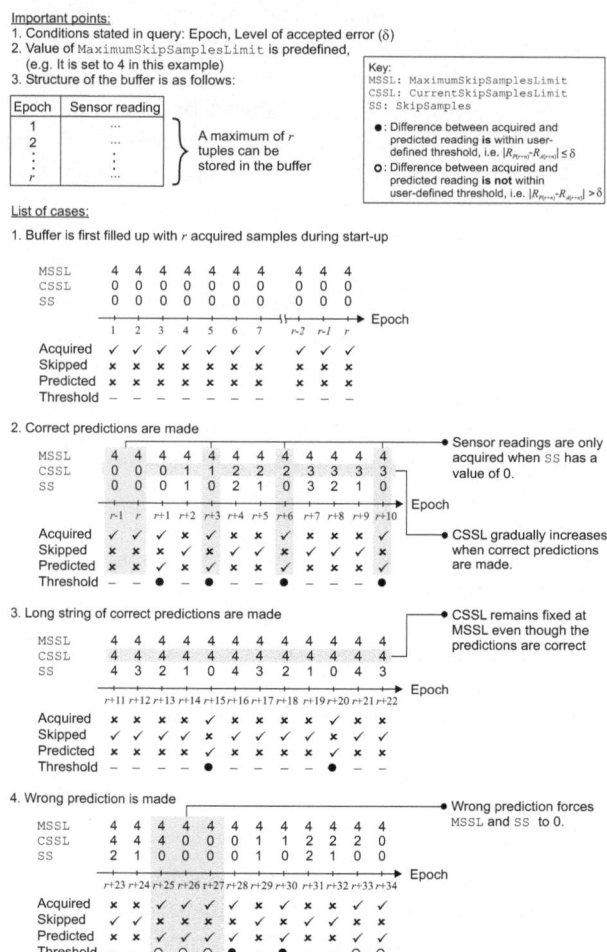

Fig. 5. Example illustrating how the values of the various variables change

Once the prediction is made, the first reading in the buffer is removed. The remaining readings are then moved back one position while the newly predicted reading (and the corresponding epoch) is placed in the rth position of the buffer. This process is repeated for all subsequent predictions (e.g. when SS>1) until a sample is taken. When a sample is taken, as shown in Figure 4(a), it is always compared with a prediction. However, instead of now placing the predicted reading in the rth position of the buffer, the actual sampled reading is placed there instead. It is obvious that the greater the number of sampled readings, the more accurate the predictions will be. Thus inserting every sampled reading into the buffer ensures that the maximum possible advantage is derived from the energy spent sampling the sensor. This is illustrated in Figure 6(a).

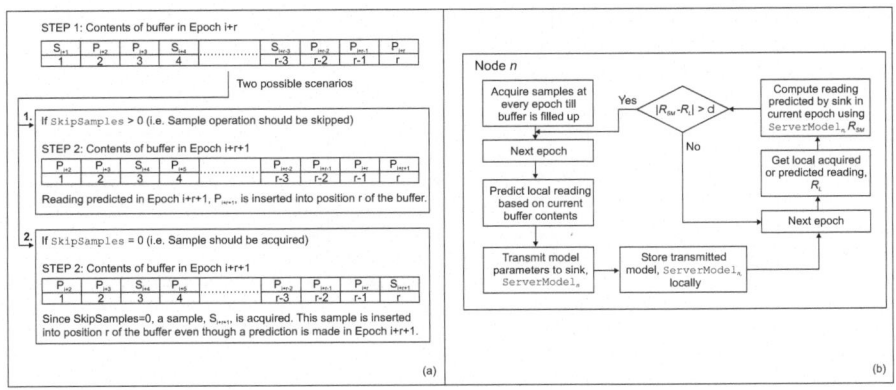

Fig. 6. (a) Updating the buffer, (b) Flowchart illustrating when model updates are transmitted to the sink by node n

Transmission of model updates to sink node. The final aim of any data collection system is to transmit the requested data to the user - who we assume in this paper to be located at the sink node. In a conventional data collection system, every node in the network would transmit a reading at every epoch (specified by the user) towards the sink node. This naturally not only results in excessive power consumption but also results in poor data quality since nodes closer to the root node may drop messages due to buffer overflows [7].

Our presented approach not only focuses on reducing the number of samples acquired but also uses the same prediction mechanism described in Section 4.1 to reduce the number of messages that need to be transmitted to the sink node. In short, instead of transmitting every acquired reading, or transmitting every time there is a change greater than the user-specified δ, we only transmit when there is a significant change in the *trend* of the measured parameter.

When the system first starts up, as mentioned earlier, a node n first samples its sensor in every epoch until the buffer gets filled up. A node does not transmit these readings to the sink node. As mentioned earlier, the first prediction is made immediately after the buffer has filled up and is based on all the inputs in the buffer. The parameters used to carry out the prediction (i.e. a_0, a_1, t, ϕ_1, X_t) are then transmitted to the sink. We refer to these model parameters of node n as ServerModel$_n$. Node n also stores ServerModel$_n$ locally. The sink node uses the received ServerModel$_n$ parameters to predict future sensor readings of node n. At every epoch after the transmission of the first ServerModel$_n$, node n uses its locally saved copy of ServerModel$_n$ to compute the sink node's prediction of node n's reading at the current epoch. If n computes that the reading predicted by the sink node differs from n's own reading by an amount greater than the user-specified δ, node n then transmits an updated ServerModel$_n$ to the sink. This update is based on the current contents of node n's buffer. In summary, node n transmits an updated ServerModel$_n$ every time the sink node's copy of the previously received ServerModel$_n$ is unable to predict node n's sensor

Table 1. Simulation settings and summary of results for the static CurrentSkipSamplesLimit scheme (The figures in brackets indicate the percentage reduction when compared to raw data collection scheme)

Data set	No. of nodes	No. of epochs	V.I.	Total no. of samples acquired			Total no. of transmissions made		
				Without prediction	**With** prediction, MIN (% **reduction**)	**With** prediction, MAX (% **reduction**)	**Without** prediction	**With** prediction, MIN (% **reduction**)	**With** prediction, MAX (% **reduction**)
LOW	100	2880	0.0411	288000	16401 (**93.4**)	98101 (**65.9**)	288000	2326 (**99.2**)	4088 (**98.6**)
MEDIUM	100	2880	0.5237	288000	101387 (**64.8**)	165636 (**42.5**)	288000	87170 (**69.7**)	150162 (**47.9**)
HIGH	100	2880	3.3769	288000	217451 (**24.5**)	253592 (**13.6**)	288000	207896 (**27.8**)	241830 (**16.0**)
NELLY	8	4936	0.0569	39488	2934 (**92.6**)	7881 (**80.0**)	39488	2629 (**93.3**)	7298 (**81.5**)
INTEL	51	2880	0.0152	146880	10731 (**92.7**)	50323 (**65.7**)	146880	4667 (**96.8**)	12868 (**91.2**)

readings accurately. This whole process of transmitting server model updates is summarized in Figure 6(b).

Effect of buffer size and MaximumSkipSamplesLimit. Values chosen for the buffer size and the MaximumSkipSamplesLimit (MSSL) can have a significant effect on the total number of samples acquired and model transmissions made by a node. In this subsection, we describe the effects of these two parameters for data sets having different V.Is. Simulations were performed using different data sets to investigate how the nodes responded when different values were used for the buffer size and MSSL. The simulation settings and results are summarized in Table 1 and the detailed results are presented in Figure 7. It can be easily observed that the results from the real data sets from Nelly Bay (Figure 7(g),(i)) and Intel (Figure 7(h),(j)) look very similar to the Low V.I. results (Figure 7(a),(d)) since they belong to the same class as far as the V.I. is concerned. Over the next few paragraphs we provide a detailed explanation of the various characteristics of the graphs in Figure 7.

Generally, when the trend of the sensor reading is fairly constant or predictable (i.e. Low V.I.), the CSSL will always reach the MSSL and remain there regardless of the value of MSSL. This is because a constant trend generally results in a long sequence of consecutive predictions which meet the users accuracy requirements. Note that MSSL can have a maximum value that is 2 less than the size of the buffer. A large CSSL effectively means that a node will sample its own sensor less frequently. Thus for data sets with a low V.I., only the value of MSSL affects the number of samples a particular sensor acquires. The size of the buffer does not have any impact.

However, the situation varies for the transmission of model updates. Recall that in the case of predicting readings used to decide on the value of CSSL, a node uses the current contents of its own buffer, which is always up-to-date since it relies on a sliding window containing the most recent r readings where r is the size of the buffer. On the other hand, when deciding whether to transmit a model update, the predictions are made using the *previously* transmitted ServerModel. Thus predictions for model updates are based on "old" data. In this case, in

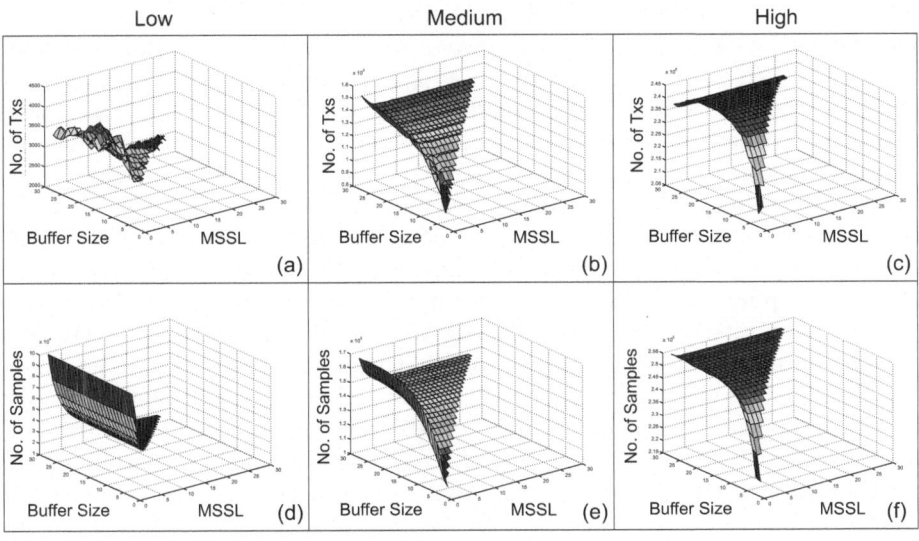

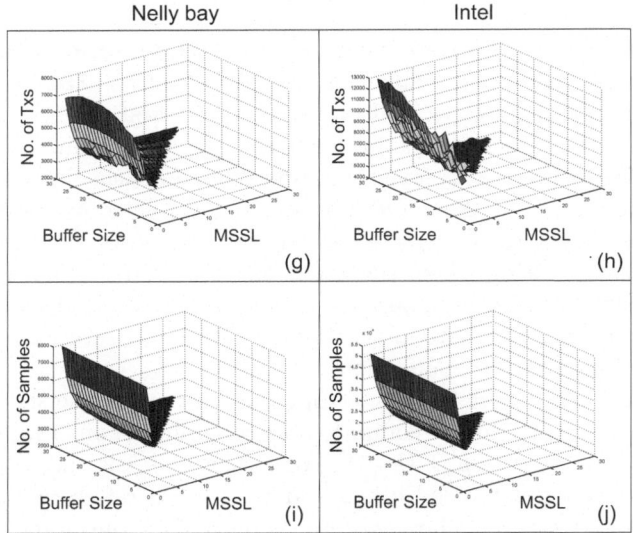

Fig. 7. Sensor sampling and model transmission results for different data sets using different buffer sizes and MaximumSkipSamplesLimit (MSSL) values (Note: Without the prediction mechanism, the number of sensor sampling and transmission operations for the different data sets would be as follows - (i) Low, Medium, High: 288000, (ii) Nelly: 39488, (iii) Intel: 146880)

addition to MSSL, the buffer size also has an impact on the number of transmitted model updates as a larger buffer size means that there are a larger number of "older" readings in every transmitted ServerModel.

For data sets with a higher V.I. the buffer size plays a more significant role than the choice of the MSSL - except when we consider smaller values of MSSL (Figures 7(b),(c),(e),(f)). Generally, MSSL hardly has a role primarily because when the trend is continuously changing, CSSL can only take up small values even if MSSL is set to a large value. Reducing the size of the buffer size has two opposing effects. Firstly, a small buffer size implies that MSSL must also have a small value since MSSL can have a maximum value that is 2 less than the buffer size. This in turn means that a node will have to sample its sensor more often. Conversely, when the trend is highly variable, a smaller buffer size would obviously help model the change in trend more accurately thus resulting in more accurate predictions. A smaller buffer size also means that only the recent past is being used for prediction. This means that the a node will sample its sensor less often. Our simulation results clearly indicate that when a smaller buffer size is used, the benefits gained from better prediction far exceed the disadvantage due to more frequent sensor sampling. Since a larger buffer size generally results in more transmissions (which in turn means more accurate data collection at the expense of higher energy consumption) we have used a buffer size of 30 for all the following simulations.

Meeting the user's coverage requirements (Dynamic MSSL). In this sub-section we see how a node can autonomously adjust its MSSL in order to meet the coverage requirements specified by the user. We first provide our definition of *coverage* and then describe the mechanism in greater detail.

When MSSL is set to m, it means that x unsampled epochs lie between any two consecutive samples. The duration of an event could span over several epochs. We consider an event to be detected as long as the sensor samples during *any one* of the epochs during which the event has occurred, i.e. our aim is not to detect an event at the earliest possible time, but to simply detect *if* at all an event has occurred. Note that this is adequate since our solution is meant only for non-critical monitoring applications.

So given that MSSL is set to m, and that the duration of the shortest possible event of interest is d epochs, we compute the probability of an event being missed by a particular node, p as $p = \frac{m+1-d}{m+1}$. Note that $m + 1$ is the total number of positions where an event could occur and $m + 1 - d$ refers to the number of uncovered event positions. Thus *coverage* is defined as the probability that an event *will* be detected, i.e. *1-p*.

If the current node has n neighbors, then we assume that an event is detected as long as any *one* of the nodes within a node's closed 1-hop neighborhood samples its sensor at any time during the occurrence of the event. The reason for this approach is that the spatial correlations between neighboring sensors in a high density network might mean that if a particular event is detected by one node, there is a high possibility of the same event being detected by an adjacent neighbor (though of course the magnitude of the measured parameter might be different). We refer the reader to our earlier work in [7] which describes typical correlation levels between adjacent sensors in a real deployment in the GBR. We also use this approach for a "distributed wake-up" mechanism where a node that detects an event asks all its adjacent neighbors to break out of the reduced sampling scheme and immediately sample a reading (by reducing SS to 0) to check if they too can detect the event. This would help improve the accuracy of the data collected. However, we do not describe this mechanism here as it falls outside the scope of this paper.

Thus when multiple nodes are involved in checking the occurrence of a certain event, the probability of missing the event is reduced. This is reflected in the following equation that computes the probability an event is missed when a node has n adjacent neighbors: $p = (\frac{m+1-d}{m+1})^{n+1}$.

Assuming that a user specifies the coverage he or she is willing to tolerate through a query injected into the network, every node can then compute the corresponding MSSL based on the number of neighbors a node has by simply rearranging the previous equation: $\text{MSSL} = \lfloor \frac{n+\sqrt[n]{p}+d-1}{1-\sqrt[n]{p}} \rfloor$

Figure 4(b) shows a plot of the above equation given that $d = 5$ and the size of the buffer is 30. It can be seen that for any user-specified p, the larger the number of neighbors a node has, the larger is the chosen value of MSSL thus allowing a node to spend more energy sampling only when the user requires it. If the computed value of m is greater than 2 less than the buffer size, m is set to $BufferSize - 2$. This explains the flat portion of the graph.

In Figures 8(a)-(c) we illustrate how the probability of missing an event varies for different event durations when data sets with different V.Is are used. Note that in the simulations, we assume that the user has predefined p to be 0.2. In Figure 8(a), when a static value of MSSL is used (e.g. 15, 28), the actual p can exceed the user-specified p. However, this is prevented in the case of Dynamic MSSL. Data sets with Medium or High Variabilities do not really require the Dynamic MSSL mechanism as CSSL values never reaches MSSL due to the frequent change in trends. Since a small MSSL implies that the nodes are sampling very frequently, they hardly ever miss any passing event. This can be clearly observed from Figures 8(b)-(c). Thus using the Dynamic MSSL mechanism is advisable if a user expects to gather data which has a low V.I. and requires a coverage guarantee.

We evaluate the overall performance of our adaptive sampling scheme by investigating the overall energy consumption of all the nodes in the network and the accuracy of the collected data. We compare our scheme with conventional raw data collection where a reading is collected from every node in every epoch and also with *approximate caching* [11]. In this algorithm, a node transmits a reading to the sink whenever its current value differs from the previously transmitted reading by an amount greater than the user-specified threshold. Note however, that while approximate caching can help reduce the number of transmissions, it cannot be used to reduce the number of sampling operations since the algorithm does not perform any predictions. Our simulations use LMAC. We use 32 slots per frame where each frame is 8 seconds long. The energy model based on the RFM TR1001 transceiver [3] and the EXCELL Salinity Sensor [4] which could typically be used in our GBR deployment. Measurements are required every 30

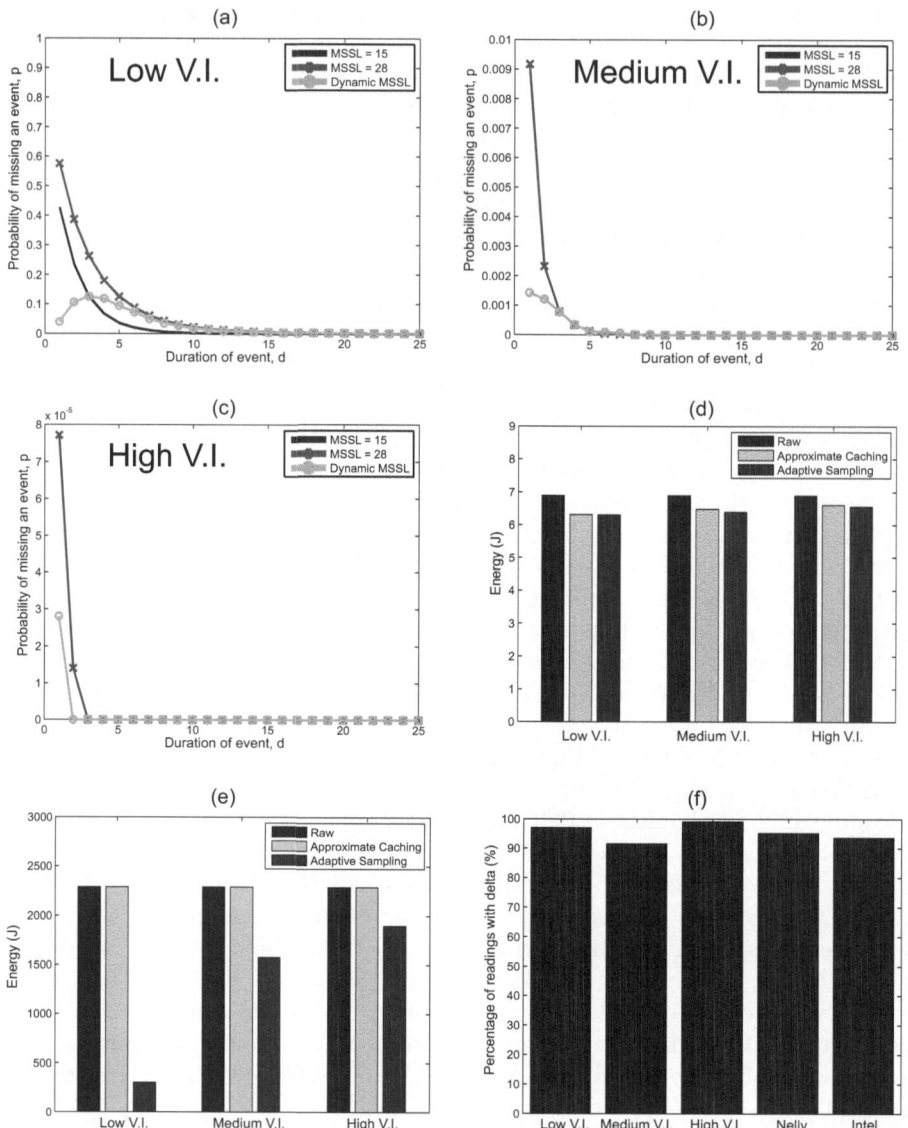

Fig. 8. (a), (b), (c): Probability of missing an event for different event durations and data sets with different V.Is: Low, Medium and High. (d) Energy consumption (Transceiver operation), (e) Overall energy consumption (Transceiver operation+Sensor sampling, (f) Accuracy).

seconds. The number of nodes used is specified in Table 1. All results have been collected over 400 frames or 53.3 minutes.

Figure 8(d) shows the energy spent operating the transceiver by all nodes in the network over the 400 frame duration. Even though there is a significant reduction in the number of messages transmitted (as shown in Table 1), the energy different between the raw, approximate caching and our adaptive sampling techniques is not very large. This is largely due to the LMAC protocol where nodes spend a lot of time listening in order to maintain synchronization even when there is no data to send. However, as shown in Figure 8(e), if we look at the overall energy consumption - which includes both the operation of the transceiver and the sensors - our scheme reduces overall power consumption by up to around 87% (Low V.I.) as compared to the other two schemes. Even though there is such a large reduction in energy consumption, in most cases, more than 90% of the collected readings are within the ±0.1 units of the actual readings. This is shown in Figure 8(f).

5   Related Work

A wide variety of techniques can be found that deal with extracting data in an energy-efficient manner. The authors in [9] describe a technique to prevent the need to sample sensors in response to an incoming query. However, the technique is not able to cope with sudden changes in the correlation models and also fails to recognize the importance of temporal fluctuations in these models. Ken [8] is able to adapt create models on the fly and thus adapt but does not describe how to deal with topology changes. Both PAQ [13] and SAF [12] are similar to our scheme in the sense that they also use time-series forecasting to identify temporal correlations within the network itself. However, all the schemes mentioned above only deal with reducing message transmissions. While this is helpful, our results in Figure 8(e) clearly indicate that a large reduction in message transmissions does not always translate into energy savings of the same magnitude largely due to the overhead of the radio (which is dependent on the MAC). This is especially true when the application uses sensors which consume a lot of energy. Note that many sophisticated sensors used for environmental monitoring also have long start-up and sampling times. This too has a large impact on energy consumption. To our knowledge, we are not aware of any other work which deals with reducing the duty cycle of the sensors themselves.

6   Conclusion

We have presented an adaptive sensor sampling scheme that takes advantage of temporal correlations of sensor readings in order to reduce energy consumption. Our localized scheme, which uses cross-layer information provided by the MAC, allows nodes to change their sampling frequency autonomously based on changes in the physical parameters being monitored or the topology of the network. Our simulations based on real and synthetic data sets, indicate a reduction in sensor sampling by up to 93%, reduction in message transmissions by up to 99% and overall energy savings of up to 87%. We also show that generally more than 90% of the readings are within the user-defined error threshold. We are currently looking into techniques that could be used to provide a bound on the error level of the acquired data.

References

1. USGS, The National Map Seamless Server, http://seamless.usgs.gov/.
2. Intel lab data (2004), http://db.csail.mit.edu/labdata/labdata.html
3. RF Monolithics, Inc, RFM TR1001 868.35MHz Hybrid Transceiver (2007), http://www.rfm.com/products/data/tr1001.pdf
4. Falmouth scientific, inc. (2008), http://www.falmouth.com/products/index.htm
5. Bondarenko, O., Kininmonth, S., Kingsford, M.: Underwater sensor networks, oceanography and plankton assemblages. In: Proceedings of the IEEE International Conference on Intelligent Sensors, Sensor Networks and Information Processing, ISSNIP 2007, Melbourne, December 2007, pp. 657–662. IEEE, Los Alamitos (2007)
6. Chatterjea, S., Kininmonth, S., Havinga, P.J.M.: Sensor networks. GeoConnexion 5(9), 20–22 (2006)
7. Chatterjea, S., Nieberg, T., Meratnia, N., Havinga, P.: A distributed and self-organizing scheduling algorithm for energy-efficient data aggregation in wireless sensor networks. ACM TOSN (to be published)
8. Chu, D., Deshpande, A., Hellerstein, J.M., Hong, W.: Approximate data collection in sensor networks using probabilistic models. In: ICDE, p. 48 (2006)
9. Deshpande, A., Guestrin, C., Madden, S., Hellerstein, J.M., Hong, W.: Model-based approximate querying in sensor networks. VLDB J 14(4), 417–443 (2005)
10. Hoesel, L.v.: Sensors on speaking terms: schedule-based medium access control protocols for wireless sensor networks. PhD thesis, University of Twente, The Netherlands (2007)
11. Olston, C., Widom, J.: Best-effort cache synchronization with source cooperation. In: SIGMOD Conference (2002)
12. Tuloen, D., Madden, S.: An energy-efficient querying framework in sensor networks for detecting node similarities. In: MSWiM 2006: Proceedings of the 9th ACM international symposium on Modeling analysis and simulation of wireless and mobile systems, pp. 191–300 (2006)
13. Tulone, D., Madden, S.: Paq: Time series forecasting for approximate query answering in sensor networks. In: EWSN, pp. 21–37 (2006)

LiveNet: Using Passive Monitoring to Reconstruct Sensor Network Dynamics

Bor-rong Chen, Geoffrey Peterson, Geoff Mainland, and Matt Welsh

School of Engineering and Applied Sciences,
Harvard University,
Cambridge MA 02138, USA
{brchen,glpeters,mainland,mdw}@eecs.harvard.edu

Abstract. We describe *LiveNet*, a set of tools and analysis methods for reconstructing the complex behavior of a deployed sensor network. LiveNet is based on the use of multiple passive packet sniffers co-located with the network, which collect packet traces that are merged to form a global picture of the network's operation. The merged trace can be used to reconstruct critical aspects of the network's operation that cannot be observed from a single vantage point or with simple application-level instrumentation. We address several challenges: merging multiple sniffer traces, determining sniffer coverage, and inference of missing information for routing path reconstruction. We perform a detailed validation of LiveNet's accuracy and coverage using a 184-node sensor network testbed, and present results from a real-world deployment involving physiological monitoring of patients during a disaster drill. Our results show that LiveNet is able to accurately reconstruct network topology, determine bandwidth usage and routing paths, identify hot-spot nodes, and disambiguate sources of packet loss observed at the application level.

1 Introduction

As sensor networks become more sophisticated and larger in scale, better tools are needed to study their behavior in live deployment settings. Understanding the complexities of network dynamics, such as the ability of a routing protocol to react to node failure, or an application's reaction to varying external stimuli, is currently very challenging. Unfortunately, few good tools exist to observe and monitor a sensor network deployment *in situ*.

In this paper, we describe *LiveNet*, a set of tools and techniques for recording and reconstructing the complex dynamics of live sensor network deployments. LiveNet is based on the use of passive monitoring of radio packets observed from one or more *sniffers* co-deployed with the network. Sniffers record traces of all packets received on the radio channel. Traces from multiple sniffers are merged into a single trace to provide a global picture of the network's behavior. The merged trace is then subject to a series of analyses to study application behavior, data rates, network topology, routing protocol dynamics, and packet loss.

S. Nikoletseas et al. (Eds.): DCOSS 2008, LNCS 5067, pp. 79–98, 2008.

Although a passive monitoring infrastructure can increase cost, we argue that in many cases this is the ideal approach to observing and validating a sensor network's operation, and in some cases is the *only* way to effectively monitor the network. LiveNet brings a number of benefits over traditional network monitoring solutions. First, LiveNet decouples packet capture from trace analysis, allowing "raw" packet traces to be studied in many different ways. In contrast, in-network monitoring relies on preconceptions of the network's operation and failure modes, and can fail when the system does not behave as expected. Second, LiveNet requires no changes to the network being monitored, which is prudent for reasons of performance and reliability. Third, the LiveNet infrastructure can be deployed, reconfigured, and torn down separately from the network under test. LiveNet can be set up on an as-needed basis, such as during the initial sensor network deployment, or during periods when unexpected behavior is observed. Finally, it is possible to use LiveNet in situations where sensor nodes are mobile or physical access to sensor nodes is unavailable. We describe the use of LiveNet during a disaster drill involving multiple mobile patient sensor nodes, fixed repeater nodes, and base stations. In this scenario, directly instrumenting each node would have been prohibitive.

Using passive monitoring to understand a sensor network's behavior raises a number of unique challenges. First, we are concerned with the coverage of the LiveNet sniffer infrastructure in terms of total number of packets observed by the system. Second, the merging process can be affected by incomplete packet traces and lack of time synchronization across sniffers. Third, understanding global network behavior requires extracting aggregate information from the detailed traces. We describe a series of analyses, including a novel *path inference* algorithm that derives routing paths based on incomplete packet traces.

We evaluate the use of LiveNet in the context of a sensor network for monitoring patient vital signs in disaster response settings [7,10]. We deployed LiveNet during a live disaster drill undertaken in August 2006 in which patients were monitored and triaged by emergency personnel following a simulated bus accident. We also perform an extensive validation of LiveNet using measurements on a 184-node indoor sensor network testbed.

Our results show that deploying the LiveNet infrastructure along with an existing sensor network can yield a great deal of valuable information on the network's behavior without requiring additional instrumentation or changes to the sensor network code. Our packet merging process and trace analyses yield an accurate picture of the network's operation. Finally, we show that our path inference algorithm correctly determines the routing path used without explicit information from the routing protocol stack itself.

2   Background and Motivation

Sensor networks are becoming increasingly complex, and correct behavior often involves subtle interactions between the link layer, routing protocol, and application logic. Achieving a deep understanding of network dynamics is extremely challenging for real sensor network deployments. It is often important to study a sensor deployment *in situ*, as well as in situations where it is impossible or undesirable to add additional instrumentation. Although simulators [6,15] and testbeds [4,17] are invaluable for development, debugging, and testing, they are fundamentally limited in their ability to capture the full complexity of radio channel characteristics, environmental stimuli, node mobility, and hardware failures that arise in real deployments. This suggests the need for *passive* and *external* observation of a sensor network's behavior "in the wild."

Several previous systems focus on monitoring and debugging live sensor deployments. Sympathy [11] is a system for reasoning about sensor node failures using information collected at *sink nodes* in the network. Sympathy has two fundamental limitations that we believe limit its applicability. First, Sympathy requires that the sensor node software be instrumented to transmit periodic *metrics* back to the sink node. However, it is often impossible or undesirable to introduce additional instrumentation into a deployed network after the fact. Second, Sympathy is limited to observe network state at sink nodes, which may be multiple routing hops from the sensor nodes being monitored. As a result, errant behavior deep in the routing tree may not be observed by the sink. However, we do believe that LiveNet could be used in conjunction with a tool like Sympathy to yield more complete information on network state.

SNMS [16] and Memento [14] are two management tools designed for inspecting state in live sensor networks. They perform functions such as neighborhood tracking, failure detection, and reporting inconsistent routing state. EnviroLog [8] is a logging tool that records function call traces to flash, which can be used after a deployment to reconstruct a node's behavior. Like Sympathy, these systems add instrumentation directly into the application code.

In contrast to these approaches, LiveNet is based on the use of passive *sniffer* nodes that capture packets transmitted by sensor nodes for later analysis. Our approach is inspired by recent work on passive monitoring for 802.11 networks, including Jigsaw [3,2] and Wit [9]. In those systems, multiple sniffer nodes collect packet traces, which are then merged into a single trace representing the network's global behavior. A series of analyses can then be performed on the global trace, for example, understanding the behavior of the 802.11 CSMA algorithm under varying loads, or performance artifacts due to co-channel interference. Although LiveNet uses a similar trace merging approach to these systems, we are focused on a different set of challenges. In 802.11 networks, the predominant communication pattern is single hop (between clients and access points), and the focus is on understanding link level behavior. In contrast, sensor networks exhibit substantially more complex dynamics, due to multihop routing and coordinated behavior across nodes. SNIF [12] is the only other passive monitoring system for sensor networks of which we are aware. Although there are some similarities, SNIF differs from LiveNet in several important respects. SNIF is focused primarily on debugging the causes of failures in sensor networks, while we are more interested in time-varying dynamics of network behavior, such as routing path dynamics, traffic load and hotspot analysis, network connectivity, and recovering the sources of path loss. As a result, SNIF requires less accuracy in terms of trace merging and sniffer coverage than we require with LiveNet. Also, SNIF uses sniffers that transmit complete packet traces to a base station via a Bluetooth scatternet. Apart from the scalability limitations of the scatternet itself, interference between Bluetooth and 802.15.4 radios commonly used by sensor networks is a concern, potentially causing the monitoring infrastructure to interfere with the network's operation.[1] In LiveNet, we decouple packet capture from trace analysis and forego the requirement of real-time processing, which we believe is less important for most uses of our system.

3   LiveNet Architecture

LiveNet consists of three main components: a *sniffer infrastructure* for passive monitoring and logging of radio packets; a *merging process* that normalizes multiple sniffer logs and combines them into a single trace; and a set of *analyses* that make use of the combined trace. While packet capture is performed in real time, merging and analysis of the traces is performed offline due to high storage and computational requirements. While providing real-time analysis of traffic captured by LiveNet would be useful in certain situations, we believe that offline trace merging and analysis meets many of the needs of users wishing to debug and analyze a network deployment.

3.1   Sniffer Infrastructure

The first component of LiveNet is a set of passive network sniffers that capture packets and log them for later analysis. Conceptually the sniffer is very simple, consisting of a sensor node either logging packets to local flash or over its serial port to an attached host. Sniffers timestamp packets as they are received by the radio, to facilitate trace merging and timing analysis.

We envision a range of deployment options for LiveNet sniffers. Sniffers can be installed either temporarily, during initial deployment and debugging, or permanently, in order to provide an unobtrusive monitoring framework. Temporary sniffers could log packets to flash for manual retrieval, while permanent sniffers would typically require a backchannel for delivering packet logs. Another scenario might involve *mobile* sniffers, each carried by an individual around the sensor network deployment site. This would be particularly useful for capturing packets to debug a performance problem without disturbing the network configuration.

3.2   Merging Process

Given a set of sniffer traces $\{S_1 \ldots S_k\}$, LiveNet's merging process combines these traces into a temporally-ordered log that represents a global view of network activity. This process must take into account the fact that each trace only

---

[1] The SNIF hardware is based on previous-generation radios operating in the 868 MHz band; it is unclear whether a Bluetooth scatternet backhaul could be used with current 802.15.4 radios.

contains a subset of the overall set of packets, with a varying degree of overlap. Also, we do not assume that sniffers are time synchronized, requiring that we normalize the timebase for each trace prior to merging. Finally, due to the large size of each trace, we cannot read traces into memory in their entirety. This requires a progressive approach to trace merging.

Our merging algorithm is somewhat similar to Jigsaw [3], which is designed for merging 802.11 packet traces. We briefly describe our algorithm here, referring the reader to a technical report [13] for further details. The merging algorithm operates in two phases. In the first phase, we compute a *time mapping* that maps the timebase in each trace S_i to a common (arbitrary) timebase reference S_0. Given a pair of traces S_i and S_j, we identify a unique packet p that appears in both traces. We calculate the timebase offset between S_i and S_j as the difference in the timestamps between the two packets in each trace, $\Delta i, j = t_i(p) - t_j(p)$. Δij represents a single edge in a *time graph* (where nodes are traces and edges are timebase offsets between traces). The offset $\Delta i, 0$ from a given trace S_i to the reference S_0 is then computed as the combined timebase offset along the shortest path in the time graph from S_i to S_0.

In the second phase, we progressively scan packets from each trace and apply the time correction $\Delta_{i,0}$. Packets are inserted into a priority queue ordered by global time and identical packets from across traces are merged. After filling the priority queue to a given window size W, we begin emitting packets, simply popping the top of the priority queue and writing each packet to a file. The algorithm alternates between scanning and emitting until all traces have been merged.

There are two potential sources of error in our algorithm. First, due to its progressive nature, it is possible that a packet p will be "prematurely" emitted before all instances of p in each trace have been scanned, leading to duplicates in the merged trace. In practice, we find very few duplicates using a window size $W = 10$ sec. Second, if link-layer ARQ has been used by the application, multiple retransmissions of the same packet will appear in the sniffer traces, complicating the merging process. For example, if 4 copies of a packet p appear in trace S_1, and 2 copies in trace S_2, it is not obvious how many transmissions of the packet occurred in total. We opt to adhere to the lower bound, since we know *at least* 4 transmissions of p occurred.

4   Trace Analysis

In this section, we describe a range of analysis algorithms for reconstructing a sensor network's behavior using the merged LiveNet trace. Several of these algorithms are generic and can be applied to essentially any type of traffic, while other analyses use application-specific knowledge.

4.1   Coverage Analysis

The most basic analysis algorithm attempts to estimate the *coverage* of the LiveNet sniffer infrastructure, by computing the fraction of packets actually transmitted by the network that were captured in the merged packet trace. Coverage can also be computed on a per-sniffer basis, which is useful for determining whether a given sniffer is well-placed. Let us define $C_i(n)$ as the *coverage* of sniffer S_i with respect to node n, which is simply the fraction of packets received by S_i that were actually transmitted by n. Estimating the number of packets transmitted by n can be accomplished using packet-level sequence numbers, or knowledge of the application transmission behavior (e.g., if the application transmits a periodic beacon packet). We assume that packet loss from nodes n to sniffers S_i is uniform, and does not depend on the contents of the packets. Note that this assumption might not be valid, for example, if longer packets are more likely to experience interference or path loss.

4.2 Overall Traffic Rate and Hotspot Analysis

Another basic analysis is to compute the overall amount of traffic generated by each node in the network, as well as to determine "hotspots" based on which nodes appear to be the source of, or destination of, more packets than others. Given the merged trace, we can start by counting the total number of packets originating from or destined to a given node n. Because LiveNet may not observe all actual transmissions, we would like to *infer* the existence of other packets. For example, if each transmission carries a unique sequence number we can infer missing packets by looking for gaps in the sequence number space. Coupled with topology inference (Section 4.3), one can also determine which nodes were likely to have received broadcast packets, which do not indicate their destination explicitly.

4.3 Network Connectivity

Reconstructing radio connectivity between nodes is seemingly straightforward: for each packet from node a to b, we record an edge $a \rightarrow b$ in the connectivity graph. However, this approach may not reconstruct the *complete* topology, since two nodes a and b within radio range may choose not to communicate directly, depending on the routing protocol in use. We make use of two approaches. First, if one assumes that connectivity is symmetric, an edge $b \rightarrow a$ can be recorded alongside $a \rightarrow b$. Although asymmetric links are common in sensor networks [1], this algorithm would establish whether two nodes are *potential* neighbors.

The second method is to inspect routing control packets. For example, several routing protocols, such as TinyOS' MultihopLQI, periodically transmit their *neighbor table* containing information on which nodes are considered neighbors, sometimes along with link quality estimates. These packets can be used to reconstruct the network connectivity from the sniffer traces. Note that this information is generally not available to a base station, which would only overhear control packets within a single radio hop.

4.4 Routing Path Inference

One of the more interesting analyses involves reconstructing the routing path taken by a packet traveling from a source node s to a destination d. The simplest

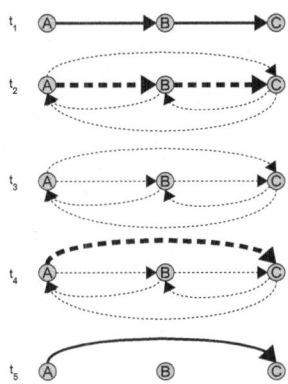

	{A,B}		{B,C}		{A,C}		{B,A}		{C,B}		{C,A}	
	Π+	Π-	Π+	Π-	Π+	Π-	Π+	Π-	Π+	Π-	Π+	Π-
t_1	**1.0**	*-0.6*	**1.0**	*-0.6*	**0.6**	*-1.0*	0	*-1.0*	0	*-1.0*	0	*-1.0*
t_2	*0.9*	*-0.7*	*0.9*	*-0.7*	*0.7*	*-0.9*	0	*-0.9*	0	*-0.9*	0	*-0.9*
t_3	*0.8*	*-0.8*	*0.8*	*-0.8*	*0.8*	*-0.8*	0	*-0.8*	0	*-0.8*	0	*-0.8*
t_4	*0.7*	*-0.9*	*0.7*	*-0.9*	*0.9*	*-0.7*	0	*-0.9*	0	*-0.9*	0	*-0.9*
t_5	*0.6*	*-1.0*	*0.6*	*-1.0*	**1.0**	*-0.6*	0	*-1.0*	0	*-1.0*	0	*-1.0*

The figure at left shows observed and inferred routing paths for a network of 3 nodes over 5 timesteps. Observed packet transmissions are shown as solid arrows; potential links are shown as dashed arrows. At time t_1, packets $A \rightarrow B$ and $B \rightarrow C$ are observed. At time t_5, packet $A \rightarrow C$ is observed. At intermediate times, the inferred path is shown in bold. At t_3, both paths $A \rightarrow B \rightarrow C$ and $A \rightarrow C$ have equal probability.

The table above shows the positive ($\Pi+$) and negative ($\Pi-$) score for each link at each timestep. Values in boldface are direct observations; those in italics are time-dilated scores based on past or future values. Here we assume a time-dilation constant $s = 0.1$.

Fig. 1. Path inference example

case involves protocols that use source-path routing, in which case the complete routing path is contained within the first transmission of a packet from the originating node. In most sensor network routing protocols, however, the routing state must be inferred by observing packet transmissions as packets travel from source to destination. However, because the merged packet trace may not contain every routing hop, there is some ambiguity in the routing path that is actually taken by a message. In addition, the routing path may evolve over time. As a worst case, we assume that the route can change between any two subsequent transmissions from the source node s.

The goal of our path inference algorithm is to determine the *most probable* routing path $P(s, d, t) = (s, n_1, ...n_k, d)$ at a given time t. We begin by quantizing time into fixed-sized windows; in our implementation, the window size is set to 1 sec. For each possible routing hop $a \rightarrow b$, we maintain a *score* $\Pi(a, b, t)$ that represents the likelihood of the hop being part of the routing path during the window containing t. $\Pi(a, b, t)$ is calculated using two values for each link: a *positive score* $\Pi^+(a, b, t)$ and a *negative score* $\Pi^-(a, b, t)$. The positive score represents any positive information that a link may be present in the routing path, based on an observation (possibly at a time in the past or future) that a message was transmitted from a to b. The negative score represents negative information for links that are *excluded* from the routing path due to the presence of other, conflicting links, as described below.

Figure 1 shows our algorithm at work on a simple example. We begin by initializing $\Pi^+(a, b, t) = \Pi^-(a, b, t) = 0$ for all values of a, b, and t. The merged packet trace is scanned, and for each observed packet transmission $a \rightarrow b$, we set $\Pi^+(a, b, t) = 1$. For each *conflicting link* $a' \rightarrow b'$, we set $\Pi^-(a', b', t) = -1$. A link conflicts with $a \rightarrow b$ if it shares one endpoint in common (i.e., $a = a'$ or $b = b'$); $b \rightarrow a$ is also conflicted by definition.

Once the scan is complete, we have a sparse matrix representing the values of Π^+ and Π^- that correspond to observed packet transmissions. To fill in the rest of the matrix, we *time dilate* the scores, in effect assigning "degraded" scores to those times before and after each observation. Given a time t for which no value has been assigned for $\Pi^+(a, b, t)$, we look for the previous and next time windows $t_{prev} = t - \delta_b$ and $t_{next} = t + \delta_f$ that contain concrete observations. We then set $\Pi^+(a, b, t) = \max(\max(0, \Pi^+(a, b, t_{next}) - s \cdot \delta_f), \max(0, \Pi^+(a, b, t_{prev}) - s \cdot \delta_b))$. That is, we take the maximum value of Π^+ time-dilated backwards from t_{next} or forwards from t_{prev}, capping the value to ≥ 0. Here, s is a scaling constant that determines how quickly the score degrades per unit time; in our implementation we set $s = 0.1$. Similarly, we fill in values for missing $\Pi^-(a, b, t)$ values, also capping them to be ≤ 0.

Once we have filled in all cells of the matrix for all links and all time windows, the next step is to compute the final link score $\Pi(a, b, t)$. For this, we set the value to either $\Pi^+(a, b, t)$ or $\Pi^-(a, b, t)$ depending on which has the greater absolute value. For links for which we have no information, $\Pi(a, b, t) = 0$. The final step is to compute the most likely routing path at each moment in time. For this, we take the acyclic path that has the highest *average* score over the route, namely: $P^\star(s, d, t) = \arg\max_{\forall P(s,d,t)} \sum_{l=\{n1,n2\} \in P(s,d,t)} \Pi(n_1, n_2, t)/|P(s, d, t)|$. The choice of this metric has several implications. First, links for which we have no information ($\Pi(a, b, t) = 0$) diminish the average score over the path. Therefore, all else being equal, our algorithm will prefer shorter paths over longer ones. For example, consider a path with a "gap" between two nodes for which no observation is ever made: $(s, n_1, \ldots? \ldots, n_2, d)$. In this case, our algorithm will fill in the gap with the direct hop $n_1 \rightarrow n_2$ since that choice maximizes the average score over any other path with more than one hop bridging the gap.

Second, note that the most likely path $P^\star$ may not be unique; it is possible that multiple routes exist with the same average score. In this case, we can use network connectivity information (Section 4.3) to exclude links that are not likely to exist in the route. While this algorithm is not guaranteed to converge on a unique solution, in practice we find that a single route tends to dominate for each time window.

4.5   Packet Loss Determination

Given a dynamic, multihop sensor network, one of the most challenging problems is understanding the causes of data loss from sources to sinks. The failure of a packet to arrive at a sink can be caused by packet loss along routing paths, node failures or reboots, or application-level logic (for example, a query timeout that causes a source node to stop transmitting). Using LiveNet we can disambiguate the sources of packet loss, since we can observe packet receptions from many vantage points, rather than only at the sink node.

Figure 2 shows the behavior of an individual node during the disaster drill described in Section 7. Each point represents a packet either observed by LiveNet or received at the base station (or both). Here, the node is transmitting packets at 1 Hz as long as the query is active; however, the query will timeout if a renewal message is not received before it expires. Each packet carries a monotonically increasing sequence number. By combining information from the LiveNet and

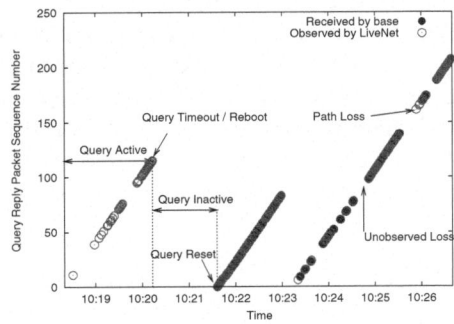

Fig. 2. Determining the causes of packet loss

base station logs, we can break down packet loss in terms of loss along the routing path to the sink, inactive query periods, and *unobserved loss*; that is, packets that were neither observed by LiveNet or the sink. For example, for the first 100 sec or so, the node is transmitting packets but none of them are received by the sink, indicating a bad routing path. Query timeouts are easily detected by a reset in the packet sequence numbers. Intervals between two subsequent queries with no observed packets indicates a period with no active query.

5 Implementation

Our implementation of LiveNet consists of three components: the sniffer infrastructure, trace merging code, and analysis algorithms. The sniffers are implemented as a modified version of the TinyOS TOSBase application, with two important changes. First, the code is modified to pass every packet received over the radio to the serial port, regardless of destination address or AM group ID. Second, the sniffer takes a local timestamp (using the SysTime.getTime32() call) on each packet reception, and prepends the timestamp to the packet header before passing it to the serial port.

We observed various issues with this design that have not yet been resolved. First, it appears that TMote Sky motes have a problem streaming data at high rates to the serial port, causing packets to be dropped by the sniffer. In our LiveNet deployment described below, a laptop connected to both a MicaZ and a TMote Sky sniffer recorded more than three times as many packets from the MicaZ. This is possibly a problem with the MSP430 UART driver in TinyOS. Second, our design only records packets received by the Active Messages layer in TinyOS. Ideally, we would like to observe control packets, such as acknowledgments, as well as packets that do not pass the AM layer CRC check.

Our merging and analysis tools are implemented in Python, using a Python back-end to the TinyOS *mig* tool to generate appropriate classes for parsing the raw packet data. A parsing script first scans each raw packet trace and emits a parsed log file in which each packet is represented as a *dictionary* mapping named keys to values. Each key represents a separate field in the packet (source ID, sequence number, and so forth). The dictionary provides an extremely flexible mechanism for reading and manipulating packet logs, simplifying the design of the merging and subsequent analysis tools. The merging code is 657 lines of code (including all comments). The various analysis tools comprise 3662 lines of code in total. A separate library (131 lines of code) is used for parsing and managing packet traces, which is shared by all of the merging and analysis tools.

6   Validation Study

The goal of our validation study is to ascertain the accuracy of the LiveNet approach to monitoring and reconstructing sensor network behavior. For this purpose, we make use of a well-provisioned indoor testbed, which allows us to study LiveNet in a controlled setting. The MoteLab [17] testbed consists of 184 TMote Sky nodes deployed over three floors of the Harvard EECS building, located mainly on bookshelves in various offices and labs. During the experiments between 120–130 nodes were active. Each node is connected to a USB-Ethernet bridge for programming and access to the node's serial port. For our validation, half of the nodes are used as sniffers and the other half used to run various applications. Although such a sniffer ratio is much larger than we would expect in a live deployment, this allows us to study the effect of varying sniffer coverage.

6.1   Sniffer Reception Rate

The first consideration is how well a single sniffer can capture packets at varying traffic rates. For these experiments, we make use of a simple TinyOS application that periodically transmits packets containing the sending node ID and a unique sequence number. Figure 3 shows the reception rate of two sniffers (a MicaZ and a TMote Sky) with up to 4 nodes transmitting at increasing rates. All nodes were located within several meters of each other. Note that due to CSMA backoff, the offered load may be lower than the sum of the transmitter's individual packet rates. We determine the offered load by computing a linear regression on the observed packet reception times at the sniffer.

As the figure shows, a single sniffer is able to sustain an offered load of 100 packets/sec, after which reception probability degrades. Note that the default MAC used in TinyOS limits the transmission rate of short packets to 284 packets/sec. Also, as mentioned in Section 5, MicaZ-based sniffers can handle somewhat higher loads than the TMote Sky. We surmise this to be due to differences in the serial I/O stack between the two mote platforms.

6.2   Merge Performance

Although LiveNet's sniffer traces are intended for offline analysis, the performance of the trace merging process is potentially of interest. For this experiment, we merge up to 25 traces containing 1000 sec of packet data on an unloaded

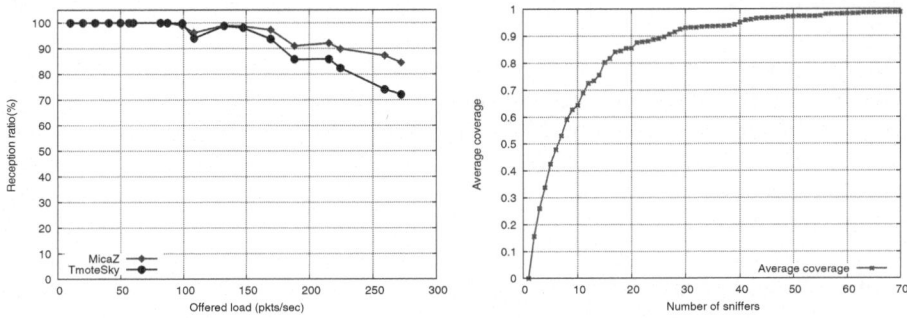

Fig. 3. Sniffer reception rate vs. offered load

Fig. 4. Sniffer coverage

2.4 GHz Linux desktop with 1 GB of memory. Merging two traces takes 301 sec, 10 traces 1418 sec, and 25 traces 2859 sec. The "break even" point where merging time takes longer than the trace duration occurs at about 8-10 traces. This suggests that for a modest number of traces, one could conceivably perform merging in real time, although this was not one of our design goals. Note that we have made no attempt to optimize the LiveNet merge code, which is implemented in Python and makes heavy use of ASCII files and regular expression matching.

6.3 Coverage

The next question is how many sniffers are required to achieve a given *coverage* in our testbed. We define coverage as the fraction of transmitted packets that are received by the LiveNet infrastructure. There are 70 sniffer traces in total for this experiment.

To compute the coverage of a random set of N traces, the most thorough, yet computationally demanding, approach is to take all $\binom{70}{N}$ subsets of traces, individually merge each subset, and compute the resulting coverage. Performing this calculation would be prohibitively expensive. Instead, we estimate the coverage of N traces by taking multiple random permutations of the traces and successively merge them, adding one trace at a time to the merge and computing the resulting coverage, as described below.

Let S_i represent a single trace and $\mathcal{S} = \{S_1 \ldots S_{70}\}$ represent the complete set of traces. Let $M(S_1 \ldots S_k)$ represent the merge of traces $S_1 \ldots S_k$, and $C(k)$ represent the coverage of these k merged traces, as defined in Section 4.1. We start by computing the coverage of the first trace S_1, yielding $C(1)$. We then merge the first two traces $M(S_1, S_2)$ and compute the coverage $C(2)$ of this merge. Next, we successively add one trace at a time to the merge, computing the resulting coverage for each trace until we have computed $C(1) \ldots C(70)$.

To avoid sensitivity to the order in which the original traces are numbered, we generate five random permutations $\mathcal{S}'$ of the original traces and compute the coverage $C(k)$ accordingly. Our final estimate of the coverage of k traces is the average of the five values of $C(k)$ computed for each of the permutations. The results are shown in Figure 4.

As the figure shows, the first 17 traces yield the greatest contribution, achieving a coverage of 84%. After this, additional sniffers result in diminishing returns. A coverage of 90% is reached with 27 sniffers, and all 70 sniffers have a coverage of just under 99%. Of course, these results are highly dependent on the physical extent and placement of our testbed nodes. The testbed covers 3 floors of a building spanning an area of 5226m². Assuming nodes are uniformly distributed in this area (which is not the case), this suggests that approximately one sniffer per 193m² would achieve a coverage of 90%. Keep in mind that sniffer locations were not planned to maximize coverage, and we are using the built-in antenna of the TMote Sky. High-gain antennas and careful placement would likely achieve better coverage with fewer nodes.

6.4   Merge Accuracy

Next, we are interested in evaluating the accuracy of the merged trace. As described earlier, our trace merging algorithm operates on fixed-length time windows and could lead to duplicate or reordered packets in the merged trace. After merging all 70 source traces from the previous experiment, we captured a total of 246,532 packets. 2920 packets are missing from the trace (coverage of 98.8%). There are a total of 354 duplicate packets (0.14%), and 13 out-of-order packets (0.005%). We feel confident that these error rates are low enough to rely on the merged trace for higher-level analyses.

6.5   Path Inference

To test the path inference algorithm described in Section 4.4, we set up an experiment in which one node routes data to a given sink node over several multihop paths. The node is programmed to automatically select a new route to the sink every 5 minutes. Since we know the routing paths in advance, we can compare the path inference algorithm against ground truth.

Space limitations prevent us from presenting complete results here, though we refer the reader to our technical report [13] for more details. In summary, the path inference algorithm correctly determined the routing path chosen by the network in all cases. When the routing path changed, the algorithm would incorrectly determine that an "intermediate" route was being used, but this would occur for no more than 1 or 2 sec until the correct route was observed.

7   Deployment Study: Disaster Drill

To evaluate LiveNet in a realistic application setting, we deployed the system as part of a live disaster drill that took place in August 2006 in Baltimore, MD, in collaboration with the AID-N team at Johns Hopkins Applied Physics Laboratory and rescue workers from Montgomery County Fire and Rescue Services.

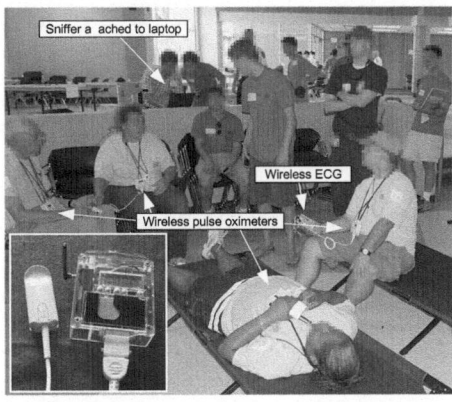

Fig. 5. The indoor treatment area of the disaster drill. *Inset shows the electronic triage tag.*

Disaster response and emergency medicine offer an exciting opportunity for use of wireless sensor networks in a highly dynamic and time-critical environment. Understanding the behavior of this network during a live deployment is essential for resolving bugs and performance issues.

The disaster drill modeled a simulated bus accident in which twenty volunteer "victims" were triaged and treated on the scene by 13 medics and firefighters participating in the drill. Each patient was outfitted with one or more sensor nodes to monitor vital signs, which formed an *ad hoc* network, relaying real-time data back to multiple laptop base stations located at the incident command post nearby. Each laptop displayed the triage status and vital signs for each patient, and logged all received data to a file. The incident commander could rapidly observe whether a given patient required immediate attention, as well as update the status of each patient, for example, by setting the triage status from "moderate" to "severe."

The network consisted of two types of sensor nodes: an *electronic triage tag* and a *electrocardiograph* (ECG) [5,10]. The triage tag incorporates a pulse oximeter (monitoring heart rate and blood oxygen saturation using a small sensor attached to the patient's finger), an LCD display for displaying vital signs, and multiple LEDs for indicating the patient's triage status (green, yellow, or red, depending on the patient's severity). The triage tags are based on the MicaZ mote with a custom daughterboard and case. The ECG node consists of a TMote Sky with a custom sensor board providing a two-lead (single-channel) electrocardiograph signal. In addition to the patient sensor nodes, a number of static repeater nodes were deployed to assist with maintaining network connectivity. The sensor nodes and repeaters all ran the CodeBlue system [7], which is designed to support real-time vital sign monitoring and triage for disaster response.

Our goal in deploying LiveNet was to capture detailed data on the operation of the network as nodes were activated, patients moved from the triage to

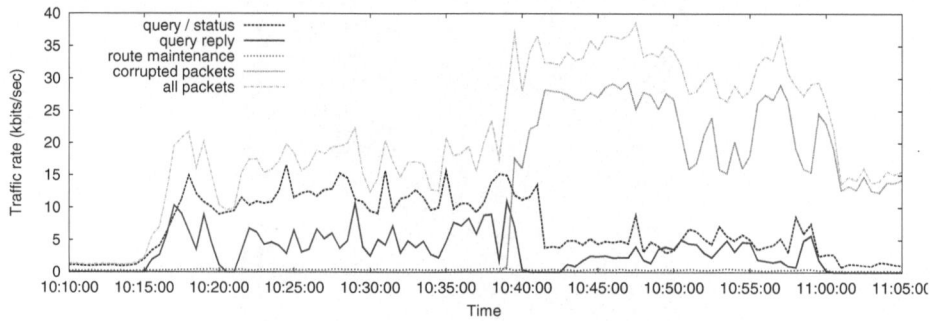

Fig. 6. Overall traffic rate during the disaster drill

treatment areas, and study the scalability and robustness of our *ad hoc* networking protocols. In this situation, it would have been impossible to record complete packet traces from each sensor node directly, motivating the need for a passive monitoring infrastructure. We made use of 6 separate sniffer nodes attached to 3 laptops (the laptops had two sniffers to improve coverage).

Figure 5 shows a picture from the drill to give a sense of the setup. The drill occurred in three stages. The first stage occurred in a parking lot area outdoors during which patients were outfitted with sensors and initial triage performed. In the second stage, most of the patients were moved to an indoor treatment area as shown in the picture. In the third stage, two of the "critical" patients were transported to a nearby hospital. LiveNet sniffers were placed in all three locations. Our analysis in this paper focuses on data from 6 sniffers located at the disaster site. The drill ran for a total of 53 minutes, during which we recorded a total of 110548 packets in the merged trace from a total of 20 nodes (11 patient sensors, 6 repeaters and 3 base stations).

8   Deployment Evaluation

In this section, we perform an evaluation of the LiveNet traces gathered during the disaster drill described in Section 7.

8.1   General Evaluation

As a general evaluation of the sensor network's operation during the drill, we first present the overall traffic rate and packet type breakdown in Figure 6 and 10. These high-level analyses help us understand the operation of the deployed network and can be used to discover performance anomalies that are not observable from the network sinks.

As Figure 6 shows, at around time $t = 10 : 39$ there is a sudden increase in corrupted packets received by LiveNet: these packets have one or more fields that appear to contain bogus data. Looking more closely at Figure 10, starting

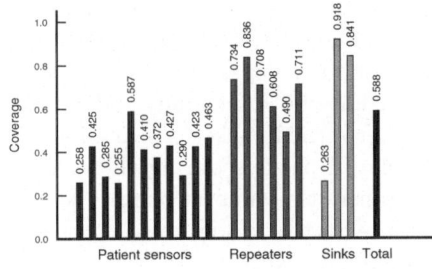

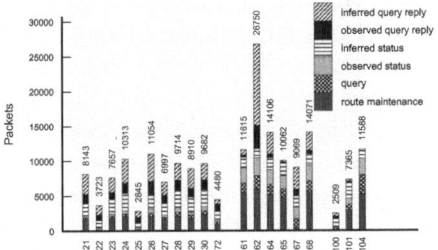

Fig. 7. Per-node coverage during the drill

Fig. 8. Inferred per-node packet traffic load during the drill

at this time we see a large number of partially-corrupted routing protocol control messages being flooded into the network. On closer inspection, we found that these packets were otherwise normal spanning-tree maintenance messages that contained bogus sequence numbers. This caused the duplicate suppression algorithm in the routing protocol to fail, initiating a perpetual broadcast storm that lasted for the entire second half of the drill. The storm also appears to have negatively affected application data traffic as seen in Figure 6.

We believe the cause to be a bug in the routing protocol (that we have since fixed) that only occurs under heavy load. Note that we had no way of observing this bug without LiveNet, since the base stations would drop these bogus packets.

8.2 Coverage

To determine sniffer coverage, we make use of periodic status messages broadcast by each sensor node once every 15 sec. Each status message contains the node ID, sensor types attached, and a unique sequence number. The sequence numbers allow us to identify gaps in the packet traces captured by LiveNet, assuming that all status messages were in fact transmitted by the node.

Figure 7 shows the coverage broken down by each of the 20 nodes in the disaster drill. There were a total of 4819 expected status messages during the run, and LiveNet captured 59% overall. We observed 89 duplicate and out-of-order packets out of 2924 packets in total, for an error rate of 3%. As the figure shows, the coverage for the fixed repeater nodes is generally greater than for the patient sensors; this is not too surprising as the patients were moving between different locations during the drill, and several patients were lying on the ground. The low coverage (26%) for one of the sink nodes is because this node was located inside an ambulance, far from the rest of the deployment.

8.3 Topology and Network Hotspots

The network topology during the drill was very chaotic, since nodes were moving and several nodes experienced reboots. Such a complex dataset is too dense to show as a figure here. However, we can discuss a few observations from analyzing the topology data. First, most nodes are observed to use several outbound links, indicating a fair amount of route adaptation. There are multiple routing trees (one rooted at each of the sinks), and node mobility causes path changes over time. Second, all but two of the patient sensors have both incoming and outgoing unicast links, indicating that patient sensors performed packet relaying for other nodes. Indeed, one of the sink nodes also relayed packets during the drill. We also observe that one of the patient sensors transmitted unicast packets to itself, suggesting corruption of the routing table state on that node; this is a protocol bug that would not have been observed without LiveNet.

Besides topology, it is useful to identify "hotspots" in the network by computing the total number of packets transmitted by each node during the drill. For several message types (query reply and status messages), we can also infer the existence of unobserved packets using sequence numbers; this information is not available for route maintenance and query messages. Figure 8 shows the breakdown by node ID and packet type. Since we are counting all transmissions by a node, these totals include forwarded packets, which explains the large variation from node to node.

The graph reveals a few interesting features. Repeater nodes generated a larger amount of traffic overall than the patient sensors, indicating that they were used heavily during the drill. Second, node 62 (a repeater) seems to have a very large number of inferred (that is, unobserved) query reply packets, far more than its coverage of 83.6% would predict. This suggests that the node internally dropped packets that it was forwarding for other nodes, possibly due to heavy load. Note that with the information in the trace, there is no way to disambiguate packets unobserved by LiveNet from those dropped by a node. If we assume that our coverage calculation is correct, we would infer a total of 4108 packets; instead we infer a total of 15017. Node 62 may have then dropped as many as $(15017 - 4108)/15017 = 72\%$ of the query replies it was forwarding. With no further information, the totals in Figure 8 therefore represent a conservative upper bound on the total traffic transmitted by the node.

8.4   Path Inference

Given the limited coverage of the LiveNet sniffers during the drill, we would not expect the path inference algorithm to provide results as clear as those in our testbed. As shown in Figure 9, we take a single patient sensor (node 22) and infer the routing paths to one of the sink nodes (node 104). The results are shown in Figure 9. As the figure shows, there are many inferred paths that lack complete observations of every hop, and the Π scores for these paths vary greatly. The path $22 \rightarrow 62 \rightarrow 104$ is most common; node 62 is one of the repeaters. In several cases the routing path alternates between repeaters 62 and 68. The node also routes packets through other repeaters and several other patient sensors. The longest inferred path is 5 hops.

This example highlights the value of LiveNet in understanding fairly complex communication dynamics during a real deployment. We were surprised to see such frequent path changes and so many routing paths in use, even for a single

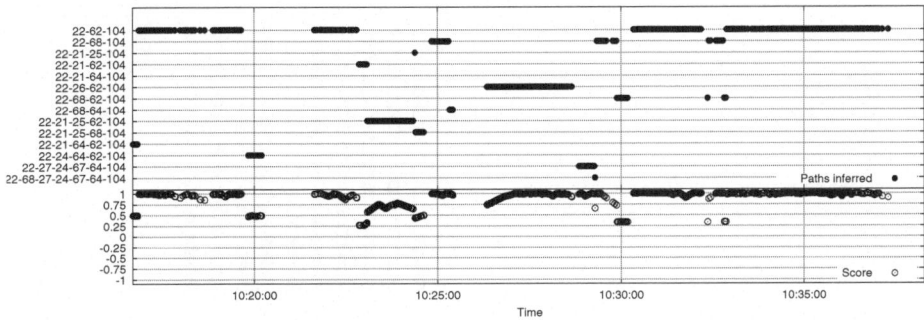

Fig. 9. Routing paths for one of the patient sensors during the disaster drill. *The upper graph shows the inferred routing path over time, while the lower graph shows the Π score for the most probable path.*

source-sink pair. This data can help us tune our routing protocol to better handle mobility and varying link quality.

8.5   Query Behavior and Yield

The final analysis that we perform involves understanding the causes of data loss from sources to sinks. We found that the overall data yield during the drill was very low, with only about 20% of the expected data transmitted by patient sensors reaching the sink nodes. There are three distinct sources of data loss: (1) packet loss along routing paths; (2) node failures or reboots, and (3) query timeouts.

During the drill, most patient nodes ran six simultaneous queries, sending two types of vital sign data at 1 Hz to each of the three sinks. Each data packet carries a unique sequence number. The CodeBlue application uses a lease model for queries, in which the node stops transmitting data after a timeout period unless the lease is renewed by the sink. Each sink was programmed to re-issue the query 10 seconds before the lease expires; however, if this query is not received by the node in time, the query will time out until the sink re-issues the query.

Using the analysis described in Section 4.5, we can disambiguate the causes of packet loss. Figure 11 shows the results. A query yield of 100% corresponds to the sink receiving packets at exactly 1 Hz during all times that the node was alive. As the figure shows, the actual query yield was 17–26%. Using the LiveNet traces, we can attribute 5–20% loss to dropped packets along routing paths, and 16–25% loss to premature query timeouts. The rest of the loss is unobserved, but likely corresponds to path loss since these packets should have been transmitted during query active periods.

This analysis underscores the value of the LiveNet monitoring infrastructure during the deployment. Without LiveNet, we would have little information to help us tease apart these different effects, since we could only observe those packets received at the sinks. With LiveNet, however, we can observe the network's

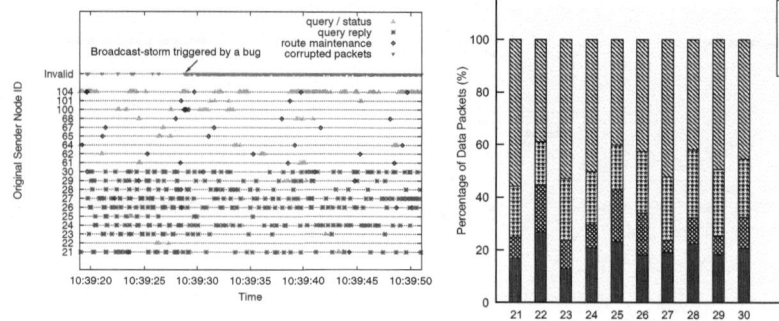

Fig. 10. Packet type breakdown for a portion of the disaster drill

Fig. 11. Query yield per node during the disaster drill

operation in much greater detail. In this case, we see that the query timeout mechanism performed poorly and needs to be made more robust. Also, routing path loss appears to be fairly high, suggesting the need for better reliability mechanisms, such as FEC.

9 Future Work and Conclusions

LiveNet provides an infrastructure for monitoring and debugging live sensor network deployments. Rather than introducing possibly intrusive instrumentation into the sensor application itself, we rely on passive, external packet monitoring coupled with trace merging and high-level analysis. We have shown that LiveNet is capable of scaling to a large number of sniffers; that merging is very accurate with respect to ground truth; that a small number of sniffers are needed to achieve good trace coverage; and that LiveNet allows one to reconstruct the behavior of a sensor network in terms of traffic load, network topology, and routing paths. We have found LiveNet to be invaluable in understanding the behavior of the disaster drill deployment, and intend to leverage the system for future deployments as well.

LiveNet is currently targeted at online data collection with offline post processing and analysis. We believe it would be valuable to investigate a somewhat different model, in which the infrastructure passively monitors network traffic and alerts an end-user when certain conditions are met. For example, LiveNet nodes could be seeded with event detectors to trigger on interesting behavior: for example, routing loops, packet floods, or corrupt packets. To increase scalability and decrease merge overhead, it may be possible to perform *partial merging* of packet traces around windows containing interesting activity. In this way the LiveNet infrastructure can discard (or perhaps summarize) the majority of the traffic that is deemed uninteresting.

Such an approach would require decomposing event detection and analysis code into components that run on individual sniffers and those that run on a backend server for merging and analysis. For example, it may be possible to perform merging in a distributed fashion, merging traces pairwise along a tree; however, this requires that at each level there is enough correspondence between peer traces to ensure timing correction can be performed. Having an expressive language for specifying trigger conditions and high-level analyses strikes us as an interesting area for future work.

Acknowledgments

The authors gratefully acknowledge the contributions of the AID-N team at Johns Hopkins Applied Physics Laboratory for their assistance with the disaster drill logistics as well as essential hardware and software development: Tia Gao, Tammara Massey, Dan Greenspan, Alex Alm, Jonathan Sharp, and David White. Leo Selavo (Univ. Virginia) developed the wireless triage tag used in the drill. Konrad Lorincz and Victor Shnayder (Harvard) contributed to the Code-Blue software platform. Geoff Werner-Allen (Harvard) developed the MoteLab sensor network testbed used for our validation studies.

References

1. Alberto Cerpa, N.B., Estrin, D.: Scale: a tool for simple connectivity assessment in lossy environments. Technical Report CENS TR-0021, UCLA (September 2003)
2. Cheng, Y.-C., Afanasyev, M., Verkaik, P., Benko, P., Chiang, J., Snoeren, A.C., Savage, S., Voelker, G.M.: Automating cross-layer diagnosis of enterprise wireless networks. In: ACM SIGCOMM Conference (2007)
3. Cheng, Y.-C., Bellardo, J., Benko, P., Snoeren, A.C., Voelker, G.M., Savage, S.: Jigsaw: Solving the puzzle of enterprise 802.11 analysis. In: ACM SIGCOMM Conference (2006)
4. Chun, B.N., Buonadonna, P., AuYoung, A., Ng, C., Parkes, D.C., Shneidman, J., Snoeren, A.C., Vahdat, A.: Mirage: A Microeconomic Resource Allocation System for SensorNet Testbeds. In: Proc. the the Second IEEE Workshop on Embedded Networked Sensors (EMNETS 2005) (May 2005)
5. Fulford-Jones, T.R.F., Wei, G.-Y., Welsh, M.: A portable, low-power, wireless two-lead ekg system. In: Proc. the 26th IEEE EMBS Annual International Conference, San Francisco (September 2004)
6. Levis, P., Lee, N., Welsh, M., Culler, D.: TOSSIM: Accurate and scalable simulation of entire TinyOS applications. In: Proc. the First ACM Conference on Embedded Networked Sensor Systems (SenSys 2003) (November 2003)
7. Lorincz, K., Malan, D., Fulford-Jones, T.R.F., Nawoj, A., Clavel, A., Shnayder, V., Mainland, G., Moulton, S., Welsh, M.: Sensor Networks for Emergency Response: Challenges and Opportunities. In: IEEE Pervasive Computing, October-December (2004)
8. Luo, L., He, T., Zhou, G., Gu, L., Abdelzaher, T., Stankovic, J.: Achieving repeatability of asynchronous events in wireless sensor networks with envirolog. In: Proc. IEEE INFOCOM Conference, Barcelona, Spain (April 2006)
9. Mahajan, R., Rodrig, M., Wetherall, D., Zahorjan, J.: Analyzing the mac-level behavior of wireless networks in the wild. In: ACM SIGCOMM Conference (2006)

10. Massey, T., Gao, T., Welsh, M., Sharp, J.: Design of a decentralized electronic triage system. In: Proc. American Medical Informatics Association Annual Conference (AMIA 2006), Washington, DC (November 2006)
11. Ramanathan, N., Chang, K., Girod, L., Kapur, R., Kohler, E., Estrin, D.: Sympathy for the sensor network debugger. In: SenSys 2005: Proceedings of the 3nd international conference on Embedded networked sensor systems (2005)
12. Ringwald, M., Römer, K., Vitaletti, A.: Passive inspection of sensor networks. In Proceedings of the 3rd IEEE International Conference on Distributed Computing in Sensor Systems (DCOSS 2007), Santa Fe, New Mexico, USA (June 2007)
13. Chen, B.r., Peterson, G., Mainland, G., Welsh, M.: Livenet: Using passive monitoring to reconstruct sensor network dynamics. Technical Report TR-11-07, School of Engineering and Applied Sciences, Harvard University (August 2007)
14. Rost, S., Balakrishnan, H.: Memento: A Health Monitoring System for Wireless Sensor Networks. In: IEEE SECON, Reston, VA (September 2006)
15. Titzer, B., Lee, D.K., Palsberg, J.: Avrora: scalable sensor network simulation with precise timing. In: Proc. Fourth International Conference on Information Processing in Sensor Networks (IPSN 2005) (April 2005)
16. Tolle, G., Culler, D.: Design of an application-cooperative management system for wireless sensor networks. In: Proc. the 2nd European Confererence on Wireless Sensor Networks (EWSN 2005) (January 2005)
17. Werner-Allen, G., Swieskowski, P., Welsh, M.: MoteLab: A Wireless Sensor Network Testbed. In: Proc. the Fourth International Conference on Information Processing in Sensor Networks (IPSN 2005) (April 2005)

Broadcast Authentication in Sensor Networks Using Compressed Bloom Filters

Yu-Shian Chen[1], I-Lun Lin[2], Chin-Laung Lei[3], and Yen-Hua Liao[4]

Department of Electrical Engineering
National Taiwan University
No. 1, Sec. 4, Roosevelt Rd., Taipei, Taiwan 106
{ethan[1],grail[2],radha[4]}@fractal.ee.ntu.edu.tw,
lei@cc.ee.ntu.edu.tw[3]

Abstract. We propose a light-weight and scalable broadcast authentication scheme, **Curtain**, for sensor network. Instead of using Merkel tree to combine multiple μTESLA instance, we apply compressed Bloom filters to multiple μTESLA. Our scheme can support longer duration and prolong the self-healing property. We greatly reduce the communication overhead at the cost of allocating a moderate space in each receiver. Combing with PKC computation like ECC, our scheme can guarantee the long-term security and also mitigate energy consumption. Moreover, our methods can be extend to the situation of multiple senders, offering efficient user addition and revocation.

Keywords: sensor networks, network security, broadcast authentication, μTESLA, Bloom filters.

1 Introduction

Broadcast authentication is probably the most critical security primitive in sensor network, since it is the most basic approach to defend against attackers in wireless environments. Especially, the central server relies on broadcast authentication to issue legitimate commands or queries to all or partial motes in the network. Broadcast authentication becomes a trivial problem when using public key cryptography (PKC). Through digital signatures, motes could verify message authenticity as communication on Internet does. However, due to the nature that most sensor motes are resource-constrained devices, PKC was thought inappropriate for a long time. Extensive research have used only symmetric primitives to achieve light-weight broadcast authentication in sensor networks. Following the prevalent μTESLA protocol [1], a diversity of schemes based on hash chain techniques are proposed [2,3]. μTESLA is a light-weight and self-healing protocol for sensor motes. By self-healing it means that a receiver who has been off-line for some period is able to recover the lost keys immediately after coming back on-line. Despite these advantages, μTESLA or μTESLA-like protocols all have their own drawbacks.

Recent research shows an 160-bits elliptic curve cryptography (ECC) [4] signature can be verified in about one second [5,6]. This dramatic improvement

S. Nikoletseas et al. (Eds.): DCOSS 2008, LNCS 5067, pp. 99–111, 2008.
© Springer-Verlag Berlin Heidelberg 2008

indicates that the protocol designers now may include PKC in their security schemes. We have the following observations.

- Nowadays public key cryptography (PKC) like ECC is no longer time-consuming operations for sensor motes, therefore it shall be included in security protocol design. However, we shall not use PKC too frequently since energy consumption is still a critical issue. PKC shall play the role to guarantee the long-term security, for example, generating digital signature on the session key used during an hour. It is reasonable for a mote to spend one second for verifying this session key. Within the middle-term session, the system shall utilize symmetric cryptography against various attacks.
- Considering energy cost, communication is more critical than computation, since computation could be improved with the advance of hardware but radio transmission would always require an minimum power. Besides, wireless environments are always unstable and difficult to predict. Therefore, practical protocols shall work on lowering down communication first, then computation.
- μTESLA-like schemes that binding multiple μTESLA instance to extend their using time all suffer their weakness. They either cause greater authentication delays or induce greater communication overhead.

Based on this observation, our design goal is to minimize the communication overhead in the μTESLA-like schemes whereas remain in the same security level. In this paper, we propose **Curtain:** an efficient and flexible broadcast authentication protocol combing μTESLA and Bloom filter techniques. Curtain extends the self-healing property across multiple key chains by storing a moderate size of Bloom bitmap. Working together with ECC, our schemes can guarantee the long-term security of broadcast authentication with lower communication overhead. Curtain highly reduces the communication overhead by compressing the bitmaps and scales up in dimension of time and senders.

Our Contribution: We employ Bloom filters technique, especially compressed Bloom filters, to broadcast authentication based on μTESLA and demonstrate that it is a viable and scalable solution compared with former research. We also point out the shortcoming and impracticality of former μTESLA-like schemes and overcome these drawbacks. Our scheme supports multiuser broadcast authentication and has great performance in updating or revoking users.

2   Preliminaries

2.1   System Model and Attack Model

System Model: In this paper, we consider the problem of long-term and scalable broadcast authentication in sensor networks. The network consists of a base station or gateway which is directly controlled by the administrator and a large number of resource-limited sensor motes which could be compromised by the adversary. The administrator is allowed to pre-load secret information into each motes before network deployment. The base station usually acts as a broadcast sender since the administrator make commands or queries to other motes through it. We assume the sensor motes are capable to verify ECC signatures in a few seconds [5,6]. There are applications where regular motes also need to broadcast authenticated messages. We will first consider the basic situation that only the base station as sender and the other motes as receivers. Later in Section 4 we will extend our approach to accommodate multiple senders in the same network. We use the same assumption that sender and all receivers have loose time synchronization with accuracy in the range of few μs as all μTESLA-like schemes did. Throughout this paper, we will use 160 bits as the length of all hash values.

Attack Model: An adversary can execute a wide range of attacks including eavesdrop, modify, forge, or replay. Basically the adversary can compromise some sensor motes and obtain their secret information. Note that denial-of-service (DoS) attack which jams the wireless channel is not our concern in this paper since there is no effective way to prevent such attack being launched due to the nature of wireless communication. We assume the adversary has computation resource equivalent to several personal computers, which means it is computationally infeasible for the adversary to break the underlying cryptographic primitives like hash and encryption. In the language of cryptanalysis, the probability that the adversary breaks the primitives in one time should be less than $2^{-\kappa}$ where κ is the security parameter. For example, the probability to find a collision in 160-bits hash function should be less than 2^{-80} (approximately $\leq 10^{-24}$). An typical attack to broadcast authentication is that an adversary intercepts all broadcast information from legitimate sender and forwards falsified message to compromised receivers.

2.2 μTESLA and μTESLA-Like Schemes

Standard μTESLA: μTESLA protocol uses only symmetric cryptographic primitives but provides very efficient solution for resource-limited sensor motes. To create the asymmetry for verification, μTESLA utilizes delay disclosure of MAC keys. Before deployment, the sender randomly chooses the last-used key K_n then generates a μTESLA key chain according to

$$K_i = F(K_{i+1}) \text{ or } K_i = F^{n-i}(K_n)$$

where F is a cryptographic one-way hash function, such as SHA-1[1] or MD5. The key chain is called a μTESLA instance and the fist-used but last-generated key K_0 is called the *commitment* of the μTESLA instance. The commitment is preloaded into each receiver's memory before deployment. The sender splits time into consecutive μTESLA *interval* I_i corresponding to the μTESLA instance. Each key

---

[1] SHA-1 is considered not secure enough since Wang et al. had greatly reduced its complexity in [7]. We could switch to the stronger alternatives such as SHA-2 for stronger security.

K_i in the chain is used for Message Authentication Code (MAC) during I_i but will not be disclosed until some later interval $I_{i+\Delta}$. (e.g. $\Delta = 2$) All broadcast messages during I_i will be attached with its MAC generated by K_i. Also the sender sends out the most recent key it can disclose $K_{i-\Delta}$ in I_i. Later in $I_{i+\Delta}$ the receivers can verify the messages queued during I_i by the new-released key.

Also μTESLA is self-healing in the sense that a receiver who has been off-line for some period is able to recover the lost keys immediately after coming back on-line. In some applications, a receiver may wait for a long time for the next broadcast message from sender. Despite that the receiver has missed many intermediate keys, he can still recursively compute hash to derive former MAC keys. Figure 1 illustrates an example of μTESLA. A typical μTESLA interval

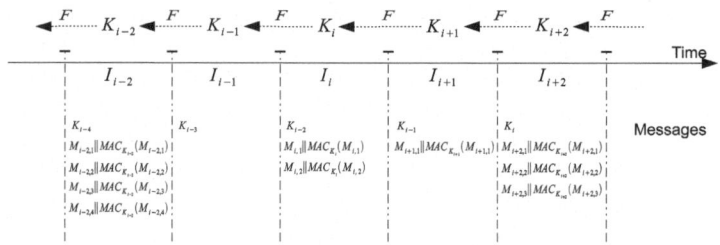

Fig. 1. An Example of μTESLA

is about 100 ms, and one key chain consists of 1000 keys, thus one key chain can sustain about 100 seconds. Following the basic μTESLA approach, there were many research working on extending the lifetime of μTESLA by combining multiple instances.

Multi-level μTESLA branches the key chains in more dimensions [2]. The higher-level μTESLA instances are used to generate lower-level instances and the lowest-level instances are used to authenticate data packets. However, it also induces multi-level authentication delays, which means even the receiver has obtained the MAC keys of those queued messages, it can not completely verify the messages because the validity of the very MAC keys relies on the higher-level key which will not be disclosed until current instance is out of usage. To be fully verify the messages and the MAC keys, the receiver has to queue at least all messages received during the instances, which is not desirable in practice.

Tree-based μTESLA: Contrary to multi-level μTESLA which binds key chains in their heads, Several research use Merkle tree to bundle multiple μTESLA instances in their tails [3,8]. The example in Figure 2 encompasses a window of $W = 8$ μTESLA instances. Every receiver is pre-loaded with the root key C_{18} instead of the last-generated keys on each μTESLA instance. To activate next instance, for example the fourth key chain, the receivers need the commitment C_4, so the sender broadcasts the commitment and all sibling keys of those nodes on the path from the commitment to the root, that is, C_4, C_3, C_{12}, and C_{58}. Then the receivers can verify whether $C_{18} = F'(F'(C_{12}\|F'(C_3\|C_4))\|C_{58})$.

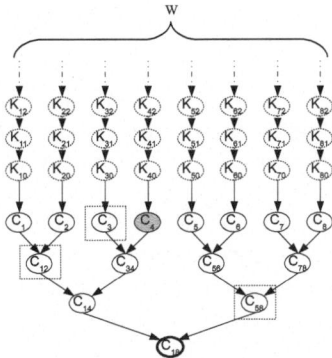

Fig. 2. Example of tree-based μTESLA. Commitments are hashed by another F' function.

Tree-based method does not have the drawback of longer authentication delays in multi-level schemes because all commitments are downward related to the root commitment. Nevertheless, the sender has to periodically broadcast authentication information for every instance, and therefore induces more traffic than naïve method. It takes $\lg W|h|$ bits to update to next instance where $|h|$ is the size of hashed key. Regarding the bundle as a whole, tree-based method's overhead will be $W \lg W|h|$ whereas the naïve method only $W|h|$.

μTESLA-like schemes all share the same philosophy: the security of all MAC keys rely on the first key, the pre-loaded *commitment*. Multi-level μTESLA bases on the framework of key generation, whereas tree-based μTESLA binds many commitments to create deeper commitments. The challenge in broadcast is how to construct the connection between keys for long-term usage. Therefore the basic principle is to generate sufficient connected keys beforehand and make the last-generated key be the loaded commitment. The whole protocol will be robust as long as the hash functions used are computationally infeasible.

2.3   Compressed Bloom Filters

Bloom filters: Bloom filters succinctly represent a set of entries in order to support membership queries with minor false positive and zero false negative rate. Bloom filters are used to represent a set $S = \{s_1, ...s_n\}$ of n entries, and consists of a bitmap with m-bits, initially all set to 0. Bloom filters use k independent random hash functions $H_1, ..., H_k$ with range $\{0, ..., m-1\}$. For each entry s_j, the bits $H_i(s_j)$ are set to 1 for $1 \leq j \leq n$ and $1 \leq i \leq k$. After all entry in S are hashed into the bitmap, the probability p that a specific bit is 0 is $(1-1/m)^{kn}$. To check if a new entry x is in S, we check whether all $H_i(x)$-th bits the bitmap are 1. If not, x is absolutely not in S. If yes, we can only mistake with a minor false positive rate $f = (1-p)^k$, depending on the values of m/n and k. Given fixed ratio m/n, the optimized f is $0.5^k = 0.6185^{m/n}$ when $k = \ln 2(m/n)$ and $p \approx 0.5$; meanwhile the bitmap is about half 1 and half 0.

Compressed Bloom filters: Mitzenmacher showed that compressed Bloom filters can achieve smaller false positive rate as a function of the compressed size over a Bloom filter without compression [9]. It based on the insight that if we allow larger size of uncompressed bitmap m and use few hash functions, then the bitmap will have extremely asymmetric distribution of 0 and 1 bits. Thus we can compress the large bitmap into a small message of size z for transmission. Moreover the same size of message using as a standard Bloom bitmap turns out to have higher false positive rate than the compressed one. Compared with original ones, compressed Bloom filters intend to optimize *transmission size* instead of false positive rate as they use bitmaps of the same size.

The essence of compressed Bloom filters is that it reduces the transmission size at the cost of using more bits to store the Bloom filters at both ends and requiring compression and decompression. Compressed Bloom filters are very efficient options when we need to transmit the bitmap as a message. In particular, compressed approach has greater improvement when we only need to transmit the updated information of a periodically changing Bloom filters. More detailed discussions and analysis please refer to [9].

3   Proposed Schemes

3.1   Overview

Figure 3 illustrates the basic of our schemes. To bundle W independent μTESLA instances, Curtain takes the last N keys as entries to be hashed into the m-bits Bloom bitmap. The bitmap replaces the last-generated key in standard μTESLA or the root key in tree-based μTESLA as the commitment. Pre-loaded by the sender to each receiver, the bitmap provides the computationally security during the usage of W μTESLA instances. For the receiver to activate next instance, it just checks the last-generated N keys are all successfully hashed into the Bloom bitmap. Before the end of the W instances, the sender broadcasts the compressed bitmap of next W instance with an ECC signature to enter next session. According to the analysis and implementation in [9], the size after compression is only a slight bigger than $z = H(p)m$ where $H(\cdot)$ is the entropy function.[2]

An adversary may try to break the system by forging key chains which can be successfully mapped unto the bitmaps. Since the adversary can not reverse hash function, it can only by brute force continuously generate a longer key chain until the last N keys are all false positive in the b bitmaps. In order to amplify the robustness against forgery, the sender can use b rather than one Bloom filter with independent hash sets to process the NW entries into b bitmaps. Figure 4 illustrates the adversary's chance to forge a valid μTESLA instance in

---

[2] $H(p) = -p \lg p - (1-p) \lg(1-p)$.

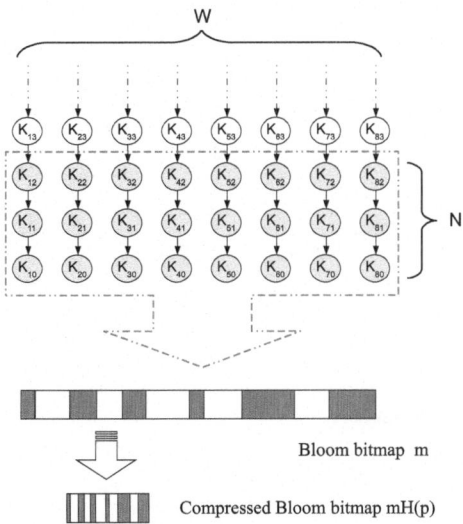

Fig. 3. Curtain. Gray μTESLA keys are entries of the Bloom filters.

different setting. Recall that the security condition of 160-bit hash function is about 10^{-24}. Therefore those points below the dashed line will be sufficient for most application. It also demonstrates that with proper parameters using Bloom filters is a viable solution. We find the chance of the adversary to success is

$$adv = f^{Nb} = (1-p)^{kNb} \tag{1}$$

Unlike tree-based method where receivers have to wait for the sender's periodical messages, Curtain spans the self-healing property of a single μTESLA instance to W instances. Even though the receiver is off-line over several instances, it is able to immediately verify the new-arrived packets.

3.2   Analysis

There are three performance metrics needed to be considered in Curtain: (1) The adversary's advantage (2) Communication overhead (3) Computation overhead. First we consider the adversary's chance to break the protocol. This advantage adv shall not be greater than $2^{-\kappa}$, which means

$$b \geq \frac{-1}{\lg(1-p)} \cdot \frac{\kappa}{kN}. \tag{2}$$

At first sight, this seems indicate that we should use more bitmaps to surpass κ. Nevertheless, there is a tradeoff between the advantage and communication overhead. The communication overhead of each session is zb, which seems to multiply significantly if we try to increase b. But z itself, like b, is also a function of k and N and interestingly they are driving in different direction.

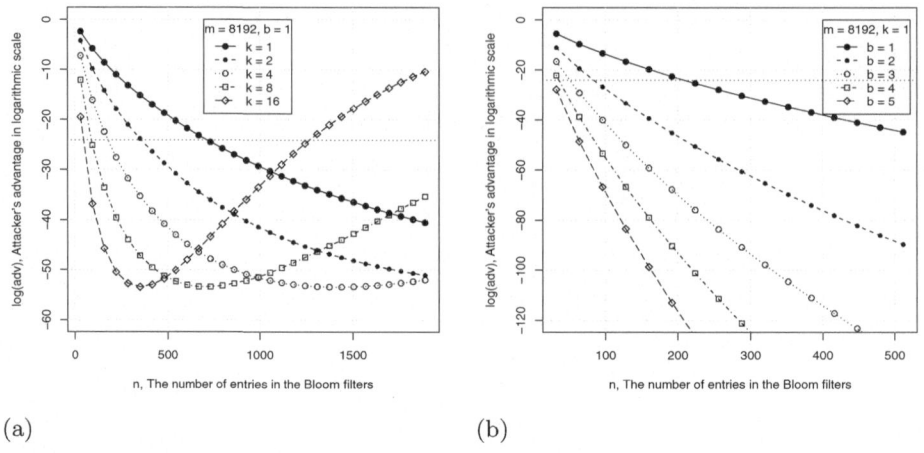

(a)   (b)

Fig. 4. The attacker's chance to break Curtain when $m = 8192$ and $W = 32$. (The dashed line indicates 10^{-24} minimum security condition.)

If we fix at specific security condition κ, then the overhead would be

$$zb = \frac{-p\lg p - (1-p)\lg(1-p)}{-\lg(1-p)} \cdot \frac{m\kappa}{kN} \tag{3}$$

Note that $p = (1-1/m)^{nk}$ is also a function of k, therefore directly analyzing the dependency between zb and k from the above formula would be a cumbersome work. We can get the intuitive idea from Figure 5(a) which shows that using more hash in Bloom filters will dramatically reduce the overhead, but eventually rise up after specific threshold. The remaining question is how b changes when we try to optimize communication overhead with respect to k.

Here a little modification will clarify the mystery. In Figure 5(a), b is treated as continuous variable. In fact, b should be like this,

$$b \geq \lceil \frac{1}{-\lg(1-p)} \cdot \frac{\kappa}{kN} \rceil \tag{4}$$

After making b to be discrete, Figure 5(a) is transformed to Figure 5(b). Looking at the optimized points in Figure 5(b) for each setting, we get an interesting result: these are the turning points where b becomes one from two. It clarifies that to achieve the same security condition and communication overhead, extra bitmaps are actually redundant. One might wonder what is lost when communication is optimized. In fact, computation overhead is exactly the third force which pulls against the other metrics. It costs Nk hash operations for each receiver to authenticate next μTESLA instance. Despite of the increment in computation, the benefit is greater than the cost. Since hash function can be executed by sensor mote efficiently and costs minor energy, it is absolutely worth to exchange more hash for less transmission. The following example reinforces this argument.

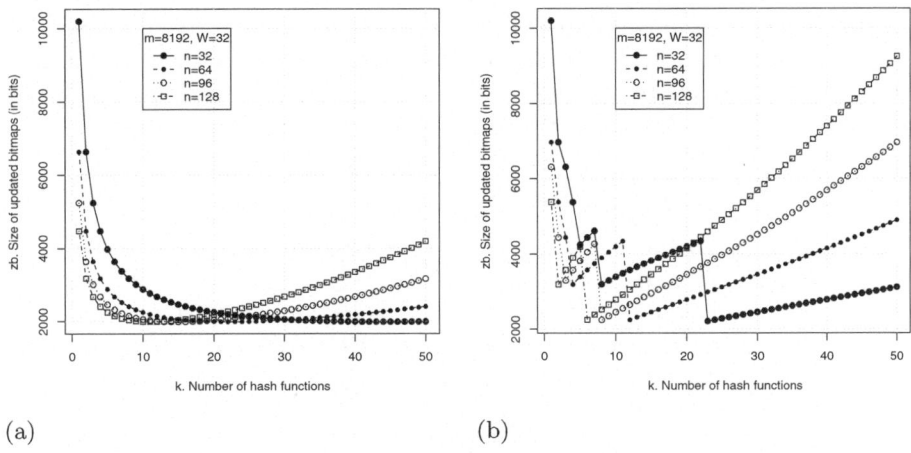

Fig. 5. Communication overhead for updating the bitmaps

3.3   An Example

By bundling $W = 32$ instances together, we can run for 32×100 seconds which is almost one hour. It is reasonable to spend 1 second on ECC computation for an hour light-weight security. We choose $m = 8192$ bits Bloom filter, which is moderate for motes to store and maintain. Let $n = 128$ and $N = n/W = 4$. The optimal value occurs when $k = 6$ and we find $z \approx 2242$ bits which can be packed into 3 4 packets according to IEEE 802.15.4 standard.Compared this example with naïve approach, which is $160 \times 32 = 5120$ bits, we save around 50%. Compared with tree-based approach, which requires the sender to broadcast extra $\lg(W)|h| = 5 \times 160 = 800$ bits every 100 seconds and totally $W \lg(W)|h| = 25600$ bits during the session, obviously Curtain is more economic. The computation overhead to authenticate a new key chain is $Nk = 24$ hash operations which costs about 24 ms.

4   Implementation

We implement Curtain on Crossbow TelosB platform which comes with 16-bit 8 MHz processor and 10KB RAM. Following the analysis from the last section, we test the case of using one bitmap with size equal to 16284, 8192, and 4096 bits, then adjust the number of hash functions to fit in with minimum security condition 2^{-80}. Table 1. shows the compressed bitmap size, decompression time, and verification time in different setting. We also include a situation where the security is not sufficient in Table 1. All hash functions are based on SHA1-160, which averagely takes about 5.48 ms each time. Further optimization on SHA-1 computation should improve the verification time. Here we only concern about the verification time of first-used key since the successive $N - 1$ keys will take additional one SHA-1 computation and the rest cost only one SHA-1 for each key. Note that the larger bitmap results in smaller compression size is due to less hash functions so less bits are marked as one in larger bitmap. We use a variant of first-order difference compression to compress the bitmap. Other compression techniques are possible to achieve the better compression ratio considering the size and entropy of different bitmaps.

Table 1. Curtain overhead with $W = 32$ and $b = 1$

Bitmap size (bits)	16284	16284	8192	4096
Number of hash per bitmap	2	4	6	12
security condition (2^{-x})	48.09	80.36	83.57	80.49
Compressed bitmap size (bytes)	270	515	745	1310
Ideal compressed bitmap size (bytes)	237	406	446	459
Decompression time (ms)	5	9	18	31
First-used key verification time (ms)	12	25	42	68

Comparison. We compare the communication overhead when bundling $W = 32$ instances using naïve, Merkle tree, and Curtain. Applying naïve way to preload all commitments takes 640 bytes. With Merkle tree it costs 3200 bytes overall by contraries. Even though the receivers cache earlier keys, the overhead will not less than 640 bytes. In our implementation with bitmap size equal to 16284 bytes, transmission overhead is only 270 bytes.

Compared with the other approaches, Curtain has better potentiality in the future. The reason is that 160 bit hash such as SHA-1 would be deprecated in near future. Researchers are gradually resorting to longer hash size like SHA-256 as security guarantee. When the time comes the expansion in hash size will cause the same expansion in naïve or Merkle methods. However Curtain does not suffer from this problem provided that the minimum security condition κ remains.

5   Extension

5.1   Curtain on Multiple Senders

Curtain possesses another advantage: it is not only scalable in terms of time but also scalable in terms of the number of senders. In simple application, the base station might be the only one doing broadcast. But there are cases where regular motes need to communicate with each other such that individual broadcast authentication from motes is required. Another scenario is that many users share the sensor network but want to use individual broadcast authentication to isolate their services. In either case, to install all commitments of different senders or services to receivers is not a feasible way.

The methodology of Curtain can be extended to bundle multiple broadcast services, and the receiver need only store one bitmap hashed from these services.

Moreover, the administrator now can add and revoke senders or services at a very low cost with Curtain. This benefit cames from the nature of compressed Bloom filters. When the network changes and the administrator need to update the user and revocation list to all receivers. Because most change are partial, the difference between the old and new bitmap is minor. According to [9,10], the administrator can exclusive-or the two bitmap, that is, the *delta* bitmap, and furthermore compress it then update. The compressed *delta* bitmap will be more succinct than compressed Bloom one. If c is the fraction of entries in Bloom bitmap that have changed, then $q = 2p(1 - p^c)$ of *delta* will be bit 0, resulting in $H(q)m$ compressed size. Figure 6 illustrates how the compression ratio varies with the fraction of change.

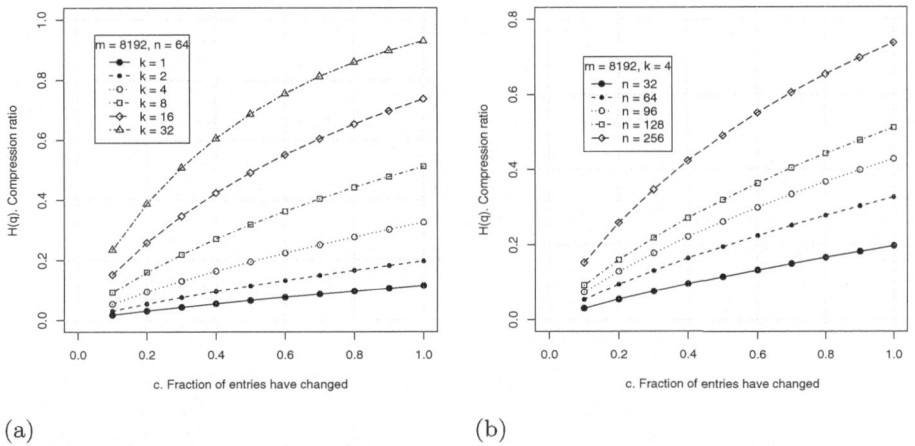

(a)   (b)

Fig. 6. The compressed ratio of *delta* with respect to the fraction of change. ($W = 32$).

5.2   Curtain with Merkle Tree

Even though we emphasize that ECC is a practical tool for authentication, there is still possibility to completely deprecate PKC in Curtain. Turning back to tree-based μTESLA, it is not ideal because the sender has to broadcast extra messages every a shot-term (100 seconds). Nevertheless we have successfully merged a middle-term period (one hour) into one commitment without more routine communication. Using Curtain as building block, we can construct Merkle tree of bitmaps at higher level, that is, now the hashes of bitmaps corresponding to their sessions will be the leaf nodes of the Merkle tree and the root node becomes the ultimate commitment. For example, a Merkle tree with 32 hashes on leafs node which encompassing a $W = 32$ session respectively will sustain more than one day, but it only needs to renew every one hour. However, the sender has to generate all the keys on all chains before determining the root, making the system not flexible. An merits of using PKC is allowing the users to dynamically decide the security policy and generate keys on the fly. We argue that the best solution for short-term broadcast authentication is Curtain, long-term security shall be left to PKC, and tree-base approaches can serve as an intermediate framework.

6 Related Work

Adapted from TESLA protocol over Internet [11], μTESLA [1] is the first lightweight broadcast authentication in sensor networks using only symmetric primitives. In [2], the authors proposed multi-level μTESLA Merkle tree methods can be found in several research [3,12,8]. MiniSec [13] is a network layer security protocol operating in two modes, one tailored for single-source communication, and another tailored for multi-source broadcast communication. In particular, MiniSec uses Bloom filters to maintain the legitimate packet information. Recently, Ren et al. [14] proposed a multi-user broadcast authentication scheme built on Bloom filters to store the users' public keys. PKC and Merkle hash tree techniques are also used in their scheme. However, it is fully based on PKC therefore the energy consumption is much higher than μTESLA series. In [8], the authors investigated the problem of broadcast authentication from various aspects and list out seven fundamental properties for designing protocols. The work in [15] discussed DoS attack which aims at exhausting the energy then increasing its response time to broadcast authentication.

7 Conclusion

We propose an efficient broadcast authentication protocol, **Curtain**, in sensor networks. It has the following advantages: (1) scalability in time and senders. (2) lower communication overhead than former schemes. (3) extending the self-healing property across multiple μTESLA instance. (4) dynamic sender addition and revocation. The implementation shows that Curtain outperforms previous tree-based methods. Curtain is an efficient way to bind the low-level μTESLA instances. But we argue that neither Curtain nor Merkle tree shall work solely without PKC. PKC is plausible in WSN and is the last resort for security. Not only in wireless sensor networks, Curtain is potential to other applications of broadcast authentication, such as in video broadcast over Internet.

Acknowledgments. The authors would like to thank the anonymous reviewers for their valuable comments. The work is partly supported by Taiwan Information Security Center (TWISC), under the grants NSC 96-2219-E-001-001 and NSC 96-2219-E-011-008, and by iCAST project under the grant NSC 96-3114-P-001-002-Y, sponsored by the National Science Council, Taiwan.

References

1. Perrig, A., Szewczyk, R., Wen, V., Culler, D., Tygar, J.D.: SPINS: Security Protocols for Sensor Networks. In: 7th ACM Annual International Conference on Mobile Computing and Networking, pp. 189–199. ACM Press, New York (2001)

2. Liu, D., Ning, P.: Multilevel μTESLA: Broadcast Authentication for Distributed Sensor Networks. Trans. on Embedded Computing Sys. 3, 800 (2004)
3. Liu, D., Ning, P., Zhu, S., Jajodia, S.: Practical Broadcast Authentication in Sensor Networks. In: 2nd Annual International Conference on Mobile and Ubiquitous Systems (MobiQuitous 2005), pp. 118–132. IEEE Computer Society, Los Alamitos (2005)
4. Menezes, A.J., Oorschot, P.C.v., Vanstone, S.A.: Handbook of Applied Cryptography, http://www.cacr.math.uwaterloo.ca/hac/
5. Eberle, H., Wander, A., Gura, N., Shantz, S.C., Gupta, V.: Architectural Extensions for Elliptic Curve Cryptography over $GF(2^m)$ on 8-bit Microprocessors. In: 16th IEEE International Conference on Application-Specific Systems, Architectures, and Processors (ASAP 2005), pp. 343–349. IEEE Computer Society, Los Alamitos (2005)
6. Gupta, V., Millard, M., Fung, S., Zhu, Y., Gura, N., Eberle, H., Shantz, S.C.: Sizzle: A Standards-Based End-to-end Security Architecture for the Embedded Internet. In: Third IEEE International Conference on Pervasive Computing and Communications, pp. 247–256. IEEE Computer Society, Los Alamitos (2005)
7. Wang, X., Yin, Y.L., Yu, H.: Finding Collisions in the Full SHA-1. In: Shoup, V. (ed.) CRYPTO 2005. LNCS, vol. 3621, pp. 17–36. Springer, Heidelberg (2005)
8. Luk, M., Perrig, A., Whillock, B.: Seven Cardinal Properties of Sensor Network Broadcast Authentication. In: Zhu, S., Liu, D. (eds.) SASN 2006, pp. 147–156. ACM Press, New York (2006)
9. Mitzenmacher, M.: Compressed Bloom Filters. IEEE/ACM Trans. Netw. 10, 604 (2002)
10. Fan, L., Cao, P., Almeida, J., Broder, A.Z.: Summary Cache: A Scalable Wide-Area Web Cache Sharing Protocol. IEEE/ACM Trans. Netw. 8, 281 (2000)
11. Perrig, A., Canetti, R., Tygar, D., Song, D.: The TESLA Broadcast Authentication Protocol. In: Cryptobytes, vol. 5(2), pp. 2–13. RSA Laboratories (2002)
12. Chang, S.-M., Shieh, S., Lin, W.W., Hsieh, C.-M.: An Efficient Broadcast Authentication Scheme in Wireless Sensor Networks. In: Lin, F.-C., Lee, D.-T., Lin, B.-S., Shieh, S., Jajodia, S. (eds.) ASIACCS 2006, pp. 311–320. ACM, New York (2006)
13. Luk, M., Ghita, M., Perrig, A., Gligor, V.: Minisec: A Secure Sensor Network Communication Architecture. In: Abdelzaher, T.F., Guibas, L.J., Welsh, M. (eds.) IPSN 2007, pp. 479–488. ACM Press, New York (2007)
14. Ren, K., Lou, W., Zhang, Y.: Multi-user Broadcast Authentication in Wireless Sensor Networks. In: 4th Annual IEEE Communications Society Conference on Sensor, Mesh and Ad Hoc Communications and Networks, pp. 223–232. IEEE Computer Society, Los Alamitos (2007)
15. Wang, R., Du, W., Ning, P.: Containing Denial-of-Service Attacks in Broadcast Authentication in Sensor Networks. In: Kranakis, E., Belding, E., Modiano, E. (eds.) MobiHoc 2007, pp. 71–79. ACM Press, New York (2007)

On the Urban Connectivity of Vehicular Sensor Networks

Hugo Conceição[1,2,3], Michel Ferreira[1,2], and João Barros[1,3,4]

[1] DCC
[2] LIACC
[3] Instituto de Telecomunicações
Faculdade de Ciências da Universidade do Porto, Portugal
[4] LIDS, MIT, Cambridge, MA, USA
{hc,michel,barros}@dcc.fc.up.pt

Abstract. Aiming at a realistic mobile connectivity model for vehicular sensor networks in urban environments, we propose the combination of large-scale traffic simulation and computational tools to characterize fundamental graph-theoretic parameters. To illustrate the proposed approach, we use the DIVERT simulation framework to illuminate the temporal evolution of the average node degree in this class of networks and provide an algorithm for computing the transitive connectivity profile that ultimately determines the flow of information in a vehicular sensor network.

1 Introduction

Mobile sensor networks are arguably substantially different from their static counterparts. First and foremost, the fact that the network topology can vary dramatically in time depending on the particular application and deployment scenario, offers richer connectivity profiles and proven opportunities e.g. to increase the system capacity [1] or to enhance the level of security [2]. Although some of the typical problems faced by static sensor networks, in particular unfair load balancing and fast energy depletion of nodes close to the sink [3], may often disappear in the presence of dynamically changing topologies, other tasks such as routing, topology control or data aggregation require non-trivial solutions capable of exploiting the existing mobility patterns and providing higher effectiveness and efficiency.

Our goal here is to develop a methodology for characterizing the aforementioned mobility patterns for vehicular sensor networks in urban environments — a task we deem crucial towards developing adequate protocols for this class of distributed systems. In sharp contrast with state-of-the-art stochastic mobility models, which frequently rely on somewhat simplistic decision models either for each sensor vehicle (e.g. the random walk or the random way point model) or with respect to the urban road network (e.g. the Manhattan grid or hexagonal lattices), we propose to use a realistic traffic simulator capable of capturing the complexity of trajectories followed by sensor nodes placed in vehicles and

S. Nikoletseas et al. (Eds.): DCOSS 2008, LNCS 5067, pp. 112–125, 2008.
© Springer-Verlag Berlin Heidelberg 2008

the dynamics of the resulting connectivity profiles when using tractable wireless communication models.

Our main contributions are as follows:

- *Modeling Methodology:* We combine large-scale traffic simulation with a basic wireless communication model in order to make inferences about the dynamic behavior of mobile sensor networks with thousands of nodes.
- *Analysis of the Average Node Degree in an Urban Environments:* Based on our large-scale vehicular network simulator, in which micro-simulated vehicles circulate in a real-life urban network, we characterize the average node degree for various traffic flow conditions, transmission radii and time intervals;
- *Connectivity Dynamics:* We present an algorithm capable of computing the transitive connectivity, which is key towards understanding information flow in a vehicular sensor network.

The remainder of the paper is organized as follows. Section 2 provides an overview of related work on modeling vehicular sensor networks. In Section 3 we describe our general methodology based on the DIVERT simulation framework and show our results for the average degree of the evolutionary connectivity graph. Section 4 then describes our algorithm and its results for the transient connectivity of the vehicular sensor network, and Section 5 concludes the paper.

2   Related Work

Vehicular sensor networks, which in the context of vehicle-to-vehicle (V2V) wireless communication are sometimes called Vehicular Ad-Hoc Networks (VANETs), consist in a distributed network of vehicles equipped with sensors that gather information related to the road. The main variables monitored are related to traffic efficiency and safety. An example of the later is the detection of hazardous locations such as ice or oil detected through the traction control devices of the vehicle. In the context of this paper traffic sensing is more relevant, because it can be exploited in the distributed computation performed by the vehicles' traffic-congestion-aware route planners. Traffic sensing is essentially done by measuring mobility (average speed) in a road segment.

Both types of information have to be propagated through the network and thus the topology of the communications network is naturally affected by the mobility of the nodes (mobility plays a dual-role in this context, as a sensing measure and as a sensor property). Realistic models to study the topology of these networks are very important, in particular for large-scale scenarios where thousands of nodes can interact in a geographically confined space, such as a city. If unrealistic models are used, invalid conclusions may be drawn [4]. Models for VANETs have two main components: modeling of mobility, which is very related to traffic modeling; and modeling of inter-vehicle communication.

According to the concept map for mobility models presented in [5], these models can be primarily categorized in analytical and simulation based. Analytical models are in general based on very simple assumptions regarding movement patterns. Typically, nodes are considered to be uniformly distributed over a given area. The direction of movement of each node is also considered to be uniformly distributed in an interval of $[0...2\pi[$, as well as speed ($[0...v_{max}[$). In some models these parameters of direction and speed remain constant until the node crosses the boundary of the given area [6], while other models allow them to change [7]. The main advantage of analytical models is the fact that they allow the derivation of mathematical expressions describing mobility metrics.

Simulation based models are much more detailed regarding movement patterns. In the context of vehicular movement these models originated from the area of transportation theory and have been an object of intense study over the past sixty years. Depending on the level of detail in which traffic entities are treated, current models can be divided into three main categories: microscopic, macroscopic and mesoscopic.

The microscopic model requires a highly detailed description where each vehicle/entity is distinguished. Each vehicle is simulated independently with its own autonomous variables, e.g. position and velocity. Microscopic models are very useful to describe the interaction between the traffic entities, braking and acceleration patterns, or navigation related parameters. There are two instances with great relevance, namely the Car-Following model (CF-model) [8] and the Cellular-Automaton model (CA-model)[9]. The former follows the principle that the variables of one vehicle depend on the vehicle (or vehicles) in front, which can be described using ordinary differential equations.

More recent than the CF-model, the CA-model divides the roads in cells with boolean states: either the cell is empty or it contains one vehicle. The simulation is done in time steps and each vehicle may move only if the next cell is free. In some CA-models the vehicles can advance more than one cell in each step. The velocity is calculated considering the number of cells one vehicle advances per step and associating values to the size of each cell (typically 7.5m) and to the duration of each step (typically 1s).

The model with the lowest-detail description is the macroscopic model, which tries to describe the traffic flow as if it was a fluid or a gas[10], thus analyzing the traffic in terms of density, flow and velocity.

As a more recent approach, the mesoscopic model can be viewed as standing between the aforementioned models in terms of detail. It combines some of the advantages of the microscopic model, such as the possibility to define individual routes for each vehicle, with the easiness to calibrate the model parameters of the macroscopic model. In the mesoscopic model the speed of the vehicle is defined by the cell in which it is, contrary to what happens in the microscopic model. This speed is the result of a speed-density function that relates the speed with the current density in each cell. Ongoing research efforts in transportation theory are targeting a hybrid model [11]. Here, the main idea is to use a microscopic approach in road segments where a high level of detail is necessary, using a mesoscopic model for the remaining parts. This approach facilitates the calibration of the simulators and saves on computational complexity.

All these models have an associated degree of randomness in parameters such as routes, direction and speed. A highly used simulation model for the study of wireless systems is the random walk model [12], where nodes can move freely anywhere in the simulation area. In vehicular simulators the randomness of the direction of movements is usually much more limited, as it is bounded by the spatial configuration of the road network. Randomness in vehicular simulators usually lies in the choice of speed and direction at road intersections. The randomness of speed is minimized in sophisticated simulators by relating it to the density of road segments or defining it by monitoring distance to the vehicles ahead. However, the randomness associated to direction decision in these simulators is quite unrealistic and does not reflect the actual mobility behaviour of vehicles in real scenarios. A more deterministic approach, where pre-defined routes are set up and their frequency is calibrated to mimic a realistic scenario is very likely to provide more coherent results.

Capturing the fundamental properties of wireless communications is also a challenging task towards realistic modeling of mobile sensor networks. Beyond the simple random geometric graph, where each node is connected to any other node within a prescribed radius, it is possible to incorporate more sophisticated elements such as shadowing and fading [13,14]. Reference [15] gives a critical overview of the most common modeling assumptions.

3   Connectivity of Vehicular Adhoc Networks

In this section we describe the study of mobile nodes connectivity in the context of a large-scale vehicular sensor network, simulated over an urban scenario using the DIVERT framework [16], which will be briefly described. We present our study abstracting the details of the spatial configuration of the road network and of vehicles and communication location into concepts from graph theory, such as node degree and transitive connectivity.

3.1   The DIVERT Simulation Environment

The DIVERT simulation environment (Development of Inter-VEhicular Reliable Telematics) has been developed in order to provide a state-of-the-art tool to study the behaviour of realistic and large-scale vehicular sensor networks, providing a testbed for the development and evaluation of distributed protocols, such as collaborative route-planning [17]. It is out-of-the-scope of this paper to describe the DIVERT framework.Other examples of traffic simulators in the spirit of DIVERT include TraNS [18], NCTUns [19] and GrooveNet [20]. We just outline the realistic features present in DIVERT, pointing out how such features can substantially affect the connectivity profile of the network. A video of a simulation is available at http://myddas.dcc.fc.up.pt/divert.

DIVERT works over real maps of cities (that can be setup by the user). The city used for the simulations presented in this paper was Porto, second largest in Portugal. It has several types of roads, from highways to residential streets, totaling 965kms. Each road has a speed limit, that is used in the simulation, and mimics the traffic signs enforcements. Each road has a different number of traffic lanes, ranging from 4 to 1, and the connectivity of each lane is discriminated in road intersections. In most roads, traffic circulates two-ways, which has clearly a crucial impact on the pattern of short-range communications. DIVERT includes a detailed traffic signs layer, in particular of priorities at intersections and traffic lights location and parametrization. This is also crucial, has our simulations show that the updating of traffic information happens primarily at road intersections, being the stop time of each vehicle a vital parameter. DIVERT includes parking location. Vehicles appear and disappear in the simulation, either because they enter/leave the city area, or because they start/stop in a parking place. This also affects connectivity. DIVERT allows defining the aimed number of vehicles in the city, dividing in sensor vehicles (vehicles that are part of the VANET), and non-communicating vehicles, which just affect the mobility of the former. This allows the realistic simulation of scenarios with different penetration rates of V2V technology. In the simulations presented in this paper, a 10% penetration rate was considered, for an aimed total of 5000 vehicles in simultaneous simulation. Regarding vehicle routes, DIVERT uses a hybrid model of pre-defined routes and randomly generated routes. For randomly generated routes, DIVERT arbitrarily selects an origin and a destination and calculates the route based on a shortest-path algorithm, either parameterized by distance or by time. Shortest-path based on time uses not only the speed limits of segments, but mainly the dynamic calibration of average mobility derived from previous simulation results. Pre-defined routes are setup by the user and have an associated frequency. In the simulations presented in this paper they have been carefully designed to approximate the simulation to our perception of traffic distribution in Porto. DIVERT allows parameterizing the transmission range of sensor vehicles, which obviously affects the network connectivity. In this paper we present results for simulations with varying transmission ranges and show its impact on the network connectivity. The DIVERT model for communication is based on the purely geometric link model.

3.2   Graphs with Mobile Nodes

We abstract the current configuration of a VANET into a graph representation as follows: each equipped vehicle corresponds to a node in the graph and we denote the set of nodes by $\mathcal{U} = \{u_1, u_2, ..., u_n\}$, mapping its latitude and longitude (corrected through map-matching) into a point in the spatial reference system of the graph; links in the graph are determined by the wireless communication model that is used, which in our case is the purely geometric link model, and thus a node has links to all nodes that are within its circular transmission range. Each link is represented by an unordered node pair $\{u_i, u_j\}$ and we denote the complete set of links by $\mathcal{L} \subseteq \mathcal{U}^2$. An example of such a graph in the initial configuration of a VANET simulation with DIVERT is illustrated in Fig. 1.

An important measure of the connectivity of a graph is its *average node degree*, which is given by:

$$d_{avg} = \frac{1}{n} \sum_{i=1}^{n} |\mathcal{N}(u_i)|$$

where $|\mathcal{N}(u_i)|$ is the cardinality of the set of nodes to which node i is connected, the *neighbors* of node u_i, denoted by $\mathcal{N}(u_i)$.

Fig. 1. Graph of the initial configuration of a VANET

As it can be seen from Fig. 1, the average node degree of the initial configuration of our VANET is very low ($\approx$ 1.6 in Fig. 1, where a transmission range of 150 meters was used), resulting in an extremely disconnected network. From the spatial configuration of the nodes, which reflects the actual geographic distribution of vehicles, we can see that even a significant increase in the transmission range would still result in a very low average node degree. However, the mobility of the nodes can significantly improve the network connectivity. If we consider the mobility of a node u_i in a time interval $[0...t]$, we can define the degree of u_i in that interval as follows:

$$d(u_i)_{[0...t]} = \left| \bigcup_{j=0}^{t} \mathcal{N}(u_i)_j \right| \tag{1}$$

where $\mathcal{N}(u_i)_j$ is the set of neighbors of u_i at instant j. In DIVERT simulations, one j step corresponds to one second.

Figure 2 shows the average node degree variation as we increase the mobility time interval, from 0 to 300 seconds. Again the transmission range used in the simulation was of 150m. Average node degree increases from 1.6 for static nodes, to a value higher than 20 when a 300s mobility time is considered.

Clearly, the timeliness of the information exchanged between nodes using a connectivity model based on mobility is different from the timeliness of the information exchanged between nodes in a static connectivity model. However,

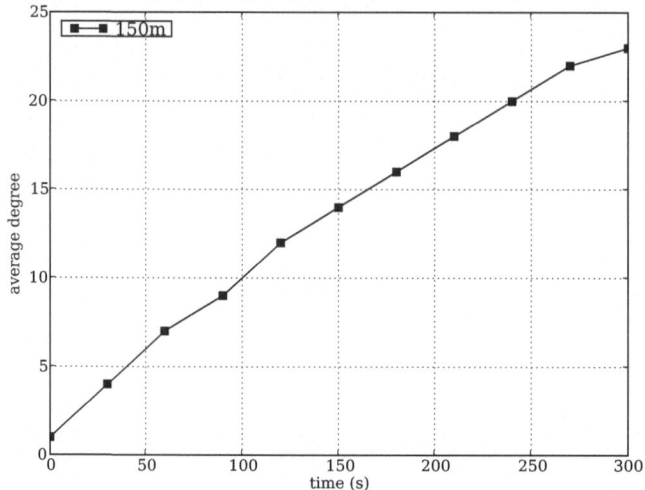

Fig. 2. Average node connectivity with varying mobility time

some of the information circulating a VANET is tolerant to a certain level of obsoleteness, such as traffic speed in road segments, where a 300-seconds-old report can still be validly used by dynamic navigation engines to avoid a traffic jam.

Information exchange in a VANET graph happens not only between nodes directly connected but also through transitive links. An example is the dissemination of traffic information, where communicating vehicles exchange information about the average speed of road segments they traveled, but also about the average speed of road segments they received from other vehicles. This transitivity of node connection is conveyed in the graph representation in Fig. 1, where node mobility has not been considered, as nodes can be reached through multi-link transitive paths. However, when mobility is considered in the connectivity of a node, simple transitive paths cannot be assumed to allow a node to reach another node. Consider for instance that $\mathcal{N}(u_1)_0 = \{u_2\}$, $\mathcal{N}(u_2)_0 = \{u_1\}$, $\mathcal{N}(u_2)_1 = \{u_3\}$ and $\mathcal{N}(u_3)_1 = \{u_2\}$. At $t = 0$, u_1 and u_2 are connected and can exchange information. At $t = 1$, u_2 and u_3 connect, and u_3 will receive information from u_2 that includes the data u_1 sent at $t = 0$, but u_1 will not received any information from u_3. In this case, we consider that $u_{3[0...1]}$ is transitively connected to $u_{1[0...1]}$ but not vice-versa. Links in the mobile graph become directed, while in the static model they were undirected.

We generalize the set of links for a mobile graph in an interval $[0...t]$ to a set of triples of the form (u_i, u_j, t'), for t' in $[0...t]$ (e.g. $(u_1, u_2, 0)$). The set of links is denoted by $\mathcal{L}' \subseteq \mathcal{U}^2 \times [0...t]$ and we consider the following rule to compute the transitive closure $\mathcal{L}'^*$ of $\mathcal{L}'$:

$$\forall u_i, u_j \in \mathcal{U}, t' \in [0...t],$$

$$(u_i, u_j, t') \in \mathcal{L}'^* \iff \begin{cases} u_j \in \mathcal{N}(u_i)_{t'} \\ \exists u_k \in \mathcal{U}, (u_i, u_k, t') \in \mathcal{L}' \wedge (u_k, u_j, t'') \in \mathcal{L}'^* \wedge t' \geq t'' \end{cases}$$

$$(2)$$

By 2, $\mathcal{L}'^* = \{(u_1, u_2, 0), (u_2, u_1, 0), (u_2, u_3, 1), (u_3, u_2, 1), (u_3, u_1, 1)\}$. Note that u_3 is linked to u_1 but not vice-versa.

After computing the transitive closure of $\mathcal{L}'$ we perform the following projection on its triples, $\Pi_{1,2}(\mathcal{L}'^*)$, obtaining the set of pairs of the graph that represents the transitive connectivity in the mobility interval $[0...t]$. Nodes are represented in their position at instant t. Figure 3 represents such as graph where an interval of 30s of node mobility was considered, maintaining the 150m transmission range. This graph is based on the simulation that generated graph in Fig. 1 at instant 0. Nodes in this graph can only exchange information through direct links. The mobility-based average node degree is thus a very good measure to quantify the graph connectivity.

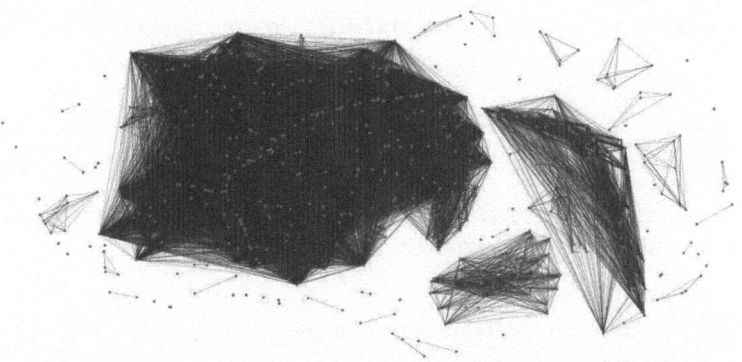

Fig. 3. Transitive closure graph considering a node mobility of 30s

Figure 4 shows the evolution of the average node degree for the transitive closure graph, in an interval from 0s to 300s. From the 1.6 average node degree at instant 0, we attain a value of almost 400 considering a mobility of one minute, resulting in an almost completely connected network. With a 150m range, the value of 60s of mobility seems to be critical in terms of connectivity.

3.3   Computing Transitive Connectivity for Mobile Nodes

The problem of computing the transitive closure of a graph is very well established. A common solution is the Floyd-Warshall Algorithm and we base our solution on it. However, since we are dealing with graphs with mobile nodes where links can change over time, a new insight was needed. Our notion of connectivity and transitivity is defined in the previous section. Let $\mathcal{L}'^*_{a,b}(n)$ be the set of links (a, b, t') resultant from applying rule 2 to paths $P_{a,b} =< (a, u_c, t'), ..., (u_d, b, t'') >$ with intermediary nodes restricted to $u_1, ..., u_n$ ($1 \leq c, d \leq n$) such that

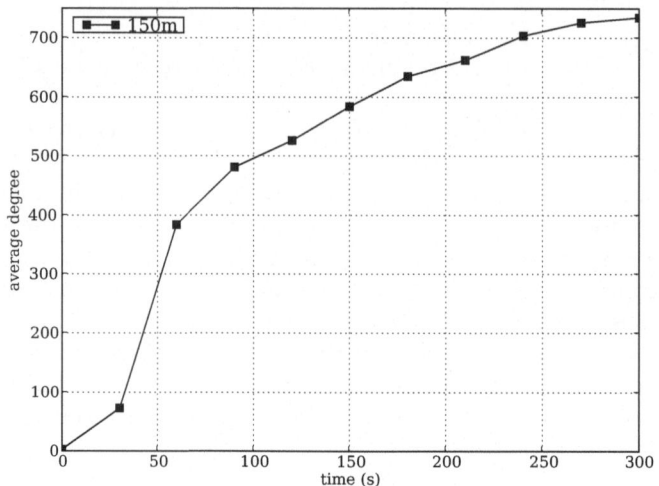

Fig. 4. Average node degree for the transitive closure graph in inteval $[0...300s]$

$$\mathcal{L'}^{*}_{a,b}(n) = \{(a,b,t') : (a,u_n,t') \in \mathcal{L'}^{*}_{a,u_n}(n-1) \wedge (u_n,b,t'') \in \mathcal{L'}^{*}_{u_n,b}(n-1), t' \geq t''\}$$

and by this, $\mathcal{L'}^{*}_{a,b}(|\mathcal{U}|) = \mathcal{L'}^{*}_{a,b}$.

Algorithm 1 is applied to $\mathcal{L'}$ in order to obtain $\mathcal{L'}^{*}$. To show the correctness of the algorithm, we need to prove the following lemma:

Algorithm 1. Compute the transitive closure

1: $\forall a, b \in \mathcal{U}, \mathcal{L'}^{*}_{a,b}(0) \leftarrow \mathcal{L'}$
2: **for all** u_k in $\mathcal{U}$ **do**
3:    **for all** u_i in $\mathcal{U}$ **do**
4:       **for all** u_j in $\mathcal{U}$ **do**
5:          $\mathcal{L'}^{*}_{u_i,u_j}(k+1) = \{(u_i,u_j,t') : (u_i,u_{k+1},t') \in \mathcal{L'}^{*}_{u_i,u_k}(k) \wedge (u_{k+1},j,t'') \in \mathcal{L'}^{*}_{u_k,u_j}(k), t' \geq t''\}$
6:       **end for**
7:    **end for**
8: **end for**

Lemma 1. *After k iterations of the outer loop,*

$$\forall a, b \in \mathcal{U}, \mathcal{L'}^{*}_{a,b}(k) =$$
$$= \{(a,b,t') : (a,u_k,t') \in \mathcal{L'}^{*}_{a,u_k}(k-1) \wedge (u_k,b,t'') \in \mathcal{L'}^{*}_{u_k,b}(k-1), t' \geq t''\}$$

if and only if there exists a path from a to b with nodes restricted to $u_1, u_2, ..., u_k$, following rule 2.

The proof is in APPENDIX A.

4   Varying Mobility Time/Transmission Range

So far in this paper we have used a transmission range of $150m$ in our DIVERT simulations. However, connectivity is clearly affected by different transmission ranges. Figure 5 shows the variation of the average node degree with varying mobility time ($[0..300s]$) and transmission range ($[50..300m]$). The graphic plots expression 1 in Section 3.2 for the 6 different transmission ranges considered. As it can be seen from Fig. 5, an average node degree of 10 is obtained after $30s$ of mobility for a $300m$ transmission range, after $60s$ for a range of $200m$ and after $120s$ for a range of $100m$. If the latency of the information is not critical, higher mobility times can achieve the same connectivity with lower transmission ranges.

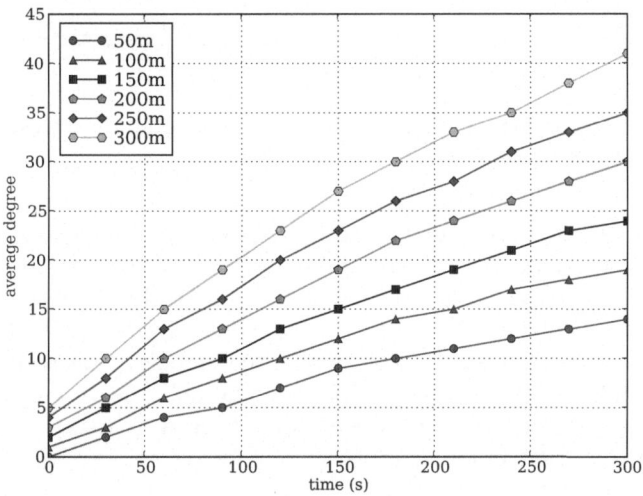

Fig. 5. Average node degree with varying mobility time and transmission range

Results for the transitive closure graph are presented in Fig. 6, considering the different transmission ranges. We now use the average percentual node degree, meaning that we divide the degree of the nodes by the total number of nodes simulated. It should be noted that in a realistic environment sensors can leave the city area, or end their trip within the city, during the mobility time being considered. On the other hand, some sensors enter or appear in the city near the end of the mobility time being considered. It is thus almost impossible to achieve a 100% connected network. Our approach is to measure connectivity in the presence of such realistic behaviour, rather than defining unrealistic border models (such as reflection) or forcing all trips to last more than the mobility time.

It is clear from Fig. 6 that the connectivity profile of the network percolates over the two dimensions considered, transmission radii and mobility time. The

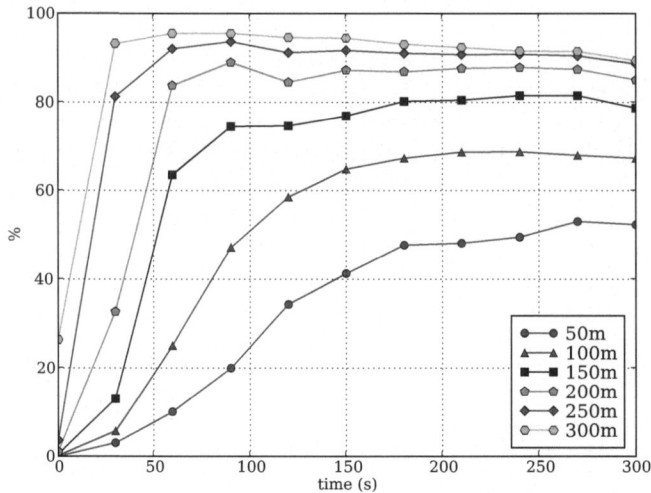

Fig. 6. Average percentual node degree (transitive closure graph) with varying mobility time and transmission range

phenomena of percolation is in fact known to govern the behaviour of wireless networks [21]. If we consider the mobility time of $30s$, percolation on the transmission radius happens around the critical threshold of $200m$. Only for the transmission ranges higher than $200m$ we achieve an average node degree higher than 50% of the total of nodes. Percolation on the mobility time seems to happen around thresholds of $60s$ or $90s$. At $90s$, average node degrees of around 50% or higher of the total of nodes are already found for all the transmission ranges except for the conservative $50m$ radius.

5   Conclusions and Further Work

Vehicular sensor networks are undoubtedly poised to become a very relevant example of distributed computation, featuring large numbers of vehicles that cooperate in a self-organized fashion to find optimal ways of exploring the road network. The deployment of such networks and the development of their protocols require a fundamental understanding of the dominant mobility and connectivity patterns — a task that in our belief must rely heavily on trustworthy computer simulations.

Focusing on one instance of this problem, we provided a preliminary characterization of the connectivity of a vehicular sensor network operating in an urban environment. Seeking a realistic model for sensor mobility, we opted for a scenario in which vehicular trajectories and velocities are bounded by the geography of the road network (in our case, a real city) and determined by detailed traffic entities, such as traffic lights, speed limits and number of lanes of a given road.

Vehicle movement in our model is also very much determined by the interaction with other vehicles and the ensuing decisions of each driver.

As part of our contribution, we were able to extract some of the most significant graph-theoretic parameters that capture the dynamics of this complex network. For this purpose, we extended concepts such as average node degree to account for mobility intervals, and developed an algorithm for computing the transitive closure of such graphs based on mobile links. Our results illuminate the relationship between transmission radius and mobility time, showing an attenuation of percolation effects when these two dimensions are viewed as complementary.

In future work, we plan to compare our results with simpler mobility models, such as the random way point or the Manhattan grid, using the same graph-based evaluation, highlighting the role of realistic mobility models towards understanding the dynamics of urban connectivity for vehicular sensor networks.

Acknowledgments

The authors gratefully acknowledge useful discussions with Prof. Luís Damas (LIACC/UP and DCC-FCUP), who contributed decisively towards the development of the DIVERT simulation framework. The first author is also grateful to Cláudio Amaral for useful discussions.

This work has been partially supported by the Portuguese Foundation for Science and Tecnology (FCT) under projects MYDDAS (POSC/EIA/59154/2004), JEDI (PTDC/EIA/66924/2006) and also STAMPA (PTDC/EIA/67738/2006), and by funds granted to LIACC and IT through the Programa de Financiamento Plurianual and POSC.

References

1. Grossglauser, M., Tse, D.: Mobility increases the capacity of ad hoc wireless networks. IEEE/ACM Transactions on Networking (TON) 10(4), 477–486 (2002)
2. Capkun, S., Hubaux, J., Buttyán, L.: Mobility helps security in ad hoc networks. In: Proceedings of the 4th ACM international symposium on Mobile ad hoc networking & computing, pp. 46–56 (2003)
3. Intanagonwiwat, C., Govindan, R., Estrin, D.: Directed diffusion: A scalable and robust communication paradigm for sensor networks. In: Proceedings of the ACM/IEEE International Conference on Mobile Computing and Networking, pp. 56–67 (2000)
4. Kurkowski, S., Camp, T., Colagrosso, M.: MANET simulation studies: the incredibles. Mobile Computing and Communications Review 9(4), 50–61 (2005)
5. Bettstetter, C.: Smooth is better than sharp: a random mobility model for simulation of wireless networks. In: Meo, M., Dahlberg, T.A., Donatiello, L. (eds.) MSWiM, pp. 19–27. ACM, New York (2001)
6. Hong, D., Rappaport, S.S.: Traffic model and performance analysis for cellular mobile radio telephone systems with prioritized and non-prioritized handoff procedures. In: ICC, pp. 1146–1150 (1986)

7. Zonoozi, M.M., Dassanayake, P.: User mobility modeling and characterization of mobility patterns. IEEE Journal on Selected Areas in Communications 15(7), 1239–1252 (1997)
8. Wilhelm, W.E., Schmidt, J.W.: Review of car following theory. Transportation Enginnering Journal 99, 923–933 (1973)
9. Hoogendoorn, S., Bovy, P.: State-of-the-art of vehicular traffic flow modelling. Proceedings of the Institution of Mechanical Engineers, Part I: Journal of Systems and Control Engineering 215(4), 283–303 (2001)
10. Hoogendoorn, S., Bovy, P.: Gas-Kinetic Model for Multilane Heterogeneous Traffic Flow. Transportation Research Record 1678(-1), 150–159 (1999)
11. Burghout, W., Koutsopoulos, H., Andréasson, I.: Hybrid Mesoscopic-Microscopic Traffic Simulation. Transportation Research Record 1934(-1), 218–255 (2005)
12. Bar-Noy, A., Kessler, I., Sidi, M.: Mobile users: To update or not to update? Wireless Networks 1(2), 175–185 (1995)
13. Bettstetter, C., Hartmann, C.: Connectivity of Wireless Multihop Networks in a Shadow Fading Environment. Wireless Networks 11(5), 571–579 (2005)
14. Chiang, C., Wu, H., Liu, W., Gerla, M.: Routing in clustered multihop, mobile wireless networks with fading channel. Proceedings of IEEE SICON 97, 197–211 (1997)
15. Kotz, D., Newport, C., Gray, R.S., Liu, J., Yuan, Y., Elliott, C.: Experimental evaluation of wireless simulation assumptions. In: MSWiM 2004: Proceedings of the 7th ACM international symposium on Modeling, analysis and simulation of wireless and mobile systems, pp. 78–82. ACM, New York (2004)
16. Conceição, H., Damas, L., Ferreira, M., Barros, J.: Large-Scale Simulation of V2V Environments. In: Proceedings of the 23rd Annual ACM Symposium on Applied Computing, SAC 2008, Fortaleza, Ceará, Brazil, March 2008. ACM Press, New York (2008)
17. Yamashita, T., Izumi, K., Kurumatani, K., Nakashima, H.: Smooth traffic flow with a cooperative car navigation system. In: AAMAS 2005: Proceedings of the fourth international joint conference on Autonomous agents and multiagent systems, pp. 478–485. ACM, New York (2005)
18. Piorkowski, M., Raya, M., Lugo, A., Papadimitratos, P., Grossglauser, M., Hubaux, J.P.: TraNS: Realistic Joint Traffic and Network Simulator for VANETs. ACM SIGMOBILE Mobile Computing and Communications Review
19. Wang, S., Chou, C., Huang, C., Hwang, C., Yang, Z., Chiou, C., Lin, C.: The design and implementation of the NCTUns 1.0 network simulator. Computer Networks 42(2), 175–197 (2003)
20. Mangharam, R., Weller, D., Rajkumar, R., Mudalige, P., Bai, F.: GrooveNet: A Hybrid Simulator for Vehicle-to-Vehicle Networks. In: Second International Workshop on Vehicle-to-Vehicle Communications (IEEE V2VCOM), San Jose, USA (July 2006)
21. Booth, L., Bruck, J., Franceschetti, M., Meester, R.: Covering Algorithms, Continuum Percolation and the Geometry of Wireless Networks. The Annals of Applied Probability 13(2), 722–741 (2003)

APPENDIX A

The following proves lemma 1:

Proof. By induction on k.

Let $\mathcal{A}(x) = \mathcal{L}'^*_{i,j}(x)$. When $k = 0$ the statement is obviously true as the paths $P_{i,j} = <i,j>$, that is, the paths from i to j without intermediary nodes, are the links from i to j and $\mathcal{L}'^*_{i,j}(0)$ is initialized with $\mathcal{L}'$.

Assume lemma 1 to be true for the first k iterations. In the $(k+1)^{th}$ iteration,

$$\mathcal{A}(k+1) = \{(i,j,t') : \{i, u_{k+1}, t'\} \in \mathcal{L}'^*_{i,u_{k+1}}(k) \wedge (u_{k+1}, j, t'') \in \mathcal{L}'^*_{u_{k+1},j}(k), t' \geq t''\}$$

and so u_{k+1} is added to the set of allowed intermediary nodes. The paths considered are of the form $<(i, u_a, tt), ..., (u_b, u_{k+1}, t'), (u_{k+1}, u_c, t''), ..., (u_d, j, tt')>$. By induction assumption the subpaths considered $<(i, u_a, tt), ..., (u_b, u_{k+1}, t')>$ and $<(u_{k+1}, u_c, t''), ..., (u_d, j, tt')>$ are valid since they are paths considered in $\mathcal{L}'^*_{i,u_{k+1}}(k)$ and $\mathcal{L}'^*_{u_{k+1},j}(k)$, respectively. Also the intermediary nodes of these subpaths are restricted to nodes $u_1...u_k$. To see that the complete path is valid it is only necessary to verify that $<(u_b, u_{k+1}, t'), (u_{k+1}, u_c, t'')>$ respects rule 2. Since it is ensured that $t' \geq t''$ the path is valid. Thus we get that in the $(k+1)^{th}$ iteration

$$\mathcal{A}(k+1) = \{(i,j,t') : \{i, u_{k+1}, t'\} \in \mathcal{L}'^*_{i,u_{k+1}}(k) \wedge (u_{k+1}, j, t'') \in \mathcal{L}'^*_{u_{k+1},j}(k), t' \geq t''\}$$

if and only if there exists a path from i to j with nodes restricted to $u_1, ..., u_{k+1}$, following rule 2. Therefore, lemma 1 holds for the $(k+1)^{th}$ iteration of the outer loop and by the principle of mathematical induction, the lemma holds for all values of k.

By considering $k = |\mathcal{U}|$ we prove that $\mathcal{L}'^*_{a,b}(|\mathcal{U}|) = \mathcal{L}'^*_{a,b}$.

FIT: A Flexible, LIght-Weight, and Real-Time Scheduling System for Wireless Sensor Platforms

Wei Dong[1], Chun Chen[1], Xue Liu[2], Kougen Zheng[1], Rui Chu[3], and Jiajun Bu[1]

973 WSN Joint Lab
[1] College of Computer Science, Zhejiang University
[2] School of Computer Science, McGill University
[3] National University of Defense Technology
dongw@zju.edu.cn, chenc@zju.edu.cn, xueliu@cs.mcgill.ca,
zkg@zju.edu.cn, rchu@nudt.edu.cn, bjj@zju.edu.cn

Abstract. We propose FIT, a flexible, light-weight and real-time scheduling system for wireless sensor platforms. There are three salient features of FIT. First, its two-tier hierarchical framework supports customizable application-specific scheduling policies, hence FIT is very **flexible**. Second, FIT is **light-weight** in terms of minimizing thread number to reduce preemptions and memory consumption while at the same time ensuring system schedulability. We propose a novel Minimum Thread Scheduling Policy (MTSP) exploration algorithm within FIT to achieve this goal. Finally, FIT provides a detailed **real-time schedulability** analysis method to help check if application's temporal requirements can be met. We implemented FIT on MICAz motes, and carried out extensive evaluations. Results demonstrate that FIT is indeed flexible and light-weight for implementing real-time applications, at the same time, the schedulability analysis provided can predict the real-time behavior. FIT is a promising scheduling system for implementing complex real-time applications in sensor networks.

1 Introduction

Recently, Wireless Sensor Networks (WSNs) have seen an explosive growth in both academia and industry [1,2]. They have received significant attention and are envisioned to support a variety of applications including military surveillance, habitat monitoring and infrastructure protection, etc.

WSNs typically consist of a large number of micro sensor nodes that self-organize into a multi-hop wireless network. As the complexities for real-world applications continue to grow, infrastructural support for sensor network applications in the form of system software is becoming increasingly important. The limitation exhibited in sensor hardware and the need to support increasingly complicated and diverse applications has resulted in the need for sophisticated system software. As a large portion of WSN applications are real-time in nature, a good real-time scheduling system plays a central role in task processing in sensor nodes.

The first emerged sensornet OS, TinyOS [3], is especially designed for resource-constrained sensor nodes. Because of its simplicity in the event-based single-threaded scheduling policy, time-sensitive tasks cannot be handled gracefully in conjunction with

S. Nikoletseas et al. (Eds.): DCOSS 2008, LNCS 5067, pp. 126–139, 2008.
© Springer-Verlag Berlin Heidelberg 2008

complicated tasks as task priority preemption is not supported. Thus as a programming hack, a long task usually has to be manually split into smaller subtasks to ensure the temporal correctness of the whole system to be met. Otherwise, a critical task could be blocked for too long hence miss its deadline. Two other notable similar sensornet OSes, Contiki [4] and SOS [5], both fall into the same category.

As an alternative solution, Mantis OS [6] uses time-sliced multi-threaded scheduling. It supports task preemption and blocking I/O, enabling micro sensor nodes to natively interleave complex tasks with time-sensitive tasks. However, Mantis OS is not very flexible and has a relative higher overhead: making changes to its scheduling policy is not an easy task, as the scheduling subsystem is tightly coupled with other subsystems; time-sliced multi-threaded scheduling incurs a higher scheduling overhead because of preemptions and extra memory consumption for thread management.

In this paper, we present a novel scheduling system, FIT, for micro sensor platforms. It is flexible through the careful design of its two-tier hierarchical scheduling framework. Under this framework, application programmers can easily implement the most appropriate application-specific scheduling policy by customizing the second-tier schedulers. It is light-weight compared with Mantis OS, as it minimizes the thread number to reduce preemptions and memory consumption while ensuring schedulability by exploiting the Minimum Thread Scheduling Policy (MTSP) exploration algorithm. In addition, FIT provides detailed real-time schedulability analysis to help the designers to check if application's real-time temporal requirements can be met under FIT's system model.

To validate FIT's efficacy and efficiency, we implemented FIT on MICAz motes, and carried out extensive evaluations. Our results show that FIT meets its design objectives. It is flexible as a series of scheduling policies in existing sensornet OSes can be easily incorporated into FIT's two-tier scheduling hierarchy (Section 6.1). It is light-weight as it effectively reduces the running thread number by using MTSP exploration algorithm (Section 6.2). It is real-time as the schedulability analysis is conducted in the MTSP exploration algorithm, thus real-time guarantee can be achieved by employing the *explored* MTSP (Section 6.3).

The rest of this paper is organized as follows: Section 2 describes related work most pertinent to this paper. Section 3 presents our flexible two-tier hierarchical scheduling framework within FIT. Section 4 details the light-weight MTSP exploration algorithm which relies on the real-time schedulability analysis presented in Section 5. Section 6 shows the evaluation results. Finally, we conclude this paper and give future directions of work in Section 7.

2   Related Work

FIT borrows heavily from three large areas of prior work: hierarchical scheduling, task scheduling on sensor nodes and real-time scheduling.

Hierarchical scheduling. Hierarchical scheduling techniques have been used in a number of research projects to create flexible real-time systems, such as PShED [7], HLS [8], etc. While their work focuses on general-purpose PC in open environment, our current work focuses on closed, static, deeply embedded sensor nodes. Regehr *et al.* [9]

describe and analyze the hierarchical priority schedulers already present in essentially all real-time and embedded systems while our current work uses the design philosophy of hierarchical scheduling to implement a two-tier flexible scheduling architecture on resource-constrained sensor nodes.

Task scheduling on sensor nodes. The event-driven scheme is commonly used in sensornet OSes. TinyOS [3], SOS [5] both fall into this category. In TinyOS, tasks are scheduled in a FIFO manner with a run-to-completion semantics. Like TinyOS, the SOS [5] scheduling policy is also non-preemptive and hence can not provide good real-time guarantee. Mantis OS [6] uses a different scheme. It supports time-sliced preemptive multi-threading. Similar to Mantis OS, Nano-RK [10] provides multi-threading support and uses priority-based preemptive scheduling. It performs static schedulability analysis to provide real-time guarantee. A large body of work was also devoted to improving the capability or the efficiency of task scheduling based on TinyOS [9,11,12,13].

FIT differs from most existing work in sensornet OSes in many important ways. Its two-tier hierarchical framework supports customizable application-specific scheduling policies, hence FIT is more flexible. FIT is also *light-weight* in terms of effectively reducing the running thread number while at the same time ensuring schedulability. Compared with [10], we adopt an event-based programming style, hence it has the opportunity to effectively reduce the running thread number while ensuring schedulability. Compared with [13], FIT automatically assign tasks to appropriate scheduling queues according to their temporal requirements, and provides detailed schedulability analysis.

Real-time scheduling. There is a large body of work devoted to real-time scheduling [14]. Priority mapping and PTS (Preemption Threshold Scheduling) are the most pertinent ones to our work. Our current work is different from traditional priority mapping [15] in that each task has a two-dimensional priority. Within the same global priority, we differentiate tasks by their local priorities. Hence, with the same number of global priorities (preemption levels), there are possibilities that MTSP exploration algorithm generates feasible assignment that those approaches cannot. We not only try to overlap the global priorities (preemption levels) but also try to overlap the local priorities. Our work is also different from PTS [16] as it has no notion of preemption threshold, thus priorities do not change at run-time. Finally, our work is different from the abovementioned ones since tasks in FIT consist of multiple jobs with run-to-completion semantics. We analyze schedulability for tasks instead of individual jobs. So the schedulability analysis is different from existing ones. Besides, we also take into account resource access time, hence the analysis is better suited to realistic situations.

3   Flexible Two-Tier Hierarchical Framework

Scheduling policies in current sensornet OSes are usually difficult to customize. Taking TinyOS-1.x for example, the tight coupling of the FIFO scheduling policy and the nesC programming language makes it hard, if not impossible, to modify. A *flexible scheduling framework* is important for easier customization of different scheduling policies. This framework should cleanly separate from other components in the system and ideally, provide lower level scheduling mechanisms (as libraries) to reduce the customization

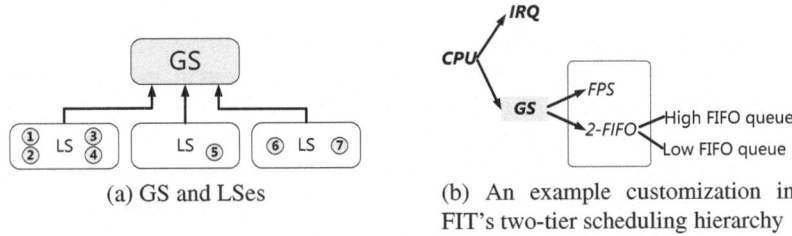

(a) GS and LSes

(b) An example customization in FIT's two-tier scheduling hierarchy

Fig. 1. Two-tier scheduling hierarchy

overhead to application programmers. Another fact in current sensornet OSes is that different scheduling policies must be employed exclusively. However, application programmers sometimes need to extend the scheduling system without affecting the current system behavior, thus different scheduling policies may need to coexist. We solve this problem through decomposition of schedulers. We propose a *two-tier scheduling hierarchy* to enhance FIT's flexibility.

3.1   Two-Tier Scheduling Hierarchy

Our two-tier scheduling hierarchy as depicted in Figure (a) comprises of two tiers of schedulers. The first-tier scheduler, GS, is designed to schedule LSes. The global scheduling policy employed by the GS is *preemptive* priority scheduling. The GS schedules LSes according to their *global priorities*. The second-tier schedulers, LSes, are designed to schedule individual tasks. Each LS is implemented as one thread and has its own thread context. The local scheduling policy employed by each LS is in a non-preemptive manner. LS schedules tasks according to their *local priorities*. Inside the LS, multiple tasks share a common thread context. The local scheduling policy depends on the number of different local priorities it needs to handle. If there is only one local priority, then FIFO (as in TinyOS) is employed. If the number of local priorities is a constant c that is lower than a threshold, then c-FIFO (which manages c FIFO queues of different priorities) is employed. In practice, we select this threshold as 3, similar to that used in the implementation of SOS [5]. The reason is that when the number of local priorities is small, using FIFOs will incur less overhead compared with maintaining a dedicated priority queue. When the number of local priorities is even larger, a priority queue is maintained and Fixed Priority Scheduling (FPS) is employed. The number of LSes and each LS's scheduling policy are customized for different applications.

Figure (b) illustrates an example customization in our two-tier scheduling hierarchy. In this example, our GS schedules two LSes (i.e., the FPS-LS and the 2-FIFO-LS) preemptively. The FPS-LS has a higher global priority than the 2-FIFO-LS (which manages 2 FIFO queues of different priorities). Thus any task in the FPS-LS can preempt tasks in the 2-FIFO-LS. The FPS-LS schedules tasks assigned to it non-preemptively with the FPS scheduling policy. The 2-FIFO-LS has a local high priority FIFO queue as well as a local low priority FIFO queue. It schedules tasks in two priority levels and uses FIFO scheduling within one priority level.

3.2   System Model

We formally define the system model and notations which will be used in the rest the paper in this subsection. As discussed in Section 3.1, in FIT system, each task is assigned a global priority and local priority. A task with higher global priority can preempt a task with lower global priority. Tasks with the same global priority are scheduled by the same LS within which they may have different local priorities. LS schedules tasks in a non-preemptive manner. Tasks with both the same global priority and local priority are scheduled in a FIFO manner.

Whenever we say task A has a higher priority than task B, we mean either task A has a higher global priority than task B or task A has a higher local priority than task B when their global priorities are equal. Task A and task B have the same priority if and only if they have the same global priority and the same local priority.

A task comprises of several jobs written with a run-to-completion semantics, i.e., they cannot suspend themselves. The basic scheduling unit of our system is a job. Because I/Os must be done in split phases, we assume *request* to be done at the end of a job while *signal* invokes the start of the next consecutive job. There is only one active job at any instant within a task.

We formally define the system model as follows: The system, Γ, consists of a set of n tasks $\tau_1, \ldots, \tau_n$. Each task τ_i is activated by a periodic sequence of events with period T_i and is specified a deadline D_i. A task, τ_i, contains $|\tau_i|$ jobs, and each job may not be activated (released for execution) until the request of the preceding job is signaled. We use J_{ij} to denote a job. The first subscript denotes which task the job belongs to, and the second subscript denotes the index of the job within the task. A job, J_{ij}, is characterized by a tuple of $\langle C_{ij}, B_{ij}, G_{ij}, L_{ij} \rangle$. C_{ij} is the worst case execution time, B_{ij} is the maximum blocking time to access the shared resource requested by the preceding job. It is the time interval from the completion of the preceding job and the start of the current job. G_{ij} is the global priority used and L_{ij} is the local priority used. The blocking time to access a shared resource, B_{ij}, consists of resource service time b_{ij} and resource waiting time b'_{ij}. The resource waiting time, b'_{ij}, is related to the specific resource scheduling scheme employed.

We make the following assumptions: (1) Task deadline is specified no longer than task period, i.e., $D_i \leq T_i$. (2) A good estimation of C_{ij} and b_{ij} is available. (3) All jobs within a task share the common period (i.e., jobs within a task are periodic albeit written with a run-to-completion semantics), and, have the same global priority and local priority, i.e., $G_{ij} = G_i \wedge L_{ij} = L_i, \forall 1 \leq j \leq |\tau_i|$.

4   Light-Weight MTSP Exploration

In the previous section, we discussed the flexible two-tier hierarchical scheduling framework can facilitate customizing different scheduling policies. In this section, we propose a method to find appropriate scheduling policies for a specific application. Specifically, we present the MTSP exploration algorithm which can effectively reduce the running thread number while ensuring system schedulability. It is worth noting that we make a distinction between thread and task. Task is from the perspective of functionality while

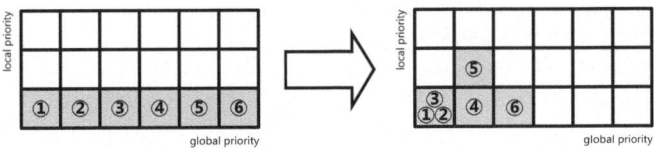

Fig. 2. Global priority and local priority assignment

thread is from the perspective of implementation. Thread has implementation and running overhead, e.g., thread context switches, thread control block, thread stack, etc. Traditional general-purpose OSes, including Mantis OS, treat each task as a separate thread. In contrast, our scheduling system tries to overlap multiple tasks so as to reduce the number of threads, and hence the implementation and running overhead, which is important for resource-constrained sensor nodes.

4.1 Problem Formulation

Because the number of global priorities maps directly to the number of LSes (which are implemented as threads), to find an MTSP, we can try to minimize the number of global priorities. The constraints are that schedulability of all tasks must be ensured.

We start from a fully preemptive Deadline Monotonic (DM) policy (the left part in Figure 2). This can be seen as the most capable scheduling policy as DM is optimal in FPS [14] conforming the assumptions under our system model. We call the initially assigned priority *natural priority*. Then we try to map the natural priorities to the one with as few global priorities as possible, thus to reduce the running thread number. Hence, memory consumption can be reduced, which is important for resource-constrained sensor nodes.

So the problem is that given a task set $\Gamma = \{\tau_1, \ldots, \tau_n\}$ with increasing natural priority. i.e., natural priority assignment is $(G_i, L_i) = (i, 0)$. Find a priority mapping $(G_i, L_i) : (i, 0) \mapsto (g_i, l_i)$, $1 \leq g_i, l_i \leq n$, such that

minimize $|G|$, where $G = \{g_i\}$ **subject to**   all tasks are schedulable.   (1)

We take *direct mapping* as the mapping rule as in [15] and [17].

Definition 1 (direct mapping). *Assignment* ① $\mapsto$ *Assignment* ② *is a direct mapping, then, if any task τ_i has higher priority than any task τ_j in Assignment* ①, *task τ_j cannot have higher priority than that of task τ_i in Assignment* ②.

As an illustrative example, let's look at the assignment in Figure 2. The left part assignment gives a separate global priority to each task. It implies that up to 6 LSes are needed which corresponds to 6 threads. This is a relatively costly scheme. It may be fine in traditional general-purpose OSes where CPU speed and memory capacity are abundant. However, for sensor platforms where resources are usually limited, the left assignment is not desirable. Using MTSP, we can perform static analysis at compile time to effectively reduce the thread number while still ensuring schedulability. The result is shown in the right part. The thread number could be reduced to 3.

Algorithm 1: MTSP exploration algorithm

Input: A task set Γ with decreasing deadlines

Output: Assignment to $(G_i, L_i) = (g_i, l_i)$ 10: $(G_i, L_i) \leftarrow (G_{i-1}, L_{i-1})$

1: **for** $i \leftarrow 1, n$ **do**   11:   **if** test_task(i) == TRUE **then**

2:    $(G_i, L_i) \leftarrow (i, 0)$ 12:   **continue**

3: **end for**   13:   **end if**

4: **if** test_all_task() != TRUE **then**   14: $L_i = L_i + 1$

5:    **print** unschedulable task set   15:   **if** test_task(i) == TRUE **then**

6:    **return**   16:   **continue**

7: **end if**   17:   **end if**

8: **for** $i \leftarrow 2, n$ **do**   18:   **restore** G_i and L_i

9:    **save** G_i and L_i 19: **end for**

4.2   MTSP Exploration Algorithm

Our MTSP exploration algorithm is presented in Algorithm 1. After initially assigning sperate global priority to each task, we explore opportunities whether a task could be overlapped with a lower global priority. After iterating for all tasks, we obtain the final result. The time complexity of MTSP exploration algorithm is $O(n)$.

Again, let's revisit Figure 2 as an example. First, we start with each task assigned a natural priority. It is worth noting that in this paper, we use a larger value to indicate a higher priority. Then we test if all tasks are schedulable with the most capable scheduling policy, i.e., preemptive DM. If that fails, we report a non-schedulable message and end up with each task assigned to the natural priority. While the task set is schedulable, we examine whether it is still schedulable with a less capable but more lightweight scheduling policy. We start from τ_2. We test whether it is schedulable when assigned with the same global priority and local priority as the previous one (τ_1 in this case). If it is, iterate for the next one (τ_3 in this case). Otherwise, we test whether τ_2 is schedulable when assigned with a higher local priority and with the same global priority as the last one τ_1. If it is, iterate for the next task. If τ_2 is not schedulable in both of the cases, we leave its global priority and local priority unchanged and iterate for the next task. After iterating for all of the tasks, we end up with the resulting global priorities and local priorities while ensuring schedulability.

In Algorithm 1, notice we only test the schedulability of τ_i in the loop. This is ensured by the following theorem[1].

Theorem 1. *When τ_i moves to a lower priority, the schedulability of all tasks are ensured as long as τ_i is schedulable.*

5   Real-Time Schedulability Analysis

In this section, we consider the real-time constraints presented in the minimization problem in Section 4.1. It is worth mention that our schedulability analysis is conducted on

---

[1] The proof of Theorem 1 and the *optimality* of MTSP are given in [18] due to space limit.

tasks instead of individual jobs. To derive the schedulability test, We analyze the processor demand [14] of each task. In FIT, the processor demand of a task consists of (1) C_i: the execution time of τ_i. (2) I_{lp}: interference time from lower priority tasks. (3) I_{sp}: interference time from the same priority tasks. (4) I_{hp}: interference time from higher priority tasks. (5) B_i: blocking time to access shared resources.

A sufficient condition to make τ_i schedulable is [14]: $\min_{0 < t \le D_i} \frac{W_i(t)}{t} \le 1$, where $W_i(t) = C_i + I_{lp} + I_{sp} + I_{hp} + B_i$.

Note that C_i, I_{lp}, I_{sp}, I_{hp} and B_i may depend on t. When there is no ambiguity occurred, we omit it here and in the following sections for the simplicity of notations.

Determining C_i. C_i is τ_i's execution time which equals to the sum of the execution time of all jobs within τ_i, i.e., $C_i = \sum_{j=1}^{|\tau_i|} C_{ij}$.

Determining I_{lp}. I_{lp} is the maximum interference time caused by lower priority tasks. In FIT, as lower global priority tasks can always be preempted by τ_i, they can never block the execution of τ_i. Thus they introduce no interference. Tasks with the same global priority but with lower local priorities, however, can block the execution of τ_i as they are scheduled non-preemptively with τ_i. We denote $lp(i)$ as the task set in which tasks have the same global priority but have lower local priority. $lp(i)$ represents the lower priority tasks which actually cause interference.

The maximum interference occurs when there is a lower priority task executing each time a job in τ_i releases. τ_i can be at most blocked for $|\tau_i|$ times as there are $|\tau_i|$ run-to-completion jobs. The blocking time each time will not exceed the *maximum* execution time of all the lower priority jobs. Hence,

$$I_{lp} \le \max_{\substack{k \in lp(i) \\ 1 \le j \le |\tau_k|}} \{C_{kj}\} \cdot |\tau_i|.$$

It is worth noting that a tighter bound exists, because the times of interference caused by a job, J_{kj}, is limited to $\lceil t/T_k \rceil$. That means the job with maximum execution time may not block τ_i for $|\tau_i|$ times (if $\lceil t/T_k \rceil < |\tau_i|$). We sort $\{C_{kj}\}$, for *all* $k \in lp(i)$, $1 \le j \le |\tau_i|$, in non-increasing order, i.e., $C_{k_1 j_1} \ge C_{k_2 j_2} \ge \ldots \ge C_{k_m j_m} \ge \ldots$, then,

$$I_{lp} \le \underbrace{C_{k_1 j_1} \left\lceil \frac{t}{T_{k_1}} \right\rceil + C_{k_2 j_2} \left\lceil \frac{t}{T_{k_2}} \right\rceil + \ldots + C_{k_m j_m} + \ldots + C_{k_m j_m}}_{\text{There are total } |\tau_i| \text{ number of } C_{kj}}. \tag{2}$$

We end up with either $|\tau_i|$ number of C_{kj} are added up or we have added all the terms under consideration.

Determining I_{sp}. I_{sp} is the maximum interference time caused by the same priority tasks. All the same priority tasks have opportunities to interfere the execution of τ_i. We denote $sp(i)$ as the task set in which tasks have the same priority as τ_i.

In FIT, tasks in $sp(i)$ are scheduled in a FIFO manner with τ_i. As with the computation of I_{lp}, τ_i can be blocked at most $|\tau_i|$ times. The blocking time each time will not exceed *all* jobs ahead of τ_i in the same priority FIFO. As there is only one active job among a task, the blocking time each time is at most $\sum_k \max_j \{C_{kj}\}$, where $k \in sp(i)$, $1 \le j \le |\tau_k|$. Hence,

$$I_{sp} \le \left(\sum_{k \in sp(i)} \max_{1 \le j \le |\tau_k|} \{C_{kj}\} \right) \cdot |\tau_i|.$$

As with the same reasoning in Section 5, a job, J_{kj}, may not block τ_i for $|\tau_i|$ times, because it is further limited by $\lceil t/T_k \rceil$. For *each* $k \in sp(i)$, $1 \le j \le |\tau_k|$, we sort $\{C_{kj}\}$ in non-increasing order, i.e., $C_{kj_1} \ge C_{kj_2} \ge \ldots \ge C_{kj_{|\tau_k|}}$, The interference caused by τ_k is

$$I_k \le \underbrace{C_{kj_1} \left\lceil \frac{t}{T_k} \right\rceil + C_{kj_2} \left\lceil \frac{t}{T_k} \right\rceil + \ldots + C_{kj_m} + \ldots + C_{kj_m}}_{\text{There are total } |\tau_i| \text{ number of } C_{kj}}. \tag{3}$$

Also, we end up with either $|\tau_i|$ number of C_{kj} are added up or we have added all the terms under consideration. Then we consider the (worst-case) overall interference, i.e., $I_{sp} \le \sum_{k \in sp(i)} I_k$.

Determining I_{hp}. I_{hp} is the maximum interference caused by higher priority tasks. As all higher priority tasks can interfere the execution of τ_i, we denote $hp(i)$ as the task set in which tasks have higher priority than τ_i. The processor demand of higher priority tasks is only limited by $I_{hp} \le \sum_{k \in hp(i)} C_k \lceil t/T_k \rceil$. where $C_k = \sum_{j=1}^{|\tau_k|} C_{kj}$.

Determining B_i. τ_i blocks $|\tau_i| - 1$ times (assume there is no blocking for the first job to execute) to access shared resources. As discussed in Section 3.2, B_{ij} denotes the maximum blocking time to access the shared resource requested by the preceding job $J_{i(j-1)}$. Each blocking time B_{ij} consists of resource service time b_{ij} and resource waiting time b'_{ij}. Hence, $B_i = \sum_{j=2}^{|\tau_i|} B_{ij} = \sum_{j=2}^{|\tau_i|} (b_{ij} + b'_{ij})$.

The resource service time b_{ij} is estimated beforehand while the resource waiting time depends on the resource scheduling scheme. We denote $rc(i,j)$ as the task set in which tasks also use the same resource as that J_{ij} blocks on (before execution). Further, we introduce $rc_m(i,j)$ to represent all the jobs in task m, that $\forall k \in rc_m(i,j)$, J_{mk} blocks on the same resource as that J_{ij} blocks on. If FIFO resource scheduling scheme is employed, then $b'_{ij} \le \sum_{m \in rc(i,j)} \max_{k \in rc_m(i,j)} \{b_{mk}\}$.

6 Evaluation

We implemented FIT on MICAz motes and carried out extensive tests to evaluate the flexibility, lightweightness and real-time performance of FIT. First, it examines whether FIT is flexible enough to incorporate scheduling policies in existing sensornet OSes; it also measures the overhead as compared to TinyOS and Mantis OS. Second, by a case study of typical workloads on sensor nodes, it examines whether FIT can effectively reduce the running thread number to reduce preemptions and memory consumption. Finally, it examines the real-time performance of FIT.

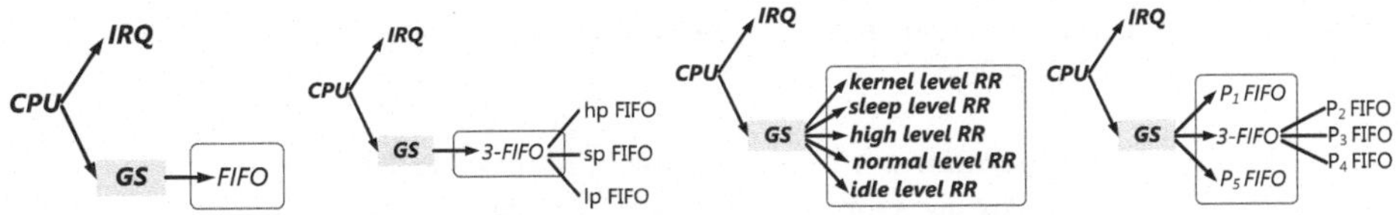

(a)   Two-tier (b) Two-tier scheduling (c) Two-tier scheduling (d) Two-tier scheduling
scheduling hierar- hierarchy for SOS   hierarchy for Mantis OS hierarchy for the PL
chy for TinyOS   scheduler

Fig. 3. Existing scheduling policies under FIT's two-tier scheduling hierarchy

6.1 Flexibility

A series of existing scheduling policies in sensornet OSes can be implemented under
FIT's two-tier scheduling hierarchy. Figure (a) shows the two-tier scheduling hierarchy
for TinyOS [3]. It has only one LS, i.e., the FIFO-LS. Whenever the GS gains the CPU
control to dispatch jobs, it always passes the control down to the FIFO-LS which schedules tasks in a non-preemptive FIFO manner. Figure (b) shows the two-tier scheduling
hierarchy for SOS [5]. Also, it has one LS, i.e., the 3-FIFO-LS, which manages three
FIFO queues of high, system and low priorities respectively. The scheduling hierarchy
of Mantis OS [6] is shown in Figure (c). It has five LSes with different global priorities.
For each LS, different from the previous two, is a *preemptive* round-robin scheduler.
Although as we have discussed in Section 3.1, each LS is assumed as a non-preemptive
scheduler, FIT's two-tier scheduling hierarchy is general and do allow each LS to employ arbitrary scheduling policy. The real-time schedulability analysis and MTSP exploration algorithm, however, should be revised once this assumption is broken. Figure (d)
shows the scheduling hierarchy of the PL scheduler [13] for TinyOS-2.x. The GS schedules three LSes in a preemptive manner. The 3-FIFO-LS manages three FIFO queues
of different local priorities and schedules tasks in a non-preemptive manner. As we can
see from Figure 3, FIT's two-tier scheduling hierarchy is very flexible as a series of
existing scheduling policies can be easily implemented under this framework.

We evaluate the implementation overhead of FIT's two-tier hierarchical scheduling
via four metrics. (1) Overhead of posting a job; (2) Overhead of scheduling consecutive
jobs in the same task which is measured as the time interval from the completion of the
preceding job to the start of the current job without considering the blocking time to
access shared resources for simplicity; (3) Overhead of LS creation which corresponds
to thread creation in Mantis OS; (4) Overhead of scheduling jobs in different LSes
which incurs context switching.

We measure the clock cycles of each operation by using Avrora [19]. The results are
reported in Table 1. The post operation of TinyOS is 42cc (cc is a shorthand for clock
cycles) while FIT has 90cc. The extra overhead lies in the fact that we make scheduling
decisions in our post operation while TinyOS does not need to as priority preemption is
not supported. When just the post operation is measured, FIT takes about 44cc which is
very close to that of TinyOS. Scheduling jobs in the same LS, however, consumes more
time than TinyOS, because FIT will relinquish the CPU control to the GS whenever a
job returns. Then the GS passes the CPU control down to the LS which schedules the

Table 1. Implementation overhead of FIT's two-tier hierarchical scheduling compared to TinyOS and Mantis OS (in clock cycles)

Operations	TinyOS	Mantis OS	FIT
1. Posting a job	42	N/A	90
2. Scheduling jobs in the same LS	63	N/A	130
3. LS/thread creation	N/A	2481	366
4. Scheduling jobs in different LSes	N/A	447	409

next job. In FIT, LS scheduling is about 55cc which is close to TinyOS; GS scheduling consumes about 67cc which is an extra overhead introduced by our two-tier hierarchical scheduling. LS/thread creation in FIT consumes 366cc, as opposed to 2481cc in Mantis OS. The significant difference stems from the fact that Mantis OS uses a dynamic memory allocation which consumes about 1655cc as well as invoking a thread dispatching routine which consumes about 447cc. Without considering these overheads, it leaves 379cc, a little larger than that of FIT. In FIT, scheduling jobs in different LSes involves context switching, which consumes 409cc, much larger than scheduling jobs in the same LS, but slightly smaller than Mantis OS. As we can see, FIT's flexible two-tier hierarchical scheduling scheme has a quite acceptable implementation overhead.

6.2 Lightweightness

To evaluate the lightweightness of FIT, we set up experiments to generate random inputs to our MTSP exploration algorithm to study how many threads are actually needed to ensure the system's schedulability. The parameters we used are as follows. The system consists of $|\Gamma| = 5$ number of tasks where all tasks fall into two categories: long executing tasks with $C_{ij} \sim U(300, 50)$ and real-time tasks with $C_{ij} \sim U(10, 8)$. In addition, real-time tasks have a more urgent deadline than long executing tasks. Each task contains 1–5 number of jobs. The default number of shared resources is $|R| = 4$ and the resource service time varies from 1–24ms. Meanwhile, FIFO resource scheduling scheme is assumed.

We are interested to see how many threads are required as the average percentage of real-time tasks increases. We generate 1000 cases of task sets for each percentage. Note some of the task sets might be schedulable while others are not. The average results are shown in Figure 4. The solid line represents the percentage of schedulable cases out of all the generated cases. The white bar represents the average number of threads out of all the generated cases while the gray bar represents the average number of threads out of all the schedulable cases. As we can see from the solid line, as the average percentage of real-time tasks increases, the percentage of schedulable cases out of all the generated cases decreases. Our MTSP exploration algorithm ends up with the maximum number of threads when it encounters an unschedulable case. Thus, the average number of threads out of all the generated cases is approaching $|\Gamma| = 5$ when the average percentage of real-time tasks reaches 80% as the majority of the cases are

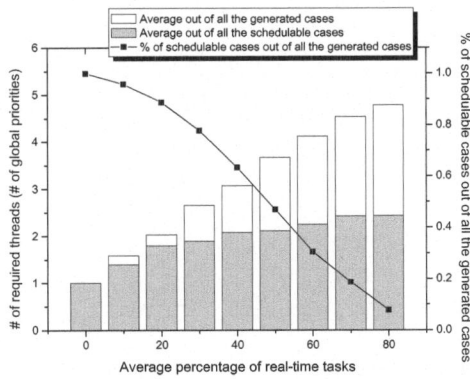

Fig. 4. Number of required threads vs. Average percentage of real-time tasks

unschedulable. To get a closer look at the cases when the task sets are schedulable, as shown by the gray bar in Figure 4, we observe the average number of threads out of all the schedulable cases never exceeds 3, reducing at least 2 threads compared with the worst case.

6.3 Real-Time Guarantee

Due to space limit, we show FIT's real-time guarantee via a case study. The system we studied has two shared resources (R_A and R_B) and their individual service time is 1ms (R_A) and 18ms (R_B) respectively. The system consists of the following tasks: (1) Task τ_1 that consists of 4 jobs. The individual job execution times are selected as 3ms, 4ms, 5ms and 3ms respectively. The shared resources accessed between consecutive jobs are chosen to be R_A, R_B and R_A. Task τ_1's period (T_1) and deadline (D_4) are set to 1s, i.e., $T_1 = D_1 = 1$s. (2) Task τ_2 with the same setting as τ_1. (3) Task τ_3 with the same setting as τ_1. (4) Task τ_4 that consists of 3 jobs. The individual job execution times are selected as 3ms, 4ms and 18ms respectively. The shared resources accessed between consecutive jobs are chosen to be R_B and R_A. This is the task under consideration and we vary $T_4 = D_4$ from 20ms to 160ms. (5) Task τ_5 that consists of 3 jobs. The individual job execution times are selected as 3ms, 36ms and 524ms. The shared resources accessed between consecutive jobs are chosen to be R_A and R_B. $T_5 = D_5 = 10$s.

Figure (a) and Figure (b) illustrate the cases when we vary the deadline D_4 within 40ms–200ms and 200ms–800ms respectively. We can see from Figure (a) that from the theoretical analysis discussed in Section 5, the system will be schedulable when D_4 reaches 120ms. When $D_4 < 120$ms, there could be missed deadlines from the theoretical analysis. The results collected from testbed show that when $D_4 \geq 100$ms, employing MTSP (with two LSes) will result in no missed deadlines in practice. The gap (between 120ms and 100ms) exists because the theoretical analysis considers the *worst* case while in practice the *runtime* interference could be smaller, thus mitigate the percentage of missed deadlines even in the face of an unschedulable system. When $D_4 < 120$ms, FIT's real-time schedulability analysis will also report τ_4 as the unschedulable task, thus designers can take various ways (e.g., relaxing its deadline, reducing concurrency,

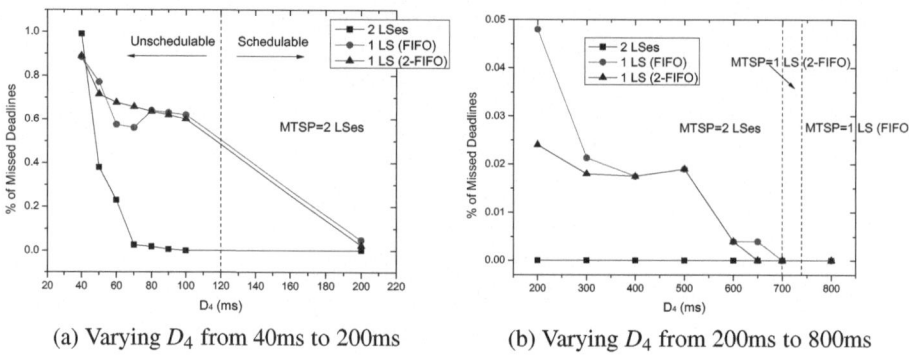

(a) Varying D_4 from 40ms to 200ms   (b) Varying D_4 from 200ms to 800ms

Fig. 5. Varying D_4 from 40ms to 800ms

etc) to redesign a schedulable system. Figure (a) also indicates that using the theoretical analysis, when $D_4 \geq 120$ms FIT can ensure that there are no missed deadlines by employing MTSP while the simple FIFO or 2-FIFO scheduling policies may produce missed deadlines in practice. Also notice that the percentage of missed deadlines may increase in Figure (a). This is because as D_4 $(= T_4)$ increases, the total number of jobs released are reduced.

Figure (b) shows when $D_4 \geq 700$ms, from the theoretical analysis, the system will be schedulable with a 2-FIFO scheme, and, when $D_4 \geq 740$ms, the system will be schedulable with a FIFO scheme. In practice, this transitional region (700ms–740ms) is located with a smaller D_4 (650ms–700ms). As we can see from Figure (b), in practice, we could employ 2-FIFO in 650ms–700ms as the 2-FIFO scheme will result in no missed deadlines while the FIFO scheme still causes a small fraction of missed deadlines. Anyway, as Figure (a) and Figure (b) indicate, FIT always selects the scheduling policy with minimum overhead while at the same time ensuring the schedulability of the system.

7   Conclusion and Future Work

In this paper, we present a novel scheduling system, FIT, for micro sensor platforms. FIT is flexible in terms of supporting customizable application-specific scheduling policies. This is achieved through the careful design of its two-tier hierarchical scheduling framework in FIT. It is light-weight by exploiting the proposed MTSP exploration algorithm which effectively reduces the running thread number to reduce preemptions and memory consumption while ensuring system schedulability. In addition, FIT provides detailed real-time schedulability analysis to predict the real-time behavior of the underlying system running on top of it, thus helps designers to check if application's temporal requirements can be met in design time.

While we have shown that FIT is a promising scheduling system for implementing complex real-time applications in sensor networks, there are several enhancements and optimizations we would like to explore. In particular, we are currently designing a new language to support programming easily under FIT's system model.

Acknowledgements. The authors would like to thank Prof. Lionel Ni and Prof. Yunhao Liu for their valuable input. This work is supported by the National Basic Research Program of China (973 Program) under grant No. 2006CB303000, and in part by an NSERC Discovery Grant under grant No. 341823-07.

References

1. Liu, Y., Li, M.: Iso-Map: Energy-Efficient Contour Mapping in Wireless Sensor Networks. In: ICDCS 2007 (2007)
2. Yang, Z., Li, M., Liu, Y.: Sea Depth Measurement with Restricted Floating Sensors. In: RTSS 2007 (2007)
3. TinyOS: http://www.tinyos.net
4. Dunkels, A., Grönvall, B., Voigt, T.: Contiki—a Lightweight and Flexible Operating System for Tiny Networked Sensors. In: EmNets 2004 (2004)
5. Han, C.C., Kumar, R., Shea, R., Kohler, E., Srivastava, M.: A Dynamic Operating System for Sensor Nodes. In: MobiSys 2005 (2005)
6. Bhatti, S., Carlson, J., Dai, H., et al.: MANTIS OS: An Embedded Multithreaded Operating System for Wireless Micro Sensor Platforms. MONET Journal, Special Issue on Wireless Sensor Networks 10, 563–579 (2005)
7. Lipari, G., Carpenter, J., Baruah, S.: A framework for achieving inter-application isolation in multiprogrammed hard real-time environments. In: RTSS 2000 (2000)
8. Regehr, J., Stankovic, J.A.: HLS: A framework for composing soft real-time schedulers. In: RTSS 2001 (2001)
9. Regehr, J., Reid, A., Webb, K., Parker, M., Lepreau, J.: Evolving real-time systems using hierarchical scheduling and concurrency analysis. In: RTSS 2003 (2003)
10. Eswaran, A., Rowe, A., Rajkumar, R.: Nano-RK: An Energy-Aware Resource-Centric RTOS for Sensor Networks. In: RTSS 2005 (2005)
11. Trumpler, E., Han, R.: A Systematic framework for evolving TinyOS. In: EmNets 2006 (2006)
12. McCartney, W.P., Sridhar, N.: Abstractions for Safe Concurrent Programming in Networked Embedded Systems. In: SenSys 2006 (2006)
13. Duffy, C., Roedig, U., Herbert, J., Sreenan, C.J.: Adding Preemption to TinyOS. In: EmNets 2007 (2007)
14. Liu, J.W.S.: Real-Time Systems. Prentice-Hall, Englewood Cliffs (2000)
15. Sathaye, S.S., Katcher, D.I., Strosnider, J.K.: Fixed Priority Scheduling with Limited Priority Levels. IEEE Trans. Comput. 44(9), 1140–1144 (1995)
16. Wang, Y., Saksena, M.: Scheduling fixed priority tasks with preemption threshold (1999)
17. DiPippo, L.C., Wolfe, V.F., et al.: Scheduling and Priority Mapping for Static Real-Time Middleware. Real-Time Systems 20(2), 155–182 (2001)
18. Dong, W., Chen, C., Liu, X.: FIT: A Flexible, Lightweight and Realtime Scheduling System for Wireless Sensors. Technical report, Zhejiang University (2007)
19. Titzer, B.L., Lee, D.K., Palsberg, J.: Avrora: Scalable Sensor Network Simulation with Precise Timing. In: IPSN 2005 (2005)

Automatic Collection of Fuel Prices from a Network of Mobile Cameras

Y.F. Dong[1], S. Kanhere[1], C.T. Chou[1], and N. Bulusu[2]

[1] School of Computer Science & Engineering, University of New South Wales,
Sydney, Australia
{ydon,salilk,ctchou}@cse.unsw.edu.au
[2] Department of Computer Science, Portland State University, USA
nbulusu@cs.pdx.edu

Abstract. It is an undeniable fact that people want information. Unfortunately, even in today's highly automated society, a lot of the information we desire is still manually collected. An example is fuel prices where websites providing fuel price information either send their workers out to manually collect the prices or depend on volunteers manually relaying the information. This paper proposes a novel application of wireless sensor networks to automatically collect fuel prices from camera images of road-side price board (billboard) of service (or gas) stations. Our system exploits the ubiquity of mobile phones that have cameras as well as users contributing and sharing data. In our proposed system, cameras of contributing users will be automatically triggered when they get close to a service station. These images will then be processed by computer vision algorithms to extract the fuel prices. In this paper, we will describe the system architecture and present results from our computer vision algorithms. Based on 52 images, our system achieves a hit rate of 92.3% for correctly detecting the fuel price board from the image background and reads the prices correctly in 87.7% of them. To the best of our knowledge, this is the first instance of a sensor network being used for collecting consumer pricing information.

Keywords: Automatic data collection, Computer-vision-based sensing, Consumer pricing information gathering, Participatory sensor networks, Vehicular sensor networks.

1 Introduction

The technology of wireless sensor networks (WSNs) ha been applied to a plethora of application domains, e.g., farming [1], structure monitoring [2], military [3], environmental monitoring [4], home health care [5], home environment control [6] etc. In this paper, we propose a novel application of using WSN to automatically collect prices of fuel[1] from still images of road-side price board of

---

[1] Unfortunately, different parts of the English speaking world use different words to refer to fuel for automobiles. In North America, it is commonly known as "gasoline" while it is called "petrol" in the United Kingdom and Australia. We have chosen to use "fuel" in this paper to avoid bias towards a particular community.

S. Nikoletseas et al. (Eds.): DCOSS 2008, LNCS 5067, pp. 140–156, 2008.

service stations. To the best of our knowledge, this is the first instance of a WSN being used for collecting consumer pricing information.

Price dispersion of homogenous goods appears to be a fixture of our economy. It is not uncommon for us to find two shops within a close proximity selling an identical item at different prices. For instance, on 18 January 2008, around 9pm, two service stations in Roseville, Sydney, that are less than 1 km apart, were selling unleaded fuel at 1.359 and 1.439 Australian dollars per litre. With the presence of price dispersion, the availability of consumer pricing information can therefore be advantageous and this has been confirmed by a number of studies by economists on the effect of information on the on-line economy. For example, the study in [7] showed that consumers who used on-line price comparison service could save up to 16% in buying electronic goods on-line. Also, the on-line economy has also been shown to drive down insurance prices in [8].

Given the existence of price dispersion in fuel, a number of websites exist to enable consumers to compare fuel prices. Examples of these include Gaswatch, GasBuddy[2] in the United States; fuelprice [3] in the UK; RACV, fuelwatch and motormouth[4] in Australia. These websites use a couple of different methods to collect their fuel price data. Some of these websites send their workers out, once or twice a day, to collect the data; however, this is labour intensive and it is difficult to track price changes since different service stations update their price at different times of the day[5]. Some other websites rely on volunteers manually relaying the fuel price information. Although the use of volunteers can provide more frequent updates, the need to manually relay the fuel price data may discourage many people from volunteering. Are there other alternatives other than manual collection? One possibility is deploying infrastructure at each service station to monitor the prices; however, the capital investment can be costly.

Given the above discussion, we therefore see the efficient collection of fuel price information (or more generally of a lot of information at the street level) to be a technical challenge. We believe that a method to overcome this technical challenge is to use the *Sensing Data Market (SenseMart)* framework that the authors of this paper have proposed in [9]. The SenseMart framework consists of two key ideas. Firstly, it leverages the existing sensing and communications infrastructure in collecting sensing data. Secondly, it provides a platform for people to exchange and share sensing data. This paper proposes a WSN architecture that allows volunteers to automatically collect, contribute and share fuel price information, thus lowering the cost or barrier for sharing. Our proposed system (which will be detailed in Sect.2) leverages on the ubiquity of cameras, mobile Internet connectivity, GPS (Global Positioning System) and GIS (Geographic Information System). Through the use of GPS and GIS, our system

---

[2] http://www.gaspricewatch.com,   http://www.gasbuddy.com/
[3] http://www.petrolprices.com/
[4] http://www.racv.com.au, http://www.fuelwatch.wa.gov.au/,
    http://motormouth.com.au/default_nf.aspx
[5] Price of fuel does change a few times a day at many service stations across Sydney.

knows that when the vehicle of a contributing user is getting close to a service station and triggers the camera automatically. These images are then processed by a computer vision algorithm to extract the fuel price, the details of which will be described in Sect.3 and 4.

There are several applications, which largely rely on the altruistic participation of users, such as Wikipedia, Youtube and BitTorrent,and yet have become hugely popular. Their success can be attributed to the following characteristics: (1) easy to use software for uploading, sharing and searching data and (2) insignificant monetary costs for uploads/downloads. The preliminary version of our system presented in this paper focusses on addressing the former. In the current incarnation, the computer vision algorithms are executed at a central server and the users are required to upload the raw images. Our ultimate goal is to accomplish all image processing tasks on the mobile device. Consequently, users would only have to upload a few bits of data, thus lowering the monetary barrier for information sharing.

This paper makes the following contributions:

- We propose a new application of WSN to collect consumer pricing information.
- We propose a novel WSN architecture — whose components include cameras, mobile phones, GPS, GIS, computer vision algorithms as well as the sharing of sensing data — that can automatically collect fuel prices from images of service station, thus lowering the barrier of sharing sensing data.
- We implemented a working prototype which automatically detects and classifies fuel prices from images of service station billboards. Based on 52 images, our system achieves a hit rate of 92.3% for correctly detecting the fuel price board from the image background and reads the prices correctly for 87.7% of them.

2   System Design

Our system has two principal modes of operation: (i) fuel price collection and (ii) user query. In this paper, we primarily focus on the former. The most important goal of our system is to automate the process of collecting the fuel prices. This is achieved by automatically triggering the mobile phones of contributing users to take pictures of roadside fuel price boards when they approach service stations while driving. Our system employs sophisticated computer vision algorithms to scan these images and retrieve the fuel prices. To reduce the complexity of the computer vision tasks, our system relies on contextual information that is made available by GPS and GIS software. Figure 1(a) presents a pictorial overview of our system. As depicted in the picture, the data collection process involves three steps: (i) capturing images of the fuel price boards, (ii) uploading the images to the central server and (iii) extracting fuel prices from the images . Each of these tasks is executed by a distinct component of the system. In the following, we discuss the design and implementation of the three main components that make up our system and describe their operation.

Camera Sensor: The primary function of the camera sensor is to automatically capture images of the price boards of approaching service stations. In our system, we have assumed that the mobile phones of contributing users will serve this purpose. Note that, almost all current mobile phones are equipped with in-built cameras. We assume that the phone is mounted on the dashboard in front of the front passenger, with the camera lens pointing towards the road, i.e. leftwards. Consequently, the camera can capture pictures of roadside objects on the left of the car (Note that in Australia we drive on the left side of the road). In the future, it is expected that cars would be fitted with cameras for implemented sophisticated ITS (Intelligent Transport System) applications. Our system can be readily interfaced with these in-car cameras. In this instance, the images captured by the cameras could be transferred wirelessly via Wifi or Bluetooth to the mobile phone.

Figure 1(b) depicts a logical representation of the various components that contribute to realizing the automated capturing of images. The control unit, which oversees the operation is implemented as a daemon in the mobile phone. The control unit periodically polls the in-built GPS receiver of the phone to obtain the current location coordinates of the car. It should be noted that a large majority of the new mobile phones are equipped with GPS receivers. Alternately, it is fairly straightforward to connect an external GPS receiver to a phone via Bluetooth. Our system also requires that GIS software such as Google Maps, TomTom or Nokia Maps is installed on the phone. GIS systems use the GPS location coordinates and a map database of the city road network to estimate the current street position of the mobile. GIS systems can also call to attention any approaching Points of Interest (POI) such as restaurants, shopping malls and service stations. For example, TomTom alerts users of approaching service stations and also indicates the provider information (i.e. brand, e.g.: Caltex, Mobil, etc). Our system takes advantage of these capabilities of the GIS software for automating the data collection process. The control unit queries the GIS system to obtain the location coordinates and the brand of the next approaching service station along the path of the car. Using this location data, the control unit first determines if the service station is on the appropriate side of the road. Recall, that as explained earlier the mobile phone is mounted with the camera lens facing the left-side of the road. Next it determines if the car is within an acceptable distance for taking pictures (a configurable threshold which is set to 10 meters) and then automatically triggers the camera sensor to take several pictures. The camera is switched off once the car has passed the service station. The images are tagged with contextual information provided by the GPS and GIS software such as the service station location coordinates, brand information and time of capture. This information significantly simplifies the complexity of the image processing algorithms that are executed at the server. The images together with the meta-data are passed on to the data upload unit, which is responsible for uploading the pictures to the central server. It is important to note that the entire process is automated and does not require any user intervention.

Data Transport: Our system leverages on the existing communication infrastructure for effecting the transfer of images to the server. Current mobile phones have ubiquitous Internet connectivity via the GSM/GPRS/3G/HSDPA cellular network and in some cases also via in-built 802.11 interfaces. The data upload unit of the mobile phone establishes a TCP connection with the central server using any of the available underlying access technologies. The images are uploaded to the server using this reliable channel. An alternative is to upload the images as multimedia SMS messages.

Central Server: The central server implements the computer vision algorithms for processing the images and extracting the fuel prices. A standard desktop computer is sufficient for implementing this functionality. We have implemented a TCP server daemon, which receives the images uploaded by the mobile users. The meta-data (location coordinates, service station brand and time) is extracted and stored separately. The images along with the fuel brand information are passed on to the image processing engine. Figure 1(c) represents the sequence of steps involved in the price extraction process. The first step involves detecting the existence of a fuel price board in the image, which by itself is a significantly challenging task. However, note that, each fuel brand uses a specific color combination for their fuel price boards. For example, BP uses a green color code, whereas Mobil boards are blue in color. To reduce the complexity of the problem, we utilize the knowledge of the brand of the service station to assist in the detection process. Recall that this information is included as meta-data with the uploaded pictures. For each brand, we employ a tailored color thresholding that is able to capture regions within the images, which have a similar color scheme as that of the price board. In certain situations, surrounding objects within the image may have colors resembling the board resulting in more than one potential candidate regions, e.g.: the blue sky in the background image may be similar to the color of the Mobil board. In this situation, we use two post-processing techniques to narrow down the search. In the first instance, we make use of certain features of the dimensions of the price boards to exclude some of the candidate regions selected by our color thresholding algorithm. Further refinement is achieved by comparing the color histogram of all candidate regions with that of a sample image of the price board of that particular brand. The detection process concludes with the identification of the precise location of the price board within the image. The image is then cropped such that it solely contains the price board. The cropped image is then normalized to a standard size and resolution. The goal of the next step is to extract the individual fuel price characters from the image. To achieve this, we first convert the color picture to a binary image and then employ connected component labeling to extract out the individual numeral characters of the fuel prices. The final step uses a Feedforward Backpropagation Neural Network algorithm for classifying the price numeral characters. The neural network utilizes a training template containing numeral characters compiled from a number of sample images of fuel price boards. A detailed description of all of the aforementioned steps can be found in the subsequent section.

Once the prices are extracted they are stored in a database which is linked to a GIS road network database populated with the locations of the service stations. This GIS database is consistent with that installed on the mobile phones. Since, each image is tagged with the time and location, the server updates the fuel prices of the appropriate service station in the database if the current image has a timestamp later than that of the stored prices. The past history of fuel prices at each service station is also recorded for analyzing pricing trends.

As discussed earlier, this paper primarily focusses on the data collection aspect of our system. We now briefly describe how our system deals with user queries. The central server implements a simple query resolver daemon for servicing pricing queries from users. A simple client program implemented on the mobile phone allows users to query for the cheapest refueling option in their vicinity. The client program obtains the current location coordinates of the user by querying the GPS receiver. The users' query tagged with his location information is then sent to the central sever. The server searches the fuel price database for the cheapest service station in the vicinity of the user and sends back a reply to the users mobile phone, which contains the price, location coordinates and brand of the chosen service station. In our future work, we also intend to implement support for more sophisticated queries such as "find the cheapest fuel station collocated with a convenience store". In addition, we plan on providing for alternate modes of access to the pricing information, such as via SMS and a web-based interface.

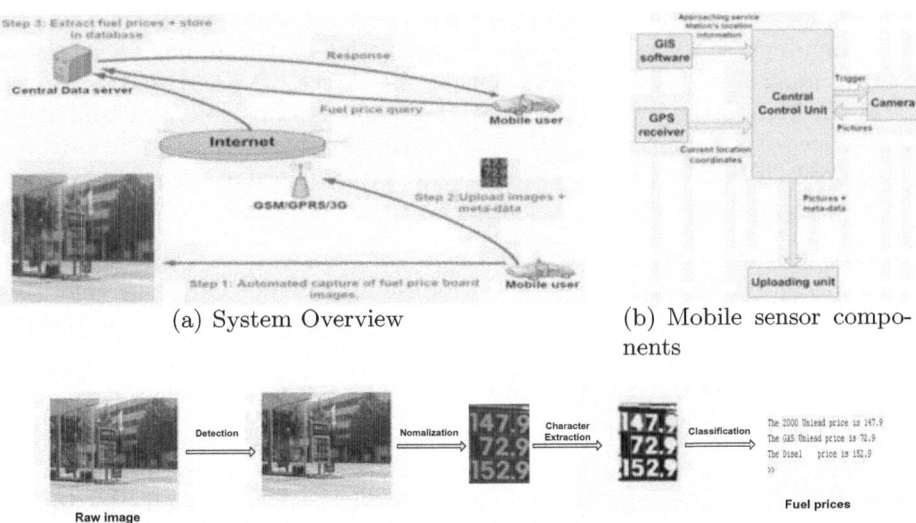

(a) System Overview

(b) Mobile sensor components

(c) Overview of the computer vision algorithm

Fig. 1. System design

3   Computer Vision Algorithm for Extracting Fuel Prices

Despite recent advances, computer vision still can not deal with complex environmental problems, though it can work accurately in controlled environments. In the following we present some of the key environmental challenges that are encountered by our system:

1. Some of the pictures may contain extraneous objects that overlap with the fuel price board. For example, in Fig.2(a) [6] a white car has masked the lower part of the board.
2. The color of some objects in the background may be similar to the color of the price board, E.g., the blue color of the wall in the background in Fig.2(b) is similar to that of the fuel price board. Consequently, the wall may be wrongly interpreted as the price board if color is used as the key feature in the detection process.
3. Since the images are captured while in motion, it is likely that some of them appear blurred or unfocussed. Figure 2(c) illustrates one such case.
4. Recognition and detection of road signs from digital images is a mature area of research [10,11,12]. However, a road sign differs significantly from a fuel price board as evidenced by comparing Fig.3(a) and Fig.3(b). A typical image of a fuel price board not only contains several small characters but also is more likely to have extraneous objects that could be wrongly interpreted as characters. For example, the cropped white label to the left of the price in the top row may be wrongly interpreted as "1". Figure 3(c) illustrates an instance where all individual characters are successfully extracted.
5. Computer vision algorithms used for extracting text from images consist of two important steps: (i) Detection of the target area and (ii) Extraction of the characters. It is likely that the quality of the images required for the successful completion of the individual tasks may vary significantly. For example, it may be fairly easy to detect the price board in a particular image. However, the text extraction may fail for the same image.

To overcome these and other challenges, our system makes use of certain contextual information by employing the assistance of GPS and GIS. Our algorithm consists of two key steps, which are described in the subsequent sub-sections.

3.1   Fuel Price Board Detection

Given an image, the task of detecting the presence of a fuel board and precisely identifying its location is quite difficult to solve with the current state-of-the-art in computer vision. Hence, our system relies on GPS and GIS to provide additional contextual information, which significantly simplifies the problem. As explained in Sect.2, the GIS system includes the service station brand as metadata with the image. Each fuel brand uses a specific color combination for its

---

[6] This and other images in Sect.3 are best viewed on a colour screen. Full size versions of these images can be found at the following link:
http://www.cse.unsw.edu.au/~ydon/dcoss08_images/

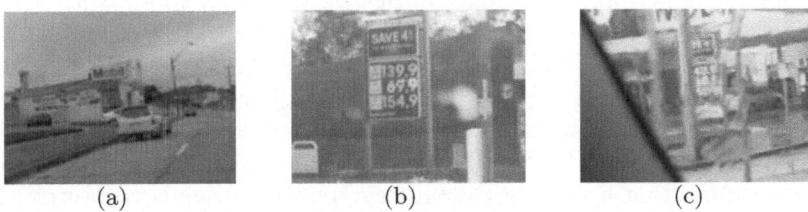

Fig. 2. Examples illustrating the challenges encountered by computer vision

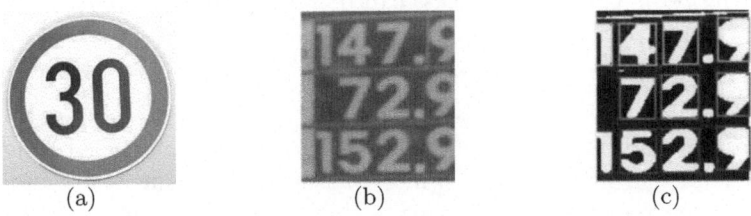

Fig. 3. Problems encountered in classifying fuel price characters

price boards, e.g.: BP price boards are green and Mobil boards use blue as the backdrop. We use this color information as a key feature for isolating objects in the picture which can be potential fuel price boards. In this paper we have only focussed on BP and Mobil, two of the largest fuel providers in Sydney. In our future work, we intend to encompass all the major providers in Australia. There are two prominent schemes for representing color images: RGB (Red, Green and Blue) and HIS (Hue, Intensity and Saturation). The HIS color space is immune to changes in the lighting condition, but is computationally intensive since it involves nonlinear transformation [10]. On the contrary, RGB is sensitive to changes in the lighting conditions, but is computationally efficient. Since our ultimate goal is to implement the computer vision algorithms directly on the mobile phones, we have chosen to use the RGB color space.

Figure 4 depicts the step-by-step evolution of the board detection algorithm for one particular example image. The original image, containing a Mobil board that serves as the input to the detection algorithm is shown in Fig.4(a). We use the Mobil board which is blue in color as an illustrative example to explain our algorithm. A similar logic was used to identify BP boards.

Since RGB is an additive color space, a blue component image can be readily obtained by boosting the blue level of each pixel and subtracting the corresponding green and red components as indicated by (1),

$$B(x,y) = 2 * f_b(x,y) - f_r(x,y) - f_g(x,y) \qquad (1)$$

where $f_b(x,y)$, $f_r(x,y)$ and $f_g(x,y)$ are, respectively, the functions representing the blue, red and green levels of each pixel in the image [10]. Figure 4(b) represents the blue component image of the original image in Fig.4(a). A similar approach can be employed for other colors.

Color thresholding [13] is one of the earliest techniques for detecting objects in digital images. Thresholding aims to classify all the pixels of an image into *object pixels*, which correspond to regions potentially containing the object and *background pixels*. Since the Mobil board is blue, we employ color thresholding with the criteria that a pixel in the blue component image is an object pixel if $B(x,y)$ exceeds a certain threshold, B_{th}. The output of the color thresholding produces a binary (i.e. black and white) image, with all object pixels taking on a value of 255 (note that, each pixel is represented by 8 bits) and background pixels being zero. The above process can be represented mathematically as,

$$B(x,y) = \begin{cases} 255, & \text{if } B(x,y) \geq B_{th}, \\ 0, & \text{else}, \end{cases} \tag{2}$$

In our first attempt at implementing color thresholding, we tried to use a singular value of the threshold B_{th}, which was calculated using a few test images. However, when we used this value for testing our dataset (details in Sect.4), we observed very poor results. This was expected given that the RGB color space is quite sensitive to lighting conditions. As an improvement we used a set of test images and manually classified them into 5 distinct groups representing different lighting conditions. We then computed the range of the average intensities (AI) of the blue component images for each group. Next, we determined the value of B_{th} for each group that achieved a 100% positive detection of the price board. Table 1 shows the range of AI values for each of the groups and the corresponding threshold value, B_{th}.

Table 1. Thresholds for images with different average intensities

Average Intensity Range	$AI \leq 3$	$3 < AI \leq 9$	$9 < AI \leq 15$	$15 < AI \leq 70$	$70 < AI$
B_{th}	0	20	50	80	120

Figure 4(c), shows the binary image produced by the color thresholding. Since, each row of the fuel prices is separated by a line, the fuel price board actually appears as three sub-divided regions in the binary image. These adjacent regions are merged together to form the complete price board, as indicated in Fig.4(d). Finally, connected component labeling [13] is employed to group the pixels into components based on pixel connectivity, i.e, all pixels in a connected component share similar pixel intensity values and are in some way connected with each other. Figure 4(e) shows the output of the the connected component labeling. One can readily observe that it contains several regions, which can be potential price boards. Our algorithm uses two post-processing techniques, which make use of certain known features (color and dimension) of the fuel price boards to narrow down the choices and correctly detect the price board.

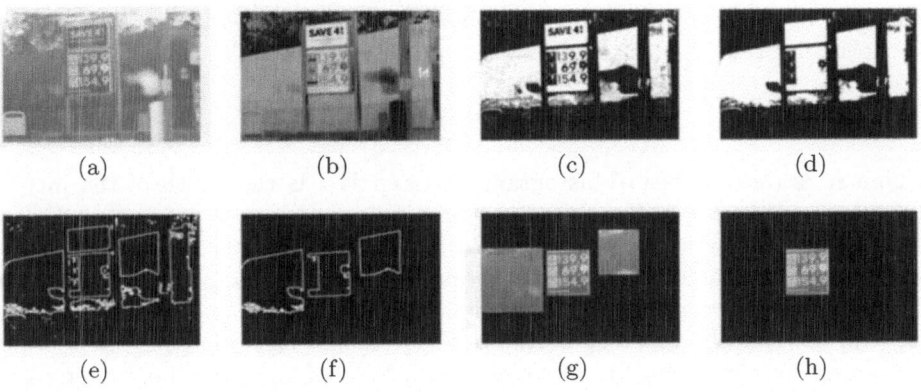

Fig. 4. Intermediate steps involved in the detection process

Dimension Comparison: The first step makes use of the a *priori* knowledge of the dimensions of a typical price board to exclude regions, which are obviously either too small or too large. However, determining a set of constraints based on dimensions can be tricky since the pictures are not captured from a static location. The camera can be at variable distances from the price boards when the pictures are captured. Further, the orientation of the camera lens relative to the price board may be different in each image. We have analyzed a set of test images of the price boards to encompass all these variables. Our test set includes close-up and far-away views of the price boards and pictures captured with the camera lens at different angles to the plane of the price board. Our analysis concludes that a candidate region of width W_i and height H_i can only qualify as a potential fuel price board, if the following constraints are met:

$$W_i \geq \frac{W_{\text{image}}}{30} \tag{3}$$

$$1 \leq \frac{H_i}{W_i} < 2.5 \tag{4}$$

where W_{image} is the width of the image containing the candidate region. Figure 4(f) is the output of the first post-processing step. As compared to the input image in Fig.4(e), the number of candidate regions have now been reduced to 3.

Color Histogram Comparison: The second post-processing technique compares the histogram distribution of the candidate regions to that of a template of the price board. The Chi square distance [14] between h_i, the histogram of the candidate region and h_j, the histogram of the price board template is computed as follows,

$$\chi^2(h_i, h_j) = \frac{1}{2} \sum_{m=1}^{K} \frac{\left[h_i(m) - h_j(m)\right]^2}{h_i(m) + h_j(m)} \tag{5}$$

where K is the number of histogram bins. As χ turns out to be a large number, we normalize this distance as follows,

$$D_{norm} = \frac{\chi}{K \times WI} \tag{6}$$

where K is the number of histogram bins and WI is the width of the image. In our system, we set K as 60, since it provides for good accuracy and low complexity. WI is set to the standard value of 480 pixels. To determine an appropriate threshold value for the normalized distance, D_{norm}, we used a test set containing 10 cropped images that include the fuel price board and 10 images that do not contain one. The normalized distance of each image in the set was computed by comparing its histogram with that of the price board template using 6. Based on the test results, we conclude that if a candidate region has $D_{norm} \leq 2.5$, then it is very likely that this region corresponds to the fuel price board. Figure 5(a) shows the template used in our tests and Fig.5(d) represents its color histogram. Figure 5(b) is a cropped image from the test set that contains the fuel price board and its color histogram is plotted in Fig.5(e). D_{norm} for this particular image is 1.6. Figure 5(c) represents an image from the test set that does not contain the price board and its corresponding histogram is shown in Fig.5(f). D_{norm} for this image is 3.9, considerably higher than the threshold of 2.5. Figure 4(h) is the output of applying the histogram comparison. The fuel price board has now been correctly identified. This concludes the detection phase. The next step is to extract the fuel prices from the board.

3.2   Fuel Price Classification

The last section describes an algorithm on detecting the fuel price board from the background of the image. This section describes the procedure to extract and classify the numeral characters which make up the fuel prices in a price board. The procedure consists of two steps: extraction of the characters and followed by classification by a neural network.

Character extraction: The fuel price boards are usually designed such that there is sufficient contrast between the characters and the background color. As a result, an image of the fuel price board does not contain excessive noise. Therefore, it is sufficient for us to use a binary image (instead of full color image) for character extraction which can significantly reduce the amount of processing. For character extraction, we employ the bounding box algorithm in [15]. The cropped (or extracted) characters are then normalised to a standard size of 50-by-70 pixels. For the construction of feature vectors for character recognition, we divide each 50-by-70 pixel character into 35 10-by-10 pixel regions and compute the average intensity in each region. Thus, each extracted character will be represented by a 35×1 feature vector.

Character Recognition: For character recognition, we employ the Feed-forward Back-propagation Neural Networks (FFBPNN) due to its simple implementation and robustness to interference [16]. The FFBPNN is first trained

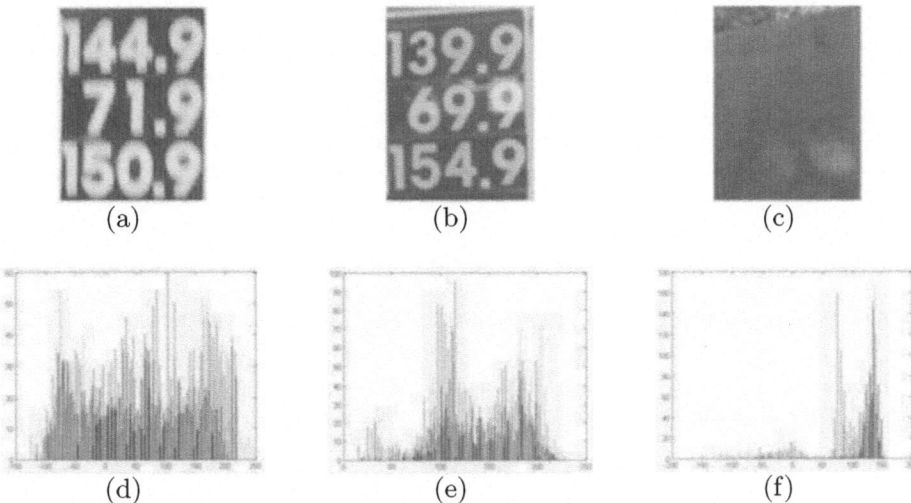

Fig. 5. Color histograms for a few example images

by using characters extracted from 20 sample fuel price boards. The feature vector construction procedure is identical to the one described in the last paragraph. Thus, each character in the training set is represented by a 35×1 feature vector.

For the identification of fuel type, we have, for the time being, chosen to use the *a priori* knowledge of the order in which different types of fuel are arranged for a particular chain of service stations. For example, all Mobil stations in Australia arrange the prices of fuel in the following order: unleaded fuel, Liquid Petroleum Gas (LPG) and diesel. Our future work will integrate fuel type identification into our system instead of depending on *a priori* knowledge.

4   Evaluation and Discussion

In order to evaluate our system, we have set up a data set with 52 images of 5 Mobil and 3 BP service stations. These images were captured by either a 5-megapixel Nokia N95 mobile phone camera or a 4-megapixel Canon IXUS 400 camera. The cameras were held by a passenger sitting in the front passenger seat of a moving car. Our data set covers many different conditions — relative distance of the price board from the vehicle (close or far away), weather conditions (sunny, cloudy) and daylight disparities — in order to test the generality of our computer vision algorithm. Each image in the data set contains one fuel price board. There are 3 fuel prices per board and on average there are 11 numeral characters per board. The size of each image is 640-by-480 pixels. Our system is implemented in Matlab and tested on a 2.0-GHz Pentium centrino duo laptop.

We first evaluate the performance of our price board detection algorithm (Sect.3.1) using our data set. We classify an image as "close" if the price board occupies more than $\frac{1}{8}$ of the size of the image; otherwise, it is classified as "far away". The performance of our detection algorithm is shown in Table 2 where a hit (or a true positive) is obtained when a price board is correctly detected; otherwise, it results in a miss or a false positive. The overall hit rate of 96% is very high considering the many possible different backgrounds that these images can have. In addition, both "close" and "far away" images include images that are taken under many different lighting conditions. This shows that our algorithm is robust against lighting variations. In our next step of evaluation, we passed

Table 2. Test results of automatic fuel price board detection on 52 images

Size of the board	No. of fuel price board	Hit Rate	Miss or False Positive
Close	31	29/31= 93.5 %	2/31= 6.5 %
Far away	21	19/21=90.5%	2/21= 9.5%
Overall	52	92.3%	7.7 %

those 48 images whose price board we have successfully detected to the fuel price classification module described in Sect.3.2. Out of these 48 images, 15 of them contain a price board that is so blurry that even human cannot read the fuel prices. We input these 15 blurry images to our algorithm and it did not return any prices, thus giving a true negative rate of 100%. We will now focus on the remaining 33 images which consist of 15 Mobil and 18 BP fuel price boards. There are a total of 330 characters and 99 fuel prices (a fuel price consists of either 3 or 4 characters) in these 33 fuel price boards. We measure the correct extraction rate and correct classification rate of both the characters and prices. The results are shown in Table 3. The overall correct classification rate for fuel prices is about 88% and is pretty high. Note that the classification result for Mobil boards is always lower than that of BP, because the classification algorithm mistook the white border of the Mobil price board as a character in some of the cases.

Table 3. Test results of neural network based character recognition method based on 43 fuel price boards

Brand	No. of fuel price boards	No. of characters	No. of fuel prices	Extraction rate of characters	Classification rate of characters	Extraction rate of fuel prices	Classification rate of fuel prices
BP	18	165	54	100%	99.3%	98.1%	91%
Mobil	15	165	45	92%	89%	91.1%	84.4%
Overall	33	330	99	96%	94%	94.6%	87.7%

5   Related Work

Wireless sensor networks: Most of the existing research work on WSNs utilises small form factor and power constrained nodes [17] for sensing and communications. However, the WSN community is beginning to appreciate the value of using vehicles, mobile phones, GPS, cameras etc to collect everyday information. A vehicular WSN aims to exploit sensors mounted on a vehicle to collect sensing data. An example of a vehicular WSN is Cartel [18] which collects information on traffic conditions, WiFi availability and potholes on the road. Other work in vehicular sensor networks focus on routing or efficient information dissemination, see [19] or references therein. The wide-spread availability of mobile phones and cameras forms the basis of Participatory Sensing [20] for collecting data on our urban environment, e.g. air pollution [21], cyclist experience [22] as well as the on-line proceeding of a recent workshop in the area[7]. In addition, the Nokia Sensor Planet[8] initiative aims to explore the use of camera phones as mobile sensing devices. Our work also leverages on the Participatory Sensing concept but with an emphasis on users contributing and sharing data [9]. Our work can make use of the methods for authenticating the location and timing of participatory sensing data discussed in [20]. However, our work differs from existing work in vehicular WSNs and Participatory Sensing in that we focus on the challenges in automatically collecting consumer pricing data.

Detection and recognition of objects from digital images: Automatic detection of road traffic signs is a mature area of research [12,23,24,25]. There have also been several recent successful attempts at detecting text from a diverse set of digital sources such as video, newspapers, advertisements and photographs [26]. However, most of these algorithms require a very large set of training samples (up to 1000 images in some cases). Consequently the training process takes up a large amount of time. Given the difficulty in populating such a large training set for our system, we have chosen to use a much simpler and computationally efficient approach, which can achieve a high success rate. The simplicity of our algorithm will also help our future direction, since we wish to migrate some of the computer vision tasks to run on the resource constrained mobile devices.

Commercial OCR (Optical Character Recognition) software has been successfully used in some instances to realize text recognition [26,24,27]. An OCR system makes certain assumptions about the source of the text and works well only with standard layouts and fonts. However, it is unable to cope with unstructured environments. Given that our system has to work in an outdoor environment, which is known to involve a large number of variables (lighting condition, weather, distance, etc), we cannot use commercial OCR software.

6   Conclusions and Future Work

This paper has proposed and described a mobile WSN that allows road-users to automatically collect, contribute and share fuel price information. It represents, to the best of our knowledge, the first WSN application for collecting consumer

---

[7] http://urban.cens.ucla.edu/sensys07/index.php?
  title=Accepted_Papers_and_Workshop_Agenda
[8] http://research.nokia.com/research/projects/sensorplanet/index.html

pricing information. Our proposed system is based on three key ideas. Firstly, it provides a platform for people to contribute and share sensing data, which can be viewed as the BitTorrent of sensing data. Secondly, our proposed system leverages on the existing sensing and communication infrastructure and as a result, lowers the barrier for a volunteer to share sensing information. Thirdly, the use of computer vision algorithms to extract the fuel price information from the contributed images. In this paper, we have focused on defining the architecture of our proposed system and describing the computer vision algorithm. Our preliminary experimental results show that our system is able to correctly classify 87.7% of the fuel prices from 33 images of BP and Mobil service stations. Our future work aims to produce a complete prototype of the proposed system which integrates GPS, GIS and automatic camera triggering. We will also investigate the effect of GPS accuracy on the performance of our system. Note that, privacy can be a concern since the users upload their location along with the pricing information. We intend to explore various options, including anonymising user identities, to address this issue. We also plan to test our algorithm on other service station chains. In addition, all the computer vision processing is currently performed on the server and we plan to test how much of the processing can be performed on the current generation of mobile devices. Lastly, we would like to point out that our proposed system is also applicable for other applications. For example, GIS providers can utilise user contributed images of street level view to update their POI instead of relying on manual update.

Acknowledgements

We would like to thank Maria Myung-Hee Kim and Chen Zhang for coding the classification algorithm, Gina Lam for taking the images, and Rajib Rana and Assad Mehmood for helpful discussion. This research is funded by the Australian Research Council Discovery Grant DP0770523.

References

1. Wark, T., et al.: The design and evaluation of a mobile sensor/actuator network for autonomous animal control. In: Proceedings of Information Processing in Sensor Networks (IPSN 2007/SPOTS 2007) (April 2007)
2. Kim, S., Pakzad, S., Culler, D., Demmel, J., Fenves, G., Glaser, S., Turon, M.: Health monitoring of civil infrastructures using wireless sensor networks. In: 6th International Symposium on Information Processing in Sensor Networks (IPSN), April 25-27, pp. 254–263 (2007)
3. Lédeczi, Á., et al.: Countersniper system for urban warfare. ACM Transactions on Sensor Networks 1(2), 153–177 (2005)
4. Hu, W., Tran, V.N., Bulusu, N., Chou, C.T., Jha, S.: The design and evaluation of a hybrid sensor network for cane-toad monitoring. In: Proceedings of Information Processing in Sensor Networks (IPSN 2005/SPOTS 2005) (April 2005)
5. Baker, C.R., et al.: Wireless sensor networks for home health care. In: 21st International Conference on Advanced Information Networking and Applications Workshops (AINAW), vol. 2, pp. 832–837 (2007)

6. Singhvi, V., Krause, A., Guestrin, C., James, H., Garrett, J., Matthews, H.S.: Intelligent light control using sensor networks. In: Proceedings of the 3rd international conference on Embedded networked sensor systems (SenSys), pp. 218–229. ACM, New York (2005)
7. Baye, M., Morgan, J., Scholten, P.: The value of information in an online consumer electronics market. Journal of Public Policy and Marketings 3, 481–507 (2003)
8. Brown, J., Goolsbee, A.: Does the internet make markets more competitive? Evidence from the life insurance industry. Journal of Political Economy 1, 17–25 (2002)
9. Chou, C.T., Bulusu, N., Kanhere, S.: Sensing data market. In: Proceedings of Poster Papers of 3rd IEEE International Conference on Distributed Computing in Sensor Systems (DCOSS 2007) (June 2007),
   http://www.dcoss.org/dcoss07/dcoss07posterproceedings.pdf
10. Garcia, M., Sotelo, M., Gorostiza, E.: Traffic sign detection in static images using matlab. In: IEEE Conference on Emerging Technologies and Factory Automation (ETFA), September 16-19, vol. 2, pp. 212–215 (2003)
11. Paulo, C.F., Correia, P.L.: Automatic detection and classification of traffic signs. In: Eighth International Workshop on Image Analysis for Multimedia Interactive Services (WIAMIS), June 6-8, pp. 11–11 (2007)
12. Bahlmann, C., Zhu, Y., Ramesh, V., Pellkofer, M., Koehler, T.: A system for traffic sign detection, tracking, and recognition using color, shape, and motion information. In: Proceedings of Intelligent Vehicles Symposium, June 6-8, pp. 255–260 (2005)
13. Gonzalez, R.C., Woods, R.E.: Digital Image Processing. Prentice-Hall, Englewood Cliffs (1992)
14. Ye, N., Borror, C.M., Parmar, D.: Scalable chi-square distance versus conventional statistical distance for process monitoring with uncorrelated data variables. Quality and Reliability Engineering International 19(6), 505–515 (2003)
15. Yuan, B., Kwoh, L.K., Tan, C.L.: Finding the best-fit bounding-boxes. Document Analysis Systems VII 3872/2006, 268–279 (2006)
16. Jenq, J.J., Li, W.: Feedforward backpropagation artificial neural networks on reconfigurable meshes. Future Generation Computer Systems 14(5-6), 313–319 (1998)
17. Karl, H., Willig, A.: Protocols and Architectures for Wireless Sensor Networks. Wiley, Chichester (2006)
18. Hull, B., Bychkovsky, V., Zhang, Y., Chen, K., Goraczko, M., Miu, A., Shih, E., Balakrishnan, H., Madden, S.: Cartel: a distributed mobile sensor computing system. In: Proceedings of the 4th international conference on Embedded networked sensor systems (SenSys), pp. 125–138. ACM, New York (2006)
19. Lee, U., Zhou, B., Gerla, M., Magistretti, E., Bellavista, P., Corradi, A.: Mobeyes: smart mobs for urban monitoring with a vehicular sensor network. IEEE Wireless Communications 13(5), 52–57 (2006)
20. Burke, J.A., Estrin, D., Hansen, M., Parker, A., Ramanathan, N., Reddy, S., Srivastava, M.B.: Participatory sensing. In: WSW 2006 at SenSys 2006. ACM, New York (October 31, 2006)
21. Paulos, E., Honicky, R., Goodman, E.: Sensing atmosphere. In: Workshop on Sensing on Everyday Mobile Phones in Support of Participatory Research (2007),
   http://urban.cens.ucla.edu/sensys07/index.php?
   title=Accepted_Papers_and_Workshop_Agenda

22. Eisenman, S.B., Miluzzo, E., Lane, N.D., Peterson, R.A., Ahn, G.S., Campbell, A.T.: The bikenet mobile sensing system for cyclist experience mapping. In: Proceedings of the 5th ACM International Conference on Embedded Networked Sensor Systems (SenSys 2007), pp. 87–101 (2007)
23. Wu, W., Chen, X., Yang, J.: Dtection of text on road signs from video. IEEE Transactions on Intelligent Transportation Systems 6, 378–390 (2005)
24. Chen, X., Yuille, A.: Detecting and reading text in natural scenes. In: Proceedings of the 2004 IEEE Computer Society Conference on Computer Vision and Pattern Recognition (CVPR), 27 June-2 July, vol. 2, pp. 366–373 (2004)
25. de la Escalera, A., Armingol, J., Mata, M.: Traffic sign recognition and analysis for intelligent vehicles. Image and Vision Computing 21(3), 247–258 (2003)
26. Jain, A., Yu, B.: Automatic text location in images and video frames. In: Fourteenth International Conference on Pattern Recognition, August 16-20, vol. 2, pp. 1497–1499 (1998)
27. Chen, X., Yang, J., Zhang, J., Waibel, A.: Automatic detection and recognition of signs from natural scenes. IEEE Transactions on Image Processing 13(1), 87–99 (2004)

Techniques for Improving Opportunistic Sensor Networking Performance

Shane B. Eisenman[1], Nicholas D. Lane[2], and Andrew T. Campbell[2]

[1] Columbia University, New York NY 10027, USA
shane@ee.columbia.edu
[2] Dartmouth College, Hanover NH 03755, USA
{niclane,campbell}@cs.dartmouth.edu

Abstract. A number of recently proposed mobile sensor network architectures rely on uncontrolled, or weakly-controlled mobility to achieve sensing coverage over time at low cost, an opportunistic sensor networking approach. However, this reliance on mobility also introduces a number of challenges. In this paper, we discuss the challenges inherent in this networking paradigm, and describe two composable techniques, sensor sharing and substitution, to make the system more robust in terms of data fidelity and delay. We present a numerical analysis of these techniques, separately and in combination, based on a simple Markov model of an opportunistic sensor network.

Keywords: Architecture, Mobility, Modeling, Performance, Wireless Sensor Networks, Opportunistic Networking.

1 Introduction

The recent integration of sensors with personal electronic devices like mobile phones has invited a number of researchers to consider appropriate architectures [4] [12] [23] [1] and applications (e.g., social [16] [25], recreational [8] [9]) for large-scale people-centric sensing systems. Generally, these systems leverage human-carried or vehicle-mounted sensors networked using short/mid-range radios (e.g., ZigBee, WiFi, Bluetooth), and an Internet gateway tier composed of tasking and collection entities. The gateway tier delivers sensing instructions to the mobile sensors on behalf of user applications and accepts incoming sensed data. These proposals rely to some extent on the mobility of humans and their vehicles to get wide area sensing coverage over time with a relatively sparse deployment of heterogeneous mobile sensors. We term sensing with this dependence on uncontrolled mobility *opportunistic sensor networking (OSN)*. While this novel OSN approach can allow large scale sensing at a lower cost compared to an ubiquitous static infrastructure of sensing devices, the opportunistic nature of sensing and communication presents challenges to the fundamental sensor networking operations. In the OSN approach, these operations can be described in terms of opportunistic tasking, opportunistic sensing and opportunistic collection. Opportunistic tasking refers to the process by which a tasking entity

S. Nikoletseas et al. (Eds.): DCOSS 2008, LNCS 5067, pp. 157–175, 2008.
© Springer-Verlag Berlin Heidelberg 2008

instructs an appropriate mobile sensor to attempt to meet a certain application request. The tasking is opportunistic since there is no guarantee that an appropriate mobile sensor will stay within the radio range of a tasking entity long enough for the tasking operation to complete. By "appropriate", we refer minimally to a mobile sensor that has the necessary sensing equipment to meet the application request, and may include other requirements (e.g., remaining energy, security clearance, inferred direction of motion). Opportunistic sensing refers to the process by which a mobile sensor that has been assigned a given application task senses the target within the preferred time frame. The sensing is opportunistic since the tasked mobile sensor may not move close enough to the target within the preferred time frame. Opportunistic collection refers to the process by which a mobile sensor that has sensed data in line with the requirements of an application request delivers this data to a collection entity. The collection is again opportunistic since the mobility of the mobile sensor that has sensed the target may not bring it within the radio range of the collection entity and keep it there long enough for the sensed data upload operation to complete.

Noting the aforementioned challenges of OSNs, in [10], we define in situ *sensor sharing* in the context of real sensing applications, and design, implement and experimentally evaluate the system performance for these application scenarios. In this paper, we augment that experimental work with a theoretical analysis of the properties (e.g., scalability and sensitivity to device heterogeneity) of sensor sharing, alone and in combination with *sensor substitution*. These two composable techniques aim to increase the robustness of the OSN paradigm, mitigating the fundamental challenges of uncontrolled human mobility and device heterogeneity to successfully and more expediently complete tasking, sensing and data collection. In particular, we effectively loosen the constraints on which mobile sensors are fit to be tasked for a particular application query. Sensor sharing does this by allowing tasked sensors without the right sensor type for a given sensing task to exploit the resources of others it encounters in the field. Sensor substitution is used in situations where one measurements from one sensor type can act as a reasonable (i.e., within fidelity bounds acceptable to the application) stand-in for another's. In Section 2, we give a more detailed description of each technique in the context of the OSN challenges it addresses. In Section 3, we provide a baseline numerical analysis of an OSN using a Markov model, and then provide an analysis of the potential improvement provided by sensor sharing and sensor substitution, first separately and then in combination, with respect to the Markov model. Section 4 discusses related work before we conclude.

2   Sensor Sharing and Substitution

To meet the requirements of the application request, a tasking entity must choose an appropriate mobile sensor from the pool of available mobile sensors in its radio range. However, this sensor selection problem is difficult for two reasons: (i) the available pool of mobile sensors is limited by the uncontrolled mobility of humans and vehicles and may not contain an appropriate mobile sensor, and (ii) it is difficult to predict whether the mobility of a given available mobile sensor will keep it within the tasking entity's radio range long enough to complete the tasking, and take it to the target region within the preferred time window. For convenience, we term these the *tasking availability problem* and the *tasking prediction problem*, respectively.

The tasking availability problem can be addressed by relaxing the requirements on what constitutes an appropriate mobile sensor, probabilistically increasing the chances that a "taskable" mobile sensor will enter the radio range of the tasking entity within the preferred time window. In particular, we propose to relax the requirement on sensing instrumentation by allowing for *sensor sharing* and *sensor substitution*. With sensor sharing, a mobile sensor A that requires sensor type α, e.g., a CO_2 sensor, to meet the application requirements, but does not itself possess this sensor type (α might be expensive, heavy or rare), can conscript another sensor B that does possess sensor type α to share its sensed data. We envision two scenarios for sensor sharing: (i) sensor A encounters sensor B at the target region, asks B to capture and share data from sensor type α, and receives this data from B, and (ii) sensor A encounters sensor B outside the target region, and leverages the approach proposed in [17] to request and receive data from sensor type α from sensor B. As an example of the former scenario, for an allergen mapping application, a mobile sensor A may be instructed to record particulates in the vicinity of a busy street intersection in the center of the city. Lacking a particle counter, when sensor A finds itself at the target intersection it might broadcast a request for a particulate reading and receive a response from sensor B which is within radio range of A and possesses a particle counter (pull-based sharing). In a more constrained case, a system might be engineered such that at least one mobile sensor Q with sensor type α is present among a group of mobile sensors that might require data from sensor type α. Here, node Q might periodically broadcast readings from sensor type α to the group (push-based sharing). An investigation of a more sophisticated communication protocol in support of sensor sharing, and algorithms to decide which of the possible neighbors is most appropriate to share from is left as future work. We assume an environment where sensors are willing to cooperate to complete sensing tasks. The cooperation might be pro bono, or quid pro quo as in social-network-based sharing [16].

With sensor substitution, a mobile sensor C that requires data from sensor type β to meet the application requirements but does not possess this sensor type, can instead use a substitute method of acquiring equivalent or similar data. We envision two scenarios for sensor substitution that we term *direct substitution* and *indirect substitution*. In direct substitution, a mobile sensor C instructed to collect data from sensor type β uses sensor type γ to collect data equivalent to that given by β, where equivalence does not necessarily include accuracy or precision specifications. For example, for a simple terrain mapping application a mobile sensor may be instructed to measure the slope of a given section of road using a three dimensional accelerometer. Lacking a three axis accelerometer, but possessing a GPS receiver (e.g., the Nokia N95), the mobile sensor can use interpolation between periodic altitude readings from the GPS receiver to calculate and report the slope of the road [26]. Note that GPS-derived altitude measurements are less accurate and less precise that those provided by a three axis accelerometer, but nonetheless for the case of road slope mapping the GPS receiver can act as a direct substitute for the three axis accelerometer. In indirect substitution, a mobile sensor C instructed to collect data from sensor type β instead reports a combination of sensed data from a number of other sensor types that can be used to generate, e.g., using inference techniques, data which is similar to the requested data. For example, for a simple location mapping application a mobile sensor may be instructed to periodically record latitude/longitude readings from a GPS receiver. Lacking a GPS receiver, but possessing a three axis magnetometer and a three axis accelerometer, the mobile sensor can report direction relative to the magnetic field of the Earth (i.e., a compass reading), and distance calculated as the double integral of the acceleration. With knowledge of at starting location, the latitude/longitude values can be approximated using the combination of the the direction and distance traveled (i.e., dead reckoning [21]). Note that localization using dead reckoning is less accurate than localization using data from a GPS receiver, but for localization over relatively short distances the combination of a three axis magnetometer and a three axis accelerometer can act as an indirect substitute for a GPS receiver. Another example is recognizing locations that can be uniquely identified by their sensor signature [13], as a substitute to GPS.

Note that the use of sensor sharing or sensor substitution is not exclusive, but rather both are composable blocks that can be used in combination to increase the probability an application request is met. The sensing action that is likely to yield the higher fidelity data is taken. Extending the previous example, if the terrain mapping application requests a slope measurement, we first check if the hardware natively supports high accuracy slope measurement via accelerometer. If not, it tries sensor sharing to see if another nearby node (with appropriate context) will share its accelerometer. If not, it tries sensor substitution to at least get a (lower accuracy) estimate of slope via GPS. Thus, success is achieved if the tasked mobile node has the required sensor, or if it has an appropriate substitute sensor, or if it meets a mobile node at the right place that is willing to share the required sensor, with the commensurate impact on data fidelity. In this way, sensor sharing and substitution help to decouple application design from hardware design (i.e., sensors, on board and opportunistically encountered as people rendezvous in the field), helping to support multiple applications on heterogeneous mobile devices.

3   Analysis and Discussion

To get a preliminary understanding of the baseline performance of an OSN and the theoretical impact of sensor sharing and sensor substitution, we model an OSN scenario using mobile sensor nodes moving according to a discrete Markov model. In the following, we develop the baseline OSN model, and then compare

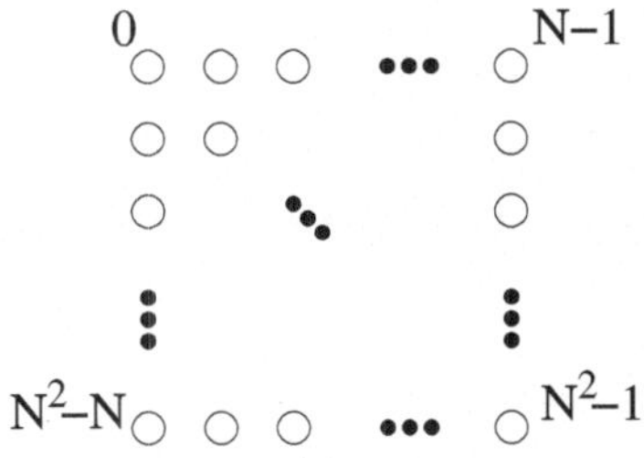

$$
\begin{array}{ccccc}
\overset{0}{\bigcirc} & \bigcirc & \bigcirc & \bullet\bullet\bullet & \overset{N-1}{\bigcirc} \\
& \bigcirc & \bigcirc & & \bigcirc \\
& \bigcirc & & & \bigcirc \\
& & \ddots & & \vdots \\
\vdots & & & & \vdots \\
\underset{N^2-N}{\bigcirc} & \bigcirc & \bigcirc & \bullet\bullet\bullet & \underset{N^2-1}{\bigcirc}
\end{array}
$$

(a)

$$
p_{i,j} = \begin{cases}
\frac{1}{5}, & j = \{\alpha + (\beta + 1) \bmod N, & \text{(right)} \\
& \alpha + (\beta + N - 1) \bmod N, & \text{(left)} \\
& (i + N) \bmod N^2, & \text{(below)} \\
& (i + N^2 - N)) \bmod N^2, & \text{(above)} \\
& i\}, & \text{(stay)} \\
0, & \text{otherwise,}
\end{cases}
$$

where $\alpha = \lfloor \frac{i}{N} \rfloor \cdot N$, and $\beta = i \bmod N$ for $N > 2$.

(b)

Fig. 1. An N^2 element Markov chain models a neighborhood where the states represent a grid of points covering the 2-D ground surface of the neighborhood. The grid points are numbered as shown in Figure 1(a). We study a toroidal scenario where nodes may move north, east, south, west, or remain stationary with equal probability (Figure 1(b)); and a more realistic scenario where transition probabilities are derived from the connectivity graph shown in Figure 2.

the performance of sensor sharing and sensor substitution with the baseline in terms of sensing success probability. We numerically evaluate the derived probability expressions to get a sense of the performance boost given by sensor sharing and substitution, and the sensitivity to the number of mobile sensor nodes and the number of sensors per node.

3.1 Model

We model a neighborhood with a Markov chain with a state space SS of size N^2. We investigate two topology scenarios. In the first, the states represent an $N \times N$ toroidal grid of points covering the 2-D ground surface of the neighborhood. The grid points are numbered as shown in Figure 1(a). We assign the transition probabilities of the Markov matrix $\mathcal{M}$ to allow mobile sensor nodes to move north, east, south, west, or remain stationary with equal probability (see Figure 1(b)). To investigate a more realistic application of the model, in the second topology scenario we overlay the $N \times N$ grid on a physical map of the northwest corner of the Columbia University Morningside campus [6] and assign transition probabilities that respect human pathways in the actual campus. To start, the map (see Figure 2) is sectioned into a 10x10 grid (solid lines), and a connectivity graph (solid squares in the center of each grid square, and dotted lines) of the sections is derived on the basis of walls, doors, pathways, etc. Then, each of the nodes in the graph is treated as a state in the Markov chain, where at each node in the graph each edge (including the implicit self-edge) is taken with equal probability.

Suppose that a query injection point at location σ receives a query from an application running on a back end server at time t for information from sensor type s at target location τ in the grid. Suppose that the request has deadline u, such that the mobile sensor node must be tasked with the request by the query injection point and arrive at the sensing target location τ by time

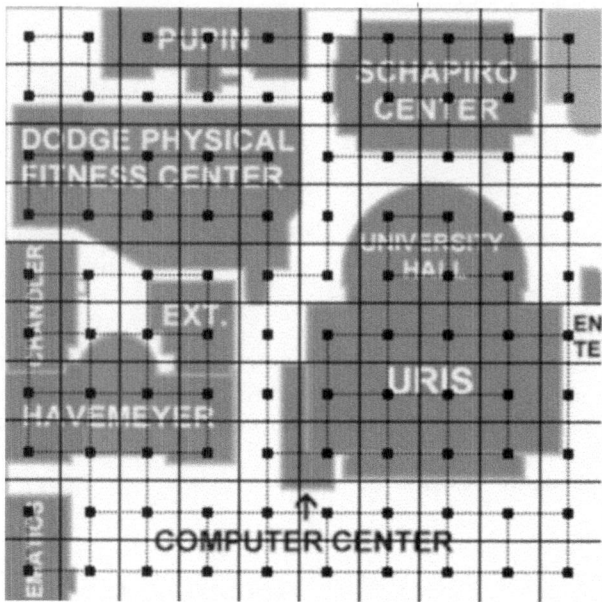

Fig. 2. A partial campus map showing the northwest corner of the Columbia University campus. The map is sectioned into a 10x10 grid (solid lines), and a connectivity graph (solid squares in the center of each grid square, and dotted lines) of the sections is derived on the basis of doors, pathways, etc. in the actual campus. This graph serves as the basis for the campus scenario probability transition matrix used in Section 3.2.

$t + u$. We assume there are sensors $\mathcal{S} = \{s_1, ..., s_z\}$ and that each mobile sensor node possesses $r \leq z$ of these sensors through random assignment (e.g., at the factory), with the constraint that the r sensors are distinct (i.e., there are $\binom{z}{r}$ sensor configurations).

3.2   Baseline OSN

We wish to determine the probability of success as a function of the node density, sensor configuration, and sensing deadline in this Markov state space. We start by determining the probability that a suitable mobile sensor node will first visit the query injection point at time $t + k, k \leq u$. We assume that t is large enough that the population is well mixed, i.e., the probability that a given mobile sensor node is in a given state i at time t is $p_i = \frac{1}{N^2}$, or equivalently that the initial probability distribution for all states i is $\nu_i = \frac{1}{N^2}$. Let $A_k(i, \sigma)$ denote the event that a single mobile sensor node starting at location i will first visit the query injection point σ at time $t+k$. The probability $F_k(i, \sigma) = Prob(A_k(i, \sigma))$ is given by the recursion

$$F_k(i, \sigma) = \begin{cases} \mathcal{M}(i, \sigma), & k = 1, \\ \displaystyle\sum_{b \in SS - \{\sigma\}} \mathcal{M}(i, b) F_{k-1}(b, \sigma) & k \geq 2. \end{cases}$$

or equivalently as

$$F_k(i, \sigma) = \mathcal{Q}^{k-1} \mathcal{M}(i, \sigma). \tag{1}$$

where $\mathcal{M}(i, j)$ represents the (i, j)th entry of the Markov probability transition matrix, and $\mathcal{Q}$ is the matrix obtained from $\mathcal{M}$ by replacing its column σ with all zeros. Summing over all starting positions i, we have

$$F_k(\sigma) = \frac{1}{N^2} \sum_{i=0}^{N^2 - 1} \mathcal{Q}^{k-1} \mathcal{M}(i, \sigma), \tag{2}$$

However, we are only interested in those mobile sensor nodes that are equipped with the sensor s that can meet the application/query request (else we assume the mobile sensor node will not be tasked by the query injection point). We model the query arrival process by assuming that a $s \in \mathcal{S}$ is chosen uniformly at random for each application query, such that $Prob(s = s_i) = \frac{1}{z}$. Let B denote the event that a given mobile sensor node's sensor configuration includes the sensor $s \in \mathcal{S}$ specified in the query. Then,

$$Prob(B) = \sum_{i=1}^{z} Prob(s = s_i) \frac{\binom{z-1}{r-1}}{\binom{z}{r}} = \frac{r}{z}. \tag{3}$$

Let C denote that event that a mobile sensor node equipped with the proper sensor first visits the query injection point at time $t + k$. Since the mobile sensor node's sensor configuration assignment and the mobile node's motion on the neighborhood grid are independent, the probability of event C is simply obtained from Equations 2 and 3 as $Prob(C) = Prob(B) \cdot F_k(\sigma)$.

Once the appropriate mobile sensor is tasked by the query injection point at location σ, it must reach the target at location τ by the sensing deadline $t + u$ in order for the mission to be successful. Thus, we can write the probability of success in the baseline OSN as

$$Prob(Success) = \sum_{l=1}^{u} Prob(C) \sum_{m=1}^{u-l} F_m(\sigma, \tau), \tag{4}$$

where l is the time when the SAP is visited.

Finally, suppose there are y mobile sensor nodes moving in the neighborhood grid. We write the probability that any of the y mobile sensor nodes succeed as

$$Prob^{(y)}(Success) = 1 - (1 - Prob(Success))^y). \tag{5}$$

As the deadline goes to infinity, the probability of success is limited by $1 - (1 - \frac{r}{z})^y$, regardless of the grid dimension N. This limitation imposed by the probability of a matching sensor configuration ($\frac{r}{z}$) strongly motivates the consideration of both sensor sharing and sensor substitution.

In Figure 3, we compare the success probability for both of the baseline OSN scenarios described previously, i.e., toroid and campus. The tasking point is placed in the lower right corner of the neighborhood (Markov state 99) and the sensing target is placed in the center of the neighborhood (Markov state 45). For both scenarios, we choose a grid granularity of $N = 10$ and use $z = 20$ based on the set of sensors currently used in personal mobile devices and personal sensing systems (viz., camera, microphone, Bluetooth, WiFi, accelerometer, GPS, and temperature samples can be taken from the Nokia N95; the Moteiv Tmote Invent includes light and humidity sensors; magnetometer, galvanic skin resistance, and CO_2 sensors are used in the BikeNet sensing system [8]; the Intel Mobile Sensing Platform includes a barometer and a gyroscope; and the Apple iPhone adds an FM transceiver and an infrared sensor). Future hardware generations will be even more sensor rich, but it is unlikely that any device will integrate all available sensors. In Figure 3(a), we plot Equation 5 versus a range of sensing deadlines u for different numbers of mobile sensor nodes y. Here the ranges of tested y values is meant to represent the rough number of participating users of the OSN in the area of campus shown in Figure 2 throughout the day (y=200) and night (y=20). We use $r = 3$ to reflect the camera, microphone, and Bluetooth link that nearly all mobile phones on the market possess. The expected trend is evident, as more time is allowed for the sensing to occur the probability of sensing success goes up. It is interesting to note, however, the effect of topology on the success probability. The campus topology from Figure 2 limits the free flow of mobile nodes as compared with the unobstructed toroid scenario, requiring a longer time to reach the same success probability. We also observe (plots omitted due to space constraints) that in the campus scenario, the performance can change substantially depending on the placement of the tasking point σ and the sensing target τ. In Figure 3(a), we again plot Equation 5, but this time versus the number of equipped sensors per node r for different numbers of mobile sensor nodes y when the sensing deadline u is fixed at 50. As intuition suggests, the more sensors each node carries, the higher the chance that a query for a random sensor type will be successful. We also see that, as r is an intrinsic property of a mobile node and not dependent on mobility or node interactions per se, the campus ensemble qualitatively matches the toroid ensemble, albeit at a lower success probability. This is due to the fixed sensing deadline of $u = 50$ used, since from 3(a) we see that the success probability rises more slowly for the campus topology than for the toroid topology.

In the following analysis of sensor sharing and sensor substitution we report only results of the toroid scenario. We do this both to avoid cluttering the figures and also, since we have seen that the placement of the tasking point and sensing target greatly impacts importance in the campus scenario, to get a more general idea of the impact of sharing and substitution.

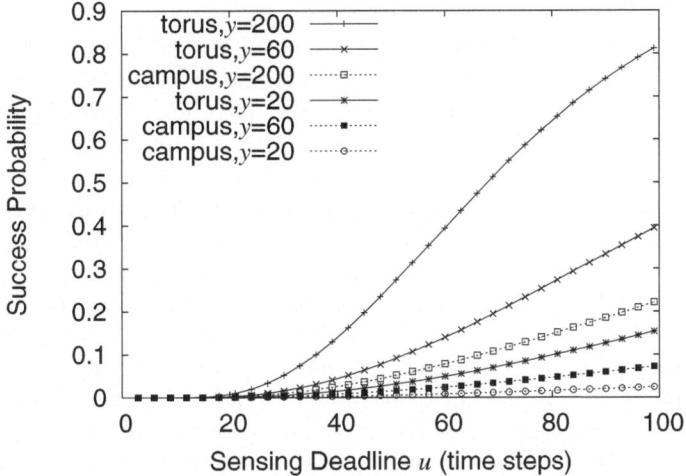

(a) Plot of the success probability versus the sensing deadline u for various values of y (number of mobile sensor nodes) for two topologies in the baseline OSN scenario. The campus topology from Figure 2 limits the free flow of mobile nodes, requiring a longer deadline to reach the same success probability.

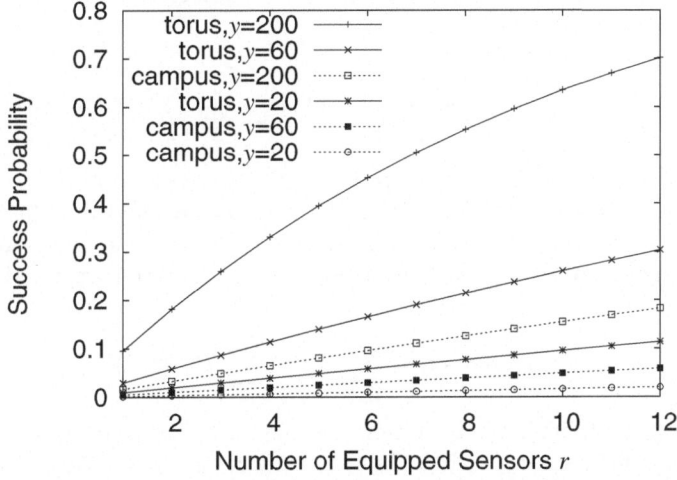

(b) Plot of success probability versus number of sensors r equipped per node for various values of y (number of nodes) for two topologies in the baseline OSN scenario. While the qualitative trends resulting from changes in r are not affected by topology (note that the campus curves follow the same trends as the toroid curves), the overall success probabilities are lower for campus than for toroid.

Fig. 3.

3.3   Sensor Sharing

To analyze sensor sharing[1], in the following development we relax the constraint that the query injection points task only those mobile sensor nodes that have the proper sensors to satisfy the application query. We allow a tasked mobile sensor node that arrives at the target region, but does not have the sensor required, to ask other mobile sensor nodes that may be at the target at the same time and do have the required sensor for their samples. In the analysis presented here we use a narrower form of sharing and do not consider the possibility of leveraging the mobile sensor node rendezvous outside of the target region [17].

In Equation 4, $F_k(\sigma, \tau)$ represents the probability that the mobile sensor node tasked by the query injection point first makes it to the sensing target location by time $t+k$. Successful sensor sampling at that point occurs with unity probability in the baseline scenario since the query injection point only tasks a mobile sensor node with the appropriate sensor equipment (as captured by $Prob(C)$ in the same equation). With sensor sharing a tasked mobile sensor node arriving at the target may not have the appropriate sensor. Let D denote the event that successful sampling occurs under these conditions. Then, the probability of D covers two cases: (i) the tasked mobile sensor node arrives at the target and has the required sensor, or (ii) the tasked mobile sensor node arrives at the target and does not have the required sensor but at least one other mobile sensor node at the target concurrently does have the sensor. Assuming there are y total mobile sensor nodes,

$$Prob(D) = F_k(\sigma, \tau) \cdot \left[\frac{r}{z} + (1 - \frac{r}{z}) \cdot \left(1 - (1 - \frac{r}{z} \cdot \frac{1}{N^2} \sum_{i=0}^{N^2-1} \mathcal{M}^k(i, \tau))^{y-1} \right) \right]. \quad (6)$$

Thus, to modify the baseline expression for success probability for sensor sharing, we substitute $Prob(D)$ in for $F_k(\sigma, \tau)$ in Equation 4. Further, since the constraint on the query injection point to task only a node with the required sensor no longer applies, $Prob(C)$ in Equation 4 is replaced with simply $F_k(\sigma)$, and we get

$$Prob(Success) = \sum_{l=1}^{u} F_k(\sigma) \sum_{m=1}^{u-l} Prob(D). \quad (7)$$

This success probability from Equation 7 is then plugged into Equation 5 to get the final result for sensor sharing. As the deadline goes to infinity, the success probability is limited by approximately $1 - (1 - (\frac{r}{z} + (1 - \frac{r}{z}) \cdot \frac{1}{N^2})^y$.

Comparison. We calculate the normalized sensing success probability improvement given by sensor sharing in our model by first evaluating Equation 5 for both the baseline and sensor sharing cases, and then calculating

---

[1] In our model we do not treat communication costs for either the baseline case or sharing, and consider only the opportunity for sensing mission success. The sensor sharing implementation in [10] minimizes this cost by limiting communications to a single wireless hop, given the increased complexity and loss probability for multi-hop routing.

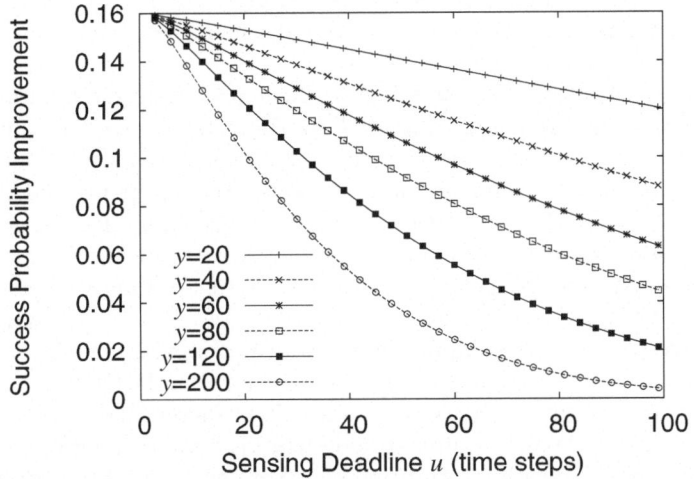

(a) Plot of the success probability improvement ratio versus sensing deadline u for various values of y (number of mobile sensor nodes) when sharing is allowed.

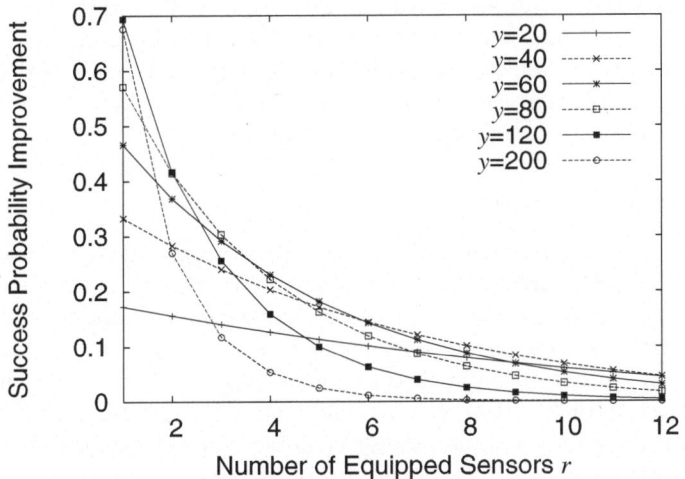

(b) Plot of success probability improvement ratio versus number of sensors r equipped per node for various values of y (number of nodes) when sharing is allowed.

Fig. 4.

$$\frac{(\text{Succ. prob. w/ sensor sharing}) - (\text{Succ. prob. w/ baseline})}{(\text{Succ. prob. w/ baseline})}.$$

We plot this success probability improvement in Figure 4(a) as a function of the sensing deadline u for various numbers of mobile sensors nodes y in the neighborhood when $r = 3$. In Figure 4(b) we plot the success probability improvement as a function of the number of sensors per configuration r for various numbers of mobile sensor nodes y in the neighborhood when $u = 50$. For both plots, we fix $N = 10$, and $z = 20$.

From Figure 4(a), we see that using sharing we can get an improvement of up to 16%, which is mainly dictated by the ratio $\frac{r}{z}$. As the sensing dead-line increases, the possible improvement decreases since even without sharing a properly equipped node will eventually go to the tasking point and then to the sensing target. Similarly, we see that with an increasing number of nodes in the system the baseline success probability also increases, reducing the space for possible improvement due to sharing. Figure 4(b) shows an improvement up to 70% when using sharing across all tested conditions. We observe the same general trend with respect to the equipped number of nodes whereby the possible improvement decreases with increasing node density. Similarly, the improvement generally decreases with increasing r, since it is increasingly likely that even without sharing a properly equipped node will visit the tasking point and sensing target before the deadline. Additionally, there exists an interesting interplay between the node density and r when r is low with a relatively small deadline u. In these situations, we conjecture, even when the node density is high, sensor sharing an offer a large improvement in success probability over the baseline case since sharing takes advantage of the higher density of sharing candidates in a shorter amount of time, while with few sensors a mobile node in the baseline case must rely on the uncontrolled mobility over time. When the deadline u is extended this effect disappears. In remains to be seen to what extent, if any, this effect persists under a different mobility model.

3.4   Sensor Substitution

Here we address direct sensor substitution (deferring a study of indirect sensor substitution to future work). In the context of the model we are developing we essentially extend the notion of a suitable sensor. We model the fact that a more sophisticated/expensive sensor can to some degree do the job of other simpler sensors, either through direct sensing (e.g., a GPS sensor substituting for a 3-axis accelerometer to measure road slope [26]) or inference (e.g., a CO_2 sensor substituting for a magnetometer to detect car density [8]). The potential for one sensor to substitute for another in some capacity, and the commensurate sensed data fidelity penalty, are dependent on the specific sensors in question. In our initial model, we abstract away these particulars and use p to denote the probability that a given sensor can act as a direct substitute for the sensor s specified by the application query, incorporating in p the probability that the corresponding loss of fidelity (if any) is within bounds acceptable to the application. A study of empirically generated correlation functions between various common sensor types will be the subject of our future work. We write

$$Prob\{i \simeq j\} = \begin{cases} 1, & i = j, \\ p, & i \neq j, \end{cases} \tag{8}$$

where $i \simeq j$ denotes that sensor i is a suitable substitute for sensor j. Letting B' denote the event that a given mobile sensor's configuration of r sensors includes a suitable (substitute) sensor, given a randomly chosen $s \in \mathcal{S}$ we have

$$Prob(B') = Prob\{i \simeq j \mid i = j\}Prob\{i = j\} + Prob\{i \simeq j \mid i \neq j\}Prob\{i \neq j\}$$
$$= 1 \cdot \frac{r}{z} + (1 - (1-p)^r)(1 - \frac{r}{z}). \tag{9}$$

The modified success probability when using direct sensor substitution is given by simply substituting $Prob(B')$ in for $Prob(B)$. Comparing Equation 9 with Equation 3, it is clear that the benefit of direct sensor substitution is given by the second term in Equation 9.

Comparison. We calculate the normalized sensing success probability improvement given by sensor substitution in our model by first evaluating Equation 5 for both the baseline and sensor substitution cases, and then calculating

$$\frac{\text{(Succ. prob. w/ sensor subst.)} - \text{(Succ. prob. w/ baseline)}}{\text{(Succ. prob. w/ baseline)}}.$$

We plot this success probability improvement in Figure 5(a) as a function of the sensing deadline u for various values of the substitution probability p when $y = 20$ and $r = 3$. In Figure 5(b) we plot the success probability improvement as a function of the number of sensors per configuration r for various numbers of mobile sensor nodes y in the neighborhood when $p = 0.02$ and $u = 50$. For both plots, we fix $N = 10$, and $z = 20$.

From Figure 5(a), we see success probability improvements ranging from 25% to 155% for relatively modest substitution probabilities, over the first 100 time steps. As before, the gain decreases over time due to the fact that eventually a properly equipped mobile sensor will visit the tasking point and then the sensing target even when sensor substitution is not used to increase the pool of "properly equipped" sensors. Figure 5(b) shows that an improvement over the baseline is achieved across all tested node densities and numbers of equipped sensors. Similar to the sensor sharing results in Section 3.3, we observe the possible improvement decreases with increasing node density, and the improvement also decreases with increasing r, since it is increasingly likely that even without substitution a properly equipped node will visit the tasking and sensing target before the deadline. The feature of overlapping curves seen in Figure 4(b) is not seen in Figure 5(b) since substitution does not involve interactions between nodes but rather operates independently on each node. Generally, the improvement given by sensor substitution is less sensitive to node density, as indicated by the tighter envelope of the curve ensemble, than is sensor sharing. On the other hand, substitution is more sensitive than sharing is to r, as indicated by the larger delta across the tested range of r (e.g., 0.28 for sensor substitution and 0.13 for sensor sharing when $y = 20$).

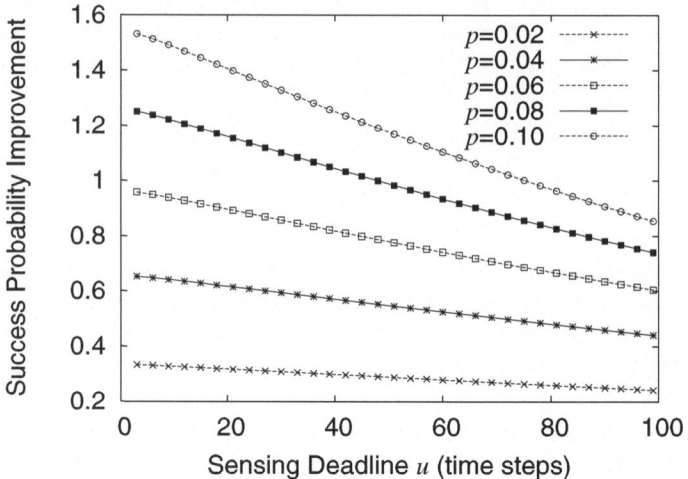

(a) Plot of success probability versus sensing deadline u for various values of p (substitute probability), when direct substitution is allowed.

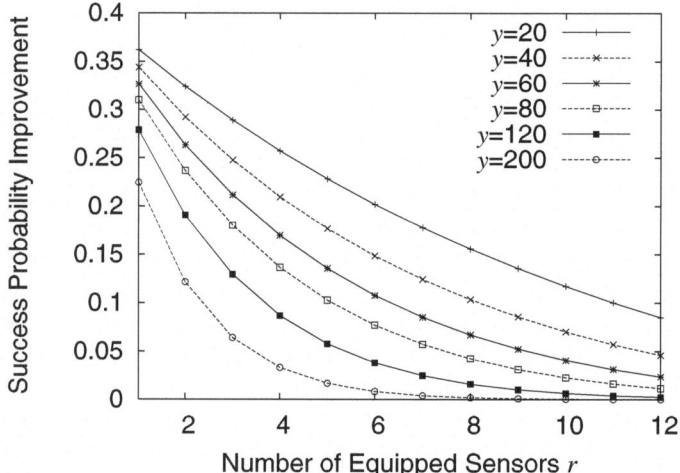

(b) Plot of success probability versus number of sensors r for various values of y (number of mobile sensors), when direct substitution is allowed.

Fig. 5.

3.5 Composing Sensor Sharing and Sensor Substitution

In the following we demonstrate the cumulative benefit of composing both sensor sharing and sensor substitution. In this scenario, again the constraint on exactly matching the preferred sensor is relaxed and all mobile sensor nodes reaching the tasking point before the deadline are tasked. Sensing success occurs if a node reaches the sensing target before the deadline, and: (i) it possesses the preferred sensor; or (ii) it does not have the preferred sensor but it is equipped with a suitable substitute sensor; or (iii) it does not have either the preferred or a substitute sensor, but it is able to share sensor readings from another mobile node at the sensing target.

We calculate the normalized sensing success probability improvement given by sensor substitution and sharing in our model by first evaluating Equation 5 for both the baseline, then applying both sensor sharing and sensor substitution, and calculating

$$\frac{(\text{Succ. prob. w/ sharing and subst.}) - (\text{Succ. prob. w/ baseline})}{(\text{Succ. prob. w/ baseline})}.$$

We plot this success probability improvement in Figure 6(a) as a function of the sensing deadline u for various values of the substitution probability p when $y = 20$ and $r = 3$. In Figure 6(b) we plot the success probability improvement as a function of the number of sensors per configuration r for various numbers of mobile sensor nodes y in the neighborhood when $p = 0.02$ and $u = 50$. For both plots, we fix $N = 10$, and $z = 20$.

From Figure 6, the benefit of enabling both sharing and substitution is clear. In Figure 6(a), the success probability improvement over the baseline is up to 270%, far outpacing the gains given by sharing (up to 16%) or substitution (up to 160%) alone, over the same range of sensing deadlines (u) (c.f. Figures 4(a) and 5(a)). Similarly, Figure 6(b) shows a success probability improvement across the same range of equipped sensors (r) over the baseline of up to 140% as opposed to only 70% for sharing and 35% for substitution alone. When both techniques are enabled, even when a node is equipped neither with the preferred sensor nor with a suitable substitute sensor, sensor sharing can probabilistically help to compensate, and vice versa.

4   Related Work

In the relatively new area of opportunistic sensor networking, there is little published work specifically addressing its challenges. However, more generally, there are alternative approaches for dealing with missing data points, both spatially and temporally. For example, Bayesian nets and other interpolation techniques can be used to infer missing data (e.g., [11]) when nodes lack access to the sensing hardware they require, at the time they require it. This approach allows for a reasonable approximation of missing data (within the accuracy of the sensor and sensed phenomenon models), but in the end is still an approximation. In contrast, both sensor sharing and substitution provide access to the real sensor samples, if supported by opportunistic rendezvous, without requiring any heavyweight computation or access to a central data store. Clearly, sharing and substitution can be used in concert with approximation techniques.

While in situ sensor sharing in opportunistic sensor networks is unaddressed in the literature, sensor sharing has been studied in the non-opportunistic networking

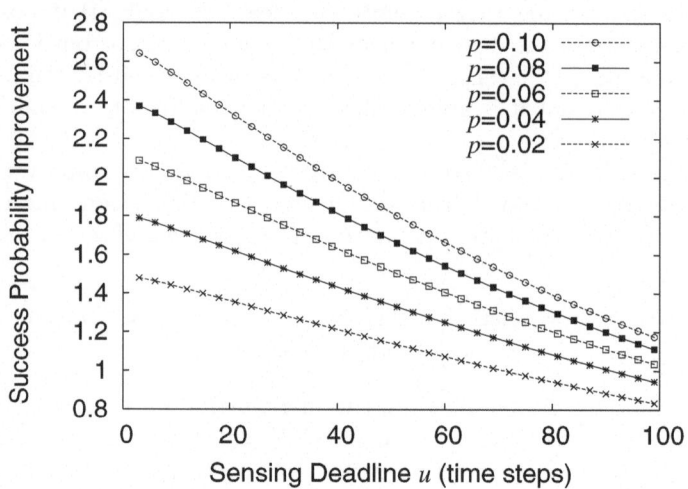

(a) Plot of success probability versus sensing deadline u for various values of p (substitute probability), when both sensor sharing and direct substitution are allowed.

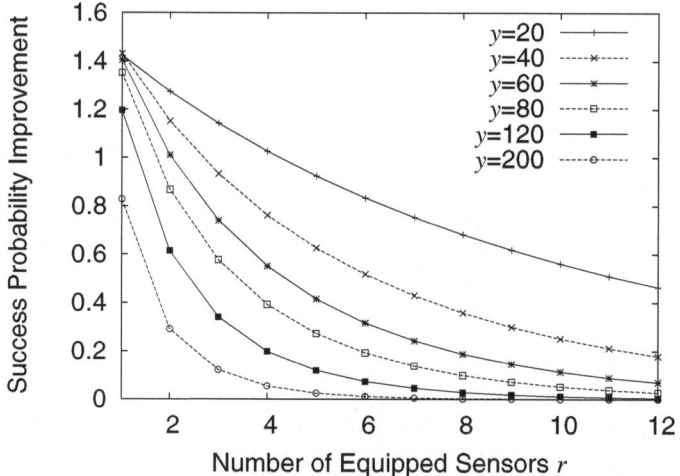

(b) Plot of success probability versus number of sensors r for various values of y (number of mobile sensors), when both sensor sharing and direct substitution are allowed.

Fig. 6.

setting. The authors of [19] present a mechanism enabling robot team members to share sensor information to achieve tightly-coupled cooperative tasks. The approach uses dynamically configured schema to route information among pre-formed groups to generate different cooperative control strategies. Conversely, our notion of sensor sharing relies only on completely opportunistic mobile node rendezvous. Analogous to mobile nodes that can carry a limited number of sensors, [5] explores the possibility of sensor sharing between integrated wired aerospace subsystems to reduce system part count. The authors of [15] explore the same design concept for reconfigurable sensor networking platforms. These subsystems are statically connected and therefore do not have to deal with the same challenges as the OSN domain we consider. When one considers the radio as a sensor, there are many examples of sensor data being shared between nodes as part of a system control mechanism. As an example of this type of sharing, in [24], the authors propose a system of mobile nodes that adjust their communication protocol parameters based on collaboratively-sensed environmental conditions. Using their radio receivers as sensors these nodes measure and share RSSI readings of WiFi AP beacons. While this sharing is in situ and the networking can be opportunistic, it is notable that in these cases each of the nodes possesses the sensor (the radio). Thus, this type of sharing is really by-design in-network information sharing, and not sensor sharing to meet an ad hoc application query. The authors of [18] and [22] have proposed a conceptual architecture and a prototype implementation to facilitate the sharing of sensor data among scientists and others once the data has been harvested to the back end. In contrast to this type of data sharing, the *sensor sharing* we propose takes place between mobile nodes in situ. These two types of sharing are complementary, as sensor sharing helps to provide the data streams that can be shared on the back end systems.

The concept of sensor substitution, though straightforward, has received little explicit attention in the sensor networking literature. This is likely due to the fact that to date the bulk of sensor network deployments have been engineered to meet the needs of a single application. We believe that both sensor sharing and sensor substitution will receive more attention as sparse, multi-application, device-heterogeneous, mobile sensor networks gain momentum. We are motivated by parallels in the health sciences domain where sensor substitution is an area of active interest for the treatment of human disabilities arising from the damage or decay of primary sensory systems. In [2], the authors summarize a study on people with balance disorders that concludes that sound may substitute, at least partially, for the lack of vestibular sensory information to control postural sway in stance.

In a broader context, methods for gaining ad hoc access to required resources (e.g., speakers, projectors, printers) have been considered by researchers in the pervasive networking community, in support of smart environments [20] and nomadic applications like "smart projector" and mobile "music service" [3].

5   Conclusion

While mobility in the context of OSNs enables sensing coverage at a lower deployment cost compared to an ubiquitous static sensor deployment, this uncontrolled mobility also poses a number of challenges related to tasking, sensing and data collection. With sensor sharing and sensor substitution we have proposed two techniques that can be used together or alone to improve the probability of successfully and more expediently completing these activities. The initial numerical evaluation is encouraging, showing non-negligible gains through low complexity techniques, warranting further study.

Acknowledgment

This work is supported in part by Intel Corp., Nokia, NSF NCS-0631289, ARO W911NF-04-1-0311, and the Institute for Security Technology Studies (ISTS) at Dartmouth College. ISTS support is provided by the U.S. Department of Homeland Security under award 2006-CS-001-000001, and by award 0NANB6D6130 from the U.S. Department of Commerce. The views and conclusions contained in this document are those of the authors and should not be interpreted as necessarily representing the official policies, either expressed or implied, of any supporting agency.

References

1. Abdelzaher, T., Anokwa, Y., Boda, P., Burke, J., Estrin, D., Guibas, L., Kansal, A., Madden, S., Reich, J.: Mobiscopes for Human Spaces. IEEE Pervasive Computing 6(2), 20–29 (2007)
2. Barclay, L., Vega, C.: Audio-Biofeedback May Improve Stability for Patients With Bilateral Vestibular Loss. In: Medscape Medical News (2005), http://www.medscape.com/viewarticle/508833
3. Basu, P., Wang, K., Little, T.D.C.: Dynamic Task-based Anycasting in Mobile Ad Hoc Networks. Mobile Networks and Applications 8(5) (October 2003)
4. Campbell, A.T., Eisenman, S.B., Lane, N.D., Miluzzo, E., Peterson, R.A.: People-Centric Urban Sensing (Invited Paper). In: Proc. of 2nd ACM/IEEE Int'l Conf. on Wireless Internet (WiCon 2006), Boston (August 2006)
5. Carlin, C.M., Hastings, W.J.: Propulsion control sensor sharing opportunities. In: Proc. of the 21st AIAA Aerospace Sciences Meeting, Reno (January 1983)
6. Columbia University Morningside campus map (retrieved January 1, 2008), http://www.columbia.edu/about_columbia/map/images/maps-large/mudd.gif
7. Deshpande, A., Guestrin, C., Hong, W., Madden, S.: Exploiting Correlated Attributes in Acquisitional Query Processing. In: Proc. of the 21st Int'l Conf. on Data Engineering (ICDE 2005), Tokyo, April 2005, pp. 143–154 (2005)
8. Eisenman, S.B., Miluzzo, E., Lane, N.D., Peterson, R.A., Ahn, G.-S., Campbell, A.T.: The BikeNet Mobile Sensing System for Cyclist Experience Mapping. In: Proc. of 5th ACM Conf. on Embedded Networked Sensor Systems, Sydney (November 2007)
9. Eisenman, S.B., Campbell, A.T.: SkiScape Sensing (Poster Abstract). In: Proc. of ACM 4th Int'l Conf. on Embedded Networked Sensor Systems, Boulder (November 2006)
10. Eisenman, S.B., Lane, N.D., Campbell, A.T.: Quintet: Orchestrating Neighborhood Sensors to Increase Sensing Fidelity through Opportunistic Sharing. Dartmouth Computer Science Technical Report (December 2007)
11. Hruschka Jr., E.R., Hruschka, E.R., Ebecken, N.F.F.: Applying Bayesian Networks for Meteorological Data Mining. In: Proc. of the 25th SGAI Int'l Conf. on Innovative Techniques and Applications of Artificial Intelligence, Cambridge (December 2005)

12. Hull, B., Bychkovsky, V., Zhang, Y., Chen, K., Goraczko, M., Miu, A., Shih, E., Balakrishnan, H., Madden, S.: CarTel: A Distributed Mobile Sensor Computing System. In: Proc. of 4th Int'l Conf. on Embedded Networked Sensor Systems, Boulder, November 2006, pp. 125–138 (2006)
13. Lane, N.D., Lu, H., Campbell, A.T.: Ambient Beacon Localization: Using Sensed Characteristics of the Physical World to Localize Mobile Sensors. In: Proc. of 4th Workshop on Embedded Networked Sensors, Cork (June 2007)
14. Lorincz, K., Welsh, M.: MoteTrack: A Robust, Decentralized Approach to RF-Based Location Tracking. Springer Personal and Ubiquitous Computing 11(6), 489–503 (2007)
15. Lymberopoulos, D., Priyantha, N.B., Zhao, F.: mPlatform: A Reconfigurable Architecture and Efficient Data Sharing Mechanism for Modular Sensor Nodes. In: Proc. of Int'l Conf. on Information Processing in Sensor Networks, Cambridge, April 2007, pp. 128–137 (2007)
16. Miluzzo, E., Lane, N.D., Eisenman, S.B., Campbell, A.T.: CenceMe – Injecting Sensing Presence into Social Networking Applications. In: Kortuem, G., Finney, J., Lea, R., Sundramoorthy, V. (eds.) EuroSSC 2007. LNCS, vol. 4793, pp. 1–28. Springer, Heidelberg (2007)
17. Miluzzo, E., Lane, N.D., Campbell, A.T.: Virtual Sensing Range (Poster Abstract). In: Proc. of 4th ACM Conf. on Embedded Networked Sensor Systems, Boulder, November 2006, pp. 397–398 (2006)
18. Parker, A., Reddy, S., Schmid, T., Chang, K., Saurabh, G., Srivastava, M., Hansen, M., Burke, J., Estrin, D., Allman, M., Paxson, V.: Network System Challenges in Selective Sharing and Verification for Personal, Social, and Urban-scale Sensing Applications. In: Proc. of HotNets-V, Irvine (November 2006)
19. Parker, L.E., Chandra, M., Tang, F.: Enabling Autonomous Sensor-Sharing for Tightly-Coupled Cooperative Tasks. In: Multi-Robot Systems. From Swarms to Intelligent Automata, vol. III, pp. 119–130. Springer, Netherlands (2005)
20. Rakotonirainy, A., Groves, G.: Resource Discovery for Pervasive Environments. In: Proc. of the 4th Int'l Symp. on Distributed Objects and Applications, October 2002, pp. 866–883 (2002)
21. Randell, C., Djiallis, C., Muller, H.: Personal Position Measurement Using Dead Reckoning. In: Proc. of 7th IEEE Int'l Symp. on Wearable Computers, October 2003, pp. 166–173 (2003)
22. Reddy, S., Chen, G., Fulkerson, B., Kim, S.J., Park, U., Yau, N., Cho, J., Heidemann, J., Hansen, M.: Sensor-Internet Share and Search–Enabling Collaboration of Citizen Scientists. In: Proc. of the ACM Workshop on Data Sharing and Interoperability on the World-wide Sensor Web, Cambridge (April 2007)
23. Srivastava, M., Hansen, M., Burke, J., Parker, A., Reddy, S., Saurabh, G., Allman, M., Paxson, V., Estrin, D.: Wireless Urban Sensing Systems. CENS Technical Report #65 (2006)
24. Troxel, G.D., Caro, A., Castineyra, I., Goffee, N., Haigh, K.Z., Hussain, T., Kawadia, V., Rubel, P.G., Wiggins, D.: Cognitive Adaptation for Teams in ADROIT (Invited paper). In: IEEE Global Communications Conference, Washington D.C (November 2007)
25. Tuulos, V., Scheible, J., Nyholm, H.: Combining Web, Mobile Phones and Public Displays in Large-Scale: Manhattan Story Mashup. In: Proc. of the 5th Int'l Conf. on Pervasive Computing, Toronto (May 2007)
26. Voyadgis, D.E., Ryder, W.H.: Measuring the Performance of Ground Slope Generation. In: Proc. of the 16th Annual ESRI User Conf. (May 1996)

On the Average Case Communication Complexity for Detection in Sensor Networks[*]

N.E. Venkatesan, Tarun Agarwal, and P. Vijay Kumar[**]

Dept. of ECE, Indian Institute of Science, Bangalore, 560012
{venkyne,tarun,vijay}@ece.iisc.ernet.in

Abstract. The problem of sensor-network-based distributed intrusion detection in the presence of clutter is considered. It is argued that sensing is best regarded as a local phenomenon in that only sensors in the immediate vicinity of an intruder are triggered. In such a setting, lack of knowledge of intruder location gives rise to correlated sensor readings. A *signal-space* viewpoint is introduced in which the noise-free sensor readings associated to intruder and clutter appear as surfaces $\mathcal{S}_\mathcal{I}$ and $\mathcal{S}_\mathcal{C}$ and the problem reduces to one of determining in distributed fashion, whether the current noisy sensor reading is best classified as intruder or clutter. Two approaches to distributed detection are pursued. In the first, a decision surface separating $\mathcal{S}_\mathcal{I}$ and $\mathcal{S}_\mathcal{C}$ is identified using Neyman-Pearson criteria. Thereafter, the individual sensor nodes interactively exchange bits to determine whether the sensor readings are on one side or the other of the decision surface. Bounds on the number of bits needed to be exchanged are derived, based on communication complexity (CC) theory. A lower bound derived for the two-party average case CC of general functions is compared against the performance of a greedy algorithm. The average case CC of the relevant *greater-than (GT)* function is characterized within two bits. In the second approach, each sensor node broadcasts a single bit arising from appropriate two-level quantization of its own sensor reading, keeping in mind the fusion rule to be subsequently applied at a local fusion center. The optimality of a threshold test as a quantization rule is proved under simplifying assumptions. Finally, results from a QualNet simulation of the algorithms are presented that include intruder tracking using a naive polynomial-regression algorithm.

1 Introduction

1.1 Background and Outline

The problem of hypothesis testing between a pair of hypotheses (H_I and H_C) based on observations $\{\theta_i\}_1^K$, has been studied in various settings. The K observations can be thought of readings from K different sensors. In particular, the

---

[*] This work was supported by the Defence Research and Development Organization (DRDO), Ministry of Defence, Government of India under a research grant on wireless sensor networks (DRDO 571, IISc).
[**] P. Vijay Kumar is on leave of absence from Department of EE-systems, University of Southern California, Los Angeles, CA 900089, USA.

S. Nikoletseas et al. (Eds.): DCOSS 2008, LNCS 5067, pp. 176–189, 2008.

problem is well studied in the case when the readings are independent conditioned upon the hypothesis. The problem of decision fusion is discussed in [1], optimal rules when sensors parley or exchange bits for the independent case are obtained in [3]. Some recent papers [10] relate communication complexity (CC) to distributed processing in sensor networks. There are very few results for the correlated case [2]. Motivated by these, we try to establish some results for a case of correlated observations that we consider.

We consider the problem of detection of an intruder in the presence of clutter[1], in a distributed manner towards building a wireless sensor network surveillance system that senses ground vibrations produced by human footsteps. The propagation model is described in Section 1.2. In Section 2 we argue that sensing is a local phenomenon and propose a *signal space* view-point of the detection problem. For this propagation model, we investigate the conditions under which a threshold test is optimal for the decision fusion problem described in Section 3. We further investigate an alternate formulation in which the sensors are allowed to exchange multiple bits and the decision test is fixed by exploiting its relation to CC in Sections 4 and 5. The motivation is to get bounds on the energy spent in detection by minimizing the number of bits exchanged between the sensors during detection. Section 6 looks at the simulation results of the algorithms in QualNet (a sensor network simulation environment).

1.2   A Propagation Model for Vibrations

A simplified propagation model [11] is given by

$$v_b = \frac{v_a}{max(1, r_b)}.$$

where

v_b : source vibration amplitude at point b
v_a : source vibration amplitude measured at 1 meter from source.
r_b : distance between source and point b

applicable for various soils. The model assumes a reference amplitude $v_a = 10^{-6}$ to 10^{-5} g [13] for a typical human footstep. We assume an AWGN model for the sensor reading, i.e the ith sensor reading θ_i is given by $\theta_i = \frac{v_a}{\max(1, d_i)} + n_i$. where d_i is the distance between the ith sensor and the vibration source and n_i is $\sim N(0, \sigma^2)$, independent across space and time. For our case, v_a takes on the value v_I for an intruder and v_C for clutter, with $v_C = \kappa v_I$, $0 \le \kappa \le 1$. We assume that the intruder or clutter can be present anywhere in the field with uniform probability. Thus, given that the source location is unknown, d_i is a random variable which brings about the correlation among the sensor readings. We also assume that the intruder or clutter are stationary for the period of detection and v_C, v_I and κ are known. Though this is a common model for propagation, it is not quite so well studied in the literature [8], [5], [9].

---

[1] Clutter is a harmless source of disturbance which is not of interest to the sensor network.

2   Sensing as a Local Phenomenon and the Signal Space

2.1   Sensing as a Local Phenomenon

With the current status of sensing technology, the value of intruder vibration is $\sim 10^{-5}$g and noise floor in sensing devices ~ 300 ng/$\sqrt{\text{Hz}}$ [12]. To keep the sensor density to affordable levels, suppose we take a sensor density of say 10,000 nodes per square km, we get an inter-sensor distance of around 10m. From Table 1, it is clear that only a few sensors in the vicinity of the vibration capture a significant SNR. The parameters assumed are Noise floor $= 300$ ng/$\sqrt{\text{Hz}}$, Bandwidth (BW) $= 500$Hz and SNR is calculated as follows, SNR $= 20\log\left(\dfrac{\frac{v_a}{r_b}}{\text{noise floor} \times \sqrt{BW}}\right)$.

Table 1. Variation of SNR with distance

v_a	Distance from source (r_b)	SNR(dB)
1e-6	5	-30.512
	10	-36.532
	15	-40.054

2.2   Signal Space View Point

Let K be the number of sensors present in the vicinity of the intruder which register a significant SNR. Consider a random placement of sensors. Using the sensor locations as vertices of triangles, the entire field can be partitioned into scalene triangles [2] (see Fig. 1(a)). Clearly, an intruder or clutter anywhere in the field, must be located inside one of these triangles. Given such a tiling, the K sensors nearest to a particular tile will collaborate to test the presence of a target in that tile and relay the decision to the base station.

Let $\underline{\theta} = (\theta_1, \theta_2, \cdots, \theta_K)$ denote the K-tuple of these readings. This K-tuple may be regarded as a point in K-dimensional *signal space* $\in \mathbb{R}^K$. Let $\mathcal{S}_I$ denote the collection of points in signal space corresponding to the set of all intruder locations within the triangle. Analogously, we get $\mathcal{S}_C$ for clutter. From Section 1.2 we have that $\mathcal{S}_C = \kappa\mathcal{S}_I$. Fig. 1(b) shows the signal space for a scalene triangle with $K = 3$, where the three sensors are at the vertices of the triangle.

Some properties of signal space are given below, proofs may be found in [9].

Lemma 1. *Given three noiseless sensor readings from sensors having known generic location, the vibration amplitude and location of the source of vibration can be uniquely determined.*

---

[2] The tiling may require sensors to know their exact locations. The tiling is also known as Delaunay triangulation. For a 2-D Delaunay triangulation, it can be shown that the closest sensor to a point in a triangle is one of the vertices of that triangle.

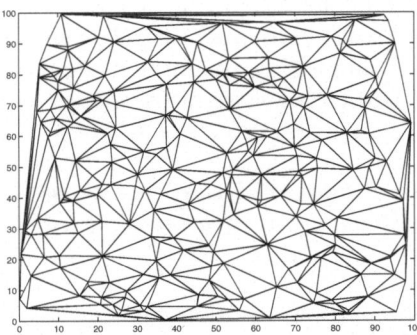 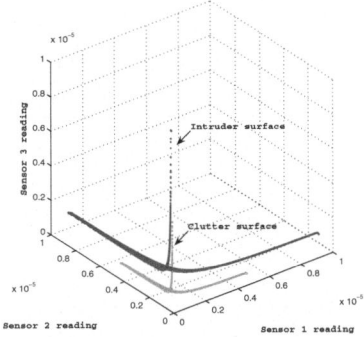

(a) Delaunay triangulation of the field by randomly placed sensors. Triangle vertices are the sensors.

(b) Three dimensional signal associated to the 3 closest sensor nodes for a scalene triangle.

Fig. 1. Delaunay triangulation and signal space

Corollary 1. *Intruder and clutter surfaces S_I, S_C are disjoint i.e., $S_I \cap S_C = \phi$.*

Lemma 2. *For the case when we would like to estimate the amplitude of the source v_a using a best-linear-unbiased-estimator, given a point in the signal space, the points closest to the origin have the maximum estimation error.*

The signal-space formulation of the detection problem can now be stated. The collection of sensor readings correspond to a point in the K-dimensional signal space. Measurement noise will case this point to drift from its origins in either S_I or S_C. The detection problem has thus been reduced to one of determining in distributed fashion, whether the current noisy sensor reading is best classified as intruder or clutter. Two approaches to distributed detection are pursued here. In the first, a decision surface separating S_I and S_C is identified using Neyman-Pearson criteria. Thereafter, the individual sensor nodes interactively exchange bits to determine whether the sensor readings are on one side or the other of the decision surface. Bounds on the number of bits needed to be exchanged are derived, based on communication complexity (CC) theory. A lower bound derived for the two-party average case CC of general functions is compared against the performance of a greedy algorithm. The average case CC of the relevant *greater-than (GT)* function is characterized within two bits. In the second approach, each sensor node broadcasts a single bit arising from appropriate two-level quantization of its own sensor reading, keeping in mind the fusion rule to be subsequently applied at a local fusion center. The optimality of a threshold test as a quantization rule is proved under simplifying assumptions.

3 Optimality of Threshold Test

For the classical decision fusion problem, where each sensor sends a bit to the fusion center, local log-likelihood ratio (LLR) decision tests are shown to be optimal for independent observations [1] and the fusion rule is a weighted sum of the bits [4]. There are very few results for the correlated case [2]; Tsitsiklis [7] shows that the problem is NP-hard. Under simplifying assumptions we prove the optimality of a threshold test at each sensor for the case of correlated readings that we are considering. The following theorem states our result.

Theorem 1. *For the case of two sensors, fixed AND/OR fusion rule at the fusion center and $\kappa = 0$, the optimal decision rule at each sensor is a threshold test on the reading.*

Remark: For the case of two sensors, the optimal decision rule can be one of AND, OR or XOR. The analysis is difficult for the XOR rule (also noted by Willet et al [2]). $\kappa = 0$ is only a sufficient condition and the result may hold for a range of κ. The proof is along the lines of that in [2] [3].

Proof. Let B_i be the bit communicated by the ith sensor $(i = 1, 2)$. Then,

$$B_i = \begin{cases} 1 \text{ if } \theta_i \in A_i \\ 0 \text{ if } \theta_i \notin A_i \end{cases}.$$

where A_i is a union of intervals (possibly infinite). It suffices to show that A_i is a single semi-infinite interval. For AND fusion rule and using a result from [6] the decision test simplifies to the following:

$$A_i = \left\{ \theta_i : L(\theta_i) \triangleq \frac{\int_{\theta_j \in A_j} p(\theta_j, \theta_i | H_I) d\theta_j}{\int_{\theta_j \in A_j} p(\theta_j, \theta_i | H_C) d\theta_j} > \tau \right\}.$$

For $\kappa = 0$ we get[4],

$$\frac{\partial L(\theta_1)}{\partial \theta_1} \geq 0.$$

Therefore, independent of the region A_2 of the second sensor, $L(\theta_1)$ is an increasing function of θ_1. Hence A_1 is a positive semi-infinite interval. Analogously, A_2 is a positive semi-infinite interval.

Similarly the OR rule simplifies to the following,

$$A_i = \left\{ \theta_i : \frac{1 - \int_{\theta_j \in A_j} p(\theta_j, \theta_i | H_I) d\theta_j}{1 - \int_{\theta_j \in A_j} p(\theta_j, \theta_i | H_C) d\theta_j} > \tau \right\}.$$

A derivation, similar to the AND rule, shows that A_1 and A_2 are single connected regions.

The above result can be generalized for $K \geq 2$ sensors.

---

[3] It should be noted that [2] gets rules for a correlated Gaussian noise setting where as in our case correlation arises due to the propagation model.
[4] Detailed derivation has been omitted for sake of brevity.

Theorem 2. *For $K \geq 2$ sensors, fixed AND/OR fusion rule at the fusion center and $\kappa = 0$, the optimal decision rule at each sensor is a threshold test on the reading.*

Proof. The proof follows the proof of Theorem 1.

Remark: Theorem 1 is clearly a special case of Theorem 2 but is of interest as in the two sensors case, AND/OR/XOR are the only possible fusion rules.

4   Communication Complexity

In this section we investigate the viewpoint wherein the decision surface is fixed apriori (based on the NP-criterion) and sensors exchange bits to determine whether the received signal point is on one or the other side of the surface. This is equivalent to evaluating $f(\theta) \geq \tau$ where f characterizes the decision surface. This is a classical function computation problem and hence motivates us to use the theory of CC. This framework would lead to a step-by-step optimal solution, wherein in the first step we optimize the detection performance and in the second step we optimize over the number of bits exchanged. A sequential probability ratio test (SPRT) could be carried out for our problem and then one could try to optimize over the bits. However this is a tough problem and we wish to focus more on distributed detection and not on the performance and hence we fix the decision surface apriori and try to optimize over the bits exchanged.

4.1   Lower Bound for the Average Case CC

The theory of CC was introduced by Yao [14] in 1979. The problem definition is as follows. X and Y are two nodes that possess values $x \in X = \{0,1\}^n$ and $y \in Y = \{0,1\}^n$ respectively. Let $(X,Y) \sim p_{XY}(x,y)$ be the joint distribution of the random variables. The objective is to calculate a Boolean function $f(x,y)$. Messages are exchanged between X and Y until both determine $f(x,y)$. Then, for a particular realization of the random variables $\{X = x\}$ and $\{Y = y\}$, let $s_P(x,y) = (m_1(x,y), ..., m_{r(x,y)}(x,y))$ be the sequence of messages transmitted under a protocol P, where $m_i(x,y)$ is the ith message sent in the protocol. Messages $m_i(x,y)$ may be varying length binary strings; $|m_i(x,y)|$ denotes the length. CC theory characterizes the bounds associated with the communication cost involved in the above problem. The reader is referred to [15] for a more detailed discussion on CC.

The deterministic average case CC of a protocol P, $D_{avg}(P)$, is defined as follows

$$D_{avg}(P) \triangleq \mathbb{E}_{x,y} |s_P(x,y)|.$$

where $|s_P(x,y)| = \sum_{i=1}^{r(x,y)} |m_i(x,y)|$ and $\mathbb{E}_{x,y}$ is the expectation over all the inputs. The deterministic average case CC of a function $f(x,y)$ is defined as

$$D_{avg}(f) = \min_P D_{avg}(P).$$

A function matrix F, associated with a function $f(x, y)$, has its rows and columns indexed by the values taken by X and Y respectively, and $F_{ij} = f(i, j)$, where (i, j) can be taken as the integer representations of the strings (x, y). A cartesian product $S \times T(S \subseteq X, T \subseteq Y)$ is called a f−monochromatic rectangle if f is constant over $S \times T$. For Boolean valued functions we can have 0-rectangles and 1-rectangles (see Fig. 2(b)). It turns out that any protocol induces a partition of F into such monochromatic rectangles [15]. The sequence of messages exchanged in a protocol uniquely identify a monochromatic rectangle, which determines the function value.

Let $R_{i,j}$ be the set of all possible monochromatic rectangles, $r_{i,j}$, in F which contain the element $\{i, j\}$. Let Γ_P be the set of all monochromatic rectangles, γ_P, induced by a protocol P in F. The probability that a sequence of messages will lead to a particular monochromatic rectangle γ_P is $p(\Gamma_P = \gamma_P) = \sum_{\{i,j\} \in \gamma_P} p_{X,Y}(i, j)$. Similarly, $p(r_{i,j})$ is defined. Thus, we have the following entropies:

$$H(\Gamma_P) = - \sum_{\gamma_P \in \Gamma_P} p(\gamma_P) \log p(\gamma_P).$$

$$H(\gamma_P) = - \sum_{\{i,j\} \in \gamma_P} \frac{p_{X,Y}(i, j)}{p(\gamma_P)} \log \frac{p_{X,Y}(i, j)}{p(\gamma_P)}.$$

$$H(r_{i,j}) = - \sum_{\{p,q\} \in r_{i,j}} \frac{p_{X,Y}(p, q)}{p(r_{i,j})} \log \frac{p_{X,Y}(p, q)}{p(r_{i,j})}.$$

The following theorem states the lower bound that we propose for the deterministic average case CC.

Theorem 3. $D_{avg}(f) \geq H(X, Y) - \sum_{i,j \in \{X,Y\}} p_{X,Y}(i, j) H(r_{i,j}^*)$, where $r_{i,j}^* = \arg \max_{r_{i,j} \in R_{i,j}} H(r_{i,j})$.

Proof. A sketch of the proof is given below. From the grouping axiom of entropy we have

$$H(X, Y) = H(\Gamma_P) + \sum_{\gamma_P \in \Gamma_P} \sum_{\{i,j\} \in \gamma_P} p_{X,Y}(i, j) H(\gamma_P)$$

From the definition of $H(r_{i,j}^*)$ we get

$$H(\Gamma_P) \geq H(X, Y) - \sum_{\{i,j\} \in \{X,Y\}} p_{X,Y}(i, j) H(r_{i,j}^*)$$

The average number of bits spent by a protocol, $D_{avg}(P)$, cannot be lesser than the entropy associated with the monochromatic rectangles into which the protocol partitions the function matrix. Therefore it is clear that for any protocol P, $D_{avg}(P) \geq H(\Gamma_P)$. Thus

$$D_{avg}(f) = \min_P D_{avg}(P) \geq H(X,Y) - \sum_{\{i,j\} \in \{X,Y\}} p_{X,Y}(i,j) H(r^*_{i,j})$$

$$\triangleq E_{LB}(f)$$

The average case CC of the greater-than function is characterized in the next section.

4.2   Average Case CC of the Greater-Than (GT) Function

The classical greater-than function is defined as follows

$$GT(x,y) = \begin{cases} 1 \text{ if } x \geq y \\ 0 \text{ otherwise.} \end{cases}$$

where $(x,y) \in ((0,1)^n, (0,1)^n)$ with each string represented by its integer value and $p_{XY}(x,y)$ is a uniform distribution. See Table 2(a) for the function matrix of this function. The randomized CC of this function was studied by Nisan [19] under the category of threshold gates. A greedy algorithm (PGA) described in the next section, provides an upper bound for this function and we have a lower bound given by $E_{LB}(GT)$. Let $N = 2^n$. We then have the following result.

X/Y	1	2	3	4
1	1	0	0	0
2	1	1	0	0
3	1	1	1	0
4	1	1	1	1

X\Y	1	2	3	4
1	0	0	0	1
2	0	0	0	1
3	0	0	0	0
4	0	1	1	1

(a) GT Function matrix.   (b) Monochromatic partition.

Fig. 2. Function matrices

Theorem 4. *The upper bound, PGA(GT), and the lower bound, $E_{LB}(GT)$, for the GT function are given by,*
$$PGA(GT) = 4 - \frac{4}{N},$$

$$E_{LB}(GT) = 2log(N) - \sum_{i=1}^{\frac{N}{2}-1} \frac{\frac{N}{2}-i}{N^2} log\left(\frac{N^2}{4} - i^2\right) + \sum_{i=1}^{\frac{N}{2}-1} \frac{\frac{N}{2}-i}{N^2} log\left(\left(\frac{N}{2}-i\right)\left(\frac{N}{2}+i+1\right)\right)$$

$$-0.25 log\left(\frac{N^2}{4}\right) - \frac{\frac{N}{2}\left(\frac{N}{2}+1\right)+\frac{N}{2}}{N^2} log\left(\frac{N^2}{4}+\frac{N}{2}\right).$$

Proof. The proofs are based on simple counting arguments.

The plot Fig.3(a), has four bounds corresponding to $PGA(GT)$, $E_{LB}(GT)$, $n+1$ (corresponds to sending the full reading, also the worst case CC of GT) and *Discrepancy* ($Disc_{uniform}(GT)$), for $\epsilon = 0$ (a previously known lower bound for GT [15]). It can be seen that $E_{LB}(GT)$ is tighter than the discrepancy lower bound. We have $\lim_{n\to\infty}PGA(GT) = 4$, $\lim_{n\to\infty}E_{LB}(GT) = \frac{3}{2\ln2} = 2.164$. Thus the average case CC for the GT function is characterized within two bits.

The relation of $E_{LB}(f)$ to *information complexity (IC)* [16] is given below.

Lemma 3. *For any function f, for the case of deterministic CC, $IC(f) \geq E_{LB}(f)$, where $IC(f)$ is the information complexity of f.*

Proof. Omitted for sake of brevity.

5   The Detection Problem and CC

From the discussions in Section 2, it is clear that the problem of detection is one of finding a decision surface separating the intruder and clutter surfaces and detecting whether the received signal point is above or below this surface. This amounts to finding whether $f(\theta_1, \theta_2, ..., \theta_K) \gtrless \tau$, where f characterizes the decision surface. For optimal detection performance f can be taken to be the LLR of the sensor readings given by $log\left(\frac{p(\theta_1, .., \theta_K | H_I)}{p(\theta_1, .., \theta_K | H_C)}\right)$. Thus the decision function is fixed here and sensors can now exchange bits to calculate this function. We shall assume $K = 2$ for our further discussions in this section. We shall restrict our attention to the class of functions f such that f is monotone increasing or monotone decreasing in both θ_1 and θ_2. Let each of the sensor readings be quantized to $N' = 2^{n'}$ levels and F' be the function matrix associated with $f(\theta_1, \theta_2) \gtrless \tau$. It can be shown that any function matrix F' of a function f satisfying the above conditions, can be simplified to an equivalent form of the greater-than function. Let N be the row and column size of the new function matrix F, obtained after simplification. We shall now look at some greedy algorithms to calculate the function.

5.1   Greedy Algorithms

These algorithms give an upper bound on the number of bits exchanged between the sensors during the detection process.

Parallel greedy algorithm(PGA): Here both the sensors transmit a bit simultaneously in each round, based on a threshold on their reading. The thresholds are chosen in such a way that the probability of a decision in that stage is maximized i.e. the probability of the monochromatic rectangles resulting in that stage is maximized.

Entropy minimization algorithm(EMA): Here the sensors transmit bits simultaneously dividing F into two monochromatic rectangles. The thresholds are chosen in such a way that the entropy associated with the remaining rectangles is minimized.

It is easy to see that $2p + 4(1 - p)$, where p is the maximum probability of decision in the first round, is a lower bound for protocols where sensors simultaneously transmit bits in each round. Similarly for serial/sequential algorithms $2p + 3(1 - p)$ is a lower bound. Note that these two lower bounds are for the specific type of protocols considered and not necessarily that of the function.

A comparison of the ROC curves and the bounds for the number of bits spent for the detection problem under consideration are shown in Fig. 3(c), 3(e) and Fig. 3(d), 3(f). A simple hyperplane decision surface is taken i.e. $f(\theta_1, \theta_2) = \theta_1 + \theta_2$. This makes the numerical analysis of the probabilities and ROC curves tractable. The sensors are placed at $(1, 0)$ and $(-1, 0)$ and the intruder/clutter is allowed to move around in a box of size 4×4 centered around the sensors. κ is taken to be 0.5 here. The intruder and clutter hypothesis are assumed to be equally likely. $n' = 5$ in these plots. The average number of bits is plotted as a function of the probability of false alarm. The curves are obtained numerically using the probability expressions.

The nature of the bits-spent curve can be explained as follows. At low PFA, the threshold τ is higher, which means that the hyperplane is closer to the intruder surface. Thus whenever there is a clutter, the reading is small and within a few bits a decision is taken. A similar argument holds when PFA is high. It is only when the hyperplane is in between the surfaces, that more bits need to be exchanged to arrive at a decision.

The performance of the PGA is two bits away from $E_{LB}(f)$. However, it is interesting to see that the nature of the bits spent curves for $E_{LB}(f)$ and the upper bound are nearly the same, which could possibly suggest that the upper bound is a constant number of bits away from the lower bound at least for these form of functions. Also note that EMA and PGA have nearly the same bit expense. One can also see that the PGA algorithm performance is within one bit of the lower bound for parallel algorithms. One should also note that these algorithms and bounds can be applied for any function whose function matrix can be reduced to an echelon form using only row and column permutations.

6   Simulation Using QualNet

The simulation results for algorithms described in Section 5, are shown in Fig 3. The plots are for the case of both intruder and clutter, for both hexagonal and random placements, as either move in the field along the tracks as shown. The histograms of the bits spent per sensor in detection are also shown. In the histogram the x-axis is the number of bits sent by each sensor in detection. The height of the histogram is the number of detections that were made using the respective x-axis' number of bits.

An intrusion scenario is simulated in Qualnet. Following are the simulation parameters. The field size is 100 m × 100 m. Sensors are placed in a hexagonal lattice with inter-sensor distance as 11 m. [9] investigates the interesting properties

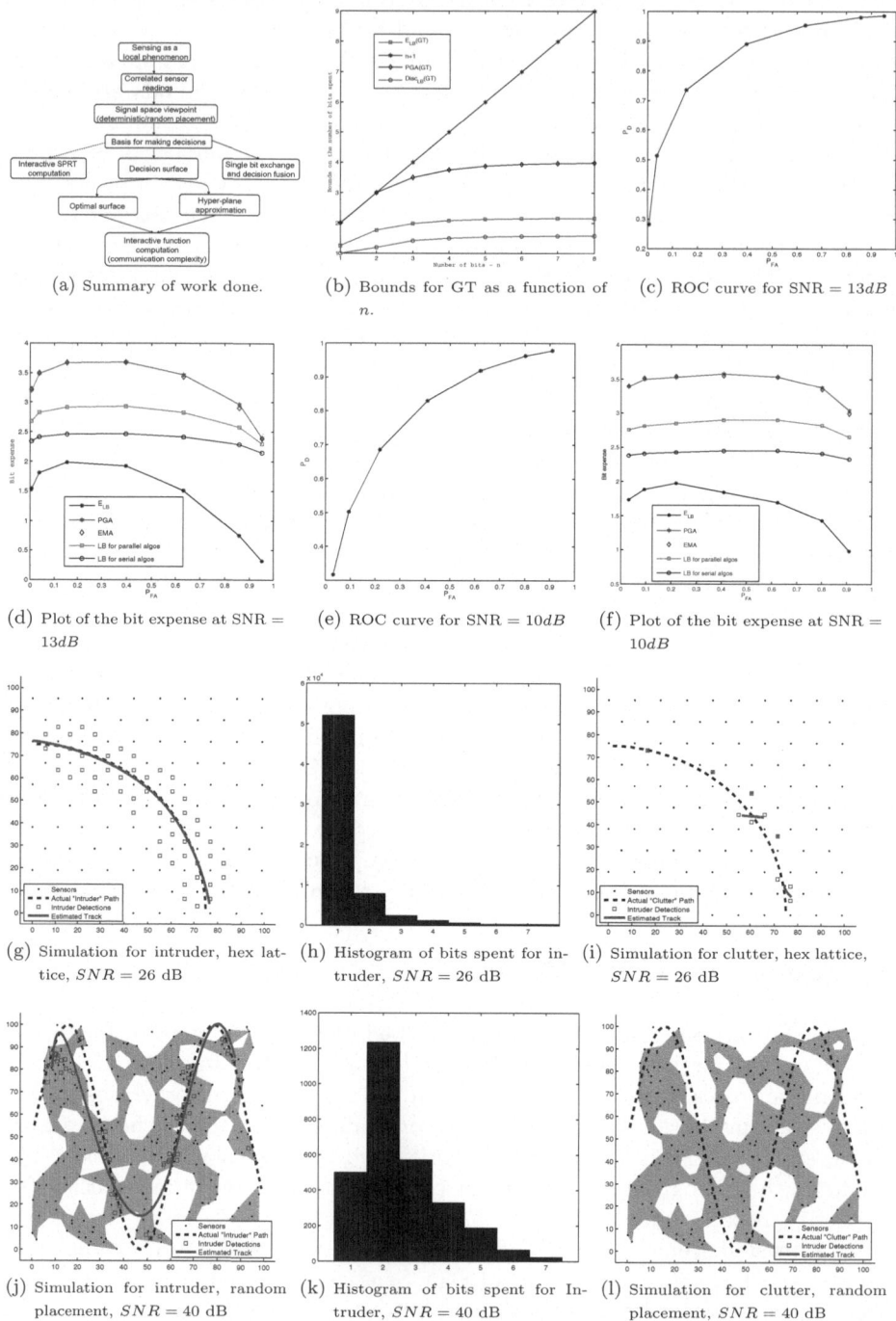

(a) Summary of work done.

(b) Bounds for GT as a function of n.

(c) ROC curve for SNR = 13dB

(d) Plot of the bit expense at SNR = 13dB

(e) ROC curve for SNR = 10dB

(f) Plot of the bit expense at SNR = 10dB

(g) Simulation for intruder, hex lattice, $SNR = 26$ dB

(h) Histogram of bits spent for intruder, $SNR = 26$ dB

(i) Simulation for clutter, hex lattice, $SNR = 26$ dB

(j) Simulation for intruder, random placement, $SNR = 40$ dB

(k) Histogram of bits spent for Intruder, $SNR = 40$ dB

(l) Simulation for clutter, random placement, $SNR = 40$ dB

Fig. 3. ROC, bit expense and simulation results

of the hexagonal grid placement[5], as relevant to sensor networks. The intruder is moved through the field at the speed of 1m/s. In the simulation plots Fig.3, the blue dotted lines represent the actual track. $v_I = 1e - 5$ g and $\kappa = 0.5$ for all simulations. For detection, a simple hyperplane test [9] is implemented in the signal space of three adjacent sensors. The hyperplane parameters are identical for any three sensor set. To minimize the number of bits spent in detection, the PGA is implemented for every set of three sensors. On positive detection, the intruder location is estimated to be at the centroid of the triangle (red squares). A sensor samples its reading at every 0.5 seconds and broadcasts the bit information, only if the reading value is above a certain threshold. To prevent collisions during transmissions, a simple random access protocol is used at the MAC layer, in which the sensor staggers its transmission by a random time. For tracking[6], a simple polynomial regression with the co-ordinates x and y taken as functions of time, is carried out for each local sets of detections. The local tracks are finally connected only if the detections are within a certain distance away. The estimated track is represented by the solid red curve. Noisy tracks can be rejected at the network level by their lack of length and direction which is again useful to separate an intruder from clutter.

A similar setting has been taken for the random placement (uniform) of sensors (Fig. 3(j), Fig. 3(l)), with a sensor density of 250 nodes per square km. The green shaded area in the background represents coverage of the field by Delaunay triangulation. A pseudo Delaunay triangulation[7] is carried out based on a local algorithm. Triangles are not formed when the sensors at the vertices of the triangle are out of radio range with respect to each other which leads to the holes shown in the field. The Delaunay triangulation for this sensor placement is shown in Fig. 1(a). It can be seen that the performance is satisfactory.

7   Conclusion

In this paper, we have discussed the problem of detection of an intruder in the presence of clutter. We introduce a signal-space viewpoint to the detection problem to tackle the case of correlated readings. We then pursue two approaches to the problem of distributed detection. In the first, a decision surface separating $\mathcal{S}_\mathcal{I}$ and $\mathcal{S}_\mathcal{C}$ is identified using Neyman-Pearson criteria and the individual sensor nodes interactively exchange bits to determine whether the sensor readings are on one side or the other of the decision surface. Bounds on the number of bits needed to be exchanged are derived, based on communication complexity (CC) theory. A lower bound derived for the two-party average case CC of general functions is compared against the performance of a greedy algorithm. In the second approach, each sensor node broadcasts a single bit arising from appropriate two-level quantization of its own sensor reading, keeping in mind the fusion rule to

---

[5] The hexagonal lattice can also be viewed as an equilateral triangle lattice which is in fact the Delaunay triangulation of the hexagonal lattice.

[6] The details of the tracking algorithm are not discussed here.

[7] Not discussed here for sake of brevity.

be subsequently applied at a local fusion center. The optimality of a threshold test as a quantization rule is proved under simplifying assumptions. QualNet simulation results are also provided.

The following are some points that need to be considered. We have assumed fixed intruder and clutter amplitudes. But in reality this may not be a good assumption (though many results in the literature assume a constant mean under the intruder hypothesis and zero mean under that of clutter). One could try to generalize the results for random amplitudes of intruder and clutter [8]. It may be more interesting to find out the minimum number of messages of a fixed length that need to be exchanged between the nodes for detection. Such a formulation is also present in CC, wherein the number of rounds of communication is minimized. The complexity of calculating $E_{LB}(f)$ may become exponential in the number of bits n, unless we are able to simplify the expression as in the case of the GT function. However for our detection application the value of n is usually not more than 8 and hence a brute force method could be used to calculate the bound. In cases where we cannot get the prior probabilities, the problem could be posed as minimizing the number of bits exchanged in any one of the hypothesis (assumed equally likely here). Analysis of the robustness of the algorithms to sensor localization errors and node failures is another interesting research direction.

References

1. Alhakeem, S., Varshney, P.K.: A unified approach to the design of decentralized detection systems. IEEE Trans. on Aerospace and Electronic Systems 31(1), 9–20 (1995)
2. Willett, P., Swaszek, P.F., Blum, R.S.: The good, bad and ugly: distributed detection of a known signal in dependent Gaussian noise. IEEE Trans. on Signal Processing 48(12), 3266–3279 (2000)
3. Swaszek, P.F., Willett, P.: Parley as an approach to distributed detection. IEEE Trans. on Aerospace and Electronic Systems 31(1), 447–457 (1995)
4. Chair, Z., Varshney, P.K.: Optimal data fusion in multiple sensor detection systems. IEEE Trans. on Aerospace and Electronic systems 22(1), 98–101 (1986)
5. Niu, R., Varshney, P.K., Moore, M., Klamer, D.: Decision fusion in a wireless sensor network with a large number of sensors. In: Proc. 7th IEEE International Conf. on Information Fusion (ICIF 2004) (2004)
6. Tang, Z.B., Pattipati, K.R., Kleinman, D.L.: A distributed M-ary hypothesis testing problem with correlated observations. In: Proc. of the 28th IEEE Conf. on Decision and Control, pp. 562–568 (1989)
7. Tsitsiklis, J., Athans, M.: On the complexity of decentralized decision making and detection problems. IEEE Trans. on Automatic Control 30(5), 440–446 (1985)
8. Tu, Z., Blum, R.S.: On the Limitations of Random Sensor Placement for Distributed Signal Detection. In: IEEE International Conf. on Comm., 2007. ICC 2007, pp. 3195–3200 (2007)

---

[8] We have carried out simulations for a uniform distribution on κ and the results were found to be satisfactory.

9. Agarwal, T., Venkatesan, N.E., Sasanapuri, M.R., Kumar, P.V.: Intruder detection over sensor placement in a hexagonal lattice. In: Proc. The 10th International Symposium on Wireless Personal Multimedia Comm., Jaipur, India (2007)
10. Giridhar, A., Kumar, P.R.: Computing and communicating functions over sensor networks. IEEE Journal on Selected Areas in Comm. 23(4), 755–764 (2005)
11. Amick, H.: A Frequency-Dependent Soil Propagation Model. In: SPIE Conf. on Current Developments in Vibration Control for Optomechanical Systems, Denver, Colorado (1999)
12. Ekimov, A., Sabatier, J.M.: Ultrasonic wave generation due to human footsteps on the ground, Acoustical Society of America(ASA) (2007)
13. Mazarakis, G.P., Avaritsiotis, J.N.: A prototype sensor node for footstep detection. In: Proc. of the Second European Workshop on Wireless Sensor Networks, pp. 415–418 (2005)
14. Yao, A.C.C.: Some complexity questions related to distributive computing (Preliminary Report). In: Proc. of the eleventh annual ACM symposium on Theory of computing, pp. 209–213 (1979)
15. Kushilevitz, E., Nisan, N.: Communication Complexity. Cambridge University Press, Cambridge (1997)
16. Bar-Yossef, Z., Jayram, T.S., Kumar, R., Sivakumar, D.: Information theory methods in communication complexity. In: Proc. 17th IEEE Annual Conf. on Computational Complexity, pp. 72–81 (2002)
17. Chakrabarti, A., Shi, Y., Wirth, A., Yao, A.: Informational complexity and the direct sum problem for simultaneous message complexity. In: Proc. 42nd IEEE Symposium on Foundations of Computer Science 2001, pp. 270–278 (2001)
18. Dietzfelbinger, M., Wunderlich, H.: A characterization of average case communication complexity. Information Processing Letters 101(6), 245–249 (2007)
19. Nisan, N.: The communication complexity of threshold gates. In: Proc. of Combinatorics, Paul Erdos is Eighty, pp. 301–315 (1993)

Fault-Tolerant Compression Algorithms for Delay-Sensitive Sensor Networks with Unreliable Links

Alexandre Guitton[1], Niki Trigoni[2], and Sven Helmer[3]

[1] Clermont University / LIMOS (CNRS), Clermont-Ferrand, France
[2] Computing Laboratory, University of Oxford, Oxford, United Kingdom
[3] Birkbeck, University of London, London, United Kingdom

Abstract. We compare the performance of standard data compression techniques in the presence of communication failures. Their performance is inferior to sending data without compression when the packet loss rate of a link is above 10%. We have developed fault-tolerant compression algorithms for sensor networks that are robust against packet loss and achieve low delays in data decoding, thus being particularly suitable for time-critical applications. We show the advantage of our technique by providing results from our extensive experimental evaluation using real sensor datasets.

1 Introduction

Large-scale multi-hop networks consisting of inexpensive battery-powered nodes that continuously monitor an environment are on the horizon. The main obstacles are the limited energy and bandwidth resources combined with unreliable wireless links. Since transmission is usually the most costly operation of a sensor node, compressing the data before transmissions seems like a good idea, as this saves bandwidth and energy. However, we cannot run just any compression algorithm: we need algorithms that are simple enough to run in an energy-conserving manner on processing-limited hardware, yet are robust enough against packet loss.

Sadler and Martonosi [17] recently proposed computationally-efficient lossless compression algorithms and studied the local energy tradeoffs for a single data-producing node performing the compression. Their technique exhibits significant energy savings at the intermediate nodes between the source and the sink, which benefit from forwarding less data. The focus was on designing sensor-adapted versions of LZW, which are tailored to nodes with limited computation and storage capabilities. Sadler and Martonosi generate the encoding/decoding table on the fly. Thus as soon as a single packet is lost during transmission, the rest of the data in a compression block (which spans several packets) cannot be decompressed anymore. The solution offered in [17] is for the receiver to send block acknowledgments and for the sender to try to retransmit any dropped packets. This causes significant delays in the decoding process especially in the

S. Nikoletseas et al. (Eds.): DCOSS 2008, LNCS 5067, pp. 190–203, 2008.
© Springer-Verlag Berlin Heidelberg 2008

case of large compression blocks[1] and intermittent/asymmetric links where not only data but also block acknowledgments can be lost.

There are environments in which delays are detrimental. For example, consider a target tracking application in which multiple sensors must combine their readings to localize a target, and to wake up sensors in the target's direction. In this case, timely dissemination and merging of their readings is critical in order not to miss the moving target. In fact it is preferred to successfully transmit a subset of the sighting readings in a timely fashion than to successfully transmit all of them after a long delay. The problem of delayed transmissions is exacerbated in networks that use a TDMA-based MAC protocol, in which sensors operate with low radio duty cycle to preserve energy and to avoid packet collisions.

Our first goal is to understand the behavior of standard compression techniques in unreliable networks, and compare their ability to deliver packets on links of varying quality. Our second goal is to develop fault-tolerant variants of compression algorithms that are robust against packet loss (i.e., all successfully delivered packets can be decompressed), while at the same time being able to adjust to changing data distributions. One of the strongest points of our approach is the ability to avoid transmission delays when faced with unreliable links. We achieve this by delaying the update of the encoding/decoding tables. In the case of reliable links, our approach exhibits communication savings comparable to those of the original algorithms. But unlike the original algorithms, our technique degrades gracefully when the conditions turn unfavorable.

Like [17], we do not consider data compression techniques that are carefully tailored to specific data distributions (where we could exploit known spatio-temporal correlations to reduce communication) or to specific query types (where for some aggregate functions, we could apply in-network aggregation to reduce the cost of result propagation). A brief overview of these techniques is presented in Section 6. Our techniques are more general, and are better suited to dynamic environments with changing data distributions, little knowledge of data correlations and a wide range of queries.

Error correcting codes (ECCs) could be used to repair bits (within a packet) flipped during transmission. Our approach is orthogonal to ECCs, as we do *not try to repair erroneous packets*, but ensure that we are able to interpret (decompress) packets that arrive successfully. That is, we strive to preserve independence between packets, so that dropped packets do not compromise our ability to decompress received packets.

The main contributions of our paper are the following: (1.) We compare standard compression techniques, like Adaptive Huffman, LZ77 and LZW, in terms of their efficiency in compressing information and sending it across lossy channels. The goal is to highlight the need for fault-tolerant compression schemes in lossy wireless environments. (2.) We study existing compression techniques that use packet retransmissions to cope with communication failures. We refer

---

[1] One the one hand, larger block sizes lead to a better compression rate, on the other hand, they also lead to longer delays.

to these algorithms as RT (ReTransmission) algorithms and we identify their shortcomings in the presence of intermittent or asymmetric links. (3.) We design novel fault-tolerant compression algorithms (FT algorithms) that overcome the shortcomings of the RT algorithms. We assess the delay and energy efficiency of the two approaches using two performance metrics defined in Section 2. For our comparison, we apply the FT and RT algorithms to real datasets of car and animal traffic data.

The remainder of this paper is organized as follows: Section 2 describes the communication model, and defines the performance metrics that we use to assess the performance of compression algorithms in sensor networks. Section 3 discusses standard compression techniques and their RT (ReTransmission) counterparts and points out the weaknesses of these existing approaches. Section 4 introduces a novel fault-tolerant mechanism that can easily be applied to various compression techniques, leading to a novel class of FT algorithms. Section 5 presents an experimental evaluation of the RT and FT algorithms, Section 6 discusses related work and Section 7 concludes the paper and identifies directions for future work.

2   Preliminaries

Communication model: It is widely accepted in the mobile and sensor network communities that radio propagation (i) is non-isotropic, i.e. connectivity is not the same in all directions from the sender at a given distance, and (ii) features non-monotonic distance decay, i.e. smaller distance does not mean better link quality, and (iii) exhibits asymmetrical links [1,5,9,12,24]. It has also been shown that there is considerable difference from link to link in the burstiness of the delivery, and most node pairs that can communicate have intermediate loss rates [1].

Our work is motivated by the difficulties that arise from applying standard compression algorithms and propagating compressed data across asymmetric links with intermediate loss rates. Depending on the link quality, interference and other environmental conditions, a number of packets will be dropped, and energy or channel considerations may not allow us to retransmit until 100% delivery is achieved. In our study, we consider two packet loss models:

- A model in which packet losses are independent and identically distributed; in this case, we vary the probability of successful packet delivery to study the performance of compression algorithms in varying error conditions.
- A first-order Markov model for the success/failure process of packet transmissions. This bursty error model, known as the Elliot-Gilbert model [8,10], is shown to accurately approximate the behavior of a fading radio channel in [26].

Performance metrics: We need to measure the amount of information that is successfully received and usable at the receiver node, as well as the delay until it becomes usable. Consider a dataset of sensor data that has been accumulated at a sender node S and must be delivered to a receiver node R along a lossy link. We compare the efficiency of various data compression techniques based on the following metrics:

Bytes of Decoded Data / Bytes Sent (BDD / BS): The metric BDD measures the number of bytes derived from successfully decoding received packets. When data is propagated without compression, this metric measures the number of bytes in successful packet transmissions. When data is propagated in a compressed manner, some packets may be successfully received, but not successfully decoded. These packets will not add to the bytes of decoded data (BDD). Hence the metric BDD/BS represents the amount of *useful* information received by the destination R, relative to the effort that the source node S puts into sending it. In an ideal channel without packet drops, this metric reflects the compression ratio achieved by an encoding scheme, i.e. the number of bytes of uncompressed data divided by the number of bytes of the compressed representation. In the context of unreliable channels, this metric combines two aspects of an algorithm's behavior: (i) its ability to reduce the size of the original sensor data as reflected by the compression rate (as in the ideal channel) and (ii) how well it encodes data into a format that can be decoded safely even in the presence of packet drops.

Delay (DEL). This metric measures the delay between the time that a packet is first sent from the sender node S and the time that it is ready to be successfully decoded by the receiver R. The delay has two components: *delivery time* and *decode waiting time.*

For packets delivered successfully the first time they are sent, the delivery time is equal to the packet transmission time: the latter consists of the delay waiting for access to the transmit channel (access time), the time needed for the packet to transit from S to R (propagation time) and the processing time required for R to pick up the packet from the channel (receive time). For simplicity, we assume that the packet transmission time is constant for all packet transmissions and lasts for 1 time unit. If a packet is dropped on the first attempt, and there are k intermediate packet transmissions (of the same or other packets) before it is retransmitted successfully, then the packet's delivery time is $k + 2$ time units.

The *decode waiting time* is the time between the arrival of a packet and its decompression. When the compression dictionary is up-to-date on receipt of a packet (or no compression is used), the decode waiting time is 0. We will show in Section 3 that existing compression algorithms require packets to be received in the order in which they were encoded, so that they can keep the compression dictionary up-to-date and decode packets successfully. In this case, packets received out of order may experience higher delays before they can be decoded, until all packets that preceded them in the encoding process are also successfully received by R.

3   Existing Algorithms

Before delving into details on how our adapted compression algorithms for sensor networks operate, we are going to briefly review some popular compression algorithms. We will demonstrate why these algorithms fail to achieve good performance when deployed in typical sensor network scenarios and show how an approach by Sadler and Martonosi [17] tries to overcome some of these problems.

3.1   Standard Compression Algorithms

LZW LZW is a refinement of the original Ziv-Lempel approach (LZ77 [25]) by Welch [22]. LZW replaces strings of symbols with (shorter) code words. The mapping of strings of symbols is done with the help of a table (or dictionary). The initial table maps individual symbols to default codes, e.g. characters to their ASCII code. Every time a new string is encountered – a substring of length $i + 1$ starting with the current symbol that is not yet found in the table – the code for the substring is output and a new code for the substring of length $i + 1$ is added to the table. Substring codes are restricted to a certain number of bits and a new code is assigned the smallest unused number. When decompressing, the coding table can be built on the fly, as long as the same initialization table is used on both the sender and the receiver side.

Adaptive Huffman Coding. Standard Huffman coding [11] uses a compression/decompression dictionary that is based on known symbol frequencies. Adaptive Huffman coding [21] determines the symbol frequencies based on the input it receives and builds the dictionary on the fly. Both sender and receiver start with an empty tree and for each symbol that is transmitted, modify the tree accordingly. Each node in Adaptive Huffman coding has a weight associated with it: the number of occurrences of a certain symbol for leaf nodes, and the sum of the weights of the children for an inner node. In addition to this, nodes are numbered in non-decreasing order by weight. This numbering scheme is called *sibling property* and reflects the order in which nodes are combined using regular Huffman coding. When updating a tree after the transmission of a symbol, it does not suffice to adjust the weights of the corresponding leaf node and its ancestors. In order to guarantee the sibling property, the algorithm may have to swap some nodes.

3.2   Weaknesses of Standard Compression Algorithms

Let us now have a look at how the algorithms described above perform in a typical sensor network environment. We simulated the transmission of road traffic data over a lossy link varying the packet drop rate of the link from 0% to 90%. Our findings are quite surprising. Figure 1 shows that, in the presence of packet loss, most of the compression algorithms (except for Huffman coding) degrade to the point where using them is worse than not compressing at all.[2]

---

[2] The value for BDD/BS is lower than 1 for no compression due to the overhead of packet headers.

What are the reasons for this? Losing packets has a devastating effect on LZ77 [25], LZW, and Adaptive Huffman coding, because these algorithms rely on previously decompressed data to successfully decompress the remainder of the data. If we lose any of this data, the algorithms will not be able to continue decoding packets successfully. Ordinary Huff-

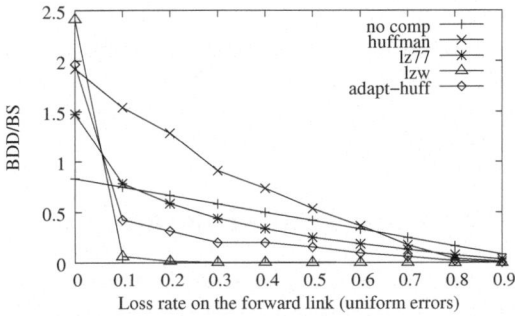

Fig. 1. Using standard compression algorithms

man coding can restart decompression after losing packets, since it does not rely on previous data. This makes it robust against packet losses, once the dictionary has been sent successfully. Nevertheless, Huffman coding also has drawbacks. It needs to know the data distribution before starting in order to yield an acceptable compression ratio. Every time this distribution changes, we have to construct a new encoding table and transmit this table to the receiver side. Doing so over an unreliable channel may result in considerable overhead.

3.3 S-LZW

Sadler and Martonosi adapted the LZW compression algorithm to sensor networks [17]. Packets are sent in blocks. In order to cope with lost packets, they keep the size of blocks containing interdependent packets small. Each block is compressed with its own dictionary, which means that a packet loss will only affect the following packets of the same block. A small block size also means that the dictionary for encoding the packets of the block will not grow very large (Sadler and Martonosi work with dictionaries having a size of 512B) keeping the memory requirements down. When looking at the effects of unreliable links on their algorithm, Sadler and Martonosi only take into consideration the energy savings of their approach. When dealing with reliability issues they rely on a RT (ReTransmission) strategy [17]: "The receiver then sends a 32B acknowledgment that tells the sender which packets it received correctly and the sender retransmits what is missing. If the transmission is not complete and the receiver does not start receiving data immediately, it assumes that the acknowledgment was lost and retransmits it. This process iterates until the receiver has successfully received and acknowledged the whole block."

However, there are no results in the paper on how much delay this causes to the data transmission. Let us illustrate this with a drastic example: assume that we transmit a block consisting of 50 packets and the first packet gets lost. All packets after the first cannot be decompressed until the first packet arrives. This retransmission will take place only after an acknowledgment has been sent, which means that there is considerable delay until the whole block is decompressed. Worse, if the connection breaks down before the first packet is resent, the whole information contained in the first block is lost. This is why Sadler and Martonosi opt for small block sizes. This, however, compromises LZW's compression rate, as LZW needs some time before reaching its full compression potential. We will demonstrate the adverse effects of the RT strategy on the speed of delivering packets in Section 5.

4   Proposed Fault-Tolerant Algorithms

In the previous section, we showed that the performance of standard compression techniques degrades significantly in the presence of packet losses, and advocated the need for fault-tolerant algorithms. The current state of the art in fault-tolerant compression consists of RT algorithms that retransmit packets in case they are dropped with the goal of achieving 100% delivery. We have shown that the RT algorithms incur high delays in packet decoding when the forward or backward channel is lossy, which may not be acceptable for delay-sensitive applications. The cause of this problem is that the RT algorithms update their dynamic dictionaries as soon as new data is being compressed or uncompressed. When no fault-tolerant mechanism is provided and some packets are lost, the sender and receiver end up using different versions of the dictionary. The RT algorithms amend this problem by retransmitting all dropped packets, but this significantly delays the process of decoding received packets.

In this section, we propose a novel mechanism that addresses these problems and can be applied to standard compression techniques, like Adaptive Huffman, LZ77 and LZW. The proposed class of algorithms, referred to as FT algorithms (e.g. FT-LZW), are robust to sudden interruptions of connectivity, and they incur considerably less delay than the RT algorithms when the forward or backward channel is lossy. Our fault-tolerant mechanism has three parts:

Block acknowledgments: The sender sends packets in blocks of n and expects to get an acknowledgment (ACK) from the receiver at the end of a block. Packets of a block are numbered from 1 to n, and when the receiver successfully receives a packet, it reads the sequence number of the packet. At the end of a block, the receiver sends an ACK that contains an n-bit vector. The i-th bit of the vector is 1 if packet i was successfully received, and 0 if it was dropped.

Periodic dictionary updates: Dictionary updates occur at the sender only immediately after receiving a block ACK. The sender reads the bit vector in the ACK to understand which packet transmissions were successful. It then modifies the dictionary by considering only the symbols encoded in the successfully delivered packets. It updates the dictionary, and uses the new updated dictionary to encode information in the next block of packets. Of course, the exact mechanism in which the dictionary is updated depends on the specifics of the algorithm. However, the idea of updating a dictionary not after each symbol, but after a block of successfully received symbols, is common to all FT algorithms and can be applied easily to a variety of compression techniques like LZW, LZ77 or Adaptive Huffman.

Countering link unreliability: Since the medium is unreliable, it is possible that the ACK sent by the receiver is lost. However, it is crucial for updating the dictionary that both the sender and the receiver agree upon which packets were successfully transmitted. Instead we propose the following simple protocol: If the sender receives a block ACK, it updates the data dictionary, otherwise the dictionary remains unchanged. Each data packet transmitted by the sender (encoder) in the next block contains a flag that denotes whether the encoder has received an ACK message for the last block and has changed its dictionary accordingly. For example, when the packets of the current block start with 1, it means that the encoder received an ACK for the previous block and changed its local dictionary based on the symbols included in the successful packets. This notifies the receiver (decoder) to also update its local version of the dictionary based on the received packets of the previous block, before using the dictionary to decode the packets of the current block.

Let us illustrate the application of our mechanism to Adaptive Huffman (the resulting algorithm being FT-Adapt-Huff). Assume that the sender sends blocks of three packets each, and the receiver sends ACKs back at the end of each block (see Figure 2 for an example). If we assume that the ACK sent for Block $(i-1)$ is dropped, the sender does not update its dictionary before starting to transmit packets of Block

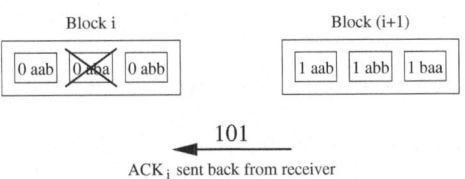

Fig. 2. Example of blockwise transmission in the FT algorithms

i. The sender denotes this by starting each packet of Block i with the flag 0. The receiver also refrains from updating the dictionary before decoding packets of Block i. At the end of Block i, the sender receives ACK 101 and updates its dictionary by inserting the stream of symbols from packets 1 and 3. As soon as the receiver picks up a packet of Block $(i+1)$ with a leading 1 bit, it also modifies its dictionary according to the symbols in the acknowledged packets of Block i and then immediately starts decompressing the packets of Block $(i+1)$ based on the updated dictionary.

Note that this fault-tolerant mechanism is applicable to several compression algorithms with dynamic dictionaries. An important feature of this mechanism is that it does not delay the transmission of data, but only the adaptation of the dictionary at the encoder and decoder. This is in contrast with the RT algorithms who delay the decoding of a received packet, until all preceding packets are successfully retransmitted. Our approach requires buffering a block's data locally at the encoder and the decoder, in order to adjust the data dictionary at the end of each block, in case an ACK is received. Our approach is fault-tolerant to losses of both data packets and ACK packets, and it is suitable for links with intermittent connectivity, that occasionally become asymmetric thus prohibiting the transmission of ACKs. It combines adaptive encoding with the ability to survive packet drops on links with intermediate or low link quality.

5   Experimental Evaluation

In Section 3, we discussed standard compression algorithms and showed that they are not suitable for sensor networks with lossy links. In particular, Figure 1 illustrated that most adaptive compression algorithms behave worse than having no compression at all (in terms of our metric: Bytes of Decoded Data / Bytes Sent), when more than 10% of the packets sent across a link get dropped. In this section, we evaluate the performance of compression algorithms that have a built-in mechanism for handling packet drops, focusing on the transmission delay. We compare the existing class of RT algorithms [17], which are based on packet retransmissions, to our novel class of FT algorithms, which are based on periodic dictionary updates, showing that we can achieve much better delay rates without compromising the energy consumption.

5.1   Simulation Setup

In our comparison, we use two performance metrics defined in Section 2: i) DEL and ii) BDD/BS. Recall that the first metric reflects the delay between the time that a packet is first sent by the sender to the time that it is ready to be decoded at the receiver, in abstract time units. The second metric reflects the communication and thus energy efficiency of an algorithm.

   In our experiments, we vary the error rate (packets dropped / packets sent) of the forward and backward direction of a lossy link. In both FT and RT algorithms, the forward link is used to send compressed data, whereas the backward link is used to propagate ACKs. When not explicitly defined, the default value of error rate at the forward and backward links is 20%. We compared existing and new algorithms using both uniform and bursty error rates. Because of limited space, we decided to include only the graphs that concern bursty errors which models the real world more closely. The Elliot-Gilbert loss model is initialized with two parameters p and r, where p is the probability that a packet transmission is successful given that the previous one was successful, and r is the probability that a packet transmission is successful given that the previous one was unsuccessful. This leads to a percentage of erroneous transmissions ϵ of $1 - r/(1 - p + r)$. To achieve a loss rate of a given ϵ, we choose $p = 1 - 0.3\epsilon$ and $r = (1 - \epsilon)(1 - p)/\epsilon$. For $\epsilon = 0.2$, this results into having $p = 0.94$ and $r = 0.24$.

   We also consider the impact of the block size on the performance of the two classes of algorithms. We set the packet size to 60B, of which 10B are the header. This explains why the value of BDD/BS is less than 1 in the no-compression algorithm in Figures 1 and 4.

   Finally, we measured the impact of different compression algorithms and data sets on the value of BDD/BS. For compression algorithms we used LZW and Adaptive Huffman, applying the FT and RT mechanisms to them. The first data

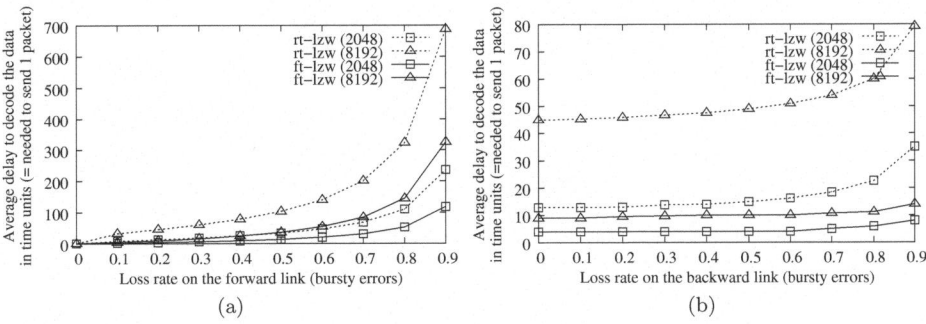

Fig. 3. Comparing RT to FT algorithms in terms of delay for forward and backward errors

set we used was the Scoot dataset, where data is generated by 118 inductive loops monitoring the flow of cars on several roads in Cambridge, UK[3]. We focus on the data generated by 68 sensors which is propagated along the most congested branch of the collection tree, from the neighbor of the gateway to the gateway. We also used a dataset from the Zebranet 2 experiment, in which several zebras wear a collar with a GPS device. For space reasons, we present the experimental results that were based on the Adaptive Huffman encoding and the Zebranet dataset in the extended version of our paper [20].

5.2 Simulation Results for Delay Rate

Impact of loss rate on the forward link: We start by applying the FT and RT mechanisms to the LZW algorithm, and compare the two approaches as we vary the rate of packet losses across the forward link. We use a bursty error model, i.e. we have bursts of successful packet transmissions followed by bursts of dropped packets. We compare FT and RT algorithms using blocks of two different sizes: 2048 bytes and 8192 bytes, corresponding to approximately 20 and 66 packets respectively. In terms of delay, our FT algorithms significantly outperform their RT counterparts. Figure 3(a) shows that as we increase the loss rate of the forward link, more compressed packets get dropped, and the average number of retransmissions per packet increases. Hence, both FT and RT algorithms experience higher delays due to retransmissions. However, the delay exhibited by the RT algorithms is 2-3 times higher than that by the FT algorithms for a given loss rate. The reason is that in the FT algorithms, packets can be decoded as soon as they are successfully received, whereas in the RT algorithms, packets can be decoded only if all preceding packets in the block are successfully received and decoded.

---

[3] http://www.cl.cam.ac.uk/research/dtg/~rkh23/time/timeloops.html

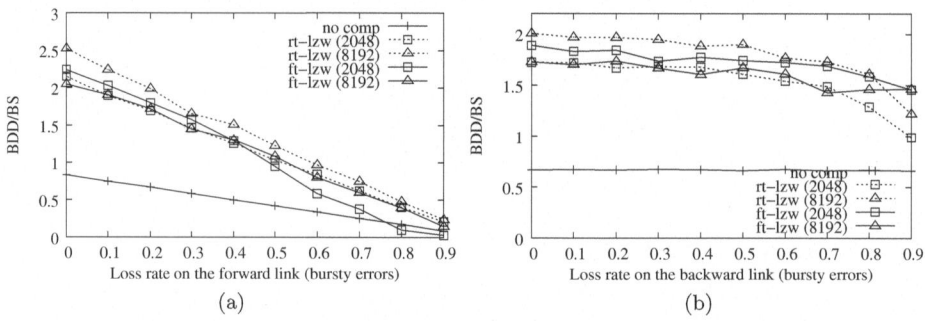

Fig. 4. Comparing RT to FT algorithms in terms of BDD/BS for forward and backward errors

Impact of loss rate on the backward link: Figure 3(b) shows another interesting feature of our FT algorithms, that makes them particularly appealing in the presence of asymmetric links. The delay of successfully decoding packets at the receiver remains almost constant as we increase the loss rate of the backward link from 0% to 90%. However, the delay of the RT algorithms rises significantly as we increase the error rate in the backward link. The reason is that the sender waits for an ACK packet before proceeding to the next block, which never arrives due to bursty errors in the backward link. Hence, our FT algorithms are particularly delay-efficient in the presence of asymmetric links with good quality of the forward link, but bursty and unreliable connectivity in the other direction.

Impact of block size: The effect of block size (2048 bytes and 8192 bytes) on the delay of the FT and RT algorithms is also illustrated in Figure 3(a) and 3(b). For a given loss rate of the forward link (or of the backward link), the higher the block size, the higher the delay in both FT and RT algorithms. The reason is that the larger the block, the longer the sender waits for an ACK that is necessary to determine which packets need retransmission. As we mentioned above, FT algorithms are always faster than RT algorithms, given a fixed block size, because the packets sent by the FT algorithms are ready to be uncompressed as soon as they arrive at the receiver, even if preceding packets have not arrived successfully.

5.3   Simulation Results for Energy Consumption

Due to space constraints, we only show a small excerpt of all the results we have obtained for the energy consumption.[4] Figure 4(a) shows that, in terms of BDD/BS, both FT and RT algorithms degrade gracefully (drop linearly) as we increase the loss rate of the forward link. They are both 2-3 times more

---

[4] A detailed analysis of these results are included in the extended version of our paper [20].

efficient than not compressing data at all, even for very lossy links. In terms of communication efficiency (BDD/BS) we observe very small differences between the FT and RT approaches.

6   Related Work

Compression, the process of finding and removing redundancy in information, is a powerful tool for data propagation in energy- and bandwidth-constrained sensor networks. A detailed discussion of trade-offs (between rate and distortion) in sensor networks can be found in [6]. However, in our paper, we focus on lossless algorithms (i.e. no distortion is allowed).

A large body of research has recently focused on distributed compression techniques for sensor networks [3,7,14]. According to the Slepian-Wolf coding theorem [19], if the joint distribution quantifying the correlation structure between two data sources is known, then, in theory, sensor nodes can compress their data independently without internode communication, by exploiting knowledge of correlations with other nodes. In our work we do not assume any spatio-temporal correlations among nodes, which might anyway change dynamically over time, but instead we focus on compressing data that is available locally at each node independently of other sensors' data.

Scaglione et al. [18] study the interdependence of routing and data compression in wireless sensor networks (WSNs). They consider the problem of collecting samples of a field at each sensor node, and using them to obtain an estimate of the entire field within a distortion threshold. Puri et al. [15] consider the problem of rate-constrained estimation of a physical process based on reliably receiving data from only a subset of sensors. Rachlin et al. [16] study the interdependence of sensing and estimation complexity in WSNs. Their goal is to estimate the underlying state of the environment based on noisy sensor observations. Xiao et al. [23] study how the signal processing capability of a WSN scales up with its size, and explore distributed signal processing algorithms with low bandwidth requirements using efficient local quantization schemes. Whereas these works focus on estimating a physical phenomenon based on unreliable noisy data, our paper aims to compare the performance of standard source coding techniques in terms of reliability, and to devise fault-tolerant versions of these techniques for WSNs. The impact of spatial correlation on routing with compression has recently been studied by Pattem et al. [13]. Unlike our work, they consider carefully selecting routes to maximize the potential of compressing spatially correlated data.

Our work is more similar to the study of Sadler and Martonosi on data compression algorithms for energy-constrained devices [17]. Like [17] we focus on compressing data locally without considering correlations with other nodes, in order to reduce the overall energy consumption of wireless data propagation. Unlike [17], which focuses on designing algorithms for nodes with limited computation and storage constraints, our goal is to tackle the problem of unreliable transmissions of compressed data. PINCO [2] buffers sensor data at each node for as long as possible, whilst adhering to user-defined end-to-end delay requirements. Energy savings are achieved by reducing available redundancy of buffered data before communicating them over the wireless medium. Barr et al. [4] measure the energy requirements of several lossless data compression schemes using a specific (Skiff) platform. They point out that energy savings can be achieved in some cases by not applying the same algorithm for both compression and decompression. Unlike our work, [2] and [4] do not address the problem of recovering from unsuccessful transmissions of compressed data.

7   Conclusions

We demonstrated that standard compression techniques behave worse in terms of communication efficiency than using no compression at all, on links with packet loss rates greater than 10%. We thus directed our attention at robust compression techniques, namely the existing RT (ReTransmission) mechanism and our novel FT (Fault-Tolerant) mechanism.

In relatively static networks the FT mechanism is comparable to the RT mechanism in terms of communication efficiency. Both variants degrade linearly in the loss rate of the forward link. However, the proposed FT algorithms are 2-3 times faster than their RT counterparts in delivering usable packets. The reason is that the FT mechanism allows each successfully received packet to be decompressed immediately without waiting for preceding packets of the same block to arrive successfully. The delay of FT algorithms remains relatively constant even for asymmetric links, i.e. cases where the backward link has a prohibitively high loss rate close to 80%.

In the future, we plan to investigate fault-tolerant compression in large-scale sensor deployments. As we propagate data along multi-hop paths, we have two choices: to compress/decompress data either hop-by-hop or only at the two ends of the path. The former may allow us to exploit correlations of data coming from neighboring nodes, but at the same time it would require more computation and memory resources. Our plan is to compare the two approaches in terms of energy-efficiency, delay and robustness. It would also be interesting to do this comparison based on a formal, probabilistic model (in order to derive bounds on the performance).

References

1. Aguayo, D., Bicket, J., Biswas, S., Judd, G., Morris, R.: Link-level measurements from an 802.11b mesh network. In: SIGCOMM, pp. 121–132 (2004)
2. Arici, T., Gedik, B., Altunbasak, Y., Liu, L.: Pinco: A pipelined in-network compression scheme for data collection in wireless sensor networks. In: ICCCN (2003)
3. Baek, S., de Veciana, G., Su, X.: Minimizing energy consumption in large-scale sensor networks through distributed data compression and hierarchical aggregation. Technical report, University of Texas (2004)
4. Barr, K., Asanovic, K.: Energy aware lossless data compression. In: MOBISYS (2003)

5. Cerpa, A., Busek, N., Estrin, D.: Scale: a tool for simple connectivity assessment in lossy environments. Tech report CENS-21, UCLA (2003)
6. Chen, M., Fowler, M.L.: Data compression trade-offs in sensor networks. In: SPIE (2004)
7. Chou, J., Petrovic, D., Ramchandran, K.: A distributed and adaptive signal processing approach to reducing energy consumption in sensor networks. In: INFOCOM (2003)
8. Elliot, E.O.: Estimates of error rates for codes on burst-noise channels. Bell Systems Technical Journal 42, 1977–1997 (1963)
9. Ganesan, D., Krishnamachari, B., Woo, A., Culler, D., Estrin, D., Wicker, S.: Complex behavior at scale: An experimental study of low-power wireless sensor networks. CSD-TR 02-0013, UCLA (February 2002)
10. Gilbert, E.N.: Capacity of a burst-noise channel. Bell Systems Technical Journal 39, 1252–1265 (1960)
11. Huffman, D.A.: A method for the construction of minimum redundancy codes. IRE 40, 1098–1101 (1952)
12. Kotz, D., Newport, C., Elliott, C.: The mistaken axioms of wireless-network research. Technical Report 467, Dartmouth College Computer Science (July 2003)
13. Pattem, S., Krishnmachari, B., Govindan, R.: The impact of spatial correlation on routing with compression in wireless sensor networks. In: IPSN (2004)
14. Pradhan, S., Kusuma, J., Ramchandran, K.: Distributed compression in a dense sensor network. IEEE Signal Processing Magazine (2002)
15. Puri, R., Ishwar, P., Pradhan, S.S., Ramchandran, K.: Rate-constrained robust estimation for unreliable sensor networks. In: AsilomarSSC, vol. 1, pp. 235–239 (2002)
16. Rachlin, Y., Negi, R., Khosla, P.: On the interdependence of sensing and estimation complexity in sensor networks. In: IPSN, pp. 160–167 (2006)
17. Sadler, C.M., Martonosi, M.: Data Compression Algorithms for Energy-Constrained Devices in Delay Tolerant Networks. In: SENSYS (November 2006)
18. Scaglione, A., Servetto, S.D.: On the interdependence of routing and data compression in multi-hop sensor networks. In: MOBICOM, pp. 140–147 (2002)
19. Slepian, D., Wolf, J.K.: Noiseless coding of correlated information sources. IEEE Trans. Inform. Theory IT-19, 471–480 (1973)
20. Trigoni, N., Guitton, A., Helmer, S.: Fault-tolerant compression algorithms for sensor networks with unreliable links. In: Technical Report BBKCS-08-01, Birkbeck, University of London (2008),
http://www.dcs.bbk.ac.uk/research/techreps/2008/
21. Vitter, J.S.: Design and analysis of dynamic Huffman codes. J. ACM 34(4), 825–845 (1987)
22. Welch, T.A.: A technique for high-performance data compression. IEEE Computer 17(6), 8–19 (1984)
23. Xiao, J.-J., Ribeiro, A., Giannakis, G.B., Luo, Z.-Q.: Distributed compression-estimation using wireless sensor networks. IEEE Signal Processing Magazine 23(4), 27–41 (2006)
24. Zhao, J., Govindan, R.: Understanding packet delivery performance in dense wireless sensor networks. In: SENSYS, pp. 1–13 (2003)
25. Ziv, J., Lempel, A.: A universal algorithm for sequential data compression. IEEE Transac. on Inf. Theory 23(3), 337–343 (1977)
26. Zorzi, M., Rao, R.R., Milstein, L.B.: On the accuracy of a first-order markov model for data transmission on fading channels. In: ICUPC, pp. 211–256 (1995)

Improved Distributed Simulation of Sensor Networks Based on Sensor Node Sleep Time

Zhong-Yi Jin and Rajesh Gupta

Dept. of Computer Science & Eng, UCSD
{zhjin,rgupta}@cs.ucsd.edu

Abstract. Sensor network simulators are important tools for the design, implementation and evaluation of wireless sensor networks. Due to the large computational requirements necessary for simulating wireless sensor networks with high fidelity, many wireless sensor network simulators, especially the cycle accurate ones, employ distributed simulation techniques to leverage the combined resources of multiple processors or computers. However, the large overheads in synchronizing sensor nodes during distributed simulations of sensor networks result in a significant increase in simulation time. In this paper, we present a novel technique that could significantly reduce such overheads by minimizing the number of sensor node synchronizations during simulations. We implement this technique in Avrora, a widely used parallel sensor network simulator, and achieve a speedup of up to 11 times in terms of average simulation speed in our test cases. For applications that have lower duty cycles, the speedups are even greater since the performance gains are proportional to the sleep times of the sensor nodes.

1   Introduction

Simulations of wireless sensor networks (WSNs) are important to provide controlled and accessible environments for developing, debugging and evaluating WSN applications. In simulations, sensor network programs run on top of simulated sensor nodes inside simulated environments. Since the simulated entities (simulation models) are transparent to simulation users, the states and interactions of sensor network programs can be inspected and studied easily and repeatedly. In addition, the properties of the simulated entities such as the locations of sensor nodes and the inputs to the sensor nodes can be changed conveniently before or during simulations.

One of the key requirements for simulating WSNs is high simulation fidelity in terms of temporal accuracy of events and actions. High fidelity simulations usually come at the cost of increased simulation time and poor scalability because significant computational resources are required for running high fidelity simulation models. Parallel and distributed simulators [1] are developed to address these issues by leveraging the combined resources of multiple processors/cores on a same computer and on a network of computers, respectively. They can significantly improve simulation speed and scalability because sensor nodes may be

S. Nikoletseas et al. (Eds.): DCOSS 2008, LNCS 5067, pp. 204–218, 2008.

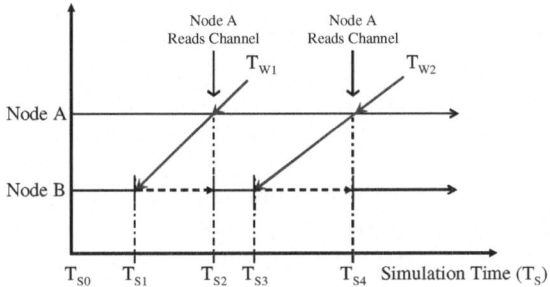

Fig. 1. The progress of simulating in parallel a wireless sensor network with two nodes that are in direct communication range of each other

simulated in parallel on different processors or computers. Since different nodes may get simulated at different speeds in this fashion, simulated nodes often need to synchronize with each other to preserve causalities and ensure correct simulation results.

Synchronizations of sensor nodes are illustrated in Figure 1 which shows the progress of simulating in parallel two sensor nodes that are within direct communication range of each other. There are two notions of time in the figure, wallclock time T_W and simulation time T_S. Wallclock time corresponds to the actual physical time while simulation time is the virtual clock time that represents the physical clock time of real sensor nodes in simulations [1]. At wallclock time T_{W1}, the simulation time of Node A (T_{SA}) is T_{S2} and the simulation time of Node B (T_{SB}) is T_{S1}. Node B is simulated slower than Node A in this case ($T_{SB} < T_{SA}$) because the thread or process that is used to simulate Node B either runs on a slower processor or receives fewer CPU cycles from an operating system (OS) task scheduler. At T_{S2}, Node A is supposed to read the wireless channel and continue its execution along different execution paths based on whether there are active wireless transmissions or not. However, despite the fact that the simulation time of Node A already reaches T_{S2} at T_{W1}, Node A probably can not advance any further because at T_{W1} Node A does not know whether Node B is going to transmit at T_{S2} or not (since $T_{SB} < T_{S2}$). There are two general approaches to handle cases like this, a conservative one, e.g., [2] and an optimistic one, e.g., [3]. With the conservative approach, Node A has to wait at T_{S2} until the simulation time of Node B reaches T_{S2}. The optimistic approach, on the other hand, would allow Node A to advance assuming there will be no transmissions from Node B at T_{S2}. However, the entire simulation state of Node A at T_{S2} has to be saved. If Node A later detects that Node B actually transmits to Node A at T_{S2}, it can correct the mistake by rolling back to the saved state and start again.

Almost all distributed WSN simulators are based on the conservative approach as it is simpler to implement and has a lower memory footprint. The conservative approach mainly involves three time-consuming steps. In the first step, the threads or processes that simulate the waiting nodes (Node A in Figure 1), have to be suspended. This would usually involve context switches to swap in other threads or processes that simulate other nodes. In the second step, the nodes that some other nodes are waiting for (Node B in Figure 1) have to notify the waiting nodes about their progresses. For example, in Figure 1, Node B must notify Node A after it advances past T_{S1} so that Node A can continue. In the last step, the waiting nodes need to be swapped back into execution once they are notified to continue. Step 1 and 3 are time-consuming as they involve context switches. Large numbers of context switches would significantly increase simulation time. Step 2 involves expensive inter-thread or inter-process communications and is generally the slowest step in distributed simulations as the notifications may have to be sent through slow networks to different computers.

Therefore, the performance gains in existing distributed WSN simulators are often compromised by the rising overheads due to inter-node synchronizations. This is reported in both Avrora [4], a cycle accurate simulator that runs over SMP (shared memory multiprocessor) computers, and DiSenS [5], a cycle accurate simulator for simulations over both SMP computers and a network of computers. In the case of Avrora, the reported time of simulating 32 nodes with 8 processors is only about 15% less than using 4 processors because of a large number of thread context switches introduced by synchronizations. In DiSenS, it is actually faster to simulate 4 nodes using 1 computer (dual-processor) than using 2 computers and simulating 2 nodes on each in most of the testing cases. This sub-linear performance in DiSenS is due to the large communication overheads in synchronizing nodes that are simulated on different computers.

In this paper, we describe a novel technique that could significantly reduce the number of synchronizations in distributed simulations of WSNs. Our technique exploits the sleep-often property of sensor network applications and uses sleep times to reduce sensor node synchronizations. It works without any prior knowledge of the sensor network applications under simulations and the number of reductions scale with both network sizes and sleep times. We demonstrate the value of our approach by its implementation in Avrora [4].

We describe our speedup technique in Section 2. Its implementation is presented in Section 3. In Section 4 we describe the results of our experiments followed by a brief overview of the related work and conclusion in Section 5 and Section 6.

2   A Novel Technique to Reduce Synchronizations in Distributed WSN Simulations

Sensor node synchronizations are required for enforcing dependencies between sensor nodes in simulations. Since the dependencies come from the interactions of sensor nodes over wireless channels, the number of required synchronizations in a distributed simulation is inversely proportional to the degree of parallelism in the WSN application under simulation [1].

To reduce the number of synchronizations, we exploit the parallelism available in WSN applications. In particular, we seek to use the information regarding

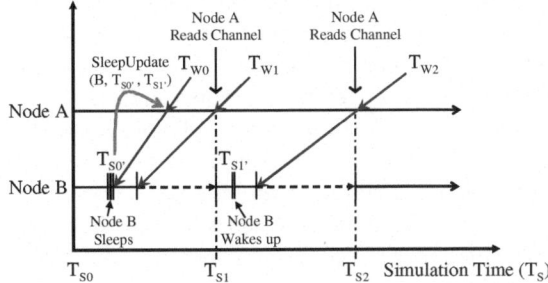

Fig. 2. The progress of simulating in parallel a wireless sensor network with two duty cycled nodes that are in direct communication range of each other

duty cycling of nodes in sensor networks to speed up simulations. Node duty cycling is common in WSN applications for power management purposes. Most WSN applications have very low duty cycles and need to carefully manage their power consumptions in order to function for an extended period of time under the constraint of limited energy supply. In other words, sensor nodes sleep most of the time and do not interact with each other frequently until certain events are detected [6]. Very little power is consumed by a sensor node in the sleep state as its wireless radio is turned off and its processor is put into a low power sleep mode.

Our speedup technique is illustrated in Figure 2 which shows the progress of simulating two duty cycled sensor nodes that are within direct communication range of each other. In the simulation, Node B enters into the sleep state at $T_{S0'}$ and wakes up at $T_{S1'}$. With existing distributed WSN simulators, Node A needs to wait for Node B at T_{S1} although Node B does not transmit anything during its sleep period. To eliminate this type of unnecessary synchronization, our technique keeps track of the time that a node enters into the sleep state and the time it wakes up. When we detect during a simulation that a node is entering into the sleep state, we immediately send both the entering time and exiting time (simulation time) in a *SleepUpdate* message to the neighboring nodes that are within direct communication range. As a result, neighboring nodes no longer need to synchronize with the sleeping node during the sleep period. For example, when we detect that Node B is entering into the sleep state at $T_{S0'}$, we immediately notify Node A that Node B will be in the sleep state from $T_{S0'}$ to $T_{S1'}$. Once Node A knows that Node B will not transmit between $T_{S0'}$ and $T_{S1'}$ and $T_{S0'} \leq T_{S1} < T_{S1'}$, it no longer needs to wait for Node B at T_{S1} and its lookahead time increases. Lookahead time is defined as the amount of simulation time that a simulated sensor node can advance freely without waiting for inputs from other simulated sensor nodes [1]. The speedup of our technique increases with the durations of sleep periods because the longer the sleep periods, the larger the lookahead time.

For the speedup technique to work, we have to be able to detect both sleep time and wakeup time in simulations. Sleep time can always be detected because a real sensor node processor has to execute some special instructions to put the node into the sleep mode. Correspondingly, sleep events, which are integral parts of any WSN simulators, are used in simulations to signal the transitions of nodes from active states to the sleep state. For example, in the case of cycle accurate sensor network simulators, sleep events are associated with the execution of specific machine instructions of the sensor node processors under simulation. However, detecting wakeup time is a challenging process as a node can be woken up by interrupts from either a timer or an external autonomous sensor such as a passive infrared sensor. Autonomous sensors are devices that can function independently without the support of a node processor and therefore may wakeup a sensor node at any time.

The wakeup time for sensor nodes that do not have autonomous sensors can always be detected. This is because the wakeup time has to be passed to a physical timer before a real sensor node is able to enter into the sleep state as the node processor can not execute any instructions during the sleep mode. For nodes equipped with autonomous sensors, if the input events to the autonomous sensors are known before a simulation starts, which is generally the case, the wakeup time can always be computed as the smaller of the time of the timers and the input events. If the input events to the autonomous sensors are not known before a node enters into the sleep mode, for example, inputs are collected in real time from real sensors [7], then the speedup technique has to be disabled on that node. However, we only need to turn off the speedup technique on those nodes receiving unpredictable input events, while other nodes can still use the technique.

3   Synchronization Algorithm and Implementation

We use Avrora [4] to evaluate the effectiveness of our approach. Avrora is a widely used cycle accurate sensor network simulator. Among all types of sensor network simulators, cycle accurate sensor network simulators [8,4,5] offer the highest level of fidelity. They provide simulation models that emulate the functions of major hardware components of a sensor node, mainly the processor. Therefore, one can run on top of them, clock cycle by clock cycle, instruction by instruction, the same binary code (images) that are executed by real sensor nodes. As a result, accurate timing and interactive behaviors of sensor network applications can be studied in details.

Avrora is written in Java and supports parallel simulations of sensor networks comprised of Mica2 Motes [9]. It allocates one thread for each simulated node and relies on the Java virtual machine (VM) to assign runnable threads to any available processors on an SMP computer. Implementing the speedup technique in Avrora mainly involves developing new code in two areas: synchronization algorithm and channel modeling. Our implementation is based on the Beta 1.6.0 code release of Avrora. Its well documented source code is publicly available.

3.1   Synchronization Algorithm

The lock-step style synchronization algorithm of Avrora is optimized for parallel simulations on SMP computers but lacks necessary features to support our

Algorithm 1. Distributed Synchronization Algorithm with Speedup Technique

Require: $nl := \{< nid, nclock >\}$ /*a list of neighboring node ids and their reported simulation time*/

Require: id /*current node ID*/, $bytetime$ /*the amount of time to transmit one byte with a wireless radio*/

1: $clock \Leftarrow 0$ /*current sim clock to 0*/, $lookahead \Leftarrow bytetime$, $intervalclock \Leftarrow 0$

2: **for** every tuple $< nid, nclock >$ in nl **do**
3:     $nclock \Leftarrow 0$ /*initialize simulation time of neighboring nodes to zero before starting the simulation*/
4: **while** $clock \leq$ user inputed simulation time **do**
5:     $waitchannel \Leftarrow false$
6:     execute $next\ instruction$
7:     **if** the $instruction$ puts a node into the sleep state **then**
8:         $exitclock = wakeuptime$, $intervalclock \Leftarrow exitclock$
9:         send a $ClockUpdate(id, exitclock)$ message to every nid node in the tuple $< nid, nclock >$ of nl
10:     **else if** the $instruction$ reads from the wireless radio **then**
11:         **if** $lookahead \geq 0$ **then**
12:             read the wireless radio
13:         **else**
14:             **if** $intervalclock \neq clock$ **then**
15:                 $intervalclock \Leftarrow clock$
16:                 send a $ClockUpdate(id, clock)$ message to every nid node in the tuple $< nid, nclock >$ of nl
17:             $waitchannel \Leftarrow true$
18:     **if** $waitchannel$ is $false$ **then**
19:         $clock \Leftarrow clock + cyclesconsumed$ /*advance clock by the clock cycles of the executed instruction*/
20:     $updated \Leftarrow false$ /*check incoming $ClockUpdate$ messages at least once per instruction*/
21:     **repeat**
22:         **for** each received $ClockUpdate(cid, cclock)$ message **do**
23:             **for** every tuple $< nid, nclock >$ in nl **do**
24:                 **if** uid equals to cid **then**
25:                     $nclock \Leftarrow cclock$
26:         $updated \Leftarrow true$
27:         $minclock = min(nclock)$ in the tuple $< nid, nclock >$ of nl /*find the neighbor with the smallest clock*/
28:         $lookahead \Leftarrow (minclock - floor(clock/bytetime) * bytetime)$ /*on byte boundaries with byte-radio*/
29:         **if** $lookahead \geq 0$ and $waitchannel$ is $true$ **then**
30:             read the wireless radio /*all neighbors have advanced past the byte boundary this node is waiting on*/
31:             $clock \Leftarrow clock + cyclesconsumed$, $waitchannel \Leftarrow false$
32:     **until** $waitchannel$ is $false$ and $update$ is $true$
33:     **if** $(clock - intervalclock) \geq bytetime$ **then**
34:         $intervalclock \Leftarrow clock$
35:         send a $ClockUpdate(id, clock)$ message to every nid node in the tuple $< nid, nclock >$ of nl

speedup technique. Our distributed synchronization algorithm is shown in Algorithm 1. It is a generic distributed synchronization algorithm similar to the one in DiSenS [5] and is suitable for both parallel and distributed simulations of WSNs.

Synchronizations are only necessary between neighboring nodes that are within direct communication range of each other. The first step before applying our algorithm is to build a neighbor node list for each node according to the locations of the sensor nodes and the maximum transmission range of their wireless radios. Mobile nodes need to be included in the neighbor node list of all other nodes. Then, a time stamp is assigned to every node (node id) in the lists to keep the last reported simulation time of that node. This list, named *nl* in Algorithm 1 is the first required input to the synchronization algorithm. There are two more inputs to the algorithm. The second input *id* is used to identify the node under simulation. The nodes in *nl* are neighbors of this node. The third input *bytetime* is the amount of time to transmit one byte with a wireless radio. It is the maximum lookahead time without synchronizations. Every node starts with that lookahead time because it takes that amount of time for one byte of data to travel from the sender to the receiver. For example, if a node starts at simulation time 0 and wants to read the wireless channel at that time, it can do so because the earliest time that a byte of data can arrive is $0 + bytetime$. Similarly, after synchronizing at time T_S, all synchronized nodes can advance freely up to $T_S + bytetime$ without any additional synchronizations. However, this approach only works if the processor and radio on a real sensor node communicate by exchanging data one byte at a time (*byte-level*).

The variable *intervalclock* in Algorithm 1 is used to ensure that *ClockUpdate* messages are sent by every simulated node once every *bytetime* if the node is not already in the sleep state. These messages update neighboring nodes about the latest simulation time of the sender and ensure neighboring nodes have the right time information to make synchronization decisions according to Condition 1. The interval chosen to send the messages will affect the performance of the algorithm as nodes may have to wait if *ClockUpdate* messages are delayed. The smallest interval one can use with *byte-level* radios is *bytetime* because the actual waiting time must fall on byte boundaries. This can be seen in Algorithm 1 where the *floor* function is used to calculate the lookahead time.

Condition 1. *If a node N_i reads data sent by a node N_s over a wireless channel C_k at simulation time T_{SN_i}, then the simulation time of node N_s, T_{SN_s}, must be greater than or equal to T_{SN_i}.*

The *SleepUpdate* message described in Section 2 is replaced in Algorithm 1 with the *ClockUpdate* message because receiving nodes only need to use the wakeup time of the sender to calculate lookahead time. However, to take advantage of a special optimization described in the future work part of Section 6, *SleepUpdate* messages must be used so the time that a node enters into the sleep state is sent as well. The time that a node enters into the sleep state is detected when the *Sleep* instruction of the ATMega128L microcontroller [10] is executed. This instruction can put the microcontroller into different sleep modes based on the values set in the *MCU Control Register*. It is critical to read that register before sending any *ClockUpdate* messages as the speedup technique only works if a microcontroller is put into *Power-Save Mode*. The reason is that the only way to wake up a microcontroller from *Power-Save Mode* is through interrupts generated by timers or external autonomous sensors. Because of that, we can find the wakeup time by keeping track of the values written to the *Timer Control Register* and using the techniques described in section 2.

The computational complexity of a synchronization algorithm is determined by the total number of synchronization messages that need to be sent and the overheads in sending and processing each of the messages [11]. Our synchronization algorithm has higher computational complexity than the one in Avrora because we choose to implement it as a generic distributed algorithm. In Avrora, a global data structure is used to keep track of the simulation time of each node and therefore a node only needs to update the global data structure once for each clock update. This centralized approach is optimized for parallel simulation over SMP computers but does not support distributed simulations over a network of computers. We implement our synchronization algorithm as a truly distributed algorithm by distributing parts of the global data structure to each node. The penalty is that a node with N nodes in direct communication range has to send a total of N messages for each clock update. However, the penalty is not significant when the number of nodes within direct communication range is not big. In fact, because the synchronization algorithm of Avora works by synchronizing a node with all other nodes regardless of whether they are within communication range or not, our distributed algorithm may even perform better when nodes are sparsely distributed. If performance is an issue in the future, we could choose to optimize our implementation for parallel simulations using a centralized approach.

3.2   Channel Modeling

With our synchronization algorithm, a node can write to a wireless channel long before the packets are read by other nodes (the transmitting code is omitted in Algorithm 1 for simplicity). Because of that, we develop a new wireless channel model that uses a circular buffer to store unread wireless packets. Our channel model uses a similar method as the original channel model in Avrora to map transmitting and receiving time into time slots that are *bytetime* apart. The slot number is used to index into the circular buffer. Note that with the original channel model, a write to a channel right after synchronizations could be dropped as the write may happen before the time slots are carried forward. Our channel implementation does not drop data.

4   Evaluation

We conduct a series of experiments to evaluate the performance of our speedup technique. The experiments are conducted on an SMP server running Linux

2.6.9. It has 4 processors (Intel Xeon 2.8GHz) and 2GByte of RAM. For comparison, the same test cases are simulated using both our modified Avrora and the original Avrora. Sun's Java 1.6.0 is used to run both simulators.

To demonstrate the effectiveness of our speedup technique, we choose two programs from the CountSleepRadio example which is a part of the TinyOS 1.1 distribution [12,13]. These two programs behave exactly like the CntToRfm and RfmToLeds programs used in the experiments of the Avrora paper and serve similar purposes. The only difference is that the new programs can put sensor nodes into sleep states to save power. (The latest TinyOS 2.0 release [14] is not used for our experiments because its radio stack is not fully compatible with the radio model in the version of the Avrora that our code is based on.)

The first program we use for our experiments is CountSleepRadio. It wakes up a node periodically from the sleep state to increase the value of a counter by one and broadcast that value in a packet. Once the packet is sent, it puts the node back into the sleep state to save power. We have to modify this program for some of our experiments because the original program has an upper bound on how long a sensor node can stay in the sleep state[1]. This is unnecessary and we work around this limitation by using the *Clock* interface of the TinyOS 1.1 directly. The problem has reportedly been fixed in TinyOS 2.0. The counterpart of our CountSleepRadio program is CountReceive. It receives packets sent by CountSleepRadio and flashes different LEDs based on the values in the packets. For simplicity, we identify nodes running CountSleepRadio and CountReceive as senders and receivers respectively in the following sections.

Both CountSleepRadio and CountReceive use the default TinyOS 1.1 CC1000 CSMA (carry sense multiple access) MAC (media access control) which is based on B-MAC [15]. Before sending a packet, the CC1000 MAC first backs off for a random amount of time and then reads its transmitting channel for ongoing transmissions. It only sends the packet if the channel is clear. Otherwise, it backs off for a random amount of time before checking the channel again. As a result, a sender in our experiments reads the wireless channel at least once before each transmission.

4.1   Performance in One-Hop Networks

In this section, the performance of our speedup technique is evaluated under various sleep times and network sizes using one-hop sensor networks. One-hop sensor networks are sensor networks set up in such a way that all sensor nodes are within direct communication range of each other. It is a common form of sensor network used in actual deployments [6].

In the one-hop sensor network experiments, nodes are laid on a 10 by 10 grid 1 meter apart and their maximum transmission ranges are set to 20 meters. A fixed node is selected as a receiver and the rest as senders. The receiver listens continuously like a gateway node [6,16] and does not enter into the sleep state. The senders are duty cycled and their sleep durations are varied for different

---

[1] The upper bound is imposed by the timer implementation of TinyOS 1.1. The timer code sets $maxTimerInteval$ to $230ms$ and the physical timers of a real sensor node can not be set to anything larger than that using the timer API.

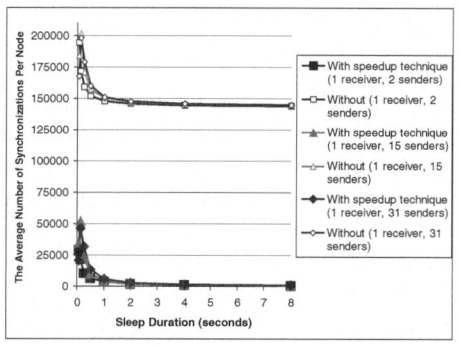

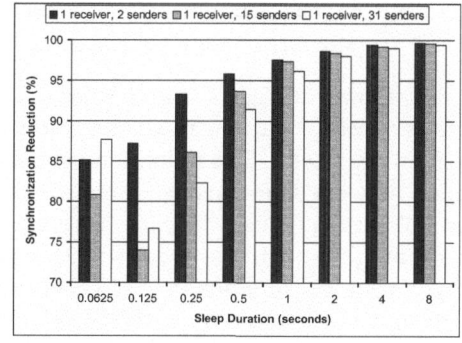

Fig. 3. Average number of synchronizations per node in one-hop networks during 60 seconds of simulation time

Fig. 4. Percentage reductions of the average number of synchronizations per node in one-hop networks during 60 seconds of simulation time

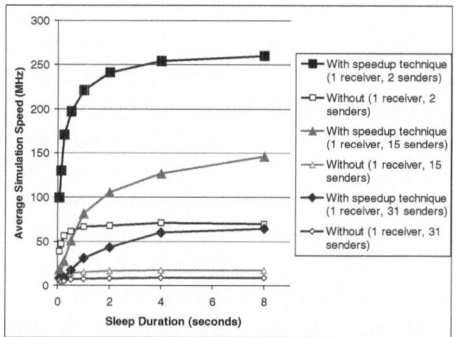

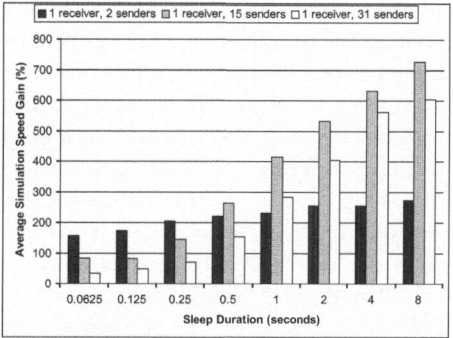

Fig. 5. Average simulation speed in one-hop networks

Fig. 6. Percentage increases of average simulation speed in one-hop networks

experiments. Sleep duration is how long a node stays in the sleep state before waking up. All results in this section are averages of three runs.

Figure 3 shows the average number of synchronizations per node in one-hop networks during 60 seconds of simulation time. Since all nodes are simulated for the same number of clock cycles (60 × clock frequency of ATMega128L) in all test cases, the average number of synchronizations per node is a good indicator to the performance of the speedup technique. The synchronization numbers are collected by logging code we add specifically for evaluation purposes. Figure 4 shows the percentage reductions of the average number of synchronizations per node in one-hop networks during the 60 seconds of simulation time. Figure 5 shows average simulation speed in one-hop networks. The average simulation speed V_{avg} is calculated using Equation 1. The percentage increases of average simulation speed in one-hop networks are shown in Figure 6.

$$V_{avg} = \frac{total \ number \ of \ clock \ cycles \ executed \ by \ the \ sensor \ nodes}{(execution \ time \ of \ the \ simulation) \times (number \ of \ sensor \ nodes)} \quad (1)$$

As shown in Figure 3 and Figure 4, the speedup technique significantly reduces synchronizations in all the test cases and the largest percentage reduction is more than 99%. The reduction percentages increase with sleep durations under fixed network sizes except for the 16 (1 receiver, 15 senders) and 32 (1 receiver and 31 sender) node test cases with $62.5ms$ sleep durations. The unusually high percentage reductions in those cases are results of using the CC1000 CSMA MAC protocol of TinyOS 1.1. When multiple senders in communication range transmit at the same time, the MAC protocol would sequence their transmission times using random backoffs. Since the senders will not return back to the sleep state until packets are successfully transmitted, the sleep times of the senders are sequenced as well. This effectively reduces synchronizations in simulations as the number of nodes that are active at a same time is reduced. The 3-node (1 receiver, 2 senders) test case is not affected by this because we randomly delay the starting time of each node between 0 and 1 second in all our experiments to prevent the nodes from artificially starting at the same time. When the number of nodes in a one-hop network decreases, the chance for concurrent transmissions decreases and the number of synchronizations increases. Similarly, the chance for concurrent transmissions also decreases when the sleep duration increases because the nodes in a one-hop network would transmit less frequently with larger sleep durations. As a result, the number of synchronizations increases with sleep durations in such cases. We can see this from Figure 7, a zoomed in view of Figure 3. When sleep duration doubles from $62.5ms$ to $125ms$, the average number of synchronizations per node actually increases for both of the 16 and 32 node test cases, regardless of whether the speedup technique is used or not. We can also see in the same test cases that the average number of synchronizations per node decreases with network size under fixed sleep durations. The speedup technique can further reduce synchronizations in cases like these because it increases the lookahead time of simulated nodes. The reduction from applying the speedup technique is greater for larger one-hop networks in those cases as there is more of this type of sequencing in larger one-hop networks under fixed sleep durations.

As shown in Figure 5 and Figure 6, the speedup technique significantly increases average simulation speed in all the test cases and the largest increase is more than 700%. Although the 3-node speedup test case has the highest percentage reduction of the average number synchronizations per node as shown in Figure 4, it does not have the largest average simulation speed increase in Figure 6. This is because the overhead in performing a synchronization is very low for the 3-node test cases. Context switches are generally not needed for synchronizations in those cases because there are more processors (4) than nodes/threads (3). We can also see in Figure 6 that the growth of the average simulation speed quickly flattens out for large sleep durations in original Avrora but continues after applying the speedup technique. We expect even better average simulation speed

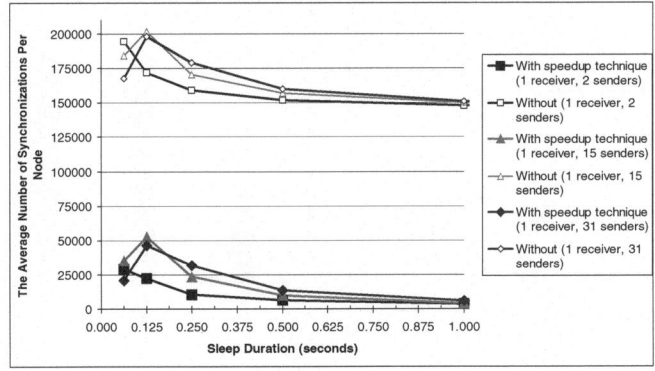

Fig. 7. Average number of synchronizations per node in one-hop networks during 60 seconds of simulation time (a zoomed in view of Figure 3)

in simulating large one-hop networks after optimizing our generic distributed synchronization algorithm for parallel simulations as discussed in Section 3.1.

The speedup technique can not completely eliminate synchronizations caused by having only limited numbers of physical processors available for simulations. We can see this from Figure 3 and Figure 7. When the sleep duration is long enough, the average number of synchronization per node with speedup is similar for all network sizes.

4.2 Performance in Multi-hop Networks

In this section, we evaluate the performance of the speedup technique using multi-hop sensor networks. Nodes are laid 20 meters apart on square grids of various sizes. Sender and receivers are positioned on the grids in such a way that nodes of the same types are not adjacent to each other. By setting a maximum transmission range of 20 meters, this setup ensures that only neighboring nodes are within direct communication range of each other. This configuration is very similar to the two dimensional topology in DiSenS [5]. Once again, only senders are duty cycled to keep the experiments simple.

Figure 8 shows the average number of synchronizations per node in multi-hop networks during 20 seconds of simulation time. The percentage reductions of the average number of synchronizations per node in multi-hop networks during the 20 seconds of simulation time is shown in Figure 9. We can see that there are significant reductions in the average number of synchronizations per node in all the test cases using the speedup technique and the reduction percentages scale with sleep durations.

Figure 10 and Figure 11 indicate that the speedup technique significantly increases average simulation speed in all multi-hop test cases. Compared to the one-hop test results in Figure 6, the speed increases scale better with network sizes in multi-hop tests. This is because our distributed synchronization

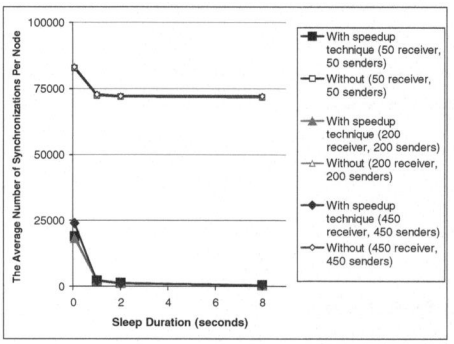

Fig. 8. Average number of synchronizations per node in multi-hop networks during 20 seconds of simulation time

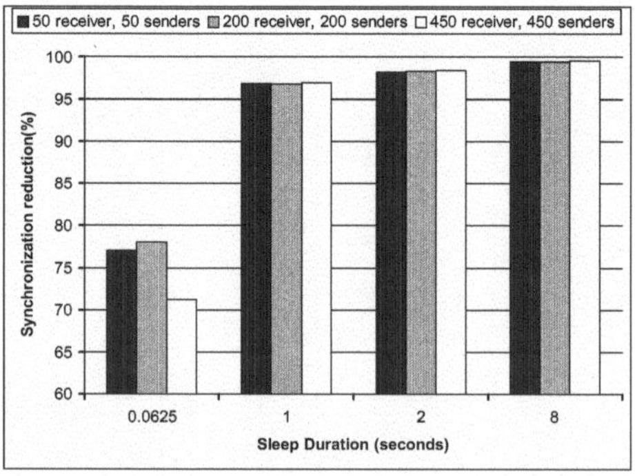

Fig. 9. Percentage reductions of the average number of synchronizations per node in multi-hop networks during 20 seconds of simulation time

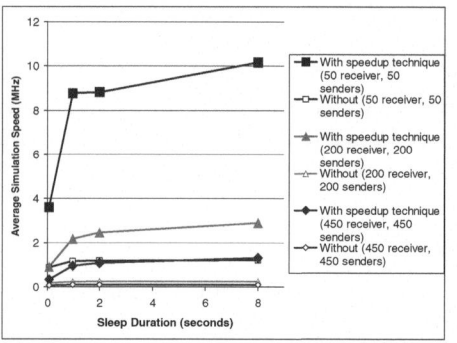

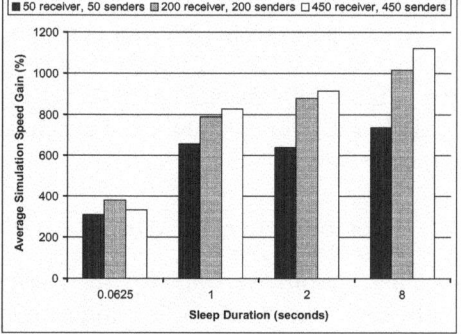

Fig. 10. Average simulation speed in multi-hop networks

Fig. 11. Percentage increases of average simulation speed in multi-hop networks

algorithm has less overhead on sensor networks that have smaller numbers of nodes within direct communication range as described in the end of Section 3.1.

5   Related Work

There is a large body of work on improving the scalability of distributed discrete event driven simulators in general. This work can be classified into two groups, those based on conservative synchronization algorithms [2] and those based on optimistic synchronization algorithms [3]. The performance of conservative approaches is bounded by worse case scenarios. The optimistic approaches do not have this limitation but they are usually very complex for implementation and require a large amount of memory to run.

Exploiting lookahead time is a very common conservative approach to improve the scalability of distributed discrete event driven simulators [17,18]. Our approach is similar to those in the sense that we also improve scalability of distributed discrete event driven simulators by increasing lookahead time. However, our technique is fundamentally different as we use different and application specific characteristics in a different context to increase lookahead time.

6   Conclusion and Future Work

We have described a speedup technique that significantly reduces sensor node synchronizations in distributed simulations of sensor networks and consequently improves average simulation speed and scalability of distributed sensor network simulators. We implemented this technique in Avrora, a widely used parallel sensor network simulator and conducted extensive experiments. The significant performance improvements with parallel simulations suggest even greater benefits in applying our technique to distributed simulations over a network of computers because of their large overheads in sending synchronization messages across computers during simulations.

As future work, we plan to merge our implementation into the latest development branch of Avrora. This would make it possible to simulate TinyOS 2.0 based applications with our speedup technique. We also plan to support an optimization that can reduce communication overheads in distributed simulations based on the speedup technique. When a sensor node is in the sleep state, its radio is off and it will not access the wireless channel at all. In other words, when a transmitting node knows that a receiving node is in the sleep state during a simulation, it no longer needs to send packets to the receiver for the entire sleep period. If the sender and receiver are simulated on different computers, the savings in terms of communication time and network bandwidth consumptions could be significant.

References

1. Fujimoto, R.M.: Parallel and distributed simulation. In: WSC 1999: Proceedings of the 31st conference on Winter simulation, pp. 122–131. ACM Press, New York (1999)
2. Chandy, K.M., Misra, J.: Asynchronous distributed simulation via a sequence of parallel computations. Commun. ACM 24(4), 198–206 (1981)
3. Jefferson, D.R.: Virtual time. ACM Trans. Program. Lang. Syst. 7(3), 404–425 (1985)
4. Titzer, B.L., Lee, D.K., Palsberg, J.: Avrora: scalable sensor network simulation with precise timing. In: IPSN 2005: Proceedings of the 4th international symposium on Information processing in sensor networks, p. 67. IEEE Press, Piscataway (2005)
5. Wen, Y., Wolski, R., Moore, G.: Disens: scalable distributed sensor network simulation. In: PPoPP 2007: Proceedings of the 12th ACM SIGPLAN symposium on Principles and practice of parallel programming, pp. 24–34. ACM Press, New York (2007)
6. Szewczyk, R., Polastre, J., Mainwaring, A.M., Culler, D.E.: Lessons from a sensor network expedition. In: EWSN, pp. 307–322 (2004)

7. Girod, L., Elson, J., Cerpa, A., Stathopoulos, T., Ramanathan, N., Estrin, D.: Emstar: a software environment for developing and deploying wireless sensor networks. In: Proceedings of the 2004 USENIX Technical Conference, Boston, MA (2004)
8. Polley, J., Blazakis, D., McGee, J., Rusk, D., Baras, J.: Atemu: a fine-grained sensor network simulator. In: IEEE SECON 2004. 2004 First Annual IEEE Communications Society Conference on Sensor and Ad Hoc Communications and Networks, October 4-7, pp. 145–152 (2004)
9. MICA2 Datasheet, Crossbow (2008)
10. ATMega128L Datasheet, Atmel (2003)
11. Nicol, D.M.: Scalability, locality, partitioning and synchronization pdes. SIGSIM Simul. Dig. 28(1), 5–11 (1998)
12. Hill, J., Szewczyk, R., Woo, A., Hollar, S., Culler, D., Pister, K.: System architecture directions for networked sensors. SIGPLAN Not. 35(11), 93–104 (2000)
13. Tinyos 1.1.15. http://www.tinyos.net/
14. Levis, P., Gay, D., Handziski, V., Hauer, J.-H., Greenstein, B., Turon, M., Hui, J., Klues, K., Sharp, C., Szewczyk, R., Polastre, J., Buonadonna, P., Nachman, L., Tolle, G., Culler, D., Wolisz, A.: T2: A second generation os for embedded sensor networks, Telecommunication Networks Group, Technische Universität Berlin. Tech. Rep. TKN-05-007 (November 2005)
15. Polastre, J., Hill, J., Culler, D.: Versatile low power media access for wireless sensor networks. In: SenSys 2004: Proceedings of the 2nd international conference on Embedded networked sensor systems, pp. 95–107. ACM, New York (2004)
16. Jin, Z., Schurgers, C., Gupta, R.: An embedded platform with duty-cycled radio and processing subsystems for wireless sensor networks. In: International Workshop on Systems, Architectures, Modeling, and Simulation (SAMOS) (2007)
17. Filo, D., Ku, D.C., Micheli, G.D.: Optimizing the control-unit through the resynchronization of operations. Integr. VLSI J. 13(3), 231–258 (1992)
18. Liu, J., Nicol, D.M.: Lookahead revisited in wireless network simulations. In: PADS 2002: Proceedings of the sixteenth workshop on Parallel and distributed simulation, pp. 79–88. IEEE Computer Society, Washington, DC (2002)

Frugal Sensor Assignment

Matthew P. Johnson[1], Hosam Rowaihy[2], Diego Pizzocaro[3],
Amotz Bar-Noy[1], Stuart Chalmers[4], Thomas La Porta[2], and Alun Preece[3]

[1] Dept. of Computer Science, Graduate Center, City University of New York, USA
[2] Dept. of Computer Science and Engineering, Pennsylvania State University, USA
[3] School of Computer Science, Cardiff University, UK
[4] Department of Computing Science, University of Aberdeen, UK

Abstract. When a sensor network is deployed in the field it is typically required
to support multiple simultaneous missions, which may start and finish at different times. Schemes that match sensor resources to mission demands thus become
necessary. In this paper, we consider new sensor-assignment problems motivated
by frugality, i.e., the conservation of resources, for both static and dynamic settings. In general, the problems we study are NP-hard even to approximate, and so
we focus on heuristic algorithms that perform well in practice. In the static setting, we propose a greedy centralized solution and a more sophisticated solution
that uses the Generalized Assignment Problem model and can be implemented
in a distributed fashion. In the dynamic setting, we give heuristic algorithms in
which available sensors propose to nearby missions as they arrive. We find that
the overall performance can be significantly improved if available sensors sometimes refuse to offer utility to missions they could help based on the value of
the mission, the sensor's remaining energy, and (if known) the remaining target
lifetime of the network. Finally, we evaluate our solutions through simulations.

1 Introduction

A sensor network deployed for monitoring applications may be tasked with achieving
multiple, possibly conflicting, missions. Although certain types of sensors, such as seis-
mic sensors, can receive data from their surroundings as a whole, other sensor types,
such as cameras, are directional. In these cases, the direction of each sensor, and thus
the mission it serves, must be chosen appropriately.

A given sensor may offer different missions varying amounts of information (because of geometry, obstructions, or utility requirements), or none at all. Missions, on
the other hand, may vary in importance (or *profit*), amount of resources they require (or
demand), and duration. In some but not all applications, it may be preferable to do one
thing well than to do many things badly, that is, to fully satisfy one mission rather than
give a small amount of utility to several. Given all currently available information, the
network should choose the "best" assignment of the available sensors to the missions.

In this paper, we examine new sensor-assignment problems motivated by frugality,
i.e., the conservation of resources. We consider two broad classes of environments:
static and dynamic. The *static* setting is motivated by situations in which different users
are granted control over the sensor network at different times. During each time period, the current user may have many simultaneous missions. While the current user

S. Nikoletseas et al. (Eds.): DCOSS 2008, LNCS 5067, pp. 219–236, 2008.
© Springer-Verlag Berlin Heidelberg 2008

will want to satisfy as many of these as possible, sensing resources may be limited and expensive, both in terms of equipment and operational cost. In some environments, replacing batteries may be difficult, expensive, or dangerous. Furthermore, a sensor operating in active mode (i.e., assigned to a mission) may be more visible than a dormant sensor, and so in greater danger of being destroyed. Therefore, we give each mission in the static problem a budget so that no single user may overtax the network and deprive future users of resources. This budget serves as a constraint in terms of the amount of resources that can be allocated to a mission regardless of profit.

Our second environment is a *dynamic* setting in which missions may start at different times and have different durations. In these cases explicit budgets may be too restrictive because we must react to new missions given our current operating environment, i.e. the condition of the sensors will change over time. Instead, we use battery lifetime as a metric to capture network lifetime and evaluate a trade-off between cost in terms of network lifetime versus mission profit before assigning sensors.

In the dynamic setting we consider two cases. First, we assume no advanced knowledge of the target network lifetime, i.e. we do not know for how long the network will be required to operate. We call this the *general dynamic setting*. Second, we consider the case in which we have knowledge of the target network lifetime, i.e. the network is needed for a finite duration. We call this the *dynamic setting with a time horizon*. By using a trade-off of lifetime versus profit instead of a hard budget, we can use the knowledge of remaining energy of the sensors and the target network lifetime, if known, to adjust the aggressiveness with which sensors accept new missions.

Our contributions. We consider several sensor-assignment problems, in both budget-constrained static settings and energy-constrained dynamic settings. In the static setting, we give an an efficient greedy algorithm and a multi-round proposal algorithm whose subroutine solves a Generalized Assignment Problem (GAP), as well as an optimal algorithm for a simplified 1-d setting. In the dynamic setting, we develop distributed schemes that adjust sensors' eagerness to participate in new missions based on their current operational status and the target network lifetime, if known. We find in the static setting that in dense networks both algorithms perform well, with the GAP-based algorithm slightly outperforming the greedy. In the dynamic setting, we find that knowledge of energy level and network lifetime is an important advantage: the algorithm given both of these significantly outperforms the algorithm using only the energy level and the algorithm that uses neither. When both the energy and target lifetime are used we can achieve profits 17% - 22% higher than if only energy is used.

The rest of this paper is organized as follows. Section 2 discusses some related work. In Section 3 we review the sensor-mission assignment problem and summarize its variations. In Sections 4 and 5 we propose schemes for the static and dynamic settings, whose performance we evaluate in Section 6. Finally, Section 7 concludes the paper.

2   Related Work

Assignment problems have received sizable attention, in both experimental and theoretical communities. In wireless sensor networks, there has been some work in defining frameworks for single and multiple mission assignment problems. For example,

[2] defines a framework for modeling the assignment problem by using the notions of utility and cost. The goal is to find a solution that maximizes the utility while staying under a predefined budget. In [7], a market-based modeling approach is used, with sensors providing information or "goods". The goal is to maximize the amount of goods delivered without exceeding the sensor's budget.

The authors of [6,8,11], for example, solve the coverage problem, which is related to the assignment problem. They try to use the fewest number of sensors in order to conserve energy. The techniques used range from dividing nodes[1] in the network into a number of sets and rotating through them, activating one set at a time [8], to using Voronoi diagram properties to ensure that the area of a node's region is as close to its sensing area as possible [11]. Sensor selection schemes have also been proposed to efficiently locate and track targets. For example, [13] uses the concept of information gain to select the most informative sensor to track a target. In [4], the author attempts target localization using acoustic sensors. The goal is to minimize the mean squared error of the target location as perceived by the active nodes. A survey of the different sensor selection and assignment schemes can be found in [9].

The Generalized Assignment Problem (GAP) [3] is a generalization of the Multiple Knapsack Problem in which the weight and value of an item may vary depending on the bin in which it is placed. There is a classical FPTAS [12] for the core knapsack problem which performs a dynamic programming procedure on a discretization of the problem input. If, for example, the knapsack budget value is not too large, then the DP can find the optimal solution in polynomial time. A stricter version of the static sensor-assignment problem was formalized as Semi-Matching with Demands (SMD) in [1,10]. In that formulation, profits are awarded only if a certain utility threshold is met, but no budgets are considered. The static problem we study here is a common generalization of these two previous problems, incorporating both budgets and a profit threshold.

3   Sensor-Mission Assignment Problems

With multiple sensors and multiple missions, sensors should be assigned in the "best" way. This goal is shared by all the problem settings we consider. There are a number of attributes, however, that characterize the nature and difficulty of the problem. In this section, we briefly enumerate the choices available in defining a particular sensor-mission assignment problem. In all settings we assume that a sensor can be assigned to one mission only, motivated e.g. by directional sensors such as cameras.

3.1   Static Setting

First consider the static setting. Given is a set of sensors $S_1, ..., S_n$ and a set of missions $M_1, ..., M_m$. Each mission is associated with a utility demand d_j, indicating the amount of sensing resources needed, and a profit p_j, indicating the importance of the mission. Each sensor-mission pair is associated with a utility value e_{ij} that mission j will receive if sensor i is assigned to it. This can be a measure of the quality of information that a sensor can provide to a particular mission. To simplify the problem we assume that the

---

[1] We use the terms *node* and *sensor* interchangeably.

Table 1. MP **P** for static setting

$$max:\ \sum_{j=1}^{m} p_j(y_j)$$
$$s.t.:\ \sum_{i=1}^{n} x_{ij}e_{ij} \geq d_j y_j,\ \text{for each } M_j,$$
$$\sum_{i=1}^{n} x_{ij}c_{ij} \leq b_j,\ \text{for each } M_j,$$
$$\sum_{j=1}^{m} x_{ij} \leq 1,\ \text{for each } S_i,$$
$$x_{ij} \in \{0,1\}\ \forall x_{ij}\ \text{and}$$
$$y_j \in [0,1]\ \forall y_j$$

Table 2. MP **P'** for dynamic setting

$$max:\ \sum_{t} \sum_{j=1}^{m} p_j(y_{jt})$$
$$s.t.:\ \sum_{i=1}^{n} x_{ijt}e_{ij} \geq d_j y_{jt},\ \text{for each } M_j \text{ and } t,$$
$$\sum_{j=1}^{m} x_{ijt} \leq 1,\ \text{for each } S_i \text{ and time } t,$$
$$\sum_{t} \sum_{j=1}^{m} x_{ijt} \leq B,\ \text{for each } S_i,$$
$$x_{ijt} \in \{0,1\}\ \forall x_{ijt}\ \text{and}$$
$$y_{jt} \in [0,1]\ \forall y_{jt}$$

utility amounts received by a mission (u_j) are additive. While this may be realistic in some settings, in others it is not; we make this simplifying assumption. Finally, a budgetary restriction is given in some form, either constraining the entire problem solution or constraining individual missions as follows: each mission has a budget b_j, and each potential sensor assignment has cost c_{ij}. All the aforementioned values are positive reals, except for costs and utility, which could be zero. The most general problem is defined by the mathematical program (MP) **P** shown in Table 1.

A sensor can be assigned ($x_{ij} = 1$) at most once. Profits are received per mission, based on its satisfaction level (y_j). Note that y_j corresponds to u_j/d_j within the range [0,1] where $u_j = \sum_{i=1}^{n} x_{ij}e_{ij}$. With strict profits, a mission receives exactly profit p_j iff $u_j \geq d_j$. With fractional profits, a mission receives a fraction of p_j proportional to its satisfaction level y_j and at most p_j. More generally, profits can be awarded fractionally, but only if a fractional satisfaction threshold T is met, i.e.:

$$p_j(u_j) = \begin{cases} p_j, & \text{if } u_j \geq d_j \\ p_j \cdot u_j/d_j, & \text{if } T \leq u_j/d_j \\ 0, & \text{otherwise} \end{cases}$$

When $T = 1$, program **P** is an integer program; when $T = 0$, it is a mixed integer program with the decision variables x_{ij} still integral. The edge values e_{ij} may be arbitrary non-negative values or may have additional structure. If sensors and missions lie in a metric space, such as line or plain, then edge values may be based in some way on the distance D_{ij} between sensor i and mission j. In the *binary sensing* model, e_{ij} is equal to 1 if distance D_{ij} is at most the sensing range, r, and 0 otherwise. In another geometric setting, e_{ij} may vary smoothly based on distance, such as $1/(1 + D_{ij})$.

Similarly, the cost values c_{ij} could be arbitrary or could exhibit some structure: the cost could depend on the sensor involved, or could e.g. correlate directly with distance D_{ij} to represent the difficulty of moving a sensor to a certain position. It could also be unit, in which case the budget would simply constrain the number of sensors. Even if profits are unit, demands are integers, edge values are 0/1 (though not necessarily depending on distance), and budgets are infinite, this problem is NP-hard and as hard to approximate as Maximum Independent Set [1]. If we also add the restriction that sensors and missions lie in the plane and that 0/1 edge utility depends on distance (i.e., the edges form a bipartite unit-disk graph), the problem remains NP-hard [1].

In certain one-dimensional settings, however, the problem can be solved optimally in polynomial-time by dynamic programming. These settings may be motivated by e.g. coastline or national border surveillance. See appendix for details.

3.2   Dynamic Setting

In the dynamic setting we have options similar to the above except that now missions do not have explicit budgets and sensor assignments do not have explicit costs. What constraints the assignment problem is the limited energy that sensors have. We also have an additional time dimension. In this setting, each sensor has a battery size B, which means that it may only be used for at most B timeslots over the entire time horizon. Also, missions may arrive at any point in time and may last for any duration.

If a sensor network is deployed with no predetermined target lifetime, then the goal may be to maximize the profit achieved by each sensor during its own lifetime. However, if there is a finite target lifetime for the network, the goal becomes to earn the maximum total profits over the entire time horizon. We assume that the profit for a mission that lasts for multiple timeslots is the sum of the profits earned over all timeslots during the mission's lifetime. The danger of any particular sensor assignment is then that the sensor in question might somehow be better used at a later time. Therefore the challenge is to find a solution that competes with an algorithm that knows the characteristics of all future missions before they arrive. The general dynamic problem is specified by the mathematical program (MP) **P'** shown in Table 2.

If *preemption* is allowed, i.e. a new mission is allowed to preempt an ongoing mission and grab some of its sensors, then in each timeslot we are free to reassign currently used sensors to other missions based on the arrival of new missions without reassignment costs. In this case, a long mission can be thought of as a series of unit-time missions, and so the sensors and missions at each timeslot form an instance of the NP-hard static (offline) problem. If preemption is forbidden, then the situation for the online algorithm is in a way simplified. If we assume without loss of generality that no two missions will arrive at exactly the same time, then the online algorithm can focus on one mission at a time. Nonetheless, the dynamic problem remains as hard as the static problem, since a reduction can be given in which the static missions are identified with dynamic missions of unit length, each starting ϵ after the previous one. In fact, we can give a stronger result, covering settings both with and without preemption.

Proposition 1. *There is no constant-competitive algorithm for the online dynamic problem, even assuming that all missions are unit-length and non-overlapping.*

Proof. First consider the problem with $B = 1$, i.e. sensors with a battery lifetime of one timeslot. Suppose at time 1, there is M_1 with $p_1 = \epsilon$ requiring S_1 (i.e. otherwise unsatisfiable); we must choose the assignment because there may be no further missions, yet at time 2 there may be a M_2 with profit $p_2 = 1$. This yields the negative result for $B = 1$. Next suppose $B = 2$. Then consider the following example: at time 1, M_1 with $p_1 = \epsilon$ requires S_1, so we must assign because there may be no future missions; at time 2, M_2 with $p_2 = \exp(\epsilon)$ requires S_1, so we must assign for the same reason; at time 3, M_3 with $p_2 = \exp(\exp(\epsilon))$ requires S_1, but it is now empty, and the algorithm fails. This construction can be extended to arbitrary B.   □

As a generalization of the static problem, the offline version of dynamic problem is again NP-hard to solve optimally; moreover, the general online version cannot be solved with a competitiveness guarantee.

4   Static Setting

In this section we describe two algorithms to solve the static-assignment problem: *Greedy* and *Multi-round Generalized Assignment Problem (MRGAP)*. The former requires global knowledge of all missions to run and hence is considered to be centralized whereas the latter can be implemented in both centralized and distributed environments which makes it more appealing to sensor networks.

4.1   Greedy

The first algorithm we consider (Algorithm 1 shown in Table 3) is a greedy algorithm that repeatedly attempts the highest-potential-profit untried mission. Because fractional profits are allowed only beyond the threshold percentage T, this need not be the mission with maximum p_j. For each such mission, sensors are assigned to it, as long the mission budget is not yet violated, in decreasing order of cost-effectiveness, i.e. the ratio of edge utility for that mission and the sensor cost. The running time of the algorithm is $O(mn(m + \log n))$. No approximation factor is given for this efficiency-motivated algorithm since, even for the first mission selected, there is no guarantee that its feasible solution will be found. This by itself is an NP-hard 0/1 knapsack problem, after all.

4.2   Multi-Round GAP (MRGAP)

The idea of the second algorithm (Algorithm 2 shown in Table 4) is to treat the missions as knapsacks that together form an instance of the Generalized Assignment Problem (GAP). The strategy of this algorithm is to find a good solution for the problem instance *when treated as GAP*, and then to do "postprocessing" to enforce the lower-bound constraint of the profit threshold, by removing missions whose satisfaction percentage is too low. Releasing these sensors may make it possible to satisfy other missions, which suggests a series of rounds. In effect, missions not making "good" progress towards satisfying their demands are precluded from competing for sensors in later rounds.

Cohen et al. [3] give an approximation algorithm for GAP which takes a knapsack algorithm as a parameter. If the knapsack subroutine has approximation guarantee $\alpha \geq 1$, then the Cohen GAP algorithm offers an approximation guarantee of $1 + \alpha$. We use the standard knapsack FPTAS [12], which yields a GAP approximation guarantee of $2 + \epsilon$. Because GAP does not consider lower bounds on profits for the individual knapsacks, which is an essential feature of our sensor-assignment problem, we enforce it using the postprocessing step.

The algorithm works as follows. The threshold is initialized to a small value, e.g. 5%. Over a series of rounds, a GAP solution is found based on the current sensors and missions. After each round, missions not meeting the threshold are removed, and their sensors are released. Any sensors assigned to a mission that has greater than 100% satisfaction, and which can be released without reducing the percentage below 100%, are released. We call such sensors *superfluous*. Sensors assigned to missions meeting the threshold remain assigned to those missions. These sensors will not be considered in the next round, in which the new demands and budgets of each mission will become the remaining demand and the remaining budget of each one of them. Finally, the threshold

Table 3. Algorithm 1: Greedy

Table 4. Algorithm 2: Multi-Round GAP

while true **do**
for each available M_j
$u_j \leftarrow \sum_{S_i(unused)} e_{ij}$;
$j \leftarrow \arg\max_j p_j(u_j)$;
if $p_j(u_j) = 0$ **break**;
$u_j \leftarrow 0$; $c_j \leftarrow 0$
for each unused S_i in decr. order of e_{ij}/c_{ij}
if $u_j \geq d_j$ or $e_{ij} = 0$
break;
if $c_j + c_{ij} \leq b_k$
assign S_i to M_j;
$u_j \leftarrow u_j + e_{ij}$;
$c_j \leftarrow c_j + c_{ij}$;

initialize set of missions $M \leftarrow \{M_1 \dots M_m\}$;
initialize global threshold $T \leftarrow 0.05$;
for $t = 0$ to T step 0.05
run the GAP algorithm of [3] on M and the
unassigned sensors;
in the resulting solution, release any
superfluous sensors;
if M_j's satisfaction level is $< t$, for any j
release all sensors assigned to M_j;
$M \leftarrow M - \{M_j\}$;
if M_j is completely satisfied OR has no
remaining budget, for any j
$M \leftarrow M - \{M_j\}$;

is incremented, with rounds continuing until all sensors are used, all missions have succeeded or been removed, or until the actual success threshold T is reached. The GAP instance solved at each round is defined by the following linear program:

$$\begin{aligned} &\textit{max: } \sum_j \sum_i p_{ij} x_{ij} \text{ (with } p_{ij} = p_j \cdot e_{ij}/\hat{d}_j) \\ &\textit{s.t.: } \sum_{S_i(unused)} x_{ij} c_{ij} \leq \hat{b}_j, \text{ for each remaining } M_j, \\ &\quad\quad \sum_{M_j(remaining)} x_{ij} \leq 1, \text{ for each unused } S_i, \text{ and } x_{ij} \in \{0, 1\} \; \forall x_{ij} \end{aligned}$$

Here $\hat{d}_j$ is the remaining demand of M_j, that is, the demand minus utility received from sensors assigned to it during previous rounds. Similarly, $\hat{b}_j$ is the remaining budget of M_j. The concepts of demand and profit are encoded in the gap model as $p_{ij} = p_j \cdot e_{ij}/\hat{d}_j$. This parameter represents the fraction of demand satisfied by the sensor, scaled by the priority of the mission. In each GAP computation, we seek an assignment of sensors that maximizes the total benefit brought to the demands of the remaining mission.

One advantage of MRGAP is that it can be implemented in a distributed fashion. For each mission there can be a sensor, close to the location of the mission, that is responsible for running the assignment algorithm. Missions that do not contend for the same sensors can run the knapsack algorithm simultaneously. If two or more missions contend for the same sensors, i.e. they are within distance $2r$ of each other, then synchronization of rounds is required to prevent them from running the knapsack algorithm at the same time. To do this, one of the missions (e.g. the one with the lowest id) can be responsible for broadcasting a synchronization message at the beginning of each new round. However, since r is typically small compared to the size of the field, we can expect many missions to be able to do their computations simultaneously.

The total running time of the algorithm depends on the threshold T and step value chosen as well as the density of the problem instance, which will determine to what extent the knapsack computations in each round can be parallelized.

5   Dynamic Setting

We have shown in Section 3.2 that the dynamic problem is NP-hard to solve and that without assuming any aditional constraints there are no competitive solutions that can provide guaranteed performance. In this section, we therefore propose heuristic-based schemes to solve the dynamic sensor-mission assignment problem. These schemes are similar in their operation to the dynamic proposal scheme that we have proposed in [10] but with a new focus. Rather than maximizing profit by trying to satisfy all available missions, we focus here on maximizing the profit over network lifetime by allowing the sensors to refuse the participation in missions they deem not worthwhile.

We deal with missions as they arrive. A *mission leader*, a node that is close to the mission's location, is selected for each mission. Finding the leader can be done using a geographic-based routing techniques [5]. The mission leaders are informed about their missions' demands and profits by a base station. Then they run a local protocol to match nearby sensors to their respective missions. Since the utility a sensor can provide to a mission is limited by a sensing range, only nearby nodes are considered. The leader advertises its mission information (demand and profit) to the nearby nodes (e.g. two-hop neighbors). The number of hops the advertisement message is sent over depends on the relation between the communication range and the sensing range of sensors.

When a nearby sensor hears such an advertisement message, it makes a decision either to propose to the mission and become eligible for selection by the leader or to ignore the advertisement. The decision is based on the current state of the sensor (and the network if known) and on potential contribution to mission profit that the sensor would be providing. We assume knowledge of the (independent) distributions of the various mission properties (namely, demand, profit and lifetime), which can be learned from historical data. To determine whether a mission is worthwhile, a sensor considers a number of factors: (1) the mission's profit, relative to the maximum profit, (2) the sensor's utility to the mission, relative to the mission's demand, (3) the sensor's remaining battery level, (4) the remaining target network lifetime, if known.

After gathering proposals from nearby sensors, the leader selects sensors based on their utility offers until it is fully satisfied or there are no more sensor offers. The mission (partially) succeeds if it reaches the success threshold; if not, it releases all sensors.

Since we assume all distributions are known, the share of mission profit potentially contributed by the sensor (i.e. if its proposal is accepted) can be compared to the expectation of this value. Based on previous samples, we can estimate the expected mission profit $E[p]$ and demand $E[d]$. Also, knowing the relationship between sensor-mission distance and edge utility, and assuming a uniform distribution on the locations of sensors and missions, we can compute the expected utility contribution $E[u]$ that a sensor can make to a typical mission *in its sensing range*. We use the following expression to characterize the expected partial profit a sensor provides to a typical mission:

$$E\left[\frac{u}{d}\right] \times \frac{E[p]}{P} \tag{1}$$

We consider two scenarios. In the first, the target network lifetime is unknown, i.e. we do not know for how long will the network be needed. In this case, sensors choose missions that provide higher profit than the expected value and hence try to last longer in anticipation of future high profit missions. In the second, the target network lifetime is known, i.e. we know the duration for which the network will be required. Sensors, in this case, take the remaining target network lifetime into account along with their expected lifetime when deciding whether to propose to a mission. In the following two subsections we describe solutions to these two settings.

5.1 Energy-Aware Scheme

In this scheme, the target lifetime of the sensor network is unknown. For a particular sensor and mission, the situation is characterized by the actual values of mission profit (p) and demand (d) and by the utility offer (u), as well as the fraction of the sensor's remaining energy (f). For the current mission, a sensor computes the following value:

$$\frac{u}{d} \times \frac{p}{P} \times f \tag{2}$$

Each time a sensor becomes aware of a mission, it evaluates expression (2). It makes an offer to the mission only if the value computed is greater than expression (1). By weighting the actual profit of a sensor in (2) by the fraction of its remaining battery value, the sensors start out eager to propose to missions but become increasingly selective and cautious over time, as their battery levels decrease. The lower a sensor's battery gets, the higher relative profit it will require before proposing to a mission. Since different sensors' batteries will fall at different rates, in a dense network we expect that most feasible missions will still receive enough proposals to succeed.

5.2 Energy and Lifetime-Aware Scheme

If the target lifetime of the network is known, then sensors can take it into account when making their proposal decisions. To do this, a sensor needs to compute what we call the *expected occupancy time*, denoted by t_α, i.e. the amount of time a sensor expects to be assigned to a mission during the remaining target network lifetime. To find this value we need to determine how many missions a given sensor is expected to see. Using the distribution of mission locations, we can compute the probability that a random mission is within a given sensor's range. If we combined this with the remaining target network lifetime and arrival rate of missions we can find the expected number of missions to which a given sensor will have the opportunity to propose. So, if the arrival rate and the (*independent*) distributions of the various mission properties are known, we can compute t_α as follows:

$$t_\alpha = \tau \times \lambda \times g \times \gamma \times E[l]$$

where:

- τ is the remaining target network lifetime, which is the initial target network lifetime minus current elapsed time.
- λ is the mission arrival rate.
- $g = \pi r^2/A$ is the probability that a given mission is within sensing range (assumes uniform distribution on missions). r is the sensing range and A is the area of the field in which sensors are deployed.

- $E[l]$ is the expected mission lifetime.
- γ is the probability that a sensor's offer is accepted. *Computing* this value would imply a circular dependency. It was chosen a priori to be 0.25.

For each possible mission, the sensor now evaluates an expression which is modified from (2). The sensor considers the ratio between its remaining lifetime and its expected occupancy time. So, if t_b is the amount of time a sensor can be actively sensing, given its current energy level, the expression becomes:

$$\frac{u}{d} \times \frac{p}{P} \times \frac{t_b}{t_\alpha} \tag{3}$$

If the value of expression (4) is greater than that of expression (1), then the sensor proposes to the mission. If the sensor's remaining target lifetime is greater than its expected occupancy time, the sensor proposes to *any* mission since in this case it expects to survive until the end of the target time. The effect on the sensor's decision of weighting the mission profit by the ration (t_b/t_α) is similar to the effect weighting the fraction of remaining energy (f) had in expression (2); less remaining energy, all things being equal, make a sensor more reluctant to propose to a mission. As the network approaches the end of its target lifetime, however, this ratio will actually increase, making a sensor more willing to choose missions with profits less than what it "deserves" in expectation. After all, there is no profit at all for energy conserved past the target lifetime.

6   Performance Evaluation

In this section we discuss the result of different experiments used to evaluate our schemes. We implemented the schemes in Java and tested them on randomly generated problem instances. We report the results of two sets of experiments. In the first set, we test the static setting, in which all missions occur simultaneously. In the second set, we consider the dynamic setting, in which missions arrive, without warning, over time and depart after spending a certain amount of time being active.

6.1   Simulation Setup

Each mission has a demand, which is an abstract value of the amount of sensing resources it requires, and a profit value, which measures its importance (the higher the profit the more important). The profit obtained from a successful mission M_j is equal to $p_j(u_j)$ as defined in Section 3. Each sensor can only be assigned to a single mission. Once assigned, the full utility of the sensor is allocated to support the mission. We consider a mission to be successful if it receives at least 50% of its demanded utility from allocated sensors (i.e. $T = 0.5$).

We assume that mission demands are exponentially distributed with an average of 2 and a minimum of 0.5. Profits for the different missions are also exponentially distributed with an average of 10 and a maximum of 100. This simulates realistic scenarios in which many missions demand few sensing resources and a smaller number demand more resources. The same applies to profit. The simulator filters out any mission that is not individually satisfiable, i.e. satisfiable in the absence of all other missions. For a sufficiently dense network, however, we can expect there to be few such impossible missions.

The utility of a sensor S_i to a mission M_j is defined as a function of the distance, D_{ij}, between them. In order for sensors to evaluate their utilities to missions, we assume that all sensors know their geographical locations. Formally the utility is:

$$e_{ij} = \begin{cases} \frac{1}{1+D_{ij}^2/c}, & \text{if } D_{ij} \leq r \\ 0, & \text{otherwise} \end{cases}$$

where r is the global sensing range. This follows typical signal attenuation models in which signal strength depends inversely on distance squared. In our experiments, we set $c = 60$ and $r = 30m$.

Nodes are deployed in uniformly random locations in a $400m \times 400m$ field. Missions are created in uniformly random locations in the field. The communication range of sensors is set to $40m$. When sensors are deployed we ensure that the network is connected (that is any sensor can communicate with any other sensor possibly over multiple hops). If a randomly created instance is not connected, it is ignored by the simulator. We do not consider communication overhead other than noting the fact that distributed schemes will typically have lower communication cost than centralized schemes and hence are preferable in that sense. We have actually studied the communication overhead of similar schemes in [10].

6.2   Static Setting

In this experiment, all missions occur simultaneously. We fix the number of sensors in the field and vary the number of missions from 10 to 100. Each sensor has a uniformly random cost in $[0, 1]$, which is the same no matter to which mission it is assigned. The associated cost can represent the actual cost in real money or a value that represents the risk of the sensor being discovered in hostile environments if it is activated, etc. Each mission has a budget, drawn from a uniform distribution with an average of 3 in the first experiment and varies from 1 to 10 in the second. In the following results we show the average of 20 runs.

The first series of results shows the fraction of the maximum mission profits achieved by the different schemes. The maximum profit is the sum of all missions profits. We show the profits for the greedy algorithm, Multi-Round GAP (MRGAP), and an estimate of the optimal value, running on two classes of sensor networks, sparse (250 nodes), and dense (500 nodes) (Figures 1 and 2, respectively). The estimate on the optimal bound is obtained by solving the LP relaxation of program **P**, in which all decision variables are allowed to take on fractional values in the range $[0, 1]$, and the profit is simply fractional based on satisfaction fraction, i.e. $p_j y_y$ for mission M_j with no attention paid to the threshold T. The MRGAP scheme, which recall can be implemented in a distributed fashion, achieves higher profits in all cases than does the greedy scheme which is strictly centralized (because missions have to be ordered in terms of profit). The difference, however, is not very large.

Figure 3 shows the fraction of the total budget each scheme spent to acquire the sensing resources it did in a network with 250 nodes. The MRGAP scheme achieves more

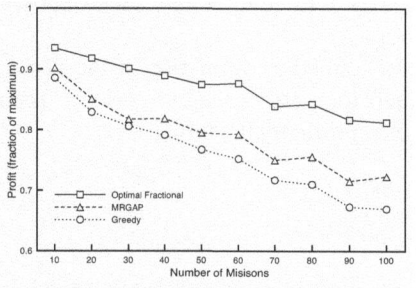

Fig. 1. % of max. profit achieved (250 nodes)

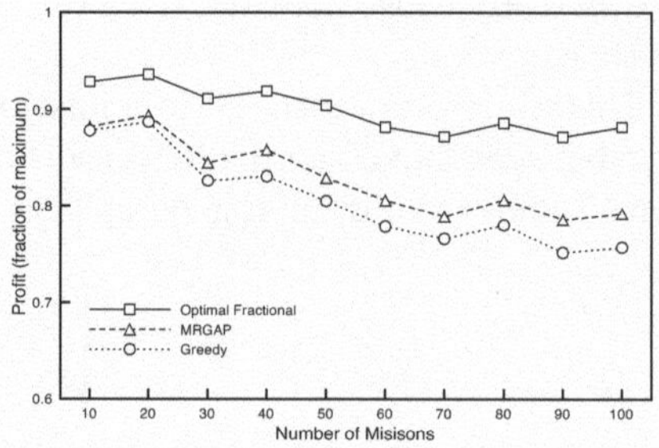

Fig. 2. % of max. profit achieved (500 nodes)

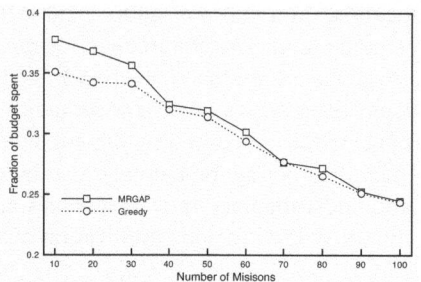

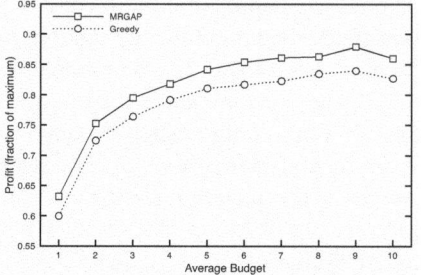

Fig. 3. Fraction of spent budget (250 nodes)

Fig. 4. Varying the average budget (250 nodes)

profits than the greedy algorithm and spends a modest amount of additional resources. The fraction of remaining budget is significant (more than 60% in all cases), which suggests that either successful missions had higher budgets than they could spend on available sensors or that unsuccessful missions had lower budgets than necessary and hence they were not able to reach the success threshold and their budgets were not spent. When the number of missions is large this can be attributed to the fact that there simply were not enough sensors due to high competition between missions. We performed another set of experiments, in which the number of missions was fixed at 50 and average budget given to missions was varied between 1 and 10. Figure 4 shows the results for a network with 250 nodes. We see that the achieved profit increases rapidly in the beginning with the budget but slows down as the issue then becomes not the limited budget but rather the competition between missions.

6.3 Dynamic Setting

In the dynamic problem, missions arrive without warning over time and the sensors used to satisfy them have limited battery lives. The goal is to maximize the total profit achieved by missions over the entire duration of the network. In this section, we test the dynamic heuristic algorithms on randomly generated *sensor network histories* in order to gauge the algorithms' real-world performance.

In addition to the assumptions made in Section 6.1, here we also assume the following. Missions arrive according to a Poisson distribution with an average arrival rate of 4 missions/hour or 8 missions/hour depending on the experiment. Each sensor starts with a battery that will last for 2 hours of continuous sensing (i.e., $B = 7200$ in seconds). We assume that this battery is used solely for sensing purposes and is different than the the the sensor's main battery which it uses for communication and maintaining its own operation. Mission lifetimes are exponentially distributed with an average of 1 hour. We limit the minimum lifetime to 5 minutes and the maximum to 4 hours. The number of nodes used in the following experiments is set to 500 nodes.

We test the performance of the dynamic schemes described in Section 5. We compare the energy-aware (*E-aware*) and the energy and lifetime-aware (*E/L-aware*) schemes with a basic scheme that does not take energy or network lifetime into account when making the decision on to which mission to propose (i.e. sensors propose to any mission in their range). For comparison purposes, we also show the performance of the network if we assume that sensors have infinite energy (i.e. $B \geq$ simulation time).

Of course, finding the true optimal performance value is NP-hard, so we cannot do an exact comparison. Even the LP relaxation of program **P'** above may be infeasible because of the number of decision variables. To provide some basis of comparison, therefore, we define a further relaxation which is feasible to solve and provides an upper-bound of the optimal. This formulation condenses the entire history into a single timeslot. The profits and demands are multiplied by the duration of the mission. Since time is elided, the sensor constraint now asserts only that a sensor be used (fractionally) for at most B timeslots, over the entire history, where B is the battery lifetime. Note that the solution value provided is the total profits over the entire history, not a time-series. We indicate its value in the results by a straight line drawn at the average profit corresponding to this total.

In both the E-aware and the E/L-aware schemes, we need to compute the expected profit of a mission (Expression 1 above), to determine whether the sensor should propose to the mission. Because we cap the distributions – limit the minimum demand and profit for missions and cap them by the available resources in the first case and by 100 in the second – the actual averages are not equal to the a priori averages of the distributions. We found an empirical average demand of $d' = 1.2$ and an average profit of $p' = 10.9$. The empirical average duration, which is used to evaluate expression (3) above, was found to be 3827.8 seconds (roughly an hour).

Figure 5 shows the achieved profit (as a fraction of maximum possible) per-timeslot. We assume the network is considered to be of use as long as this value stays above 50% (shown with a fine horizontal line). The target network lifetime is 3 days (shown as a fine vertical line) and the simulations are run for one week of time. Knowledge of the target network lifetime is used by the E/L-aware and in the LP. The other schemes assume that network will potentially have infinite duration.

From Figure 5 we see that the profits of all schemes stay above the 50% threshold for the target lifetime. The basic scheme achieves most of its profits in the beginning and then profits go down (almost linearly) as time progresses. The E-aware scheme tries to conserve its resources for high profit missions. Because it ignores the fact that we care more about the first 3 days than anytime after that it, becomes overly conservative and ignores many missions. Such a scheme is better suited to the case when there is no known target lifetime for the network and we want the network to last as long as

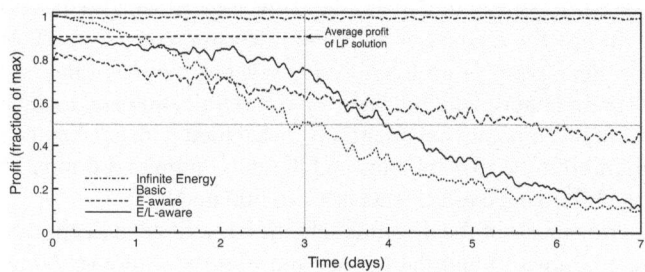

Fig. 5. Fraction of achieved profits. (arrival rate = 4).

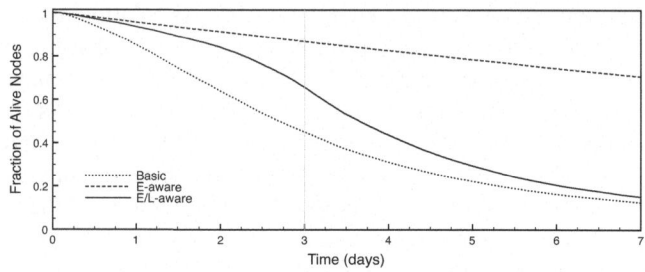

Fig. 6. Fraction of alive nodes (arrival rate = 4)

possible. We see that the profit for E-aware does not fall below the 50% threshold until the end of the sixth day.

In the E/L-aware scheme, nodes will initially be aggressive in accepting missions that might not provide their expected value but become more cautious as their energy is used. However, unlike E-aware, as their remaining expected occupancy time approaches their remaining lifetime, sensors will again accept missions with lower and lower profits. The curves for E-aware and E/L-aware cross by the middle of the fourth day, at which point the E-aware becomes better. When compared to the average LP solution, we see that E/L-aware is very close with a difference (on the average) of few percentage points. We also found that in terms of the sum of all profits during the target network lifetime (i.e. the area under the curve for the first 3 days), the E/L-aware achieves about 84% of the profits compared to 72% for the E-aware. This means that E/L-aware achieves close to 17% higher profits. If the sensors battery lifetime is increased from 2 to 3 hours the percentage increase becomes about 22%.

The fraction of remaining alive sensors over time is shown in Figure 6. Because in the basic scheme sensors propose to any mission within its range no matter how low the profit is, nodes start dying rapidly. By the the end of the third day, only half the nodes can be used for sensing and by the end of the seventh day this falls below 15%. In E-aware, nodes become very cautious as their batteries run low, which helps the network to last for longer without significant sacrifice of achieved profits per timeslot. By the end of the 7 days, about 72% of the nodes remain living. For E/L-aware, sensors accept more missions, and hence are used at a higher rate, as the target lifetime of the network approaches. In the figure, we can see this happening by the second day, when the curve of E/L-aware diverges from that of E-aware. By the end of the seventh day, it has used nearly as much energy as the basic scheme.

One thing to note is that we assume E/L-aware acts like the basic scheme once the target lifetime of the network has passed, i.e. sensors propose to all nearby missions. If this behavior were changed to emulate the E-aware, we expect the energy usage to slow down which would conserve a larger fraction of alive nodes. A similar effect is expected to happen to profit. As the fraction of alive nodes will be higher, the decrease in profit after target network lifetime will slow down.

We omit figures showing the fraction of achieved profit and fraction of alive nodes over time, with twice the previous arrival rate (8 missions/hour). Due to the increased number of missions, sensors are used more rapidly and hence both the profit and fraction of alive nodes decrease quickly. The basic scheme passes the 50% profit line by the middle of the second day and both E-aware and E/L-aware pass that point in the beginning of the fourth day. But by that point, E/L-aware achieves significantly higher profits than E-aware. Similar effects are seen on the fraction of alive nodes.

7 Conclusion and Research Directions

In this paper, we defined new sensor-assignment problems motivated by frugality and conservation of resources, in both static and dynamic settings. We proposed schemes to match sensing resources to missions in both settings and evaluated our these schemes through simulations. In the static case, we found that the multi-round GAP scheme, which can be implemented in a distributed fashion, outperformed a centralized greedy solution. In the dynamic setting, we found that overall performance can be significantly improved if sensors are allowed to sometimes refuse to offer utility to "weak" missions. Performance improves when sensors make this decision based both on their remaining energy level and the remaining target network lifetime.

More sophisticated utility models, taking into consideration not just physical attributes such as distance but also some sort of semantics, could be developed. Above, we based the potential utility of a sensor-missing pairing on their separating distance. However, in the case of video sensors, for example, the closest sensor to an event might not be the best candidate for selection because its view of the field is obstructed or it cannot provide sufficient frame rate. Also, regarding the value of utility from multiple sensors, we make the simplifying assumption of additive utilities, which is realistic in some but clearly not all scenarios. In 3-d reconstruction, for example, the benefit of having two viewing angles separated by 45 degrees may be much greater than two views from essentially the same location. In ongoing work, we are investigating models that allow for such non-additive joint utility.

Acknowledgements. This research was sponsored by US Army Research laboratory and the UK Ministry of Defence and was accomplished under Agreement Number W911NF-06-3-0001. The views and conclusions contained in this document are those of the authors and should not be interpreted as representing the official policies, either expressed or implied, of the US Army Research Laboratory, the US Government, the UK Ministry of Defence, or the UK Government. The US and UK Governments are au-

References

1. Bar-Noy, A., Brown, T., Johnson, M., La Porta, T., Liu, O., Rowaihy, H.: Assigning sensors to missions with demands. In: ALGOSENSORS 2007 (2007)
2. Byers, J., Nasser, G.: Utility-based decision-making in wireless sensor networks. In: Proceedings of MOBIHOC 2000 (2000)
3. Cohen, R., Katzir, L., Raz, D.: An efficient approximation for the generalized assignment problem. Inf. Process. Lett. 100(4), 162–166 (2006)
4. Kaplan, L.: Global node selection for localization in a distributed sensor network. IEEE Transactions on Aerospace and Electronic Systems 42(1), 113–135 (2006)
5. Karp, B., Kung, H.: Greedy perimeter stateless routing for wireless networks. In: Proceedings of MobiCom 2000, Boston, MA, August 2000, pp. 243–254 (2000)
6. Lu, J., Bao, L., Suda, T.: Coverage-Aware Sensor Engagement in Dense Sensor Networks. In: Yang, L.T., Amamiya, M., Liu, Z., Guo, M., Rammig, F.J. (eds.) EUC 2005. LNCS, vol. 3824, pp. 639–650. Springer, Heidelberg (2005)
7. Mullen, T., Avasarala, V., Hall, D.L.: Customer-driven sensor management. IEEE Intelligent Systems 21(2), 41–49 (2006)
8. Perillo, M., Heinzelman, W.: Optimal sensor management under energy and reliability constraints. In: WCNC 2003 (2003)
9. Rowaihy, H., Eswaran, S., Johnson, M., Verma, D., Bar-Noy, A., Brown, T., La Porta, T.: A survey of sensor selection schemes in wireless sensor networks. In: SPIE Defense and Security Symposium (2007)
10. Rowaihy, H., Johnson, M., Brown, T., Bar-Noy, A., La Porta, T.: Assigning Sensors to Competing Missions. Technical Report NAS-TR-0080-2007, Network and Security Research Center, Department of Computer Science and Engineering, Pennsylvania State University, University Park, PA, USA (October 2007)
11. Shih, K., Chen, Y., Chiang, C., Liu, B.: A distributed active sensor selection scheme for wireless sensor networks. In: Proceedings of the IEEE Symposium on Computers and Communications (June 2006)
12. Vazirani, V.V.: Approximation Algorithms. Springer, Heidelberg (2001)
13. Zhao, F., Shin, J., Reich, J.: Information-driven dynamic sensor collaboration. IEEE Signal Processing Magazine 19(2), 61–72 (2002)

Appendix: 1-d Static Setting

In 1-d settings such as border surveillance, sensors and missions lie on the line, and edge weights e_{ij} depend in some way on the distance between S_i and M_j. If $e_{ij} = 1/(1 + D_{ij})$, for example, then the 1-d problem is strongly NP-hard, even with no budget constraint, since the known 3-Partition reduction [10] also works in 1-d. To adapt the reduction to 1-d, place all missions at one point on the line, and choose the sensor locations so that the resulting edge weights equal the corresponding 3-Partition element values. If we adopt a binary sensor model, however, then the budget-constrained 1-d SMD problem is in P. We give a polynomial-time algorithm for a 1-dimensional, budget-constrained version of SMD in which edge-weights are 0/1 based on sensing range and the budget limits the *total number of sensors* that may be used. In this setting, $e_{ij} = 1$ if $D_{ij} \leq r$ and 0 otherwise. We may therefore assume without loss of generality that the demands d_j are integral, but profits p_j may be arbitrary positive values. Assignments and profits are both non-fractional. For now, assume budgets are infinite.

We first observe that if r is large enough so that $e_{ij} = 1$ for all i, j then this is an instance of the 0/1 knapsack problem, solvable in polynomial time (in terms of *our problem*'s input size, not knapsack's). In this case, sensors are simply the units in which knapsack weights are expressed, with the knapsack budget equal to n, so the knapsack DP runs in time $O(mn)$. In general, though, we will have $e_{ij} = 0$ for some i, j. We can solve the problem by a modification of the standard 0/1 knapsack DP [12]. We construct an $m \times n$ table of optimal solution values, where $v(j, i)$ is the optimal value of a subinstance comprising the first j missions and the first i sensors. Row 1 and Column 1 are initialized to all zeros (optimal profits with the zero-length prefix of missions, and the zero-length prefix of sensors, respectively). We fill in the entries row-by-row, using a value d'_j defined below:

$$v(j, i) = \begin{cases} v(j - 1, i), \text{if } M_j \text{ is not satisfiable with } S_1, ..., S_i \\ \max(v(j - 1, i), \; p_j + v(j - 1, i - d'_j)), \text{o.w.} \end{cases}$$

The two entries of the *max* correspond to satisfying M_j or not. The first option is clear. If we do satisfy M_j in a solution with only sensors $1, ..., i$ available, the important observation is that we may as well use for M_j the rightmost d_j sensors possible, since in this sub-instance M_j is the rightmost mission. (The DP will consider satisfying M_j when solving sub-instances with fewer sensors.) Note that we might be unable to assign the rightmost d_j sensors, since some of these may be too far to the right for M_j to be within their sensing range. Let d'_j equal d_j plus the number of sensors among $S_1, ..., S_i$ too far to the right to apply to M_j. Any sensor too far to the right to reach M_j is also too far to the right to reach M_k for $k < j$. What remains among this sub-instance will therefore be missions $1, ..., j - 1$ and sensors $1, ..., i - d'_j$. This completes the recursion definition.

Proposition 2. *With unbounded budgets, the 1-d problem can be solved optimally in time*
$O(mn)$.

Proof. Optimality is by induction. The running time results from the table size.

In the formulation of the infinite-r setting above, the number of sensors n becomes the knapsack budget, whereas in the general setting the sensors must be assigned to missions in range. We now introduce an explicit budget into the problem. Let the model be the same as above, except now also given is an integral budget $B \leq n$, which is the total number of sensors that may be used. First delete any missions with $d_j > B$. Then the budgeted version is solved by creating a 3-d DP table, whose third coordinate is budget. We initialize this table by filling in all nm entries for budget 0 with value 0. Then for layer b, we again fill in the entries row-by-row.

Notice that the remaining budget to be spent on the "prefix" instance is $b - d_j$, not $b - d'_j$, since only d_j sensors are actually assigned to mission M_j. Thus we conclude:

Table 5. Algorithm 3: 1-d 0/1 dynamic program

$v(0, j, i) \leftarrow 0$ **for** $0 \leq j \leq m, 0 \leq i \leq n$
for $b = 1$ to B **do**
  $v(b, 0, i) \leftarrow 0$ **for** $0 \leq i \leq n$
  for $j = 1$ to m **do**
    for $i = 0$ to n **do**
      if M_j is not satisfiable with b sensors among $1..i$ **then**
        $v(b, j, i) \leftarrow v(b, j - 1, i)$
      else
        $v(b, j, i) \leftarrow \max(v(b, j - 1, i),\ p_j + v(b - d_j, j - 1, i - d'_j))$

Proposition 3. *Algorithm 3 is optimal and runs in time $O(mn^2)$.*

Tug-of-War: An Adaptive and Cost-Optimal Data Storage and Query Mechanism in Wireless Sensor Networks

Yuh-Jzer Joung and Shih-Hsiang Huang

Dept. of Information Management, National Taiwan University, Taipei 10617, Taiwan

Abstract. We propose *Tug-of-War* (*ToW*) for data storage and query mechanism in sensor networks. ToW is based on the concept of *data-centric storage* where events and queries meet on some rendezvous node so that queries for the events can be sent directly to the node without being flooded to the network. However, rather than fixing the rendezvous point, we dynamically replicate the point and float them in between query nodes and event sensor nodes according to the relative frequencies of events and queries. By carefully calculating the optimal setting, we are able to minimize the total communication costs while adapting to the environment.

1 Introduction

The main role of sensor nodes in a wireless sensor network is to interact with the physical environment to detect relevant event data, which are then to be queried by interested clients. The problem of how to efficiently and effectively gather these data back to the clients for further processing then arises. There are two canonical ways: *External Storage* and *Local Storage* [1]. In External Storage, whenever sensor nodes detect events, the data are sent directly to an external node, referred to as *sink*, where the data are stored and processed. Obviously queries can be very efficient as they need not be disseminated into the network. However, much communication cost is wasted if the stored data are less frequently queried.

In contrast, in the Local Storage mechanism event information is stored at the detecting nodes. When the client needs the data, it sends query messages to collect them. Compared to External Storage, data is transmitted only when needed, thus avoiding unnecessary transmission. The performance of the system then is primarily influenced by the way how queries are disseminated into the network. Because nodes in a wireless sensor network are usually deployed in a random manner and the client has no knowledge about where the events might occur in the network, flooding is perhaps the most simple and intuitive way to disseminate queries. Flooding, however, generates too much traffic, and so some remedies have been proposed to reduce the costs, e.g., directed diffusion [2], data aggregation [3,4,5], TTL-based expanding search [6], and in-network processing [7].

S. Nikoletseas et al. (Eds.): DCOSS 2008, LNCS 5067, pp. 237–251, 2008.

An intermediate approach is to find a rendezvous point for events and queries. Every detected event is sent to and stored at the rendezvous node so that queries for the event can be sent directly to the node without being flooded to the network. Another advantage is that different types of events can have different rendezvous nodes to balance the load. This concept has indeed been adopted in *Data-Centric Storage (DCS)* [1]. Basically, DCS operates like many distributed hash tables (DHTs) [8,9] in peer-to-peer networks. Every event is given a name according to its type, and the name is hashed to obtain a key for storage and retrieval of the event. A representative example of DCS is Geographic Hash Table (GHT) [10]. Note that in DCS some routing mechanism must be used to send a message to a destination. GHT uses geographic location of each node to assist routing. Some follow-ups improve upon GHT/DCS by removing the need of exact geographic location for each node [11], adding support of range queries [12], and increasing failure resilience [13].

In this paper we continue this direction and focus on how to dynamically choose the rendezvous point to optimize the communication cost. Intuitively, if the rendezvous point is close to the sensor node, then storing an event to the rendezvous point takes less communication cost, while a query message needs to traverse a longer distance to reach the rendezvous point. So if events occur more often than queries, then the total communication cost can be reduced by moving the rendezvous point to the sensor node. Conversely, we can reduce the communication cost by moving the rendezvous point towards the query source if queries occur more often than events. The query source and the sensor node can be viewed as two end nodes that battle for the rendezvous point. Our contribution is to devise a mechanism, which we referred to as **Tug-of-War (ToW)**, that can adjust the rendezvous point on the fly and on an optimal basis, thereby to minimize the communication cost according to different event and query frequencies.

2   Network Model and Definitions

We represent a wireless sensor network as a graph $G = (V, E)$, where V is the set of sensor nodes and E is the set of links. Each sensor node $s \in V$ has a unique identification, and knows its position in a coordinate system. Each node can (directly) communicate with nodes within radio transmission range. We assume that all nodes have the same transmission range. Let r be the radio transmission range. Thus, for every $s_i, s_j \in V$, $(s_i, s_j) \in E$ iff $\delta(s_i, s_j) \leq r$, where δ is the Euclidean distance function. Note that edges are bi-directional. We assume that G is connected. So every node can communicate with every other node in V through hop-by-hop message routing.

The main job of the sensor nodes is to detect events and accept queries of events. We assume that events are categorized into classes, e.g., soldiers, tanks, cars, etc. Event classes can be identified by the sensors detecting them. Each event is modeled as $e = (x, y, d, t, c)$, where x and y are coordinates of the event at which it occurs, d is the radius of influence area, t is the occurring timestamp, and c is the event class of e. The event e is detected by the set of nodes $V_e \subseteq V$, where $s \in V_e$ iff $\delta(e, s) \leq d$.

Queries of events are of the form: $q = (s, c, \Delta)$, where s is the querying node, c is the target event class (type), and Δ is the time interval during which the query is interested. Only events occurring in the interval will be collected. We assume that queries can be disseminated by any node in the network, thereby allowing us to model mobile information gathering. The results in the paper can be easily extended to the case where queries can only originate from some fixed external sinks.

We also assume that events and queries are the main concerns of the system. Each detected event is to be stored at some node which can be queried by others. Therefore, the system cost is comprised of *event storage* and *query dissemination and reply*. Because the energy consumption of radio transmission in sensor networks is typically much higher than that of storage and memory access, we count only radio transmissions when measuring the system costs. Radio transmissions will be measured in terms of the number of hops a message has traversed.

3   Tug-of-War

ToW is based on GHT [10], but improves it to support an adaptive mechanism to minimize the communication cost. We begin by a brief introduction to GHT.

3.1   Geographic Hash Table (GHT)

The core step in GHT is the hashing of event types into geographic coordinates to determine the rendezvous point for events and queries of a given type. Each event type is associated with a key, and the storage and retrieval of events of the type are performed according to the key. Because GHT stores all events with the same key in the same node (called the *home node* of the events), if too many events with the same key are detected, the key's home node may become a hotspot, both for communication and storage. To alleviate the node's load, GHT applies *structured replication (SR)* to distribute the events to other nodes. To see the idea, let us call the point a key maps to the *root point* of the key. SR divides the network's geographical space into $2^d \cdot 2^d = 4^d$ subspaces with a given *hierarchy depth d*, and assigns a mirror image of the root point in each subspace. Fig. 1 shows an example of mirror images for $d = 2$ and root point at $(5, 5)$. A node that detects an event with the key in consideration can store the event at the mirror closest to its location. This reduces the storage cost from $O(\sqrt{n})$ to $O(\sqrt{n}/2^d)$, where n is the number of nodes in the network. However, a query to events of the key should be routed throughout all mirrors. In SR, this is done recursively: the query first routes to the root point (node), then from the root node to three level-1 mirror nodes. Each level-1 node in turn forwards the query to the three level-2 mirrors associated with them, and so on until all mirrors are reached. The total query cost thus increases from $O(\sqrt{n})$ to $O(2^d\sqrt{n})$. That is, SR reduces storage cost at the expense of queries. The hierarchy depth d in SR

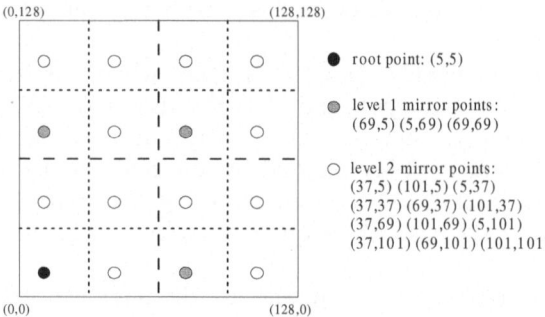

Fig. 1. Structured replication in GHT

is assumed to be determined statically by the event name, and is known to every node. However, how the depth should be calculated is not addressed in [10].

3.2 Motivation

The system cost in a sensor network is comprised primarily of event storage and query dissemination. Intuitively, if events incur more cost than queries, then the system may benefit by moving the rendezvous point to the event detecting nodes. However, if events (of the same type) may occur at any where in the network, then multiple rendezvous points must be employed, as a rendezvous point close to some node may be far away from other nodes.

We observe that the SR mechanism in GHT, although it was originally proposed to alleviate a node's load, can also effectively reduce the event storage cost, as a node needs only to store detected events to the nearest mirror image. However, since a query node has no idea which image may have the data, it needs to query all the images, thereby increasing the query cost. So the decrease of event storage cost may be offset by the increase of query cost. A natural question then arises: can the system still benefit from this bias towards event storage, and if so, what is the optimal setting?

Similarly, if queries incur more cost than events, then we can reduce query cost by letting a node send its query message only to the nearest image. For this image to have the needed data, events must be stored to all images if queries can originate from anywhere in the network. Again, the decrease of query cost is at the expense of event storage. So we need to know whether the system can still benefit by this event storage and query dissemination strategy, and if so, how to maximize the benefit.

To answer the questions we need to calculate the routing cost of events and queries, which in turn depends on how images are chosen.

3.3 Selection of Mirror Images

Like GHT, ToW assigns each event class c a home node and a set of $4^r - 1$ mirror images, where r is similar to the hierarchy depth in GHT's structured replication. Here we call it the *system resolution*. The coordinates of the home node is obtained by hashing the name of c using a globally known hash function h. Let $(h_x(c), h_y(c))$ denote the home node coordinates in binary representation. Let m be the size of the binary strings used in the coordinate system. Then the $4^r - 1$ mirror images of the home node are at the coordinates (x, y) such that x and $h_x(c)$ have identical suffix of $m - r$ bits, and similarly for y and $h_y(c)$. For a given r, we will refer to the home node of an event class and all its mirror images as the *level-r images* of the class. Note that a level-$(r - 1)$ image of c is by definition also a level-r image of c.

3.4   Operating Modes

ToW operates in two basic modes: one allows a sensor node to store an event to its nearest image, while the other allows a querying node to query only its nearest image. In the former, queries must be disseminated to all images, as the querying node does not know where an event might occur. Conversely, in the latter events must be stored at all images if we assume that queries can originate from anywhere in the network. We refer to the two modes as **write-one-query-all** and **write-all-query-one**, respectively. For now, let us assume that both querying nodes and event detecting nodes know which mode to operate. Later in Section 3.6 we will discuss how the nodes know which mode to operate.

In each mode, we also need to determine a system resolution r to find the images of an event class so that queries and the target events will meet at some image. The resolution is determined according to the relative frequencies of events and queries so as to optimize the system cost. We will return to this issue in Section 3.6. When the resolution r is determined, in write-one-query-all, a node detecting an event simply stores the event data to the nearest level-r image of the event class. A query of the event class will then be routed to all level-r images of the class. On the other hand, in the write-all-query-one mode, detected events need to be stored to all level-r images so that queries can be routed to the images that are closest to the query nodes.

3.5   Routing Algorithms

In both the write-one-query-all and write-all-query-one modes we need to route a message to a nearest image as well as to the set of all level-r images of an event class. Like in GHT, the former can be done using the GPSR algorithm [14]. GHT also presents a "**hierarchical**" routing algorithm for the latter: first routes the message to the home node. The home node then forwards the message to the three level-1 images. Each of the level-1 image in turn forwards the query to the three level-2 images associated with them, and so on until all images are reached. Fig. 2(a) illustrates the routing algorithm. Each edge represents a message delivery path, with the delivery order labeled.

Rather than using the hierarchical routing algorithm, here we present another algorithm, referred to as **combing**, that yields lower cost than the hierarchical algorithm. The combing algorithm works by first letting the originating node of a

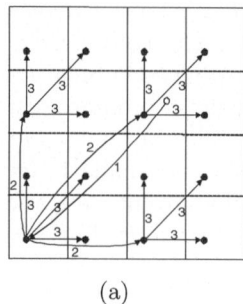   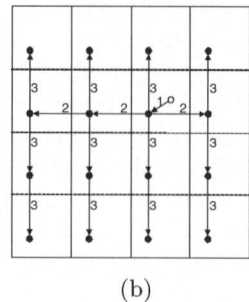

(a)   (b)

Fig. 2. Routing algorithms: (a) hierarchical, and (b) combing

message route the message to the nearest level-r image, say u, using the ordinary GPSR routing algorithm. Starting from u, all the other level-r images can be visited by first forwarding the message along the horizontal line (or alternatively, the vertical line) that contains u. So all the level-r images with the same y-coordinate as u can be visited along this horizontal line. For each level-r image w on the line, a vertical search path is then traversed to discover all the level-r images with the same x-coordinate as w. This routing scheme is illustrated in Fig. 2(b).

By comparing with the hierarchical routing algorithm, we can see that the combing algorithm takes less communication cost. For example, in Fig. 2(a), the message must first be routed to the root image (the home node), while in Fig. 2(b) it can be sent to the nearest one.

To see the actual communication cost in each algorithm, assume that the network consists of n sensor nodes uniformly deployed in a square region, so each side consists of roughly $\sqrt{n}$ nodes. For simplicity the routing cost between two nodes is their geographical distance. Recall the hierarchical algorithm, whose routing cost can be expressed by $CR_{hierarchical} = \delta(s,h)$ when $r = 0$, and $CR_{hierarchical} = \delta(s,h) + S_1 + S_2 + \ldots + S_r$ when $r \geq 1$, where $\delta(s,h)$ denotes the distance between the source node s and the root image h, and each S_i denotes the total cost for each level-$(i-1)$ image to forward a message to the three level-i images associated with it, $i \geq 1$. To calculate S_i, observe that there are 4^{i-1} level-$(i-1)$ images, and the distance between each level-$(i-1)$ image and the associated three level-i images are $\frac{1}{2^i}\sqrt{n}$, $\frac{1}{2^i}\sqrt{n}$, and $\frac{\sqrt{2}}{2^i}\sqrt{n}$. So $S_i = 4^{i-1}(\frac{2}{2^i} + \frac{\sqrt{2}}{2^i})\sqrt{n} = \frac{(2+\sqrt{2})\sqrt{n}}{2} \times 2^{i-1}$. Then, $S_1 + S_2 + \ldots + S_r = \frac{(2+\sqrt{2})\sqrt{n}}{2}(2^r - 1)$. Therefore,

$$CR_{hierarchical}$$
$$\simeq \delta(s,h) + \left[(1 + \tfrac{\sqrt{2}}{2})2^r - (1 + \tfrac{\sqrt{2}}{2})\right]\sqrt{n} \qquad (1)$$

Note that the above holds for both $r = 0$ and $r \geq 1$.

On the other hand, in the combing algorithm, the cost consists of a routing path from the message source to the nearest image, and a horizontal path of length $\sqrt{n}(1 - \frac{1}{2^r})$ followed by 2^r vertical paths of the same length. For the first part, observe that the distance d from the message source to the nearest image is $d = \delta(s, h)$ when $r = 0$ (i.e., the root image h is the nearest image), and is halved for every increment of r. So $d = \frac{\delta(s,h)}{2^r}$. So the total routing cost of the combing algorithm is

$$
\begin{aligned}
CR_{combing} &\simeq \frac{\delta(s, h)}{2^r} + (2^r + 1)(1 - \frac{1}{2^r})\sqrt{n} \\
&= \frac{\delta(s, h)}{2^r} + (2^r - \frac{1}{2^r})\sqrt{n}
\end{aligned}
\tag{2}
$$

3.6   Optimizing the System Mode and Resolution

Nodes in ToW need to know which mode to operate, and in the selected mode, which resolution to use to minimize the total communication cost. To see how this can be determined, assume that the system consists of n sensor nodes uniformly deployed in a square region, and f_e and f_q respectively denote the event frequency and query frequency for some event class c in consideration. Every event of the class has the same influence range, and will be detected by k nodes. We use $\overline{\delta}\sqrt{n}$ to denote the average distance between a querying node and the home node of c.[1] The estimation of δ depends on the query model and the selection of a home node. For example, if events and queries can originate randomly from anywhere in the square, and the home node of c is also chosen randomly from the network, then $\overline{\delta} \simeq 0.52$. This is because two random points have an average distance of $\frac{1}{15}[\sqrt{2} + 2 + 5\ln(1 + \sqrt{2})] \simeq 0.52$ in a unit square.[2] On the other hand, if the home node is fixed at the origin, then $\overline{\delta} \simeq 0.76$.[3]

Assume that the GPSR algorithm is used to route a message to the nearest level-r image, while the combing routing algorithm is used for routing a message to all level-r images. Let us first calculate the total cost in the write-one-query-all mode with a given resolution r. Observe that storing an event to its home node with resolution $r = 0$ costs $\overline{\delta}\sqrt{n}$. An increment of r will shorten the distance by half, as the event needs only be stored to the nearest level-r image. So, the event storage cost C_e per unit interval is

$$
C_e \simeq f_e \cdot k \cdot \frac{\overline{\delta}\sqrt{n}}{2^r}
$$

By (2), the cost C_q in delivering query messages per unit interval is

$$
C_q \simeq f_q\left[\frac{\overline{\delta}\sqrt{n}}{2^r} + (2^r - \frac{1}{2^r})\sqrt{n}\right]
$$

---

[1] More precisely, this distance should be $\overline{\delta}(\sqrt{n} - 1)$, as the side of the network square has length $\sqrt{n} - 1$ when measured by hop count. To avoid complicating the equations too much, in this section we use $\sqrt{n}$ to approximate the side length. Later in Section 4 the more accurate length will be used in order to obtain a precise bound.

[2] See, e.g., http://mathworld.wolfram.com/SquareLinePicking.html

[3] http://mathworld.wolfram.com/SquarePointPicking.html

Suppose every node receiving a query message needs to reply to the query. So the total query cost is twice the value of C_q. All together, the total communication cost $C_{W_1 Q_{all}}$ per unit interval is

$$C_{W_1 Q_{all}} \simeq C_e + 2C_q$$

$$= f_e \cdot k \cdot \frac{\overline{\delta}\sqrt{n}}{2^r} + 2f_q\sqrt{n}\left(2^r - \frac{1-\overline{\delta}}{2^r}\right) \tag{3}$$

Given k, f_e, and f_q, $C_{W_1 Q_{all}}$ is minimized by the following value of r:

$$\Upsilon_{W_1 Q_{all}}(f_e, f_q) \simeq \frac{1}{2}\log\left(\frac{\overline{\delta}}{2} \cdot k \cdot \frac{f_e}{f_q} - (1-\overline{\delta})\right) \tag{4}$$

We can compare how ToW improves over GHT with no structured replication. It is easy to see that the communication cost of GHT per query is

$$C_{GHT} = \frac{f_e}{f_q}k\overline{\delta}\sqrt{n} + 2\overline{\delta}\sqrt{n}$$

By using an optimal r, the cost of ToW per query is

$$C_{W_1 Q_{all}} \simeq 2\sqrt{2k\overline{\delta}}\sqrt{\frac{f_e}{f_q}}\sqrt{n} \tag{5}$$

So we have

$$\frac{C_{W_1 Q_{all}}}{C_{GHT}} \simeq \frac{2\sqrt{2k\overline{\delta}}\sqrt{\frac{f_e}{f_q}}\sqrt{n}}{\frac{f_e}{f_q}k\overline{\delta}\sqrt{n}} \simeq 2\sqrt{\frac{2f_q}{k\overline{\delta}f_e}} \tag{6}$$

From (6) we see that the larger the $\frac{f_e}{f_q}$ ratio, the smaller the cost ratio of ToW vs. GHT, and therefore the more the improvement of ToW over GHT. To illustrate, when $\frac{f_e}{f_q} = 8$ and $k = 8$, we have $\frac{C_{W_1 Q_{all}}}{C_{GHT}} \simeq 0.49$; i.e., ToW can improve nearly 50% of the cost of GHT. Note that the improvement does not depend on the network size. However, (6) can be applied up to the case when the f_e/f_q ratio in (4) yields the maximal resolution—$\log\sqrt{n}$ (as the number of images 4^r cannot exceed the total number of nodes n). When the maximal resolution is reached, ToW reduces to Local Storage, where the cost $C_{W_1 Q_{all}}$ consists of only query cost from the query source to all other nodes plus the replies; i.e., $2n - 2$. So $\frac{C_{W_1 Q_{all}}}{C_{GHT}} \simeq \frac{2f_q\sqrt{n}}{k\overline{\delta}f_e}$.

On the other hand, when write-all-query-one mode is used, by (2) the cost for storing detected events to all level-r images per unit time interval is

$$C'_e \simeq f_e \cdot k\sqrt{n}\left(2^r - \frac{1-\overline{\delta}}{2^r}\right),$$

while the cost in delivering query messages to their nearest images is

$$C'_q \simeq f_q \cdot \frac{\overline{\delta}\sqrt{n}}{2^r}$$

So the total cost per unit interval is

$$C_{W_{all}Q_1} \simeq C'_e + 2C'_q$$

$$= f_e \cdot k\sqrt{n}(2^r - \frac{1-\overline{\delta}}{2^r}) + 2f_q \cdot \frac{\overline{\delta}\sqrt{n}}{2^r} \qquad (7)$$

$C_{W_{all}Q_1}$ is minimized by the following value of r:

$$\Upsilon_{W_{all}Q_1}(f_e, f_q) \simeq \frac{1}{2}\log(\frac{2\overline{\delta}}{k} \cdot \frac{f_q}{f_e} - (1 - \overline{\delta})) \qquad (8)$$

By using an optimal r, the cost of ToW in the mode per event is

$$C_{W_{all}Q_1} \simeq 2\sqrt{2k\overline{\delta}}\sqrt{\frac{f_q}{f_e}}\sqrt{n} \qquad (9)$$

So the cost of ToW in the write-all-query-one mode relative to GHT is as follows:

$$\frac{C_{W_{all}Q_1}}{C_{GHT}} \simeq \frac{2\sqrt{2k\overline{\delta}}\sqrt{\frac{f_q}{f_e}}\sqrt{n}}{k\overline{\delta}\sqrt{n} + 2\overline{\delta}\frac{f_q}{f_e}\sqrt{n}} = \frac{2\sqrt{2k\frac{f_q}{f_e}}}{(k + 2\frac{f_q}{f_e})\sqrt{\overline{\delta}}} \qquad (10)$$

Again, we see that the improvement increases as f_q/f_e increases, regardless of the network size. Also, if f_q/f_e is large relatively to k, then the cost ratio of ToW versus GHT is approximately in proportion to $\sqrt{f_e/f_q}$. So the larger the f_q/f_e ratio, the larger the improvement of ToW over GHT. Also note that (10) can be applied up to the case when the f_q/f_e ratio in (8) yields the maximal resolution; and in that case, ToW reduces to External Storage.

Equations (4) and (8) determine an optimal resolution for each mode. To determine which mode to operate, we observe that resolution must be non-negative. For a given k, f_e, and f_q, if a nonzero optimal r exists in one mode, then no nonzero optimal r can exist in the other mode. Moreover, the two modes converge at $r = 0$ (and at which $C_{W_1Q_{all}} = C_{W_{all}Q_1}$). So, to determine the best mode, it suffices to see in which mode a nonzero optimal r exists for a given k, f_e, and f_q. Otherwise, $r = 0$ is the best choice for both modes, making it no difference in selecting which mode to use. For example, by (4) and (8) we see that when $f_e \geq f_q$, write-one-query-all is the best mode. Even if $f_e < f_q$, the mode may still be preferred if k is large.

3.7   Query and Event Frequency Estimation

Up to now we have assumed that every node in ToW has the knowledge of the event-query ratio so that they can determine which query and storage mode to operate, and in which resolution. For ToW to be adaptive, we let each node estimate the frequencies based on the number of events and queries it has encountered or learned from other nodes.

To estimate the frequencies, every level-r image of an event class c maintains two counts, the number of events c_e and the number of queries c_q addressed to the image in some fixed interval Γ. Assume that events and queries may originate uniformly at random at any point in the network. So in the write-one-query-all mode with resolution r, a recording of an event at an image node v must be compensated by another $4^r - 1$ events that may be addressed to other level-r images in order to obtain the total number of events occurred network-wise. Similarly, in query-one-write-all, each query recorded at an image must also be compensated by another $4^r - 1$ queries. The ratio of the two counts can then be used to estimate the event-query ratio.

For non-image nodes to learn the ratio, when an image node replies to a query, the ratio is timestamped and then piggybacked in the reply message. So every node in the radio range along the replying path can learn of the ratio and update their information. To disseminate new information more quickly across the network (even though queries are less frequently made), each node also attaches its event/query ratio to each message it sends so that nodes in the radio range can also learn of the ratio.

3.8   Coping with Inconsistency

The actual selection of r must be integer. So by rounding the optimal value to integer, we see some leeway for maintaining view consistency of the event-query ratio by different nodes. Similarly, since the two modes write-one-query-all and write-all-query-one converge at $r = 0$, there is also some leeway before the system needs to switch to different mode. So even if two nodes have a different view of the event-query ratio, they may still use the same mode and the same resolution. Nevertheless, as r is discrete, an inconsistent view of the ratio may lead to different system resolution. For example, when $k = 8$ and $f_e/f_q = 4$, we have $\Upsilon_{W_1 Q_{all}}(f_e, f_q) = 1.49$, yielding the round value to 1. However, when $f_e/f_q = 4.1$, $\Upsilon_{W_1 Q_{all}}(f_e, f_q) = 1.5$ and so the round value is 2.

To cope with inconsistent view of r, observe that all level-r images of an event class c are also level-$(r + 1)$ images of c. Therefore, a "conservative" approach to deal with a possible different view of r is to use level-r images rather than level-$(r + 1)$ images when a node is not sure whether the exact resolution is r or $r + 1$. Specifically, in the write-one-query-all mode, a sensor that sees a "low confident" resolution $r + 1$ may store event data to the nearest level-r image, instead of a level-$(r+1)$ image. A resolution $r+1$ is "low confident" if the original resolution $\Upsilon_{W_1 Q_{all}}(f_e, f_q)$ before rounding is $r+0.5 \leq \Upsilon_{W_1 Q_{all}}(f_e, f_q) < r+0.5+\eta$ for some $\eta \leq 0.5$. As such, a query node that views the system resolution as r or $r + 1$ is still able to retrieve the data. Note that query nodes still use the regular rounding function to convert $\Upsilon_{W_1 Q_{all}}(f_e, f_q)$ to integer. So, for example, when $\eta = 0.2$, a query node that views $\Upsilon_{W_1 Q_{all}}(f_e, f_q)$ as 1.4 and a sensor node that views $\Upsilon_{W_1 Q_{all}}(f_e, f_q)$ as 1.6 will be able to meet as both will use level-1 images. Even if the query node sees $\Upsilon_{W_1 Q_{all}}(f_e, f_q)$ also as 1.6 and use level-2 images, it is still able to find the data stored by the sensor node as the level-2 images include the level-1 image the sensor node uses. Similarly, in the write-all-query-one mode, a query node will query the nearest level-r image if it's confidence on resolution $r + 1$ is low.

4 Related Work

Recently there has been much work to study query strategies in between Local Storage and External Storage. Intuitively, in Local Storage a query source needs to "pull" relevant results out of the network, while in External Storage nodes "push" results out to the query source. A hybrid approach is to use both "push" and "pull" strategies simultaneously. For example, rumor routing [15] propagates queries and events via random walk with some TTL to allow them to meet with high probability. An in-depth analysis in [16] shows that a combined use of push-pull strategies in random walk has the best query success rate over time. Propagating queries and events along a tree of sensor nodes with root being the query source is analyzed in [17]. The result also shows that a hybrid push-pull query processing strategy is often preferred to pure approaches. Push-pull-based directed diffusion routing protocols are analyzed in [18] to determine how well they match different application scenarios.

The work by Liu, et al. [19] is most close to our work. They study push-pull event dissemination and query strategy in a grid network, and propose a deterministic mechanism, referred to as **Comb-Needle (CN)**, to guarantee query results. Like ToW, CN also considers an optimal setting to minimize communication cost, and adapts to different frequencies of queries and events. In the rest of the section we give a thorough comparison of CN and ToW to see how the two mechanisms differ when both claim an optimal setting. Note that ToW is based on DCS, but adds on a mechanism to dynamically adjust the rendezvous point. In fact, we have just learned that a comparative study of DCS with CN has recently been studied in [20]. Our analysis here complements the studies, hoping to provide more insight into hybrid push-pull query strategies in two query processing paradigms in sensor networks—*structured* (of DCS) vs. *unstructured* (of non-DCS) [20].

As CN is presented mainly in a grid environment, here we assume that the system layout is a grid. Furthermore, we assume events dominate queries; the query-dominated case can be analyzed in a similar way as we did below. CN can be illustrated by Fig. 3. When a sensor x detects an event, it disseminates the information, say vertically, to a path of length l (called the needle length). Then when some node y wishes to query the event, it can issue several horizontal search paths, each of which s distance apart, to the network. So if $l + 1 \geq s$, then one of the query paths must intersect with the event dissemination path, thereby to guarantee resolving the query. So when the event dissemination path is long, fewer number of query search paths need to be issued. The dissemination path corresponds to event cost, while the search paths correspond to query cost. So event cost can be reduced at the expense of query, and vice versa. Therefore, we can minimize the total communication cost by dynamically adjusting the event dissemination path length according to the relative frequencies of events and queries.

Although both ToW and CN use a dynamic strategy for information dissemination and gathering, there are some subtle differences. First, ToW is based on Data-Centric Storage, where events and queries meet at some globally known

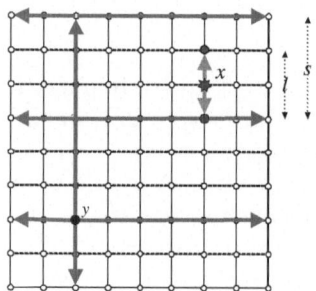

Fig. 3. The Comb-Needle information dissemination and gathering mechanism

rendezvous point. However, rather than using a fixed rendezvous point, ToW adjusts the rendezvous position according to the relative frequencies of events and queries, thereby to minimize the total communication cost.

CN, on the other hand, is not data centric, and so a query source has no prior knowledge about which specific node to gather the information. So flooding is essentially the search technique.[4] CN avoids flooding to the entire network by letting queries and events walk through orthogonal paths to ensure intersection.

To compare ToW with CN, consider a network of n nodes laid out as a grid like Fig. 3. Note that each side of the grid has length $\sqrt{n} - 1$. Consider the case where event frequency f_e is larger than query frequency f_q, and we compare the total communication cost per query. (The case for $f_e < f_q$ can be similarly analyzed.) For simplicity, we assume that the average number of nodes k that may simultaneously detect an event is 1, and the query source is fixed at the lower left corner of the grid (i.e., the origin).

Like ToW, the communication cost of CN consists of three parts: C_e for event dissemination, C_{qd} for query dissemination, and C_{qr} for query replies. It can be seen that $C_e = (s-1) \cdot \frac{f_e}{f_q}$, $C_{qd} = 2(\sqrt{n}-1) + \frac{(\sqrt{n}-1)^2}{s}$, and $C_{qr} = (\sqrt{n}-1)\frac{f_e}{f_q}$ (see [19]). By setting s to be the optimal value $s = (\sqrt{n}-1)\sqrt{\frac{f_q}{f_e}}$, we have the following total communication cost for CN:

$$C_{CN} = C_e + C_{qd} + C_{qr}$$

$$\simeq \sqrt{n}\frac{f_e}{f_q} + 2\sqrt{n} + 2\sqrt{n}\sqrt{\frac{f_e}{f_q}} \qquad (11)$$

For ToW, recall from Section 3.6 that the total communication cost is $C_{W_1 Q_{all}} = C'_e + 2C'_q$, where C'_e and C'_q are the event and query dissemination costs,

---

[4] The fact that DCS uses a globally known hash function to determine the rendezvous point and that CN (and previous work on push-pull query strategies) have no such prior knowledge can be analogized with *structured* vs. *unstructured* peer-to-peer (P2P) networks: in structured P2P networks data placement is deterministic and usually hash-based, while in unstructured P2P networks there is no prior knowledge of the data location. Query then can be done in a more efficient way in DCS than in non-DCS, as the latter basically resorts to an exhaustive search of the network.

respectively. Rather than directly using (3), for the purpose of estimating the maximum value of $C_{W_1 Q_{all}}$ shortly, below we use a more accurate estimate of C'_e and C'_q. For C'_e, the cost (per query) is

$$C'_e \simeq \frac{f_e}{f_q} \cdot \overline{\delta_{eh}} \cdot (\frac{\sqrt{n}}{2^r} - 1)$$

where $\overline{\delta_{eh}}$ denotes the average distance between an event detecting node and the home node of the event class in a unit grid. Note that here we use $\frac{\sqrt{n}}{2^r} - 1$ to represent the hop-count length of one side of a grid of $\frac{\sqrt{n}}{2^r} \times \frac{\sqrt{n}}{2^r}$ nodes. In contrast, in Section 3.6, we simply use $\frac{\sqrt{n}}{2^r}$. Similarly,

$$C'_q \simeq \overline{\delta_{qh}}(\frac{\sqrt{n}}{2^r} - 1) + (2^r - \frac{1}{2^r})\sqrt{n}$$

where $\overline{\delta_{qh}}$ denotes the average distance between a query node and the home node of the event class in a unit grid.

To estimate $\overline{\delta_{eh}}$ and $\overline{\delta_{qh}}$, we first note that by virtue of hash function we assume that the home node can be an arbitrarily node in the network. As events can arise from anywhere in the network, $\overline{\delta_{eh}}$ represents the average distance of two random points in grid. Since routing in grid uses Manhattan distance, by the fact that the average distance between two random points in a unit length is $1/3$, we have $\overline{\delta_{eh}} \simeq 2/3$. For $\overline{\delta_{qh}}$, since here we assume that the query source is fixed at the origin, we have $\overline{\delta_{qh}} \simeq 1$.

So the total communication cost per query in ToW is

$$C_{W_1 Q_{all}} = C'_e + 2C'_q \simeq \frac{2}{3}\frac{f_e}{f_q}(\frac{\sqrt{n}}{2^r} - 1) + 2(2^r \sqrt{n} - 1) \qquad (12)$$

By using an optimal $r = \log \left(\frac{1}{3}\frac{f_e}{f_q}\right)^{\frac{1}{2}}$, we have

$$C_{W_1 Q_{all}} \simeq \frac{2}{3}\sqrt{n}\sqrt{\frac{3f_e}{f_q}} + 2\sqrt{n}\sqrt{\frac{f_e}{3f_q}} - \frac{2}{3}\frac{f_e}{f_q} \qquad (13)$$

By comparing (11) and (13), we see that ToW yields lower cost than CN, as the former has the cost in the order of $O(\sqrt{n}\sqrt{\frac{f_e}{f_q}})$, while the latter is of $O(\sqrt{n}\frac{f_e}{f_q})$.

In the above we considered CN where every event yields one reply path, and so the total query reply cost is $C_{qr} = (\sqrt{n} - 1)\frac{f_e}{f_q}$. One can see that when f_e/f_q is large, it may be worth to combine reply paths by letting the end node in each horizontal query path issue a return message along the query path to gather the results hop by hop. In this way, the communication cost of CN can be estimated by

$$C'_{CN} = C_e + 2C_{qd} \simeq 4\sqrt{n} + 3\sqrt{n}\sqrt{\frac{f_e}{f_q}} \qquad (14)$$

By comparing (14) and (13), one can still see that ToW have smaller dominating constant ($2.3\sqrt{\frac{f_e}{f_q}}\sqrt{n}$ of ToW vs. $3\sqrt{\frac{f_e}{f_q}}\sqrt{n}$ of CN, approximately 23% difference). Note that in this scheme, although the gap of communication cost between CN and ToW is shortened, the query latency of CN increases: for CN it takes $2\sqrt{n}-2$ transmission time to complete a query, while for ToW on average the latency is only half of CN, i.e., $\sqrt{n}-1$.

To see the maximum cost of ToW, recall that the maximal resolution in ToW is $\frac{1}{2}\log n$, and in this case ToW reduces to the Local Storage scheme, where each node detecting an event simply stores the event locally. By setting $r = \log\left(\frac{1}{3}\frac{f_e}{f_q}\right)^{\frac{1}{2}} = \frac{1}{2}\log n$, we can obtain the threshold of f_e/f_q that turns ToW into Local Storage. So for f_e/f_q larger than the threshold, the communication cost per query is the cost to send a message from one node to all other nodes in the network plus replies from them—that is, $2n - 2$.

Similarly, CN may reduce to Local Storage if f_e/f_q is sufficiently large. This occurs when $s = (\sqrt{n} - 1)\sqrt{\frac{f_q}{f_e}} = 1$. In this case, the event dissemination cost C_e is zero, while the query dissemination cost is $C_{qd} = n - 1$. By aggregating query results along the return path, we can also have $C_{qr} = n - 1$. So in total the maximum cost of CN can also be bounded by $2n - 2$.

5   Conclusions

We proposed ToW, an adaptive, cost-optimal, data-centric storage and query mechanism for sensor networks. In data-centric storage like GHT, events and queries meet on some rendezvous node so that queries of events can be sent directly to the node without being flooded to the network. However, rather than fixing the rendezvous point, we dynamically replicate the point and float them in between query nodes and event sensor nodes to minimize the total communication costs. A theoretical analysis indicates that the adaptive mechanism of ToW can significantly reduce the communication costs of GHT. The cost ratio of ToW vs. GHT is approximately in proportion to the square root of the event/query frequency ratio, regardless of the network size.

We have also conducted a thorough comparison between ToW and a recent proposal Comb-Needle (CN), which also uses a dynamic strategy for information dissemination and gathering in sensor networks. ToW and CN represent two query mechanisms for two query processing paradigms in sensor networks: DCS (Data-Centric Storage) and non-DCS. Compared to ToW, CN is not data centric, and so a query node has no prior knowledge about which specific node to gather the information. To avoid flooding, query and event messages walk through orthogonal paths to ensure intersection. CN works best in gird, but still, costs much more than ToW.

The performance of ToW and its comparisons with other systems have also been studied via simulation. Due to space limitation, the results will be provided in the full paper.

References

1. Shenker, S., Ratnasamy, S., Karp, B., Govindan, R., Estrin, D.: Data-centric storage in sensornets. SIGCOMM Computer Communication Review 33(1), 137–142 (2003)
2. Intanagonwiwat, C., Govindan, R., Estrin, D., Heidemann, J., Silva, F.: Directed diffusion for wireless sensor networking. IEEE/ACM Transactions on Networking 11(1), 2–16 (2003)
3. Madden, S., Szewczyk, R., Franklin, M.J., Culler, D.: Supporting aggregate queries over ad-hoc wireless sensor networks. In: WMCSA 2002, pp. 49–58 (2002)
4. Madden, S., Franklin, M.J., Hellerstein, J.M., Hong, W.: TAG: a tiny aggregation service for ad-hoc sensor networks. SIGOPS Operating Systems Review 36(SI), 131–146 (2002)
5. Shrivastava, N., Buragohain, C., Agrawal, D., Suri, S.: Medians and beyond: new aggregation techniques for sensor networks. In: SenSys 2004, pp. 239–249 (2004)
6. Chang, N., Liu, M.: Revisiting the TTL-based controlled flooding search: optimality and randomization. In: MobiCom 2004, pp. 85–99 (2004)
7. Rabbat, M., Nowak, R.: Distributed optimization in sensor networks. In: IPSN 2004, pp. 20–27 (2004)
8. Ratnasamy, S., Francis, P., Handley, M., Karp, R., Shenker, S.: A scalable content-addressable network. In: SIGCOMM, pp. 161–172 (2001)
9. Stoica, I., Morris, R., Karger, D.R., Kaashoek, M.F., Balakrishnan, H.: Chord: A scalable peer-to-peer lookup service for Internet applications. In: SIGCOMM 2001, pp. 149–160 (2001)
10. Ratnasamy, S., Karp, B., Yin, L., Yu, F., Estrin, D., Govindan, R., Shenker, S.: GHT: a geographic hash table for data-centric storage. In: WSNA 2002, pp. 78–87 (2002)
11. Newsome, J., Song, D.: GEM: graph embedding for routing and data-centric storage in sensor networks without geographic information. In: SenSys 2003, pp. 76–88 (2003)
12. Greenstein, B., Estrin, D., Govindan, R., Ratnasamy, S., Shenker, S.: DIFS: a distributed index for features in sensor networks. In: SNPA 2003, pp. 163–173 (2003)
13. Ghose, A., Grossklags, J., Chuang, J.: Resilient Data-Centric Storage in Wireless Ad-Hoc Sensor Networks. In: Chen, M.-S., Chrysanthis, P.K., Sloman, M., Zaslavsky, A. (eds.) MDM 2003. LNCS, vol. 2574, pp. 45–62. Springer, Heidelberg (2003)
14. Karp, B., Kung, H.T.: GPSR: greedy perimeter stateless routing for wireless networks. In: MobiCom 2000, pp. 243–254 (2000)
15. Braginsky, D., Estrin, D.: Rumor routing algorithm for sensor networks. In: WSNA 2002, pp. 22–31 (2002)
16. Shakkottai, S.: Asymptotics of query strategies over a sensor network. In: INFOCOM 2004, pp. 548–557 (2004)
17. Trigoni, N., Yao, Y., Demers, A.J., Gehrke, J., Rajaraman, R.: Hybrid push-pull query processing for sensor networks. GI Jahrestagung (2), 370–374 (2004)
18. Krishnamachari, B., Heidemann, J.: Application-specific modelling of information routing in sensor networks. In: MWN 2004, pp. 717–722 (2004)
19. Liu, X., Huang, Q., Zhang, Y.: Combs, needles, haystacks: balancing push and pull for discovery in large-scale sensor networks. In: SenSys 2004, pp. 122–133 (2004)
20. Kapadia, S., Krishnamachari, B.: Comparative analysis of push-pull query strategies for wireless sensor networks. TR CENG-2005-16, USC (2005)

Towards Diagnostic Simulation in Sensor Networks

Mohammad Maifi Hasan Khan, Tarek Abdelzaher, and Kamal Kant Gupta

Department of Computer Science
University of Illinois at Urbana-Champaign
mmkhan2@uiuc.edu,zaher@cs.uiuc.edu,kkgupta2@ad.uiuc.edu

Abstract. While deployment and practical on-site testing remains the ultimate touchstone for sensor network code, good simulation tools can help curtail in-field troubleshooting time. Unfortunately, current simulators are successful only at evaluating system performance and exposing manifestations of errors. They are not designed to diagnose the root cause of the exposed anomalous behavior. This paper presents a *diagnostic simulator*, implemented as an extension to TOSSIM [6]. It (i) allows the user to ask questions such as "why is (some specific) bad behavior occurring?", and (ii) conjectures on possible causes of the user-specified behavior when it is encountered during simulation. The simulator works by logging event sequences and states produced in a regular simulation run. It then uses sequence extraction, and frequent pattern analysis techniques to recognize sequences and states that are possible root causes of the user-defined undesirable behavior. To evaluate the effectiveness of the tool, we have implemented the directed diffusion protocol and used our tool during the development process. During this process the tool was able to uncover two design bugs that were not addressed in the original protocol. The manifestation of these two bugs were same but the causes of failure were completely different - one was triggered by node reboot and the other was triggered by an overflow of timestamps generated by the local clock. The case study demonstrates a success scenario for diagnostic simulation.

Keywords: Sensor network, diagnostic simulation, frequent pattern mining.

1 Introduction

Simulation is widely used as an effective tool for rapid prototyping and evaluation of new applications and protocols. In this paper, we present the concept of diagnostic simulation where a simulator is augmented with a data mining backend such that it can not only perform performance evaluation but also be used as an automated troubleshooting tool to diagnose root causes of failures and anomalous behavior. Our architecture can extend prior simulation tools such as TOSSIM [6], Atemu [9], Avrora [11], EmStar [3], DiSenS [13], and S^{2}DB [12], to provide automated analysis and failure diagnosis.

S. Nikoletseas et al. (Eds.): DCOSS 2008, LNCS 5067, pp. 252–265, 2008.

A diagnostic simulator can answer questions such as "Why does the network suffer unusually high communication delays?", "Why is throughput low despite availability of resources?", or "Why does this leader election fail to elect a unique leader?". In general, the user supplies a condition they deem undesirable. The diagnostic simulator then conjectures on the root cause of the undesirable behavior.

Diagnostic simulation, as presented in this paper, is intended to uncover failures in *distributed protocols*. It may not necessarily be the right tool to uncover function-level bugs (i.e., those contained within one function), such as an improperly dereferenced pointer or a misspelled variable. Traditional debugging tools are better at stepping through lines of code within a function until an error is encountered. Instead, we focus on finding system-level or design bugs that arise from (improper) interaction between multiple components. Generally, the more complex the interaction sequence that leads to the problem, the more suitable the diagnostic simulator is at isolating that sequence. Informally, a distributed computation can be thought of as a state machine in which some states are "bad". The goal of the diagnostic simulator is to find out why a system enters a particular bad state. Generally, such state is entered because a sequence of events transformed the system from a good state to the bad one. It is desired to uncover that unlucky sequence of events, which is a data mining problem. We designed and implemented a modular, extensible framework that uses frequent pattern analysis techniques to identify sequences correlated with user-specified manifestations of errors.

To address diagnosing protocol bugs in a real deployment, we previously developed SNTS [5], a passive logging-based tool that is deployed on separate nodes in the network alongside real application nodes and records all radio communications. Those data are uploaded to a PC for automated analysis. The techniques used in the diagnostic simulator described in this paper offer two contributions over SNTS. First, by virtue of reading the state of a (simulated) system directly, the diagnostic simulator has cheap access to a much larger variety of events, as opposed to message transmission events only. This significantly enriches its diagnostic capability. Second, our diagnostic simulator uses sequence mining algorithms that can identify *event sequences* leading to failure. SNTS, in contrast, was restricted only to classification algorithms that find correlations between *current system state* and the occurrence of failures. Often failure is a function of history or a *chain of events* and not only current conditions. For example, out-of-order message delivery is a sequence of at least two events occurring in an anomalous order. If a problem is caused by a particular out of order delivery sequence, such a problem can be detected by our diagnostic simulator but not by SNTS.

The rest of this paper is organized as follows. In section 2, we describe recent work on troubleshooting sensor network applications as well as representative simulation techniques. In section 3, we describe the design of the diagnostic simulator and the challenges with our solution for such a system. To show the effectiveness of our tool, in section 4, we provide a case study by diagnosing failure cases in the directed diffusion protocol. Finally, section 5 concludes the paper.

2   Related Work

Simulation is an effective way to test sensor network applications before deployment. Among all the simulators, TOSSIM [6] is one of the most widely used in the sensor network community. One of the biggest advantages of TOSSIM is that it leverages the existing TinyOS code and thus makes it easy to transfer simulated code to real hardware. The event driven execution framework makes the simulation a realistic representation of sensor hardware behavior.

Among the existing debugging tools, ATEMU [9] provides XATDB, a GUI based debugger that can be used to manually debug the code at a line-level. DiSenS [13] uses a distributed memory based parallel cluster system for simulation of large systems. S^2DB [12] provides debugging abstractions at different levels, such as the node level and network level, by leveraging the DiSenS simulator. S^2DB provides debugging points so the developer can set breakpoints and watchpoints to access the internal device states being simulated by DiSenS. S^2DB also introduces the concept of parallel debugging where a user can set breakpoints among multiple coordinated devices to perform distributed debugging. With all these attractive features, the major shortcoming is that the process of debugging requires manual intervention and inspection.

Marionette [14] is an interactive tool that allows developers to communicate with real sensor nodes from a PC at run time. It lets users access functions at the node level and inspect and change the variables dynamically. Clairvoyant [15] is another tool that lets the user debug programs remotely using breakpoints and watchpoints. Neither offer automated diagnosis.

Sympathy [10] is a debugging tool primarily designed for distributed data gathering applications. It localizes the source of failure to a specific node or route and infers failure based on the quantity of data (rather than data content) in an automated way. The assumption of the tool is that the application being debugged generates some amount of traffic at regular intervals which limits its applicability for arbitrary application. Sympathy does not aim to diagnose arbitrary failures or anomalies.

To troubleshoot sensor network applications in a real deployment, we previously developed SNTS [5], a passive listening based diagnostic tool that can be used to diagnose protocol bugs. SNTS does not assume anything about the application nature itself and requires minimal input from the user. These features make the tool broadly applicable. SNTS, however, cannot uncover event *sequences* that lead to failure. It can only classify states (i.e., *unordered sets of conditions*) in which failures are observed with a high probability.

Our approach is inspired by advances in data mining techniques applied successfully for software bug detection [2,8,7]. In a similar manner, we use frequent sequence mining to diagnose protocol design bugs by analyzing run time event sequences rather than analyzing source code.

Finally, there is a large body of work on model-checking and formal verification of protocol properties. We do not survey this work as we take an orthogonal approach that relies on diagnosis of observed errors as opposed to proofs of absence of errors. Formal techniques are preferred in mission-critical software, such as avionics, where advance correctness guarantees are required (e.g., for certification). In contrast, diagnostic simulation is intended for softer embedded applications, where certification is not needed, and therefore the rigorous and costly process of formal specification and verification is not warranted.

3   Design and Implementation

In this section, we present the main idea of diagnostic simulation, explain the notions of events and event sequences, which are central to the operation of our tool, highlight the algorithm we borrow for detecting common sequences (that are correlated with failures or anomalous behavior), and explain some of the challenges we face in applying that algorithm to sensor network troubleshooting. Finally, we present implementation details.

3.1   Main Idea

The diagnostic simulator is motivated by the idea that, in a distributed computing environment, nodes have to interact with each other in some manner defined by their distributed protocols in order to perform tasks correctly. These interaction patterns are the concatenation of distributed sequences of events on multiple nodes. In a correctly functioning system, these sequences of events follow a path that the protocol is designed to handle. Occasionally, design flaws or omissions lead to sequences that the protocol designer did not envision, potentially causing the protocol to fail or manifest a bug. The challenge is to identify this special sequence of events which is responsible for the failure among hundreds of other common sequences that are logged during the execution but are unrelated to the failure.

At a high level, our analysis tool first identifies the time z in the log where the bug manifests. The tool then analyzes the log up to that point ignoring what happens after the bug manifestation. Note that, whatever happens after that point can not be the cause of the failure but rather the after effect of failure. Once the failure point is identified, the tool splits the log into two parts. The first part contains the sequence of all logged events from time x (by default x is set to the start time) to a time y where $x < y < z$. This part is treated as a representative of "good behavior". The second part contains all logged events from time y to time z. It is assumed to contain the sequence of events that caused "bad behavior". The experiment is repeated multiple times with different random seeds, and the "good" and "bad" parts are grouped across the runs into one "good" set and one "bad" set of sequences.

At the next step, the frequent sequence pattern mining algorithm is going to produce two separate lists of frequent patterns; one is generated from the "good" data set and the other is from the "bad" data set. Next, the tool performs differential analysis between these two lists, removing patterns common to both. The remaining patterns are those that occur predominantly in one data set but not the other. In particular, it is of interest to uncover sequences that occur in the bad data set but not the good. These sequences must be correlated with the occurrence of failure. They are sorted based on how many times they are encountered in the data set (which we call *support*). The sequences with support that is of the order of the number of observed failures are reported. The architecture of the tool is given in Figure 1.

3.2   Events and Sequences

To uncover sequences correlated with failures, the data log generated by the simulator must be analyzed. The main element in the simulation log is the single *event*. An event is defined as a tuple of the form:

$< NodeId, EventId, attribute_1, attribute_2, ...attribute_n >$

where $NodeId$ is the ID of the node which generated the event, $EventId$ denotes the ID of an event type (e.g., message receive event, message send event, flash read, flash write, function call, timer interrupt, etc). The developer defines the different types of events to be logged for his/her application, giving them descriptive labels. A header file describing the log entry format for the different types of logged events will be generated from the user's specifications and fed as input to the data mining tool. At the end of simulation the event log will consist of the sequence of encountered events, formatted as specified in the aforementioned header file. Since the simulator has a global notion of time, the log collected from the simulated distributed system is fully ordered and globally timestamped. A (sub)sequence of events is simply a list of such events. For example, $(< a >, < a >, < b, 1, 2 >, < b, 1, 2 >, < a >)$ is a sequence of five events. Our tool counts the frequency of occurrence of each subsequence in the log. The one with the higher frequency count is ranked higher. For example, in the sequence of events $(< a >, < a >, < b, 1, 2 >, < b, 1, 2 >, < a >)$, one frequent subsequence is $(< a >, < b, 1, 2 >)$. It occurs two (overlapped) times, and hence is said to have a frequency of two.

3.3   Frequent Pattern Generation

To generate the list of frequent patterns (we use "pattern" and "subsequence" interchangeably), we use the Apriori algorithm [1]. The main idea of the algorithm is as follows. This algorithm has an input parameter **minimumSupport**. At the first iteration, the frequency (number of occurrences) of every distinct event is counted and if the frequency of an event is larger than or equal to **minimumSupport**, it will be kept as a frequent pattern of length 1. Now assume that the set of frequent pattern of length 1 is C_1. At the next iteration, the candidate set of frequent patterns of length 2 is generated, which is $C_1 \times C_1$. Here "×" represents the Cartesian product. Next, the algorithm will prune all patterns from $C_1 \times C_1$ that have a frequency smaller than **minimumSupport**

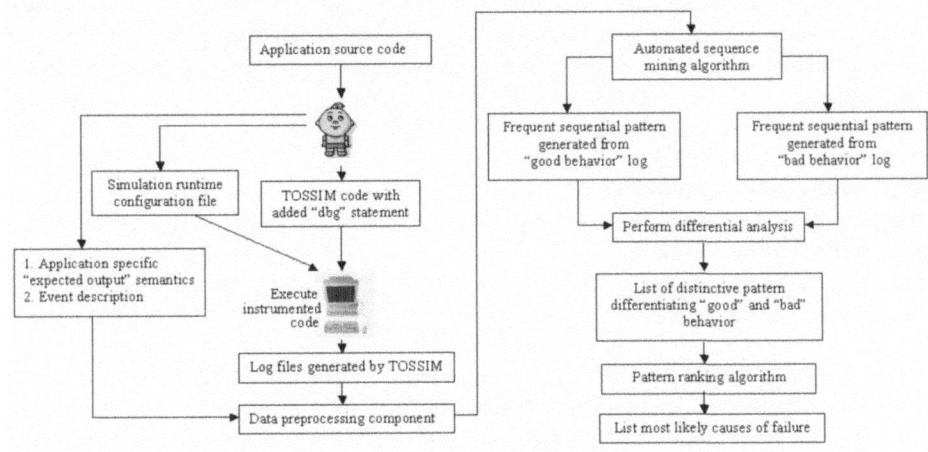

Fig. 1. Architecture of the diagnostic simulator

and generate the set of frequent patterns of length 2, which is C_2. At the next iteration, the candidate set of frequent patterns of length 3 is generated which is $C_2 \times C_1$. Again, the algorithm will prune all patterns from $C_2 \times C_1$ that have a frequency smaller than **minimumSupport** and generate the set of frequent patterns of length 3, which is C_3. This iterative process is repeated until the algorithm cannot generate any new patterns with frequency larger than or equal to **minimumSupport**. To speed up the process, the algorithm uses the **apriori property** which states that *"for an item to be frequent, all nonempty subsets of that item must also be frequent"*.

3.4   Challenges

In this section, we discuss the design challenges in using the above algorithm for purposes of finding root causes of problems, along with our proposed solutions.

Handling Multi Attribute Events. As different event types can have a different number of attributes in the tuple, mining frequent patterns becomes much more challenging as the mining algorithm has no prior knowledge of which attributes in a specific event are correlated with failure and which are not. For example, consider the following sequence of events:

```
<msg_sent,nodeid=1,msgtype=2,nodetype=1>
<msg_sent,nodeid=2,msgtype=2,nodetype=m>
```

In the above pattern, we do not know which of the attributes are correlated with failure (if any are related at all). It could be *nodeid*, or *msgtype*, or a combination of *msgtype* and *nodetype* and so on. One trivial solution is to try all possible permutations. However, this is exponential in the number of attributes and becomes unmanageable very quickly. Rather, we split such multi attribute events into a sequence of single attribute events, each with only one attribute of the original multi-attribute event. The converted sequence for the above example is as follows:

```
<msg_sent, nodeid=1>
<msg_sent, msgtype=2>
<msg_sent, nodetype=1>
<msg_sent, nodeid=2>
<msg_sent, msgtype=2>
<msg_sent, nodetype=m>
```

We can now apply simple (uni-dimensional) sequence mining techniques to the above sequence. As before, the user will be given the resulting sequences (that are most correlated with failures). In such sequences, only the relevant attributes of the original multidimensional events will likely survive. Attributes irrelevant to the occurrence of failure will likely have a larger spread of values (since these values are orthogonal to the failure) and hence a lower support. Sequences containing them will consequently have a lower support as well. The top ranking sequences are therefore more likely to focus only on attributes of interest, which is what we want to achieve.

Handling Continuous Data Types. When logged parameters are continuous, it is very hard to identify frequent patterns as there are potentially an infinite number of possible values for them. To map continuous data to a finite set, we simply discretize into a number of categories (bins).

Optimal Partitioning of Data. It is challenging to separate the log into a part that contains only good behavior and a part the includes the cause of failure. In particular, how far back from the failure point should we consider when we look for the "bad" sequence? If we go too far back, we will mix the sequences that are responsible for failure with some sequences that are unrelated to failure. If we go back too little, we may miss the root cause of failure. As a heuristic, we start with a small window size (default is set to 200 events) and iteratively increase it if necessary. After applying the sequence mining algorithm with that window size, the tool performs differential analysis between good patterns and bad patterns. If after the analysis the bad pattern list is empty, it tries a larger window.

Occurrence of Multiple Bugs. It is possible that there may be traces of multiple different bugs in the log. Fortunately, this does not impact the operation of our algorithm. As the algorithm does not assume anything about the number or types of bugs, it will report all plausible causes of failure as usual. In the presence of multiple bugs manifested due to different bad event sequences, all such sequences will be found and reported.

3.5   Implementation

As a specific instantiation of diagnostic simulation, we interface a data-mining backend to TOSSIM. Our implementation is divided into multiple independent components, namely: (i) the interface to TOSSIM, (ii) the interface to the data analysis backend, and (iii) the frequent sequence analysis algorithm.

Interface to TOSSIM. For specifying network topology and radio models we use standard TOSSIM file format. In an input file, the user specifies the location of the configuration files, duration of simulation, errors to induce at runtime (for testing fault-tolerant protocols), and the list of variables to change across different simulation runs. Based on this input file, the simulation is executed multiple times as defined by the user. If there are application specific parameters that the developer wishes to change between simulation runs, he/she needs to specify in the configuration file the parameter name, the range of values to try and the path to the header file where this parameter is declared. Based on that information, the application will be executed for each of the possible parameter values. The user can log any event inside the application using TOSSIM's "dbg" statement as in $dbg("Channel", "%d : %d : %d...", NodeId, EventId, attr1, attr2, ...)$. The simulation output is redirected to a file which is processed later for entry into the data mining tool.

Interface to Data Analysis Backend. As every application needs to log different types of events with a different number of attributes, the user needs to provide a file that describes the events logged during simulation. Using this information, the logged events are preprocessed and the required data discretization and multi attribute conversion is performed.

As every application has a different "good behavior" semantic, we provide a template of a java function that the user has to implement that defines the conditions of bad behavior. One can think of this function as a user-defined assert statement. It is the way the user defines the problem symptom in terms of logged data. This function will be called in a loop with a single event (as a parameter) at a time. If a failure condition is detected, the function should return 1. Otherwise, it should return 0. The user can check for multiple error conditions in that function and report the first violation it detects.

Data Analysis Component. The data analysis component is implemented in java. Once a failure point is detected, the log is partitioned into a "good" part and a "bad" part as described earlier. The frequent sequence mining component is called with each of the two parts separately. The configurable parameters for the frequent pattern mining algorithm are (i) **minimumSupport** (default is 1) (ii) **maximumSupport** (default is 1000, this parameter is useful if the user is not interested in sequences that are "too" frequent), and (iii) **maxSequenceLength** (if user is not interested in sequences that are "too" long).

4   A Case Study

To show the effectiveness of the tool, we implemented the directed diffusion protocol [4], which is a distributed data centric communication protocol. During the process of development and testing, we observed cases where a large number of messages that matched an interest from a sink were not delivered to that sink despite good communication. The question posed to the diagnostic simulator was to find out why these losses occurred. The tool was able to uncover two design bugs that were not addressed in the original protocol that were responsible for the message loss. The interesting fact about these two bugs was that their manifestation was same, but the causes of failure were completely different in each case. One was triggered by node reboot (and subsequent loss of state) and the other was triggered by overflow of timestamps generated by a local clock. For completeness, we briefly describe the working of the protocol in the following section, then present the use of our tool.

4.1   Directed Diffusion Protocol

In the directed diffusion protocol, data are kept as (attribute, value) pairs. A node interested in specific data sends out an interest message. Receiving a new interest, a node saves it in its local "interest cache" and forwards the interest to all of its neighbors. If a node receives the same interest from multiple neighbors, it sets up a gradient towards each neighbor from which it received the interest. This gradient information is later used to forward the data along each path. If a node matches an interest, it will mark itself as a "source" node. It forwards data matching that interest to all its neighbors towards which it has a gradient. Timestamping is used to prevent looping of the old data messages. The data will eventually be reported to the source node (possibly along multiple paths). After receiving the first report, the source node may send out "positive reinforcement" messages along a single path to increase the reporting frequency along that path. If a node receives a data message that has no matching interest cache entry, it silently drops it.

4.2   Logged Events

We have implemented the directed diffusion protocol for TinyOS 2.0. We decided to log the events shown in Table 1. One question is "How do we decide which event types to log?". As the accuracy of the tool depends on the available data, we need to decide carefully what information is important for debugging. For example, the outcome of all condition checking is important. Suppose the user can drop messages for three reasons: (1) if there is no matching interest, (2) if the data are old, which is determined based on timestamps, or (3) if data are not addressed to the node in question. In this scenario, we would like to know the reason why a message was dropped along with the event that a message was dropped. We choose to log communication events such as radio message transmission, along with relevant information such as senderId, radio

Table 1. Logged events for directed diffusion protocol

MSG_RECEIVED_EVENT, MsgType, SenderId, TimeStamp
DATA_MSG_SENT_EVENT, MsgType, ReceiverId, TimeStamp
BOOT_EVENT
STARTING_DATA_SOUCE_TIMER
INTEREST_CACHE_EMPTY
DATA_CACHE_EMPTY
MSG_DROPPED, ReasonToDrop
MSG_ENQUEUED, DataType, TimeStamp
NEW_INTEREST_INSERTION, SenderId
OLD_INTEREST_DROPPED
OLD_INTEREST_NEW_INTERESTED_PARTY, SenderId
INTEREST_MSG_SENT_EVENT, MsgType, ReceiverId, TimeStamp
REINFORCEMENT_MSG_SENT_EVENT, MsgType, ReceiverId, TimeStamp

message reception, radio message drops, reason for drop, operations on critical data structures (e.g., the interest cache and the data cache), such as insertion of data into the data cache, insertion of new interests, dropping of old interests, status of data cache, etc. Since we also wanted to experiment with tolerance with respect to node failures, we logged reboot events.

4.3   Failure Scenario I

After implementing the protocol, we ran the simulation five times and data was logged in the corresponding files. After logging all the events, we applied the application specific error detection function to check whether the observed behavior conformed to the expected behavior. For directed diffusion, if the protocol is working, the reported incidents by a source node should be received by the sink node eventually. We checked for violations of this condition. The function reported an error if it saw that the sink missed 5 consecutive messages. Next, we analyzed the logs upto each failure point. The tool generated frequent patterns for each "good" data set and "bad" data set independently. The most frequent patterns in the "bad" data that did not exist in the "good" data included the following sequences:

```
1. <DATA_MSG_SENT_EVENT:TimeStamp:255>
   <DATA_MSG_SENT_EVENT:Node_Id:4>
   <DATA_MSG_SENT_EVENT:TimeStamp:0>

2. <DATA_MSG_SENT_EVENT:TimeStamp:255>
   <MSG_DROPPED:Node_Id:2>
   <DATA_MSG_SENT_EVENT:TimeStamp:0>

3. <DATA_MSG_SENT_EVENT:TimeStamp:255>
   <MSG_RECEIVED_EVENT:Sender_Id:1>
   <DATA_MSG_SENT_EVENT:TimeStamp:0>
```

```
4.  <DATA_MSG_SENT_EVENT:TimeStamp:255>
    <MSG_RECEIVED_EVENT:Msg_Type:52>
    <MSG_DROPPED:ReasonToDrop:SAME_DATA_WITH_OLDER_TIMESTAMP>
```

The first three sequences indicate that a message with time stamp 255 was followed by one with timestamp 0, which seems consistent with a timestamp counter overflow. Thus, diagnosis suggests that a timestamp counter overflowed somewhere and it was somehow responsible for the loss of messages. The last sequence includes the event:

<MSG_DROPPED:ReasonToDrop:SAME_DATA_WITH_OLDER_TIMESTAMP>.

Combining that with the overflow, the picture becomes clear. After the timestamp overflow, messages have an "older timestamp" than previously received messages. Hence, they are dropped. Indeed, according to the design of directed diffusion, the message *TimeStamp* needs to be monotonically increasing. As the local clock has a limited resolution, after it wraps around, *TimeStamp* becomes zero. The receiving neighbors drop the most recent packet silently thinking that it is an old packet.

We illustrate the failure in figure 2 where the range of timestamps is from 0 to 255. In (a), the timestamp of the last message received by all node is 255. In (b), local clock at source node wraps around and sends out messages with timestamp 0. In (c), node 3 silently drops the message with timestamp 0 assuming it is an old message.

Observe that the above set of patterns could use some improvement. Ideally, we would have liked the tool to return the single pattern:

```
<DATA_MSG_SENT_EVENT:TimeStamp:255>
<...>
<DATA_MSG_SENT_EVENT:TimeStamp:0>
<...>
<MSG_DROPPED:ReasonToDrop:SAME_DATA_WITH_OLDER_TIMESTAMP>
```

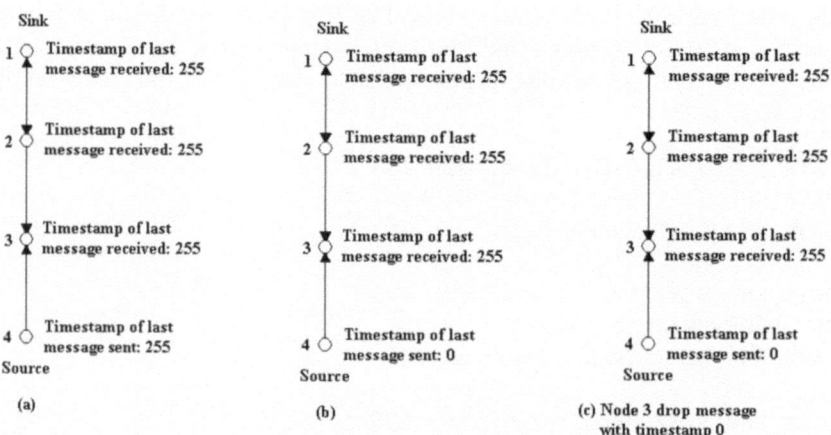

Fig. 2. Failure caused by timestamp overflow

This pattern would have explained the whole story in one step. The reason we did not see such a pattern returned is because different parts of it had a different frequency of occurrence in the log. Nevertheless, the set of subsequences returned is still very indicative of the nature of the problem.

4.4 Failure Scenario II

As a stress test, we ran the protocol and reboot a node (for our example, node 3 was rebooted) at a random time. The failure was induced at different points in time for different simulation runs. As before we ran the simulation five times and logged all events. We applied the same error detection function to check if the observed behavior conformed to the expected behavior (i.e., message sent were received). Based on the reported error point, we analyzed the logs upto that point. The frequent patterns in the "bad" data (but not "good" data) included the following:

1. `<INTEREST_CACHE_EMPTY:Node_Id:3><DATA_CACHE_EMPTY:Node_Id:3>`
   `<DATA_MSG_SENT_EVENT:TimeStamp:20>`
2. `<INTEREST_CACHE_EMPTY:Node_Id:3><DATA_CACHE_EMPTY:Node_Id:3>`
   `<DATA_MSG_SENT_EVENT:Node_Id:4>`
3. `<INTEREST_CACHE_EMPTY:Node_Id:3><DATA_CACHE_EMPTY:Node_Id:3>`
   `<DATA_MSG_SENT_EVENT:Msg_Type:52>`
4. `<INTEREST_CACHE_EMPTY:Node_Id:3>`
   `<MSG_DROPPED:ReasonToDrop:DATA_WITH_NO_MATCHING_INTEREST>`
   `<MSG_DROPPED:TimeStamp:20>`

The first three say DATA_CACHE_EMPTY and INTEREST_CACHE_EMPTY have something to do with failure. Indeed, upon a reboot, the interest and data cache is wiped out. The 4th sequence includes the event:
`<MSG_DROPPED:ReasonToDrop:DATA_WITH_NO_MATCHING_INTEREST>`.

This event explains the relation between INTEREST_CACHE_EMPTY and message loss. Indeed, if the cache is wiped out, no data matches to any interest as the interest cache is empty. One of the design choices in directed diffusion is that if there is no matching interest cache entry, the node will silently drop the message. This explains the message loss that results after a reboot.

As before, ideally, we would have liked our tool to return the pattern:

`<INTEREST_CACHE_EMPTY:Node_Id:3><DATA_CACHE_EMPTY:Node_Id:3>`
`<...>`
`<MSG_DROPPED:ReasonToDrop:DATA_WITH_NO_MATCHING_INTEREST>`

However, different subsets of the above sequence had different support and hence ended up in different returned subsequences. Nevertheless, the cause of failure is clear. Observe that although in both failure cases the manifestation of bugs were same, the causes were completely different and unrelated. Yet by logging the same set of events and using the same analysis technique, we were able to pinpoint both causes of failure. This shows the effectiveness and generality of the tool.

4.5   Performance of Frequent Pattern Mining

The configurable parameters for the frequent pattern mining algorithm are (i) **minimumSupport**, (ii) **maximumSupport**, and (iii) **maxSequenceLength**. We show the execution time of the frequent pattern mining algorithm when applied on a list of 200 events (logged from our experiment) in Table 2. From the table, we can see that increasing minimum support has a drastic impact in reducing execution time while increasing sequence length also increases the execution time drastically. If we want to mine event lists longer than 200 events, we do that in chunks of 200 events so it increases the execution time proportionally rather than exponentially. The rationale behind this strategy is that the events that are apart by more than 200 events are most likely not related strongly.

Table 2. Execution time of sequence mining algorithm

MinimumSupport	MaximumSupport	MaxSequenceLength	ExecutionTime(Sec)
1	1000	3	130.704
5	1000	3	5.172
5	1000	4	775.912
10	1000	4	0.75

5   Conclusion

This paper introduced a diagnostic simulator, which can be a useful tool for sensor network developers. To take advantage of the information hidden in the large amount of data that can be logged by a simulator, we developed an automated tool that can be used with minimal feedback from the user to diagnose root causes of failures or performance anomalies encountered during simulation. The tool presented in this paper was demonstrated using a case study on directed diffusion. Two examples of failures and their causes were uncovered. The concept of using a simulator as a diagnostic debugging tool can be very effective in the software development process of non-safety-critical applications (safety-critical applications need a more formal approach to ensure freedom from error). It can help detect a class of bugs early in the process with minimal effort.

Acknowledgements. This work was funded in part by NSF grants CNS 06-13665 and CNS 06-26342.

References

1. Agrawal, R., Srikant, R.: Fast algorithms for mining association rules. In: VLDB conference, Santiago,chile (1994)
2. Di Fatta, G., Leue, S., Stegantova, E.: Discriminative pattern mining in software fault detection. In: Proceedings of the Third International Workshop on Software Quality Assurance (SOQUA), Portland, USA (November 2006)

3. Girod, L., Elson, J., Cerpa, A., Stathopoulos, T., Ramanathan, N., Estrin, D.: Emstar: a software environment for developing and deploying wireless sensor networks. In: Proceedings of USENIX General Track (2004)
4. Inatanagonwiwat, C., Govindan, R., Estrin, D.: Directed diffusion: A scalable and robust communication paradigm for sensor networks. In: Mobicom 2000, Boston, MA, USA (2000)
5. Khan, M.M.H., Luo, L., Huang, C., Abdelzaher, T.: Snts: Sensor network troubleshooting suite. In: International Conference on Distributed Computing in Sensor Systems (DCOSS), Santa Fe, New Mexico, USA (2007)
6. Levis, P., Lee, N., Welsh, M., Culler, D.: Tossim: Accurate and scalable simulation of entire tinyos applications. In: First International Conference on Embedded Networked Sensor Systems (SenSys 2003) (November 2003)
7. Liu, C., Yan, X., Fei, L., Han, J., Midkiff, S.P.: Sober: Statistical model-based bug localization. In: ACM SIGSOFT Symp. Foundations Software Eng (FSE), Lisbon, Portugal (2005)
8. Liu, C., Yan, X., Han, J.: Mining control flow abnormality for logic error isolation. In: SIAM International conference on data mining (SDM), Bethesda, MD (April 2006)
9. Polley, J., Blazakis, D., McGee, J., Rusk, D., Baras, J.S.: Atemu: A fine-grained sensor network simulator. In: First International Conference on Sensor and Ad Hoc Communications and Networks (October 2004)
10. Ramanathan, N., Chang, K., Kapur, R., Girod, L., Kohler, E., Estrin, D.: Sympathy for the sensor network debugger. In: SenSys 2005, UCLA Center for Embedded Network Sensing,San Diego, California, USA (2005)
11. Titzer, B., Lee, D., Palsberg, J.: Avrora: Scalable sensor network simulation with precise timing. In: IPSN (2005)
12. Wen, Y., Wolski, R., Gurun, S.: S^2db: A novel simulation-based debugger for sensor network applications. UCSB 2006 (2006-01)
13. Wen, Y., Wolski, R., Moore, G.: Disens: Scalable distributed sensor network simulation. In: PPoPP (March 2007)
14. Whitehouse, K., Tolle, G., Taneja, J., Sharp, C., Kim, S., Jeong, J., Hui, J., Dutta, P., Culler, D.: Marionette: Using rpc for interactive development and debugging of wireless embedded networks. In: IPSN (April 2006)
15. Yang, J., Soffa, M.L., Selavo, L., Whitehouse, K.: Clairvoyant: A comprehensive source-level debugger for wireless sensor networks. In: SenSys, Australia (November 2007)

Sensor Placement for 3-Coverage with Minimum Separation Requirements[*]

Jung-Eun Kim[1], Man-Ki Yoon[1], Junghee Han[2], and Chang-Gun Lee[1]

[1] The School of Computer Science and Engineering,
Seoul National University, Seoul, Korea
{deep0314, ymk3337, cglee}@snu.ac.kr
[2] Samsung Electronics Co. Ltd, Suwon, Korea
{junghee0917.han}@samsung.com

Abstract. Sensors have been increasingly used for many ubiquitous computing applications such as asset location monitoring, visual surveillance, and human motion tracking. In such applications, it is important to place sensors such that every point of the target area can be sensed by more than one sensor. Especially, many practical applications require *3-coverage* for triangulation, 3D hull building, and etc. Also, in order to extract meaningful information from the data sensed by multiple sensors, those sensors need to be placed not too close to each other—*minimum separation requirement*. To address the 3-coverage problem with the minimum separation requirement, this paper proposes two methods, so called, *overlaying method* and *TRE-based method*, which complement each other depending on the minimum separation requirement. For these two methods, we also provide mathematical analysis that can clearly guide us when to use the TRE-based method and when to use the overlaying method and also how many sensors are required. To the best of our knowledge, this is the first work that systematically addresses the 3-coverage problem with the minimum separation requirement.

Keywords: sensor placement, 3-coverage, minimum separation requirement, coverage redundancy.

1 Introduction

Sensors have been increasingly used for many ubiquitous computing applications such as asset location monitoring, visual surveillance, and human motion tracking. In such applications, especially indoor applications where we have the full control of sensor positioning, it is important where to put the sensor nodes so that the target area can be fully covered using the minimum number of sensors. To address this issue, many studies have proposed various strategies [1,2,3]. In these previous studies, it is attempted to guarantee that every single point of

[*] This work was supported in part by Korean Ministry of Knowledge Economy grant 10030826 and in part by IITA through the IT Leading R&D Support Project. The corresponding author is Chang-Gun Lee.

S. Nikoletseas et al. (Eds.): DCOSS 2008, LNCS 5067, pp. 266–281, 2008.
© Springer-Verlag Berlin Heidelberg 2008

a target area should be sensed by at least one sensor (hereafter referred to as "1-coverage").

However, such an 1-coverage is not sufficient for many practical applications. Triangulation-based tracking of moving objects is one example. In this example, wherever a target object moves in the target area, at least three distance sensors should be able to sense it to figure out its position through triangulation. Also for visual surveillance applications, a target object needs to be sensed by at least three CCD cameras to build a facing-independent 3D hull of the target object [4]. From this perspective, we extend an 1-coverage problem to a general K-coverage problem, where every point of a whole target area should be covered by at least K sensors. In this paper, we specifically address a 3-coverage problem, which is meaningful for many practical applications.

When placing sensor nodes while ensuring the 3-coverage, one important issue overlooked by most researchers is the minimum separation requirement among sensors. This minimum separation requirement is motivated by our observation that sensors placed extremely close to each other are very likely to provide redundant information from the same point of view and thus fusing them does provide nothing but the one obtained by a single sensor. Figure 1 shows the measured average error of locating an object by triangulating its sensed distances from three Cricket ultrasonic sensors [5] as increasing the separation of the three sensors. When the separation of sensors is extremely small, say less than 20 cm, the error is unacceptably high. On the other hand, when the three sensors are separated enough, say more than 20 cm, the error becomes constant and it is acceptably low, say lower than 5 cm. Thus, in this application, the minimum required separation among three sensors for the 3-coverage is 20 cm.

This paper proposes sensor placement algorithms to guarantee the full 3-coverage ensuring the minimum separation requirement. Note that the communication connectivity of sensor nodes is not our concern assuming that the placed sensors can be connected through wired lines or, if wireless, the radio communication range is much larger than the sensing range, which is the case in many practical situations.

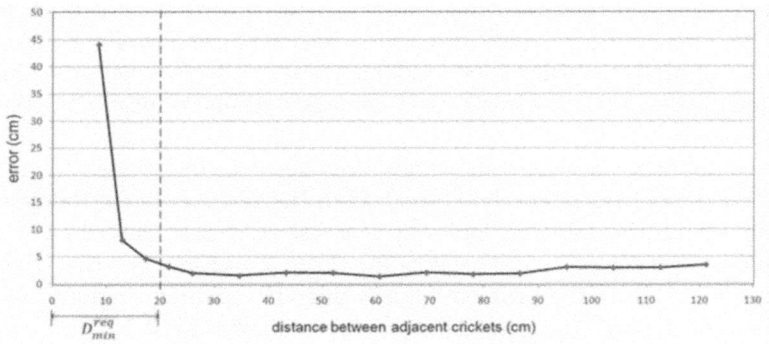

Fig. 1. D_{min}^{req} of a Cricket ultrasonic sensor

The first algorithm overlays the 1-coverage optimal placement solution [1] three times ensuring minimum separation among the sensors in different layers. However this simple algorithm has only a limited freedom of sensor placement since all sensors in the same layer must have a fixed relative distance to adjacent sensors. To enjoy more freedom when placing each individual nodes, we propose the second algorithm which first forms a 3-covered region called TRE (Triple-Rounded-Edge area) using three sensors and then places the TREs repeatedly to cover the whole target area. Another important contribution of this paper is to geographically analyze the performance of both algorithms. This analysis can answer the spectrum of minimum separation requirement for which each algorithm can find a feasible solution. Also, it can answer the over-use of sensors by each algorithm for a given minimum separation requirement. Thanks to such analysis, we can make a mix-up algorithm that uses the better way out of the two algorithms once the minimum separation requirement is given.

The rest of this paper is organized as follows: The next section summarizes the related work. Section 3 formally describes the 3-coverage problem with the minimum separation requirement. Section 4 proposes our two sensor placement algorithms. It also geographically analyzes the two proposed algorithms to suggest a guideline for combining them depending on the given minimum separation requirement. Finally, Section 5 concludes the paper with some remarks on the future work.

2   Related Work

Most of related works mainly focus on the 1-coverage problem. A historical paper written by Richard Kershner [1] presents the optimal algorithm for placing circles so that every point of the target area can be covered by at least one circle with the minimum number of circles. This is a generic solution for the 1-coverage problem without being specifically limited to sensor applications. As sensor systems become ubiquitous, many recent works have started addressing the 1-coverage problem for sensor applications. [2] presents algorithms to select the minimum number of sensors out of already densely deployed sensors to provide the 1-coverage. However, they do not address how to place sensors from the beginning. On the other hand, [6] provide sensor placement algorithms for the 1-coverage of discrete grid points. [3,7,8] address the 1-coverage problem for the continuous target area subject to the communication connectivity constraints among sensors.

For the applications where the target area needs to be covered by more than one sensor, [9,10] present methods to check if the K-coverage of the target area is possible with the already deployed sensors. However, they do not address the problem of where to place sensors for the K-coverage. For the K-coverage, many studies address the problem of selecting the minimum number of sensors out of densely deployed sensors. This problem is known to be NP-hard [6]. Thus, [11,12,13] present approximated algorithms of minimal sensor selection for the

K-coverage. However, these works again do not directly answer the sensor placement problem for the K-coverage.

The most closely related work is [14] that gives a method to place sensors for the K-coverage. However, it covers only discrete grid points not the whole target area. Also, it does not consider the minimum separation among sensors for the effective K-coverage.

To the best of our knowledge, there has been no work on the sensor placement for the 3-coverage of the continuous target area satisfying the minimum separation requirement.

3   Problem Description

To geographically model the sensor placement problem, we model the sensing coverage of a single sensor as a circle centered at the sensor position (x, y) with radius R, which is denoted by $C(x, y, R)$. This means that any target point within the circle $C(x, y, R)$ can be sensed by the sensor located at (x, y) with a high enough accuracy. As an example, Figure 2 shows our measured accuracy of distance sensing for various target positions surrounding a Cricket ultrasonic sensor [5]. For the white area, the probability that the distance sensing error is less than 10 cm is 90% or higher, and for the light gray area, that probability is 80% ~ 90%, and so on. If the required distance sensing accuracy is given as "less than 10 cm error with higher than 90%", for example, it is reasonable to assume the incircle of the white area with radius R as the sensing coverage, even though the white area does not form an exact circle. We also assume the same radius R for the sensing coverage of every sensor. The area size of each sensing coverage is denoted by $Size(C)$ dropping x, y, and R from the circle notation $C(x, y, R)$ for the notational simplicity.

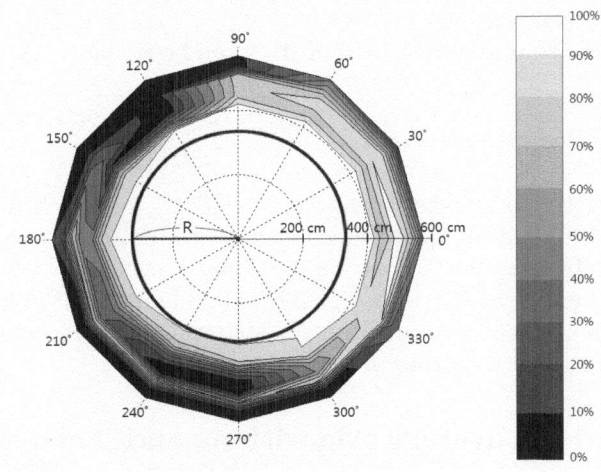

Fig. 2. Sensing coverage of a Cricket ultrasonic sensor

The target area denoted by TA is defined as an area of any shape in the 2D plane that we want to cover with the sensors. The size of TA is denoted by $Size(TA)$. In order to eliminate the effect of target boundaries when evaluating the sensor placement algorithms, we assume that the size of the target area is sufficiently larger than the size of sensing coverage of each individual sensor, that is, $Size(TA) >> Size(C)$.

For the 3-coverage of a target point (x, y), there should exist at least three sensors that cover (x, y). In addition, those three sensors should be separated at least D_{min}^{req} apart. Otherwise, the target point (x, y) is not considered as 3-covered. The minimum separation requirement D_{min}^{req} is given by applications considering the physical characteristics of the employed sensors.

With this setting, the feasibility of a sensor placement solution is formally defined as follows:

Feasibility Definition: *A sensor placement solution is defined as feasible if and only if for every point (x, y) in TA there exist at least three sensors that can cover (x, y) and they are at least D_{min}^{req} apart from each other.*

With this feasibility definition, our goal now can be formally described as follows:

Problem Description: *Find a feasible sensor placement solution with the minimum number of sensors.*

In order to evaluate the performance of the sensor placement algorithms, we use two criteria: (1) *coverage redundancy* and (2) *feasible spectrum of D_{min}^{req}*. First, the coverage redundancy reflects how much sensing coverage is unnecessarily overused compared to the ideal 3-coverage solution. The ideal 3-coverage solution is to conceptually melt the circle-shaped sensing coverage into liquid and paint the target area TA with the liquid exactly three times without any redundant painting. For this conceptually ideal solution, the total sum of the ideally used sensing coverage is exactly the same as $Size(TA) \times 3$. On the other hand, if an actual algorithm $Algo$ uses N_{Algo} sensors for feasible 3-coverage, the total sum of the actually used sensing coverage is $N_{Algo} \times Size(C)$. Thus, the coverage redundancy of the solution of Algorithm $Algo$ can be formally defined as follows:

$$CoverageRedundancy(Algo) = \frac{N_{Algo} \times Size(C) - Size(TA) \times 3}{Size(TA) \times 3}. \quad (1)$$

The second criterion reflects the spectrum of D_{min}^{req} for which Algorithm $Algo$ can find a feasible solution. The wider is the spectrum, the more generally can Algorithm $Algo$ be used for various applications.

Considering these two criteria, our goal is to build an algorithm that can find a feasible sensor placement solution with the lowest possible coverage redundancy for the widest possible spectrum of D_{min}^{req}.

4   Proposed 3-Coverage Algorithms and Their Analysis

In this section, we present two algorithms—1) *1-coverage based overlaying method* and 2) *TRE-based method*—to effectively deploy sensors for the 3-coverage of the whole target area with complying the minimum separation requirement D_{min}^{req}. The 1-coverage based overlaying method overlays the 1-coverage optimal placement solution [1] three times ensuring D_{min}^{req} among the sensors in different layers. On the other hand, the TRE-based method first forms a 3-covered region so called a TRE (Triple-Rounded-Edge area) using three sensors and then places the TREs repeatedly to cover the whole target area. Since these two algorithms enjoy different freedom of sensor placement, i.e., a group of entire single layer sensors vs. a group of three sensors, they are complementary in the sense of coverage redundancy and feasible spectrum of D_{min}^{req}.

4.1 Overlaying 1-Coverage Optimal Solutions

Our first algorithm for the 3-coverage is based on the 1-coverage optimal placement presented in [1], which is proven to be optimal in the sense of minimizing the number of sensors required for the full 1-coverage of the target area. In the 1-coverage optimal placement, all the sensors are regularly placed with the same inter-sensor distance of $\sqrt{3}R$ as shown in Figure 3.

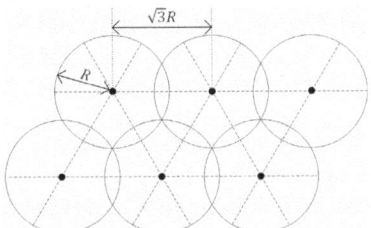

Fig. 3. Structure of 1-coverage optimal solution

Our first solution for the 3-coverage is to carefully overlay the 1-coverage optimal placement in Figure 3 three times with considering D_{min}^{req} across the three layers. Figure 4 depicts the idea of overlaying the first, second, and third layers. For the first layer, we simply use the 1-coverage optimal placement as in Figure 4(a) without worrying about D_{min}^{req}. In this figure, the black dots marked by 1 are the positions of the 1st layer sensors and solid line circles are their sensing coverages. Connecting positions of adjacent three sensors, we form an equilateral triangle whose length of each edge is $\sqrt{3}R$.

When overlaying the second layer sensors, we place a sensor of the second layer at the center of an equilateral triangle of the first layer as shown in Figure 4(b). Once we determine the placement of a single sensor of the second layer, all other sensors of the same layer are automatically placed following the rule of 1-coverage optimal placement. Such determined positions of the second layer sensors are marked by 2 in Figure 4(b) where omit the coverage circles of second layer sensors to avoid the confusion. This placement makes a sensor of the second layer farthest from its closest first layer sensor. The distance between a

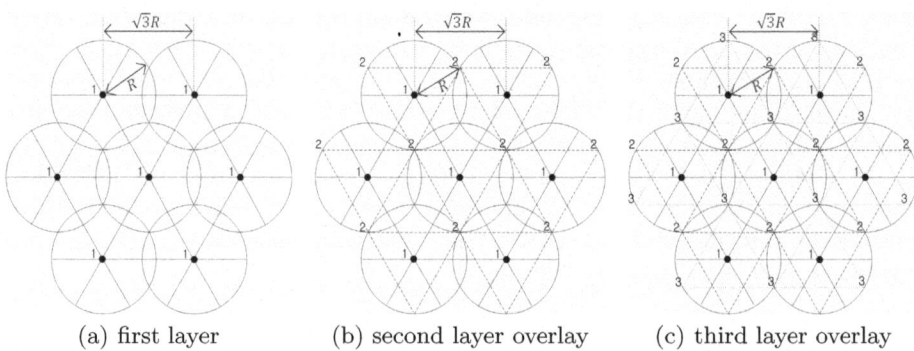

(a) first layer   (b) second layer overlay   (c) third layer overlay

Fig. 4. Overlaying of 1-coverage optimal solutions

second layer sensor and its closest first layer sensor is R. Thus, we can claim that a sensor of the second layer is at least R apart from any of first layer sensors.

Similarly, when overlaying the third layer sensors, we place sensors of the third layer at the centers of remaining equilateral triangles of the first and second layers, as marked by 3 in Figure 4(c). This way, we can make any sensor of the third layer at least R apart from any of the first and second layer sensors.

Since this overlaying method makes the sensors of one layer as far as possible from those of other layers, it is optimal in the sense of maximizing the feasible spectrum of D_{min}^{req} as we will see in the following.

A. Analysis of feasible spectrum of D_{min}^{req}

Our overlaying method has the following nice property in the sense of finding the 3-coverage placement while meeting the minimum separation requirement D_{min}^{req}.

Lemma 1. *If we place sensors using our overlaying method, each point of the target area is covered by at least three sensors which are at least R apart.*

Proof. Since each layer provides the full 1-coverage of the target area, for any point of the target area, there exists at least one sensor from each layer that covers the point. We denote such sensors from the first, second, and third layers by S_1, S_2, and S_3, respectively. Since our overlaying method ensures that any sensor of one layer is at least R apart from any sensor of other layers, S_1, S_2, and S_3 are at least R apart from each other. Thus, the above lemma holds.

Theorem 1. *Our overlaying method can find a feasible 3-coverage solution up to $D_{min}^{req} = R$.*

Proof. The theorem trivially holds due to Lemma 1.

In order to find the maximum D_{min}^{req} for which any overlaying-based method can find a feasible solution, consider the circles with radius D_{min}^{req} and centered at sensor positions of the first layer as shown in Figure 5(a). Such circles are the areas to be avoided when placing sensors of second and third layers to meet the

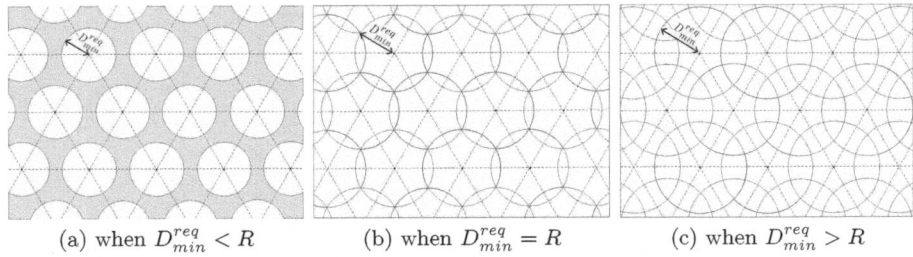

(a) when $D_{min}^{req} < R$ (b) when $D_{min}^{req} = R$ (c) when $D_{min}^{req} > R$

Fig. 5. Feasible D_{min}^{req} for any overlaying-based method

D_{min}^{req} requirement. In other words, sensors of the second and third layers can be placed only in the gray area to provide a feasible 3-coverage solution meeting D_{min}^{req}. When $D_{min}^{req} > R$ as in Figure 5(c), however, those circles completely cover the entire space without leaving any gray area. Therefore, whatever overlaying methods are used, there is no way to overlay the second and third layers meeting the D_{min}^{req} requirement. Thus, beyond $D_{min}^{req} = R$, no other overlaying-based method can find a feasible 3-coverage solution. In this sense, our overlaying method is the optimal in the sense of maximizing the feasible spectrum of D_{min}^{req}.

B. Analysis of coverage redundancy

In the case of 1-coverage problem, the coverage redundancy by Algorithm $Algo_1$ is defined as

$$CoverageRedundancy(Algo_1) = \frac{N_{Algo_1} \times Size(C) - Size(TA)}{Size(TA)} \qquad (2)$$

where N_{Algo_1} is the number of sensors required by $Algo_1$.

The 1-coverage optimal algorithm denoted by $Algo_{1opt}$ gives the coverage redundancy of 0.209 [1], that is,

$$CoverageRedundancy(Algo_{1opt}) = \frac{N_{Algo_{1opt}} \times Size(C) - Size(TA)}{Size(TA)} = 0.209.$$
$$\qquad (3)$$

Note that our overlaying method uses the 1-coverage optimal placement three times and thus the number of required sensors $N_{ourOverlay}$ is three times of $N_{Algo_{1opt}}$. Therefore, our coverage redundancy for the 3-coverage problem is

$$CoverageRedundancy(ourOverlay) = \frac{N_{ourOverlay} \times Size(C) - Size(TA) \times 3}{Size(TA) \times 3}$$
$$= \frac{3 \times N_{Algo_{1opt}} \times Size(C) - Size(TA) \times 3}{Size(TA) \times 3} = 0.209. \qquad (4)$$

Figure 6 summarizes the feasible spectrum of D_{min}^{req} and coverage redundancy achieved by our overlaying method.

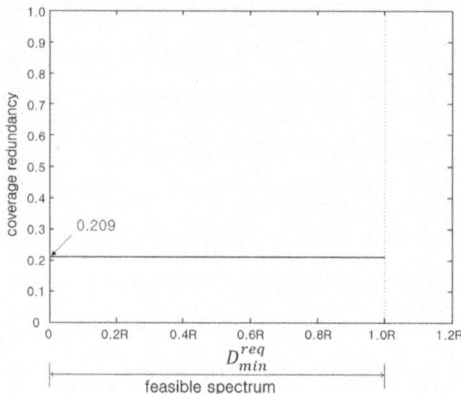

Fig. 6. Coverage redundancy according to feasible D_{min}^{req}

4.2   TRE-Based Approach

Although the aforementioned overlaying method is simple and effective, it has a limited freedom of sensor placement. In the original 1-coverage optimal method, all sensors are connected in the form of triangular grid with the fixed distance of $\sqrt{3}R$. Thus, once we place a sensor, the positions of the rest of the sensors are fixed without any freedom. Thus, the overlaying-based method has a freedom only at the level of controlling the fixed texture of entire layer, not individual sensors.

Our second method gives more freedom of controlling individual sensor positions to further improve the coverage redundancy especially when D_{min}^{req} is small relative to the sensing range R. In this method, we first form a 3-covered area, called a TRE (*Triple-Rounded-Edge* area), which is an intersection of coverage circles of three sensors equally separated by d from each other as depicted in Figure 7. The lines connecting the centers of three circles form a triangle, called a *TRE triangle*, whose sides' lengths are equally d. The d value is our engineering parameter that controls the size of a TRE—as increasing d, the TRE size decreases. Intuitively, we prefer a small d, that is, a big *TRE*, since we can make a large 3-covered area, i.e., TRE, with the same number of sensors. However, d should be larger than D_{min}^{req} to meet the minimum separation requirement. Further, there exists a minimum bound of d for the full 3-coverage as we will revisit later.

Note that any point within a TRE is covered by three sensors separated by d. Thus, our TRE-based method patches the TREs over the entire target area to provide the full 3-coverage. In this patching process, we need to minimize the overlap among TREs to optimize the coverage redundancy. Thus, we place a TRE such that its one rounded-edge just contacts that of another TRE as shown in Figure 8. If we do so, the TRE triangle of one TRE is exactly mirrored with that of the other relative to the contact point and the distance between the center of the two TRE triangles is $2 \times (R - \frac{d/2}{\cos 30°})$. If we repeat this placement of

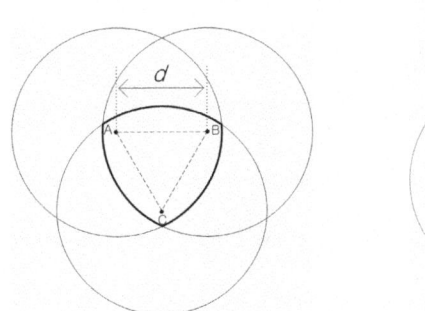

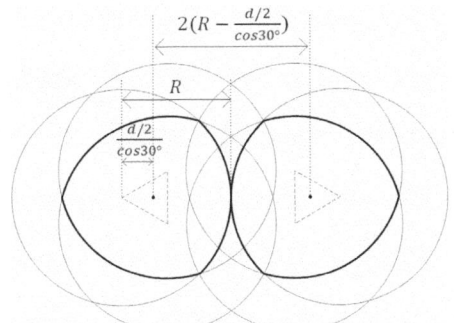

Fig. 7. A TRE (Intersection of three circles)

Fig. 8. Patching two TREs

TREs six times, the sixth TRE just contacts the first TRE as shown in Figure 9. The lines connecting the centers of the six TRE triangles form a hexagon whose sides' lengths are equally $2 \times (R - \frac{d/2}{\cos 30°})$. These six TREs are called *outer TREs*. To fill the hole surrounded by the six outer TREs, we add one more TRE called an *inner TRE* as in Figure 9. By repeating this as shown in Figure 10, we can patch TREs all over the target area with their minimum overlap.

With this TRE-based method, it is trivial to observe that any point covered by a TRE is covered by "d-separated" three sensors that form the TRE. However, there exist areas not covered by any TRE, which are depicted as gray areas in Figure 9. Such areas, however, can be covered by at least three sensors from different TREs when d is larger than a threshold, which we denote by d_{min}. This can be observed in the magnified view of a gray area in Figure 9. All the sub-areas of the gray area are 3 or more covered by surrounding sensors—see the coverage numbers of the sub-areas. However, if we decrease d, three circles forming each TRE start collapsing to one circle moving the circles toward the boundaries (thick solid) of the gray area as indicated by arrows. This can make the dark-gray sub-area first violate the 3-coverage. To find such a condition, we consider three circles (thick dashed) that form the boundaries of the dark-gray area. As decreasing d, such condition starts happening when the three circles meet at one point. For a given d, note that the relative (x, y) coordinates of the centers of the three circles are known as functions of d. Thus, we can find the d value, denoted by d_{min}, that makes the three circles meet at one point. Such found d_{min} is $0.196R$. If d is smaller than $d_{min} = 0.196R$, the gray area has a 3-coverage hole. Thus, d should be controlled above $d_{min} = 0.196R$.

Using this TRE-based patching method with $d \geq d_{min} = 0.196R$, the entire target area can be fully 3-covered without any hole. Moreover, d should be larger than or equal to D_{min}^{req} to meet the minimum separation requirement. Therefore,

$$d \geq \max(d_{min}, D_{min}^{req}). \tag{5}$$

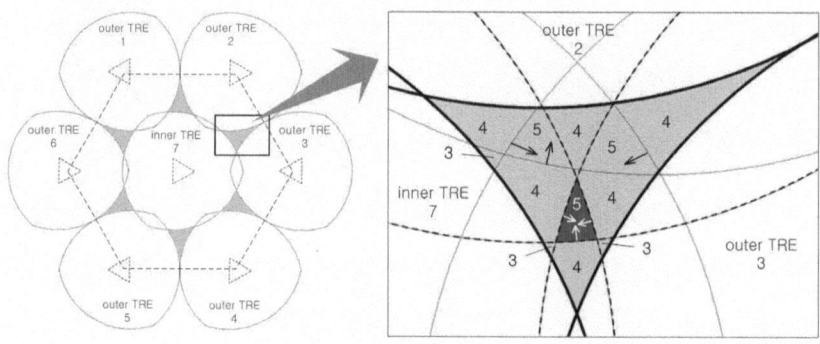

Fig. 9. Patching six TREs

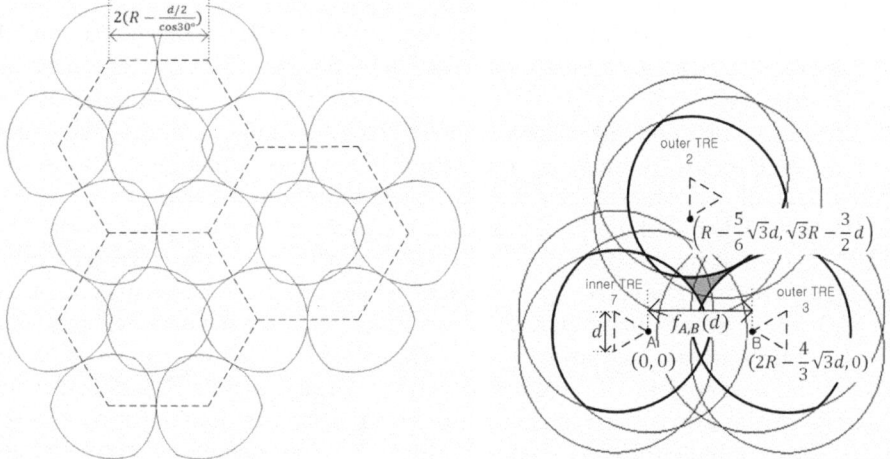

Fig. 10. Patching TREs all over the target **Fig. 11.** Meeting D_{min}^{req} in a gray area
area

Also, as we will see later, the coverage redundancy is a monotonic increasing function of d. Thus, to minimize the coverage redundancy, our TRE-based method picks the smallest possible d as follows:

$$d = \max(d_{min}, D_{min}^{req}). \qquad (6)$$

A. Analysis of feasible spectrum of D_{min}^{req}

Using our TRE-based approach, any point under a TRE is certainly covered by three sensors d apart from each other and d is larger than or equal to D_{min}^{req} by Equation (6). Therefore, the areas under TREs have no problem when checking the feasibility of our TRE-based solution for a given D_{min}^{req}. However, the gray areas in Figure 9 may be a problem, since they are 3-covered by circles from different TREs as shown in the magnified view of a gray area. As we can see in the figure, each sub-area of a gray area is covered by a combination of more than 3 sensors from different TREs, that is, at least three out of 9 sensors from three surrounding TREs. Therefore, we have to check whether those three sensors are at least D_{min}^{req} apart from each other.

This is the case up to $D_{min}^{req} = 0.6R$ as the following lemma says.

Lemma 2. *If we place sensors using our TRE-based method, the 9 sensors from three surrounding TREs of a gray area are at least D_{min}^{req} apart when $D_{min}^{req} \leq 0.6R$.*

Proof. For a gray area, consider the 9 sensors of three surrounding TREs as shown in Figure 11. For a given d value, the relative coordinates of the 9 sensors are known as functions of d. For example, assuming one sensor position of the inner TRE at $(0,0)$, other two closest sensor positions of two outer TREs are $(R - \frac{5}{6}\sqrt{3}d, \sqrt{3}R - \frac{3}{2}d)$ and $(2R - \frac{4}{3}\sqrt{3}d, 0)$ as shown in Figure 11. Therefore, the pairwise distance between two sensors A and B from two different TREs is also known as a function of d, denoted by $f_{A,B}(d)$. Note that as d getting larger, A and B are getting closer, that is, $f_{A,B}(d)$ is getting smaller. Therefore, with $\bar{d}_{A,B}$ denoting d that satisfies

$$d = f_{A,B}(d),$$

when $d \leq \bar{d}_{A,B}$, the distance between A and B, i.e., $f_{A,B}(d)$ is larger than or equal to d. If we solve the above equality for every pair of A and B from two different TREs and take the minimum of $\bar{d}_{A,B}$, that minimum value is $0.6R$. Thus, if $d \leq 0.6R$, the distance between any pair of two sensors from two different TREs is larger than or equal to d. Therefore, the 9 sensors that cover the gray area are at least d apart from each other–we call it *d-separation*, as long as $d \leq 0.6R$. When the given D_{min}^{req} is less than $d_{min} = 0.196R$, we use $d = d_{min} = 0.196R > D_{min}^{req}$ by Equation (6). Thus, the d-separation among the 9 sensors also means their D_{min}^{req}-separation. When the given D_{min}^{req} is in the range of $(d_{min} = 0.196R, 0.6R)$, we use $d = D_{min}^{req}$ by Equation (6). Thus, in this case as well, the d-separation among the 9 sensors also means their D_{min}^{req}-separation. In conclusion, for any given $D_{min}^{req} \leq 0.6R$, the 9 sensors covering a gray area are at least D_{min}^{req} apart from each other. Thus, the lemma holds.

Theorem 2. *Our TRE-based method can find a feasible 3-coverage solution up to $D_{min}^{req} = 0.6R$.*

Proof. The areas under TREs are 3-covered by sensors separated by d which is larger than or equal to D_{min}^{req} due to Equation (6). Also, the gray areas are 3-covered by sensors separated at least by D_{min}^{req} as long as $D_{min}^{req} \leq 0.6R$ by Lemma 1. Thus, the theorem holds.

B. Analysis of coverage redundancy

By our TRE-based method, the entire target area is patched by regular patterns of the same hexagons as shown in Figure 10. For each hexagon, the coverage redundancy is the same. Thus, it suffices to compute the coverage redundancy

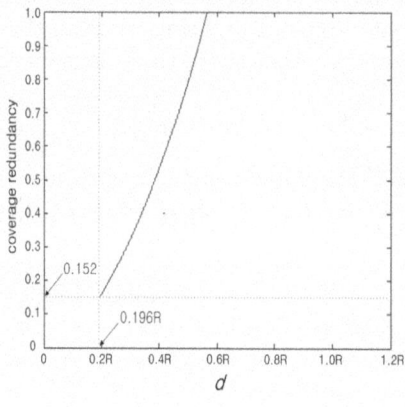
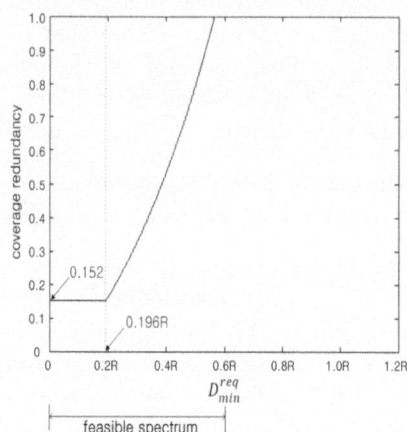

Fig. 12. Coverage redundancy according to d

Fig. 13. Coverage redundancy according to feasible D_{min}^{req}

for one hexagon. That is, the coverage redundancy of our TRE-based method can be given as follows:

$$\text{CoverageRedundancy(ourTRE)} = \frac{N_{ourTRE}(TA) \times Size(C) - Size(TA) \times 3}{Size(TA) \times 3}$$

$$= \frac{N_{ourTRE}(hexagon) \times Size(C) - Size(hexagon) \times 3}{Size(hexagon) \times 3} \quad (7)$$

where $N_{ourTRE}(TA)$ and $N_{ourTRE}(hexagon)$ are the numbers of required sensors used for covering TA and one hexagon, respectively. For covering a hexagon, we need 6 outer TREs and one inner TRE. For an outer TRE, only 1/3 portion of it is contributed to one hexagon while the inner TRE contribute its whole portion. Thus, $N_{ourTRE}(hexagon)$ can be calculated as $(3 \times \frac{1}{3}) \times 6 + 3 \times 1$. Also, since the length of the hexagon's side is $2 \times (R - \frac{d/2}{\cos 30°})$, $Size(hexagon)$ can be calculated as $\frac{\sqrt{3}}{4}(2 \times (R - \frac{d/2}{\cos 30°}))^2 \times 6$. As a result, the coverage redundancy of our TRE-based method using d is

$$\text{CoverageRedundancy(ourTRE)} = \frac{9 \times R^2\pi - (\frac{\sqrt{3}}{4}(2 \times (R - \frac{d/2}{\cos 30°}))^2 \times 6) \times 3}{(\frac{\sqrt{3}}{4}(2 \times (R - \frac{d/2}{\cos 30°}))^2 \times 6) \times 3} \quad (8)$$

Figure 12 shows such coverage redundancy as a function of our control parameter d, which is monotonically increasing. This justifies our previous decision that d should be as small as possible subject to the constraint $d \geq d_{min}$ and $d \geq D_{min}^{req}$, which derives Equation (6).

Since we use $d = d_{min} = 0.196R$ until D_{min}^{req} reaches $d_{min} = 0.196R$ and then we use $d = D_{min}^{req}$ afterward, the coverage redundancy of our TRE-based method according to D_{min}^{req} is as shown in Figure 13. The figure also shows the feasible spectrum of D_{min}^{req} by our TRE-based method.

4.3   Guideline for Combining the Two Algorithms Depending on D_{min}^{req}

From the analysis above, we can notice that our TRE-based method gives a better coverage redundancy than our overlaying method when $D_{min}^{req} \leq 0.232R$ while the latter is better than the former otherwise. Thus, we suggest to use our TRE-based method if the given minimum separation requirement D_{min}^{req} is less than or equal to $0.232R$ and to use our overlaying method if D_{min}^{req} is larger. Figure 14 shows such a combined result according to different minimum separation requirement D_{min}^{req}.

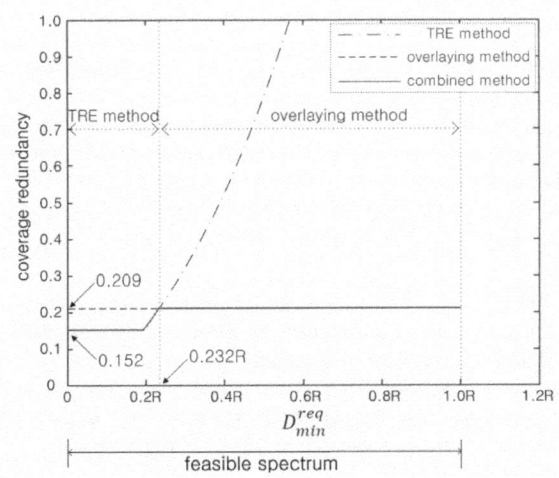

Fig. 14. Combined coverage redundancy of two approaches

Note that, in many practical applications, the minimum separation requirement D_{min}^{req} is much smaller than the sensing range R. For example, if we use the Cricket distance sensors [5] for localizing objects in a ubiquitous system [15,16], the minimum separation requirement D_{min}^{req} is 20 cm as shown in Figure 1 while the sensing range R is 320 cm as shown in Figure 2, that is, $D_{min}^{req} = 0.0625R$. In this case, our suggestion is to use our TRE-based method. The resulting coverage redundancy is 0.152 meaning that we can feasibly 3-cover the entire target area with only 15% more sensors than the ideally required sensors.

5   Conclusion

In many sensing applications, every target point needs to be sensed by more than one sensor separated by at least a certain distance, for us to extract useful information. This paper addresses a problem of placing sensor nodes to provide 3-coverage of the entire target area satisfying the minimum separation requirement. We propose two methods, i.e., overlaying method and TRE-based method, which complement each other depending on the minimum separation requirement. We also provide mathematical analysis of them in both terms of coverage redundancy and the spectrum of the minimum separation requirement for which the methods can find feasible solutions. This mathematical analysis clearly guides us when to use the TRE-based method and when to use overlaying method and also allows us to expect the number of required sensors.

In the future, we plan to address a K-coverage problem in general to find a sensor placement algorithm that works for broader applications. Another interesting direction of future research is to extend towards 3D environments.

References

1. Kershner, R.: The Number of Circles Covering a Set. American Journal of Mathematics 61(3), 665–671 (1939)
2. Tian, D., Georganas, N.D.: A Coverage-Preserving Node Scheduling Scheme for Large Wireless Sensor Networks. In: Proceedings of ACM Workshop on Wireless Sensor Networks and Applications (WSNA), pp. 32–41 (2002)
3. Bai, X., Kumar, S., Yun, Z., Xuan, D., Lai, T.H.: Deploying wireless sensors to achieve both coverage and connectivity. In: Proceedings of ACM MobiHoc, pp. 131–142 (2006)
4. Esteban, C.H., Schmitt, F.: Multi-Stereo 3D Object Reconstruction. In: Proceedings of the first International Symposium on 3D Data Processing Visualization and Transmission (3DPVT), pp. 159–166 (2002)
5. Crossbow: MCS Cricket Series (MCS410), http://www.xbow.com
6. Chakrabarty, K., Iyengar, S.S., Qi, H., Cho, E.: Grid Coverage for Surveillance and Target Location in Distributed Sensor Networks. IEEE Transactions on Computers 51(12), 1448–1453 (2002)
7. Wang, Y.C., Hu, C.C., Tseng, Y.C.: Efficient Deployment Algorithms for Ensuring Coverage and Connectivity of Wireless Sensor Networks. In: Proceedings of IEEE Wireless Internet Conference (WICON), pp. 114–121 (2005)
8. Iyengar, R., Kar, K., Banerjee, S.: Low-coordination topologies for redundancy in sensor networks. In: Proceedings of the 6th ACM international symposium on Mobile ad hoc networking and computing(MobiHoc), pp. 332–342 (2005)
9. Huang, C.F., Tseng, Y.C.: The Coverage Problem in a Wireless Sensor Network. In: Proceedings of ACM Workshop on Wireless Sensor Networks and Applications (WSNA), pp. 115–121 (2003)
10. Huang, C.F., Tseng, Y.C., Lo, L.C.: The Coverage Problem in Three-Dimensional Wireless Sensor Networks. In: Proceedings of IEEE GLOBECOM, pp. 3182–3186 (2004)
11. Yang, S., Dai, F., Cardei, M., Wu, J.: On Connected Multiple Point Coverage in Wireless Sensor Networks. Journal of Wireless Information Networks 13(4), 289–301 (2006)
12. Hefeeda, M., Bagheri, M.: Randomized k-Coverage Algorithms for Dense Sensor Networks. In: Proceedings of IEEE INFOCOM, pp. 2376–2380 (2007)

13. Wang, X., Xing, G., Zhang, Y., Lu, C., Pless, R., Gill, C.: Integrated coverage and connectivity configuration in wireless sensor networks. In: Proceedings of the 1st international conference on Embedded networked sensor systems(SenSys), pp. 28–39 (2003)
14. Xiaochun, X., Sartaj, S.: Approximation Algorithms for Sensor Deployment. IEEE Transactions on Computers 56(12), 1681–1695 (2007)
15. Nam, M.Y., Al-Sabbagh, M.Z., Kim, J.E., Yoon, M.K., Lee, C.G., Ha, E.Y.: A Real-time Ubiquitous System for Assisted Living: Combined Scheduling of Sensing and Communication for Real-Time Tracking. IEEE Transactions on Computers (to appear, 2008)
16. Nam, M.Y., Al-Sabbagh, M.Z., Lee, C.G.: Real-Time Indoor Human/Object Tracking for Inexpensive Technology-Based Assisted Living. In: Proceedings of IEEE Real-Time Systems Symposium (RTSS) (2006)

Power Assignment Problems in Wireless Communication: Covering Points by Disks, Reaching few Receivers Quickly, and Energy-Efficient Travelling Salesman Tours

Stefan Funke[1], Sören Laue[1], Rouven Naujoks[1], and Zvi Lotker[2]

[1] Max-Planck-Institut für Informatik, Saarbrücken, Germany
[2] Ben Gurion University, Beer Sheva, Israel

Abstract. A fundamental class of problems in wireless communication is concerned with the assignment of suitable transmission powers to wireless devices/stations such that the resulting communication graph satisfies certain desired properties and the overall energy consumed is minimized. Many concrete communication tasks in a wireless network like broadcast, multicast, point-to-point routing, creation of a communication backbone, etc. can be regarded as such a power assignment problem.

This paper considers several problems of that kind; the first problem was studied before in [1,6] and aims to select and assign powers to k out of a total of n wireless network stations such that all stations are within reach of at least one of the selected stations. We show that the problem can be $(1+\epsilon)$ approximated by only looking at a small subset of the input, which is of size $O\left(\frac{k^{\frac{2d}{\alpha}+1}}{\epsilon^d}\right)$, i.e. independent of n and polynomial in k and $1/\epsilon$. Here d denotes the dimension of the space where the wireless devices are distributed, so typically $d \leq 3$ and α describes the relation between the Euclidean distance between two stations and the power consumption for establishing a wireless connection between them. Using this *coreset* we are able to improve considerably on the running time of $n^{((\alpha/\epsilon)^{O(d)})}$ for the algorithm by Bilo et al. at ESA'05 ([6]) actually obtaining a running time that is *linear* in n. Furthermore we sketch how outliers can be handled in our coreset construction.

The second problem deals with the energy-efficient, bounded-hop multicast operation: Given a subset C out of a set of n stations and a designated source node s we want to assign powers to the stations such that every node in C is reached by a transmission from s within k hops. Again we show that a coreset of size independent of n and polynomial in $k, |C|, 1/\epsilon$ exists, and use this to provide an algorithm which runs in time linear in n.

The last problem deals with a variant of non-metric TSP problem where the edge costs are the squared Euclidean distances; this problem is motivated by data aggregation schemes in wireless sensor networks. We show that a good TSP tour under Euclidean edge costs can be very bad in the squared distance measure and provide a simple constant approximation algorithm, partly improving upon previous results in [5], [4].

S. Nikoletseas et al. (Eds.): DCOSS 2008, LNCS 5067, pp. 282–295, 2008.

1   Introduction

Wireless network technology has gained tremendous importance in recent years. It not only opens new application areas with the availability of high-bandwidth connections for mobile devices, but also more and more replaces so far 'wired' network installations. While the spatial aspect was already of interest in the wired network world due to cable costs etc., it has far more influence on the design and operation of wireless networks. The power required to transmit information via radio waves is heavily correlated with the Euclidean distance of sender and receivers. Hence problems in this area are prime candidates for the use of techniques from computational geometry.

Wireless devices often have limited power supply, hence the energy consumption of communication is an important optimization criterion. In this paper we use the following simple geometric graph model: Given a set P of n points in $\mathbb{R}^2$, we consider the complete graph $(P, P \times P)$ with edge weight $\omega(p, q) = |pq|^\alpha$ for some constant $\alpha > 1$ where $|pq|$ denotes the Euclidean distance between p and q. For $\alpha = 2$ the edge weights reflect the exact energy requirement for free space communication. For larger values of α (typically between 2 and 4), we get a popular heuristic model for absorption effects.

A fundamental class of problems in wireless communication is concerned with the assignment of suitable transmission powers to wireless devices/stations such that (1) the resulting communication graph satisfies a certain connectivity property Π, and (2) the overall energy assigned to all the network nodes is minimized. Many properties Π can be considered and have been treated in the literature before, see [8] for an overview. In this paper we consider several definitions of Π to solve the following problems:

k-Station Network/k-disk Coverage: *Given a set S of stations and some constant k, we want to assign transmission powers to at most k stations (senders) such that every station in S can receive a signal from at least one sender.*

k-hop Multicast: *Given a set S of stations, a specific source station s, a set of clients/receivers $C \subseteq S$, and some constant k, we want the communication graph to contain a directed tree rooted at s spanning all nodes in C with depth at most k.*

TSP under squared Euclidean distance: *Given a set S of n stations, determine a permutation $p_0, p_1, \ldots p_{n-1}$ of the nodes such that the total energy cost of the TSP tour, i.e. $\sum_{i=0}^{n-1} |p_i p_{(i+1) \bmod n}|^\alpha$ is minimized.*

1.1   Related Work

The *k-Station Network Coverage* problem was considered by Bilo et al. [6] as a k-disk cover, i.e. covering a set of n points in the plane using at most k disks such that the sum of the areas of the disks is minimized. They show that obtaining an exact solution is $\mathcal{NP}$-hard and provide a $(1 + \epsilon)$ approximation to this problem in time $n^{((\alpha/\epsilon)^{O(d)})}$ based on a plane subdivision and dynamic programming. Variants of the k-disk cover problem were also discussed in [1].

The general broadcast problem – assigning powers to stations such that the resulting communication graph contains a directed spanning tree and the total amount of energy used is minimized– has a long history. The problem is known to be $\mathcal{NP}$-hard ([7,8]), and for arbitrary, non-metric distance functions the problem can also not be approximated better than a log-factor unless $\mathcal{P} = \mathcal{NP}$ [14]. For the Euclidean setting in the plane, it is known ([2]) that the minimum spanning tree induces a power assignment for broadcast which is at most 6 times as costly as the optimum solution. This bound for a MST-based solution is tight ([7], [16]). There has also been work on restricted broadcast operations more in the spirit of the *k-hop multicast* problem we consider in this paper. In [3] the authors examine a *bounded-hop* broadcast operation where the resulting communication graph has to contain a spanning tree rooted at the source node s of depth at most k. They show how to compute an optimal k-hop broadcast range assignment for $k = 2$ in time $O(n^7)$. For $k > 2$ they show how to obtain a $(1 + \epsilon)$-approximation in time $O(n^{O(\mu)})$ where $\mu = (k^2/\epsilon)^{2^k}$, that is, their running time is triply exponential in the number of hops k and this shows up in the exponent of n. In [10], Funke and Laue show how to obtain a $(1+\epsilon)$ approximation for the k-hop broadcast problem in time doubly exponential in k based on a coreset which has size exponential in k, though.

The classical travelling salesperson problem is NPO-complete for arbitrary, non-metric distance functions (see [13]). However, if the metric satisfies the relaxed triangle inequality $\mathrm{dist}(x, y) \leq \tau(\mathrm{dist}(x, z) + \mathrm{dist}(z, y))$ for $\tau \geq 1$ and every triple of points x, y and z as in our case, then a 4τ approximation exists [5].

General surveys of algorithmic range assignment problems can be found in [8,17,12].

1.2   Our Contribution

In Section 2 we show how to find a coreset of size **independent of** n and polynomial in k and $1/\epsilon$ for the $k-$Station Network Coverage/k-Disk cover problem. This enables us improve the running time of the $(1 + \epsilon)$ approximation algorithm by Bilo et al.[6] from $n^{((\alpha/\epsilon)^{O(d)})}$ to $O\left(n + \left(\frac{k^{\frac{d}{\alpha}+1}}{\epsilon^d}\right)^{\min\{\,2k,\ (\alpha/\epsilon)^{O(d)}\,\}}\right)$,

that is, we obtain a running time that is *linear* in n. We also present a variant that allows for the senders to be placed arbitrarily (not only within the given set of points) as well as a simple algorithm which is able to tolerate few outliers and runs in polynomial time for constant values of k and the number of outliers.

Also based on the construction of a (different) coreset of small size, we show in Section 3 how to obtain a $(1 + \epsilon)$ approximate solution to the k-hop multicast problem with respect to a constant-size set C of receivers/clients. Different from the solution for the k-hop broadcast problem presented in [10] we can exhibit a coreset of size *polynomial* in $k, 1/\epsilon$ and r. The approach in [10] requires a coreset of size exponential in k.

Finally, in Section 4 we consider the problem of finding energy-optimal TSP tours. The challenge here is that the edge weights induced by the energy costs do not define a metric anymore; a simple example shows that an optimal solution to the Euclidean TSP can be a factor $\Omega(n)$ off the optimum solution. We present an $O(1)$-approximation for the TSP problem with powers α of the Euclidean distance as edge weights. For small α we improve upon previous work by Andreae [4] and Bender and Chekuri [5].

2   Energy-Minimal Network Coverage or: "How to Cover Points by Disks"

Given a set S of points in $\mathbb{R}^d$ and some constant k, we want to find at most k d-dimensional balls with radii r_i that cover all points in S while minimizing the objective function $\sum_{i=1}^{k} r_i^{\alpha}$ for some power gradient $\alpha \geq 1$. We distinguish two cases: the discrete case in which the ball centers have to be in S and the non-discrete case where the centers can be located arbitrarily in $\mathbb{R}^d$.

2.1   A Small Coreset for k-Disk Cover

In this section we describe how to find a coreset of size $O\left(\frac{k^{d/\alpha+1}}{\epsilon^d}\right)$, i.e. of size independent of n and polynomial in k and in $1/\epsilon$.

For now let us assume that we are given the cost of a λ-approximate solution S^λ for the point set S. We start by putting a regular d-dimensional grid on S with grid cell width δ depending on S^λ. For each cell C in the grid we choose an arbitrary representative point in $S \cap C$. We denote by R the set of these representatives. We say that C is *active* if $S \cap C \neq \emptyset$. Note that the distance between any point in $S \cap C$ and the representative point of C is at most $\sqrt{d} \cdot \delta$. In the following we write P^{OPT} for an optimal solution for a point set $P \subseteq S$. Now we obtain a solution R_S^{OPT} by increasing the disks in an optimal solution R^{OPT} by an additive term $\sqrt{d} \cdot \delta$. Since each point in S has a representative in R with distance at most $\sqrt{d} \cdot \delta$, R_S^{OPT} covers S. In the following we will show that (i) the cost of R_S^{OPT} is close to the cost of an optimal solution S^{OPT} to the original input set S and (ii) the size of the coreset R is small.

Theorem 1. *We have in the*

$$
\begin{aligned}
\textit{non-discrete case:} \quad & \mathrm{cost}(R_S^{OPT}) \leq (1+\epsilon)^{\alpha} \cdot \mathrm{cost}(S^{OPT}) \\
\textit{discrete case:} \quad & \mathrm{cost}(R_S^{OPT}) \leq (1+\epsilon)^{2\alpha^2} \cdot \mathrm{cost}(S^{OPT})
\end{aligned}
$$

$$
\textit{for} \quad \delta := \frac{1}{\sqrt{d}} \cdot \delta' \quad \textit{where} \quad \delta' := \frac{\epsilon}{k^{1/\alpha}} \left(\frac{\mathrm{cost}(S^\lambda)}{\lambda}\right)^{1/\alpha}
$$

Proof. Suppose R^{OPT} is given by k balls $(C_i)_{i \in \{1,\dots,k\}}$ with radii $(r_i)_{i \in \{1,\dots,k\}}$. Then

$$
\frac{\mathrm{cost}(R_S^{OPT})}{\mathrm{cost}(S^{OPT})} = \frac{\sum_{i=1}^{k}(r_i + \sqrt{d} \cdot \delta)^{\alpha}}{\mathrm{cost}(S^{OPT})} =: (I)
$$

One can easily show that this term is maximized when $r_1 = r_i \; \forall \; i \in \{2, \ldots, k\}$ (simply by differentiation). Thus we have

$$\frac{\text{cost}(R_S^{OPT})}{\text{cost}(S^{OPT})} \leq k \cdot \frac{\left(\frac{\text{cost}(R^{OPT})^{1/\alpha}}{k^{1/\alpha}} + \frac{\epsilon}{k^{1/\alpha}} \left(\frac{\text{cost}(S^\lambda)}{\lambda} \right)^{1/\alpha} \right)^\alpha}{\text{cost}(S^{OPT})}$$

$$= \frac{\left(\text{cost}(R^{OPT})^{1/\alpha} + \epsilon \left(\frac{\text{cost}(S^\lambda)}{\lambda} \right)^{1/\alpha} \right)^\alpha}{\text{cost}(S^{OPT})}$$

$$\leq \left(\frac{\text{cost}(R^{OPT})^{1/\alpha}}{\text{cost}(S^{OPT})^{1/\alpha}} + \epsilon \right)^\alpha := (II)$$

Now let us distinguish between the non-discrete and the discrete case. In the non-discrete case $(II) \leq (1+\epsilon)$ following from the monotonicity of the problem, i.e. that for all subsets $P \subseteq S : \text{cost}(P^{OPT}) \leq \text{cost}(S^{OPT})$. This can easily be seen, since each feasible solution for S is also a feasible solution for P. In the discrete case $\text{cost}(R^{OPT})$ can be bigger than $\text{cost}(S^{OPT})$ - but not much as we will see: Given an optimal solution S^{OPT}. We transform S^{OPT} into a feasible solution S_P^{OPT} for R by shifting the ball centers to their corresponding representative point in R. Since some points can be uncovered now we have to increase the ball radii by an additional $\sqrt{d} \cdot \delta$. Following exactly the same analysis as above, in this case $\text{cost}(S_R^{OPT}) \leq (1+\epsilon)^\alpha \cdot \text{cost}(S^{OPT})$. Hence $\text{cost}(R_S^{OPT}) \leq ((1+\epsilon)^\alpha + \epsilon)^\alpha \leq (1+\epsilon)^{2\alpha^2} \cdot \text{cost}(S^{OPT})$.

Knowing that the coreset is a good representation of the original input set S we will show now that R is also small.

Theorem 2. *The size of the computed coreset R is bounded by*

$$O\left(\frac{k^{\frac{d}{\alpha}+1} \cdot \lambda^{d/\alpha}}{\epsilon^d} \right)$$

Proof. Observe that the size of R is exactly given by the number of cells that contain a point in S. The idea is now to use an optimal solution S^{OPT} to bound the number of active cells. We can do so because any feasible solution for S covers all points in S and thus a cell C can only be active if such a solution covers fully or partially C. Thus the number of active cells $\#cc$ cannot be bigger than the volume of such a solution divided by the volume of a grid cell. To ensure that also the partially covered cells are taken into account we increase the radii by an additional term $\sqrt{d} \cdot \delta$. Thus consider S^{OPT} given by by k balls $(C_i)_{i \in \{1, \ldots, k\}}$ with radii $(r_i)_{i \in \{1, \ldots, k\}}$. Then

$$\#cc \leq \sum_{i=1}^{k} \frac{2^d \cdot \left(r_i + \sqrt{d} \cdot \delta\right)^d}{\delta^d} \quad = \quad (2\sqrt{d})^d \cdot \sum_{i=1}^{k} \left(1 + \frac{r_i}{\delta'}\right)^d$$

$$\overset{(*)}{\leq} (2\sqrt{d})^d \cdot \sum_{i=1}^{k} \left(1 + \frac{(k \cdot \lambda)^{1/\alpha}}{\epsilon} \cdot \left(\frac{\text{cost}(S^{OPT})}{\text{cost}(S^{\lambda})}\right)^{1/\alpha}\right)^d$$

$$\leq (2\sqrt{d})^d \cdot k \cdot \left(1 + \frac{(k \cdot \lambda)^{1/\alpha}}{\epsilon}\right)^d \leq (4\sqrt{d})^d \cdot \left(\frac{k^{\frac{1}{\alpha}+\frac{1}{d}} \cdot \lambda^{1/\alpha}}{\epsilon}\right)^d$$

$$\in O\left(\frac{k^{\frac{d}{\alpha}+1} \cdot \lambda^{d/\alpha}}{\epsilon^d}\right)$$

where inequality $(*)$ follows from the fact that $r_i \leq \text{cost}(S^{OPT})^{1/\alpha}$.

Note that it is easy to see that any $(1 + \omega)^{\alpha}$-approximation algorithm yields a $(1 + \epsilon)$-approximation algorithm by setting $\epsilon := \frac{\omega}{2\alpha}$. Doing so increases our coreset by an another *constant factor* of $(2\alpha)^d$, i.e. the size of R does not change asymptotically.

2.2   Algorithms

Still we have to show how to approximate $\text{cost}(S^{OPT})$ for the construction of the grid. In [15] Feder et al. show how to compute deterministically a 2-approximate solution for the so called k-center problem in $O(n \log k)$ time. Furthermore Har-Peled shows in [11] how to obtain such an approximation in $O(n)$ expected time for $k = O(n^{1/3}/\log n)$. The k-center problem differs from the k-disk coverage problem just in the objective function which is given by $\max_{i=1..k} r_i^{\alpha}$ where the discs have radii r_i. Since $\max_{i=1..k} r_i^{\alpha} \leq \frac{1}{k} \cdot \sum_{i=1}^{k} r_i^{\alpha}$ a 2-approximation for the k-center problem is a $2k$-approximation for the k-disk coverage problem. Using such an approximation the size of our coreset becomes $O\left(k^{\frac{2d}{\alpha}+1}/\epsilon^d\right)$. Now we will show how to solve the coreset.

Discrete Version

Via Bilo et al.:   Note that the discrete version of the k-disc cover problem can be solved by the approach of Bilo et. al. [6]. Recall that their algorithm runs $n^{((\alpha/\epsilon)^{O(d)})}$ time.

Via Exhaustive Search: Alternatively we can find an optimal solution in the following way. We consider all k-subsets of the points in the coreset R as the possible centers of the balls. Note that at least one point in R has to lie on the boundary of each ball in an optimal solution (otherwise you could create a better solution by shrinking a ball). Thus the number of possible radii for each ball is bounded by $n - k$. In total there are $(n-k)^k \cdot \binom{n}{k} \leq n^{2k}$ possible solutions. Hence

Corollary 1. *The running time of our approximation algorithm in the discrete is*

$$O\left(n + \left(\frac{k^{\frac{2d}{\alpha}+1}}{\epsilon^d}\right)^{\min\{\,2k,\ (\alpha/\epsilon)^{O(d)}\,\}}\right)$$

Non-Discrete Version Via Exhaustive Search: Note that on each ball D of an optimal solution there must be at least three points (or two points in diametrical position) that define D - otherwise it would be possible to obtain a smaller solution by shrinking D. Thus for obtaining an optimal solution via exhaustive search it is only necessary to check all k-sets of 3- respectively 2-subsets of S which yields a running time of $O(n^{3k})$. Hence

Theorem 3. *A $(1+\epsilon)$-approximate solution of the non-discrete k-disk coverage problem can be found in*

$$O\left(n + \left(\frac{k^{\frac{2d}{\alpha}+1}}{\epsilon^d}\right)^{3k}\right)$$

2.3 k-Disk Cover with Few Outliers

Assume we want to cover not all points by disks but we relax this constraint and allow a few points not to be covered, i.e. we allow let's say c outliers. This way, the optimal cover might have a considerably lower power consumption/cost.

Conceptually, we think of a k-disk cover with c outliers as a $(k+c)$-disk cover with c disks having radius 0. Doing so, we can use the same coreset construction as above, replacing k by $k + c$. Obviously, the cost of an optimal solution to the $(k + c)$-disk cover problem is a lower bound for the k-disk cover with c outliers. Hence, the imposed grid might be finer than actually needed. So snapping each point to its closest representative still ensures a $(1+\epsilon)$-approximation. Constructed as above, the coreset has size $O(\frac{(k+c)^{\frac{2d}{\alpha}+1}}{\epsilon^d})$.

Again, there are two ways to solve this reduced instance, first by a slightly modified version of the algorithm proposed by Bilo et al. [6] and second by exhaustive search.

We will shortly sketch the algorithm by Bilo et al. [6] which is based on a hierarchical subdivision scheme proposed by Erlebach et al. in [9]. Each subdivision is assigned a level and they together form a hierarchy. All possible balls are also assigned levels depending on their size. Each ball of a specific level has about the size of an ϵ-fraction of the size of the cells of the subdivision of same level. Now, a cell in the subdivision of a fixed level is called relevant if at least one input point is covered by one ball of the same level. If a relevant cell S' is included in a relevant cell S and no larger cell S'' exists that would satisfy $S' \subseteq S'' \subseteq S$, then S' is called a child cell of S and S is called the parent of S'. This naturally defines a tree. It can be shown that a relevant cell has at most a constant number of child cells (the constant only depending on ϵ, α and d). The key ingredient for the algorithm to run in polynomial time is the fact that there exists a nearly optimal solution where a relevant cell can be covered by only a constant number of balls of larger radius. The algorithm then processes all relevant cells of the hierarchical subdivision in a bottom-up way using dynamic programming. A table is constructed that for a given cell S, a given configuration P of balls having higher level than S (i.e. large balls) and an integer $i \leq k$ stores the balls of level at most the level of S (i.e. small balls) such that all input points in S are covered and the total number of balls is at most i. This is done for a cell S by looking up the entries of the child cells and iterating over all possible ways to distribute the i balls among them.

The k-disk cover problem with c outliers exhibits the same structural properties as the k-disk cover problem without outliers. Especially, the local optimality of the global optimal solution is preserved. Hence, we can adapt the dynamic programming approach of the original algorithm. In order for the algorithm to cope with c outliers we store not only one table for each cell but $c+1$ such tables. Each such table corresponds to the table for a cell S where $0, 1, \ldots, c$ points are not covered. Now, we do not only iterate over all possible ways to distribute the i balls among its child cells but also all ways to distribute $l \leq c$ outliers. This increases the running time to $n^{((\alpha/\epsilon)^{O(d)})} \cdot c^{((\alpha/\epsilon)^{O(d)})} = n^{((\alpha/\epsilon)^{O(d)})}$. Hence running the algorithm on the coreset yields the following result:

Corollary 2. *We can compute a minimum k-disk cover with c outliers $(1 + \epsilon)$ approximately in time*

$$O\left(n + \left(\frac{(k + c)^{\frac{2d}{\alpha} + 1}}{\epsilon^d}\right)^{(\alpha/\epsilon)^{O(d)}}\right).$$

For the exhaustive search approach we consider all assignments of k disks each having a representative as its center and one lying on its boundary. For each such assignment we check in time $O(kn)$ whether the number of uncovered points is at most c. We output the solution with minimal cost.

Corollary 3. *We can compute a minimum k-disk cover with c outliers $(1 + \epsilon)$ approximately in time*

$$O\left(n + k\left(\frac{(k + c)^{\frac{2d}{\alpha} + 1}}{\epsilon^d}\right)^{2k+1}\right).$$

3   Bounded-Hop Multicast or: "Reaching Few Receivers Quickly"

Given a set S of points (stations) in $\mathbb{R}^d$, a distinguished source point $s \in S$ (sender), and a set $C \subset S$ of client points (receivers) we want to assign distances/ranges $r : S \to \mathbb{R}_{\geq 0}$ to the elements in S such that the resulting communication graph contains a tree rooted at s spanning all elements in C and with depth at most k (an edge (p, q) is present in the communication graph iff $r(p) \geq |pq|$). The goal is to minimize the total assigned energy $\sum_{p \in S} r(p)^\alpha$. This can be thought of as the problem of determining an energy efficient way to quickly (i.e. within few transmissions) disseminate a message or a data stream to a set of few receivers in a wireless network.

As in the previous Section we will solve this problem by first deriving a coreset R of size independent of $|S| = n$ and then invoking an exhaustive search algorithm. We assume both k and $|C| = c$ to be (small) constants. The resulting coreset will have size *polynomial* in $1/\epsilon$, c and k. For few receivers this is a considerable improvement over the exponential-sized coreset that was used in [10] for the k-hop broadcast.

3.1   A Small Coreset for k-Hop Multicast

In the following we will restrict to the planar case in $\mathbb{R}^2$, the approach extends in the obvious way to higher (but fixed) dimensions. Assume w.l.o.g. that the maximum distance of a point $p \in S$ from s is exactly 1. We place a square grid of cell width $\delta = \frac{1}{\sqrt{2}} \frac{\epsilon}{kc}$ on $[-1, 1] \times [-1, 1] \subset \mathbb{R}^2$. The size of this grid is $O(\frac{(kc)^2}{\epsilon^2})$. Now we assign each point in P to its closest grid point. Let R be the set of grid points that had at least one point from S snapped to it, C' the set of grid points that have at least one point from C snapped to it.

It remains to show that R is indeed a coreset. We can transform any given valid range assignment r for S (wrt receiver set C) into a valid range assignment r' for R (wrt receiver set C'). We define the range assignment r' for S as

$$r'(p') = \max_{p \text{ was snapped to } p'} r(p) + \sqrt{2}\delta.$$

Since each point p is at most $\frac{1}{\sqrt{2}}\delta$ away from its closest grid point p' we certainly have a valid range assignment for R. It is easy to see that the cost of r' for R is not much larger than the cost of r for S. We have:

$$\sum_{p' \in R} (r'(p'))^\alpha = \sum_{p \in S} (\max_{p \text{ was snapped to } p'} r(p) + \sqrt{2}\delta)^\alpha$$

$$\leq \sum_{p \in S} (\max_{p \text{ was snapped to } p'} r(p) + \frac{\epsilon}{kc})^\alpha$$

$$\leq \sum_{p \in S} (r(p) + \frac{\epsilon}{kc})^\alpha.$$

The relative error satisfies

$$\frac{\text{cost}(r')}{\text{cost}(r)} \leq \frac{\sum_{p \in S}(r(p) + \frac{\epsilon}{kc})^\alpha}{\sum_{p \in S}(r(p))^\alpha}.$$

Notice, that $\sum_{p \in P} r(p) \geq 1$ and r is positive for at most kc points p (each of the c receivers must be reached within k hops). Hence, the above expression is maximized when $r(p) = \frac{1}{kc}$ (which can be seen simply by differentiation) for all points p that are assigned a positive value. Thus

$$\frac{\text{cost}(r')}{\text{cost}(r)} \leq \frac{(kc) \cdot (\frac{1}{(kc)} + \frac{\epsilon}{(kc)})^{\alpha}}{(kc) \cdot (\frac{1}{(kc)})^{\alpha}} = (1 + \epsilon)^{\alpha}.$$

On the other hand we can transform any given valid range assignment r' for R into a valid range assignment r for S as follows. We select for each grid point $g \in R$ one representative g_S from S that was snapped to it. For the grid point to which s (the source) was snapped we select s as the representative. If we define the range assignment r for S as $r(g_S) = r'(g) + \sqrt{2}\delta$ and $r(p) = 0$ if p does not belong to the chosen representatives, then r is a valid range assignment for S because every point is moved by at most $\delta/\sqrt{2}$. Using the same reasoning as above we can show that $\text{cost}(r) \leq (1 + \epsilon)^{\alpha} \text{cost}(r')$. In summary we obtain the following theorem:

Theorem 4. *For the k-hop multicast problem with c receivers there exists a coreset of size* $O(\frac{(kc)^2}{\epsilon^2})$.

3.2 Solution Via a Naive Algorithm

As we are not aware of any algorithm to solve the k-hop multicast problem we employ a naive brute-force strategy, which we can afford since after the coreset computation we are left with a 'constant' problem size. Essentially we consider every kc-subset of R as potential set of senders and try out the $|R|$ potential ranges for each of the senders. Hence, naively there are at most $\binom{\frac{kc^2}{\epsilon^2}}{kc}$.

$\left(\frac{kc^2}{\epsilon^2}\right)^{kc}$ different range assignments to consider at all. We enumerate all these assignments and for each of them check whether the range assignment is valid wrt c'; this can be done in time $|R|$. Of all the valid range assignments we return the one of minimal cost.

Assuming the floor function a coreset R for an instance of the k-hop multicast problem can be constructed in linear time. Hence we obtain the following corollary:

Corollary 4. *A $(1 + \epsilon)$-approximate solution to the k-hop multicast problem on n points in the plane can be computed in time* $O(n + \left(\frac{kc}{\epsilon}\right)^{4kc})$.

As we are only after an approximate solution, we do not have to consider all $|R|$ potential ranges but can restrict to essentially $O(\log_{1+\epsilon}\frac{kc}{\epsilon})$ many, the running time of the algorithm improves accordingly:

Corollary 5. *A $(1 + \epsilon)$-approximate solution to the k-hop multicast problem on n points in the plane can be computed in time* $O\left(n + \left(\frac{(kc)^2 \log \frac{kc}{\epsilon}}{\epsilon^3}\right)^{kc}\right)$.

4   Information Aggregation Via Energy-Minimal TSP Tours

While early wireless sensor networks (WSNs) were primarily data collection systems where sensor readings within the network are all transferred to a central computing device for evaluation, current WSNs perform a lot of the data processing *in-network*. For this purpose some nodes in the network might be interested in periodically *collecting* information from certain other nodes, some nodes might want to *disseminate* information to certain groups of other nodes. A typical approach for data collection and dissemination as well as for data aggregation purposes are tree-like subnetwork topologies, they incur certain disadvantages with respect to load-imbalance as well as non-obliviousness to varying initiators of the data collection or dissemination operation, though. Another, very simple approach could be to have a *virtual token* floating through the network (or part thereof). Sensor nodes can attach data to the token or read data from the token and then hand it over to the next node. Preferably the token should not visit a node again before all other nodes have been visited and this should happen in an energy-optimal fashion, i.e. the sum of the energies to hand over the token to the respective next node should be minimized. Such a scheme has some advantages: first of all none of sensor nodes plays a distinguished role – something that is desirable for a system of homogeneous sensor nodes – furthermore every sensor node can use the same token to initiate his data collection/dissemination operation. Abstractly speaking we are interested in finding a *Travelling Salesperson tour* (TSP) of minimum energy cost for (part of) the network nodes. Unfortunately, the classical TSP with non-metric distance function is very hard to solve (see [13]). However, for metrics satisfying the relaxed triangle inequality $\text{dist}(x, y) \leq \tau(\text{dist}(x, z) + \text{dist}(z, y))$ for $\tau \geq 1$ and every triple of points x, y and z a 4τ approximation exists [5].

In this section we show that the 'normal' Euclidean TSP is not suitable for obtaining an energy-efficient tour and devise a 6-approximation algorithm for TSP under squared Euclidean metric.

4.1   Why Euclidean TSP Does Not Work

Simply computing an optimal tour for the underlying Euclidean instance does not work. The cost for such a tour can be a factor $\Omega(n)$ off from the optimal solution for the energy-minimal tour. Consider the example where n points lie on a slightly bent line and each point having distance 1 to its right and left neighbor. An optimal Euclidean tour would visit the points in their linear order and the go back to the first point. Omitting the fact that the line is slightly bent this tour would have a cost of $(n - 1) \cdot 1^2 + (n - 1)^2 = n(n - 1)$ if the edge weights are squared Euclidean distances. However, an optimal energy-minimal tour would have a cost of $(n - 2) \cdot 2^2 + 2 \cdot 1^2 = 4(n - 1) + 2$. This tour would first

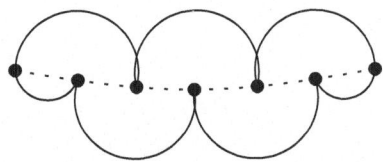

Fig. 1. An optimal energy-minimal tour for points on a line

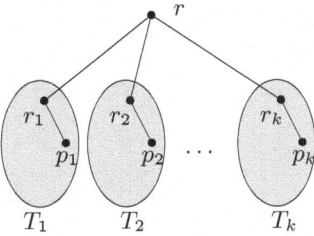

Fig. 2. Tree T and its children trees $T_1, T_2, \ldots, T_k$

visit every second point on the line and on the way back all remaining points as in figure 1.

4.2 A 6-Approximation Algorithm

In this section we will describe an algorithm which computes a 6-approximation for the TSP under squared Euclidean distance. Obviously, the cost of a minimum spanning tree is a lower bound for the optimal value OPT of the tour. Consider a non-trivial minimum spanning tree T for a graph with node set V and squared Euclidean edge weights. We denote the cost of such a tree by $\mathrm{MST}(T)$. Let r be the root of T and p be one child of T.

We define two Hamiltonian paths $\pi^a(T)$ and $\pi^b(T)$ as follows. Let $\pi^a(T)$ be a path starting at r, finishing at p that visits all nodes of T and the cost of this path is at most $6\,\mathrm{MST}(T) - 3\|rp\|^2$. Let $\pi^b(T)$ be defined in the same way but in opposite direction, i.e. it starts at p and finishes at r.

Now, if we have such a tour $\pi^a(T)$ for the original vertex set V we can construct a Hamilton tour by connecting r with p. The cost of this tour is clearly at most $6\,\mathrm{MST}(T) - 3\|rp\|^2 + \|rp\|^2 \leq 6\,\mathrm{MST}(T) \leq 6\,\mathrm{OPT}$. It remains to show how to construct such tours π^a and π^b. We will do this recursively.

For a tree T of height 1, i.e. a single node r, $\pi^a(T)$ and $\pi^b(T)$ both consist of just the single node. Conceptually, we identify p with r in this case. Obviously, the cost of both paths is trivially at most $6\,\mathrm{MST}(T) - 3\|rp\|^2$.

Now, let T be of height larger than 1 and let $T_1, \ldots, T_k$ be its children trees. Let r denote the root of T and r_i the root of T_i and p_i be a child of T_i as in figure 2. Then we set $\pi^a(T) = (r, \pi^b(T_1), \pi^b(T_2), \ldots, \pi^b(T_k))$.

The cost of the path $\pi^a(T)$ satisfies

$$
\begin{aligned}
\mathrm{cost}(\pi^a(T)) &= \|rp_1\|^2 + \mathrm{cost}(\pi^b(T_1)) + \|r_1p_2\|^2 + \mathrm{cost}(\pi^b(T_2)) + \dots \\
&\quad + \|r_{k-1}p_k\|^2 + \mathrm{cost}(\pi^b(T_k)) \\
&\leq (\|rr_1\| + \|r_1p_1\|)^2 + \mathrm{cost}(\pi^b(T_1)) \\
&\quad + (\|r_1r\| + \|rr_2\| + \|r_2p_2\|)^2 + \mathrm{cost}(\pi^b(T_2)) \\
&\qquad \vdots \\
&\quad + (\|r_{k-1}r\| + \|rr_k\| + \|r_kp_k\|)^2 + \mathrm{cost}(\pi^b(T_k)) \\
&\leq 2\|rr_1\|^2 + 2\|r_1p_1\|^2 + \mathrm{cost}(\pi^b(T_1)) \\
&\quad + 3\|r_1r\|^2 + 3\|rr_2\|^2 + 3\|r_2p_2\|^2 + \mathrm{cost}(\pi^b(T_2)) \\
&\qquad \vdots \\
&\quad + 3\|r_{k-1}r\|^2 + 3\|rr_k\|^2 + 3\|r_kp_k\|^2 + \mathrm{cost}(\pi^b(T_k)) \\
&\leq 6\sum_{i=1}^{k}\|rr_i\|^2 + 3\sum_{i=1}^{k}\|r_ip_i\|^2 + \sum_{i=1}^{k}\mathrm{cost}(\pi^b(T_i)) - 3\|rr_k\|^2 \\
&\leq 6\sum_{i=1}^{k}\|rr_i\|^2 + 6\sum_{i=1}^{k}\mathrm{MST}(T_i) - 3\|rr_k\|^2 \\
&= 6\,\mathrm{MST}(T) - 3\|rr_k\|^2.
\end{aligned}
$$

In the above calculation we used the fact that $\left(\sum_{i=1}^{n} a_i\right)^\alpha \leq n^{\alpha-1} \cdot \sum_{i=1}^{n} a_i^\alpha$, for $a_i \geq 0$ and $\alpha \geq 1$, which follows from Jensen's inequality and the fact that the function $f : x \mapsto x^\alpha$ is convex. The path $\pi^b(T)$ is constructed analogously.

In fact, the very same construction and reasoning can be generalized to the following corollary.

Corollary 6. *There exists a $2 \cdot 3^{\alpha-1}$-approximation algorithm for the TSP if the edge weights are Euclidean edge weights to the power α.*

The metric with Euclidean edge weights to the power α satisfies the relaxed triangle inequality with $\tau = 2^{\alpha-1}$. A short computation shows that our algorithm is better than previous algorithms [5,4] for small α, i.e. for $2 \leq \alpha \leq 2.7$.

References

1. Alt, H., Arkin, E.M., Brönnimann, H., Erickson, J., Fekete, S.P., Knauer, C., Lenchner, J., Mitchell, J.S.B., Whittlesey, K.: Minimum-cost coverage of point sets by disks. In: SCG 2006: Proc. 22nd Ann. Symp. on Computational Geometry, pp. 449–458. ACM Press, New York (2006)
2. Ambühl, C.: An optimal bound for the mst algorithm to compute energy-efficient broadcast trees in wireless networks. In: ICALP, pp. 1139–1150 (2005)
3. Ambühl, C., Clementi, A.E.F., Ianni, M.D., Lev-Tov, N., Monti, A., Peleg, D., Rossi, G., Silvestri, R.: Efficient algorithms for low-energy bounded-hop broadcast in ad-hoc wireless networks. In: STACS, pp. 418–427 (2004)

4. Andreae, T.: On the traveling salesman problem restricted to inputs satisfying a relaxed triangle inequality. Networks 38(2), 59–67 (2001)
5. Bender, M.A., Chekuri, C.: Performance guarantees for the TSP with a parameterized triangle inequality. Information Processing Letters 73(1–2), 17–21 (2000)
6. Bilò, V., Caragiannis, I., Kaklamanis, C., Kanellopoulos, P.: Geometric clustering to minimize the sum of cluster sizes. In: European Symposium on Algorithms (ESA), pp. 460–471 (2005)
7. Clementi, A.E.F., Crescenzi, P., Penna, P., Rossi, G., Vocca, P.: On the complexity of computing minimum energy consumption broadcast subgraphs. In: STACS, pp. 121–131 (2001)
8. Clementi, A.E.F., Huiban, G., Penna, P., Rossi, G., Verhoeven, Y.C.: Some recent theoretical advances and open questions on energy consumption in ad-hoc wireless networks. In: Proc. 3rd Workshop on Approximation and Randomization Algorithms in Communication Networks (ARACNE), pp. 23–38 (2002)
9. Erlebach, T., Jansen, K., Seidel, E.: Polynomial-time approximation schemes for geometric graphs. In: Proceedings of the 12th ACM-SIAM Symposium on Discrete Algorithms (SODA 2001), Washington, DC, pp. 671–679 (2001)
10. Funke, S., Laue, S.: Bounded-Hop Energy-Efficient Broadcast in Low-Dimensional Metrics Via Coresets. In: Thomas, W., Weil, P. (eds.) STACS 2007. LNCS, vol. 4393, pp. 272–283. Springer, Heidelberg (2007)
11. Har-Peled, S.: Clustering motion. Discrete Comput. Geom. 31(4), 545–565 (2004)
12. Kirousis, L.M., Kranakis, E., Krizanc, D., Pelc, A.: Power consumption in packet radio networks. Theor. Comput. Sci. 243(1-2), 289–305 (2000)
13. Orponen, P., Mannila, H.: On approximation preserving reductions: Complete problems and robust measures. Technical Report, University of Helsinki (1990)
14. Guha, S., Khuller, S.: Improved methods for approximating node weighted steiner trees and connected dominating sets. Information and Computation 150, 57–74 (1999)
15. Feder, T., Greene, D.: Optimal algorithms for approximate clustering. In: Proc. 20th ACM Symp. on Theory of Computing (1988)
16. Wan, P.-J., Calinescu, G., Li, X., Frieder, O.: Minimum-energy broadcast routing in static ad hoc wireless networks. In: INFOCOM, pp. 1162–1171 (2001)
17. Wieselthier, J.E., Nguyen, G.D., Ephremides, A.: On the construction of energy-efficient broadcast and multicast trees in wireless networks. In: INFOCOM, pp. 585–594 (2000)

Distributed Activity Recognition with Fuzzy-Enabled Wireless Sensor Networks

Mihai Marin-Perianu[1], Clemens Lombriser[2], Oliver Amft[2],
Paul Havinga[1], and Gerhard Tröster[2]

[1] Pervasive Systems, University of Twente, The Netherlands
{m.marinperianu, p.j.m.havinga}@utwente.nl
[2] Wearable Computing Lab, ETH Zürich, Switzerland
{lombriser, amft, troester}@ife.ee.ethz.ch

Abstract. Wireless sensor nodes can act as distributed detectors for recognizing activities online, with the final goal of assisting the users in their working environment. We propose an activity recognition architecture based on fuzzy logic, through which multiple nodes collaborate to produce a reliable recognition result from unreliable sensor data. As an extension to the regular fuzzy inference, we incorporate temporal order knowledge of the sequences of operations involved in the activities. The performance evaluation is based on experimental data from a car assembly trial. The system achieves an overall recognition performance of 0.81 recall and 0.79 precision with regular fuzzy inference, and 0.85 recall and 0.85 precision when considering temporal order knowledge. We also present early experiences with implementing the recognition system on sensor nodes. The results show that the algorithms can run online, with execution times in the order of 40ms, for the whole recognition chain, and memory overhead in the order of 1.5kB RAM.

1 Introduction

Sensor miniaturization and advances in wireless communication made real visionary technologies such as wearable computing, Wireless Sensor Networks (WSN) and the Internet of Things. As a result, we witness today a rapidly growing uptake of these technologies in various industrial and business related fields [5]. In particular, providing context-aware assistance to workers in industrial environments [1] has a number of potential benefits: (1) improves the productivity at work by supporting the workers with just-in-time, relevant information, (2) increases the reliability by supervising safety-critical process steps and (3) accelerates the learning curve of new, unskilled workers.

In this paper, we focus on the concrete application of assembling and testing car body parts at a car manufacturing site [16,3]. The mechanical assembly activities are supervised by a distributed system composed of wireless sensor nodes worn by the workers, embedded into the tools and deployed within the infrastructure. Having the technology integrated in the usual work environment ensures a simplified, unobtrusive human-machine interaction, which is important to foster user acceptance [6]. The same system can additionally support the

S. Nikoletseas et al. (Eds.): DCOSS 2008, LNCS 5067, pp. 296–313, 2008.

training of new employees. The most important technical objective of the system is to recognize the user activities with high accuracy and in real time, in order to provide prompt and exact assistance. We can expect several major difficulties related to: the inaccurate and noisy sensor data, the very limited resources (processing, memory and energy), the high variability of the data (e.g. for sensors placed on different parts of the body) and the unreliable, low bandwidth wireless communication. To overcome these problems, we propose a distributed activity recognition system based on fuzzy-enabled WSN. Fuzzy logic represents nowadays a well-established field with an impressive number of successful applications in the industry and other connected areas [12]. Fuzzy inference systems (FIS) constitute an effective tool for WSN because (1) they can be implemented on limited hardware and are computationally fast, and (2) they can handle unreliable and imprecise information (both numeric and linguistic information), offering a robust solution to decision fusion under uncertainty [14]. In addition, fuzzy logic reduces the development time; it is possible to have a running system based only on the common-sense description of the problem, and then tune it using expert knowledge or automatic training and learning methods.

In our approach we decompose the activity recognition process in three steps: (1) the detection of action events using sensors on the body, tools, and the environment (based on the previous work of Amft et al. [3]), (2) the combination of events into basic operations and (3) the final activity classification. For performance evaluation, we use the experimental data from 12 sensor nodes with 3D accelerometers, obtained during the car assembly trial conducted in [3] (see Fig. 1). Sec. 3 presents the overview of the distributed architecture. Sec. 4 analyzes the properties of the experimental data and derives the difficulties in achieving an accurate recognition. The detailed design of the FIS and the performance results are given in Sec. 5 and 6. As an extension to the normal fuzzy inference, we show how temporal order knowledge can improve the overall accuracy for the activities executed in sequences of operations. In order to study the feasibility of our solution, we implement the recognition algorithms on the Tmote Mini platform. The results from Sec. 7 show that the algorithms can run online even on resource constrained sensor nodes. Sec. 8 discusses several important factors that can affect the performance of fuzzy-enabled WSN. Finally, Sec. 9 concludes the paper.

2   Related Work

Activity recognition is a topic of high interest within the machine vision community. In particular, we can trace back the fundamental idea of dividing the recognition problem into multiple levels of complexity. In the "Inverse Hollywood Problem", Brand [4] uses coupled hidden Markov models (HMM) to visually detect causal events and fit them together into a coherent story of the ongoing action. Similarly, Ivanov and Bobick [7] address the recognition of visual activities by fusing the outputs of event detectors through a stochastic context-free grammar parsing mechanism. Wren et al. [18] show that even low resolution

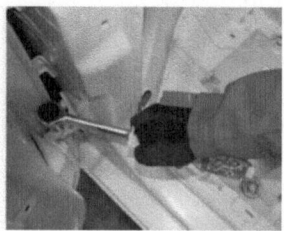

Fig. 1. A car assembly worker performing several activities. Sensors are both worn by the worker and attached to the tools and car parts.

sensors (motion detectors and ultrasonic sensors) can provide useful contextual information in large buildings.

From a different perspective, distributed event detection has received considerable attention in the field of WSN. The research focuses on fundamental issues of WSN, such as reducing the number of messages needed for stable decisions [11], minimizing the effects of data quantization and message losses [13], and using error correcting codes to counteract data corruption [17]. The sensor nodes are typically assumed to sample at low data rates and to implement simple binary classifiers based on maximum likelihood estimates. However, recent work [15] shows that sensor nodes are able to execute more elaborate tasks, such as the classification of sound data. This result opens promising perspectives for using WSN in complex activity recognition processes.

The general problem of inferring complex events from simple events is also considered in middleware systems, such as DSWare [8] or Context Zones [10]. Still, there is a gap between the inference process in these systems and the algorithms operating with low-level sensor signals. In particular, it is difficult to evaluate the influence of different levels on the overall recognition performance.

In this paper, we evaluate a distributed recognition system that can run entirely on sensor nodes and classify the activities online. The resource constraints (computation, energy, memory, communication) are extreme and make impractical the usage of state-of-the-art HMM or Bayesian inference methods. Instead, we propose a lightweight fuzzy inference engine that can run on limited hardware and proves stable to inaccurate sensor data.

3   Solution Overview

In this section we start with a brief overview of fuzzy logic, and then present in detail the architecture of the fuzzy-based activity recognition system.

3.1   Overview of Fuzzy Inference Systems

The fuzzy inference process maps crisp inputs to crisp outputs by executing four basic steps. First, the inputs are *fuzzified* through the membership functions. In contrast with crisp logic, the membership functions allow an input variable to

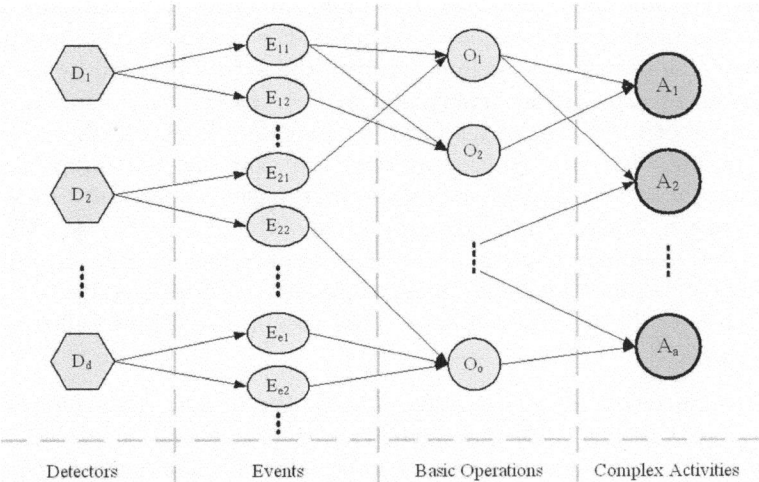

Detectors   Events   Basic Operations   Complex Activities

Fig. 2. Distributed activity recognition architecture. The three main steps are: the detection of simple events (see also [3]), the combination of events into basic operations and the final activity classification.

belong to different fuzzy sets, with partial degrees of confidence (for example, the acceleration can be fuzzified as 0.7 *High* and 0.2 *Medium*). Second, the fuzzified inputs are processed according to the *rule base*, which is a collection of IF-THEN statements with fuzzy propositions as antecedents and consequences. The rule base is derived from expert knowledge or generated automatically from numerical data. Third, the evaluation of the rules is *aggregated* into a fuzzy output. Finally, the fuzzy output is *defuzzified* to a crisp value, which is the final FIS result.

3.2 Distributed Activity Recognition Architecture

The analysis of the car assembly scenario resulted into the following architecture design issues to be considered:

1. Multiple *detectors* are available: sensors worn on the body, attached on the objects and tools, or deployed within the infrastructure. Although heterogeneous, most of these detectors have limited capabilities and low accuracy.
2. Despite their limitations, the detectors can use simple methods to extract online the relevant features of the activities from the sampled data. Online processing is a design choice imposed by the low WSN bandwidth, which makes it unfeasible to transport large amounts of data to a central fusion point. Since the features are computed over a sliding time window and reported asynchronously by different detectors, we refer to them as *events* (see [3] for the detailed algorithms and performance of the event detectors).
3. A single event from one detector gives just an estimation of the real activity going on as it is perceived by the sensor at this location. However, fusing the information from several detectors reporting similar (or correlating) events

within the same timeframe leads to a high confidence in recognizing that the user is performing a certain *basic operation*. For example, the basic operation "Mount the screws" can be inferred from the events signaled by the sensors on the user's arm, on the screwdriver and on the car part being assembled.

4. Sequences of basic operations form *complex activities*, which represent the final output of the recognition system. For example, the complex activity "Mount the brake light" consists of three basic operations: "Pick-up the brake light", "Insert the brake light" and "Mount the screws". We notice from this example that the order of the operations can be an important parameter for distinguishing among different activities (see Sec. 5.2 for details on how the execution order is used to improve the overall recognition performance).

Fig. 2 summarizes the observations above. We denote the set of detectors by $\mathcal{D} = \{D_1, D_2, ..., D_d\}$. Each detector D_i can produce the set of events $\mathcal{E}_i = \{E_{i1}, E_{i2}, ..., E_{ie_i}\}$. Events from various detectors are combined to obtain the basic operations $\mathcal{O} = \{O_1, O_2, ..., O_o\}$. The identification of the basic operations represents the first level of data fusion. The aim is to reduce the number of inputs and consequently the classification complexity. The second level of data fusion performs the combination of multiple observed basic operations into more complex activities $\mathcal{A} = \{A_1, A_2, ..., A_a\}$, where each A_i is composed of a set of operations $\{O_{i_1}, O_{i_2}, ..., O_{i_k}\}$, $1 \leq i \leq a$, $1 \leq k \leq o$.

There are two types of problems that adversely affect the performance of the recognition system. First, the low accuracy of the detectors may result into reporting false events (also referred to as *insertions*) or missing events (also referred to as *deletions*). Second, there are overlaps among different basic operations and also among different complex activities (i.e. the same events are involved in more operations and the same operations are performed during various activities), which can lead eventually to a misclassification due to confusion.

Fuzzy logic can be used throughout the whole activity recognition chain. The detectors usually identify the events with a certain confidence and cannot distinguish perfectly among similar events. This situation maps well to the notion of membership functions, as explained in Sec. 3.1. Likewise, combining events into basic operations is similar to fuzzy majority voting, where different weights are assigned to the detectors based for example on their placement. Rule-based fuzzy inference is appropriate for the final activity classification, by checking the occurrence of the right operations. An interesting extension is to include the time sequence of the operations in the inference process (see Sec. 5.2).

4   Experimental Data

In order to evaluate the performance of our system, we utilize the experimental data obtained by Amft et al. [3] from a car assembly trial. Throughout the trial, 12 sensors with 3D accelerometers were used as detectors. Their placement is shown in Table 1. The detectors could spot a total number of 49 distinct events. We use the confidences of recognizing the events, ranging between (0; 1], as inputs

Table 1. Placement of the detectors

Detector	Placement	Detector	Placement
D_1	Right arm	D_7	Trunk door
D_2	Left arm	D_8	Screwdriver 1
D_3	Upper back	D_9	Screwdriver 2
D_4	Front light	D_{10}	Front door
D_5	Brake light	D_{11}	Back door
D_6	Hood	D_{12}	Socket wrench

Table 2. Activity A_1-"Mount the front door"

Basic operation	Events (Detectors)
O_1-Pick-up the front door	$9(D_1)$, $15(D_2)$, $39(D_{10})$
O_2-Attach the front door	$40(D_{10})$
O_3-Mount the screws	$6(D_1)$, $20(D_3)$
O_4-Pick-up the socket wrench	$7(D_1)$, $21(D_3)$, $47(D_{12})$
O_5-Use the socket wrench	$10(D_1)$, $48(D_{12})$
O_6-Return the socket wrench	$49(D_{12})$
O_7-Close the front door	$16(D_2)$

to our fuzzy-based recognition system. As explained in Sec. 3.2, we first combine the events corresponding to a certain basic operation, then perform the complex activity classification. The following example will best explain this process.

Let us consider the complex activity A_1-"Mount the front door".[1] Table 2 lists in the left column the operations involved in this activity, O_1 to O_7. The right column enumerates the events (represented as numbers) and the corresponding detectors (given in brackets). We distinguish the following cases:

– *Operations identified by only one detector/event, such as O_2.* The confidence of executing such a basic operation is given by the associated detector/event confidence.
– *Operations identified by several detectors on the body, such as O_3.* Compared to the previous case, the combined confidence is higher, but still solely dependent on the accuracy of the gesture recognition. For computing the confidence of the basic operation, the average of the event confidences can be used. If multiple detectors are available, we can apply a fuzzy majority measure [9].
– *Operations identified by several detectors on the body and on the tools (e.g. O_4) or the car parts (e.g. O_1).* This is the best case, as we can fuse information from different sources and produce a more reliable detection. The fusion can be done through a simple weighted average or a fuzzy majority measure, if more detectors of the same type are available.

---

[1] The complete description of the activities can be found in [3].

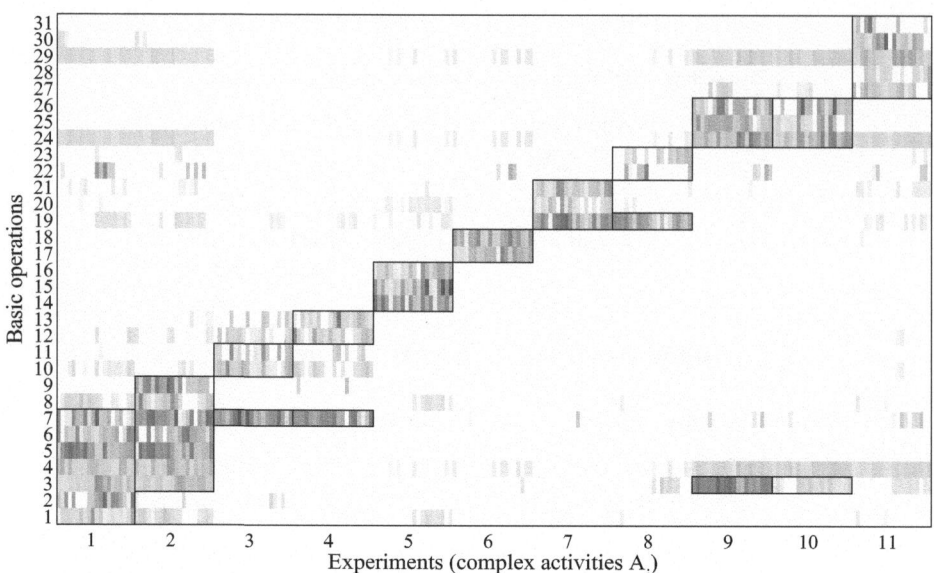

Fig. 3. Basic operations observed by the detectors in the car assembly experiment. The gray spots represent confidence values, where a darker spot means a higher confidence. Operations relevant for the activity are marked by rectangles.

Fig. 3 depicts a grayscale map of the whole car assembly experiment. On the horizontal axis we have 11 complex activities, each of them performed 10 times by each of two subjects. The total recorded data amounts to approximately 5 hours. On the vertical axis we have the basic operations derived from the events reported by the detectors. By using a first level of data fusion, we reduced the number of input variables from 49 events to 31 operations. The gray spots indicate the confidence levels of the basic operations; higher confidences correspond to darker regions. The solid line rectangles mark the real operations involved in each of the 11 activities.

We observe that the fuzzy-based classification method has to cope with the following difficulties:

- Activities with significant overlapping definitions, such as A_1 and A_2, which have the operations O_3 to O_7 in common.
- Activities such as A_3 and A_4, which have theoretically only O_7 in common, but in practice record a large amount of false detections (insertions) causing overlaps.
- Constantly high-confidence insertions, for example O_{29} during A_1 and A_2.
- Missed detections (deletions), such as O_{22} during A_8 or O_{31} during A_{11}. The continuity of the deletions in these examples suggests that they are caused by packet losses at the associated detectors.

5   Fuzzy-Based Activity Recognition

In this section we present in detail the design the fuzzy-based recognition system. We first analyze several alternatives for the main FIS components, then extend the inference for sequences of operations ordered in time.

5.1   Fuzzy System Components

The experimental data described in Sec. 4 is used as input to the fuzzy inference for the final activity classification. There are three important components of the FIS: the membership functions, the rule base and the defuzzifier. Making the right choices for these components is the most difficult task in fuzzy system design. However, the difficulties can be leveraged if (1) there is expert knowledge about the process to be modeled and/or (2) there is a collection of input/output data that can be used for learning and training. The car assembly scenario matches both conditions. Therefore, we base the FIS design on the following list of facts derived from the actual process:

1. The number of output classes is known, equal to the number of complex activities (11).
2. The mapping of inputs (basic operations) to outputs is known from the definition of each activity.
3. The input data is noisy and contains insertions (see Sec. 4) that generate confusion among output classes.
4. The erroneous wireless communication causes deletions in the input data (see Sec. 4), which translate into non-activation of the fuzzy rules.

Fact 2 allows us to define the rule base. Each complex activity is selected by a rule taking the inputs corresponding to the activity definition. For example, activity A_7 generates the rule:

$$\text{IF} \quad O_{19} \text{ is } High \text{ AND } O_{20} \text{ is } High \text{ AND } O_{21} \text{ is } High$$
$$\text{THEN} \quad A_7 \text{ is } High \tag{1}$$

The next step is to choose the membership functions for the fuzzy sets. For computational simplicity, we use trapezoidal membership functions. The first problem is to choose the upper threshold of the membership functions. For each input O_i, we compute the threshold as the mean value $m(O_i)$ over the training set. Consequently, confidences higher than $m(O_i)$ will be fuzzified to 1. The second problem arises from Fact 4: deletions (i.e. zero value inputs) determine the non-activation of the appropriate rules when using the usual *max-min* or *sum-product* inference. To alleviate this effect, we lower the zero threshold of the membership functions, so that zero input values are fuzzified to strictly positive numbers. The downside of this approach is that the level of confusion among output classes increases. We will see however in Sec. 6 that this has no major impact on the overall performance.

The last component to discuss is the defuzzifier. The standard centroid-based defuzzification is the most time-consuming operation on a sensor node, taking

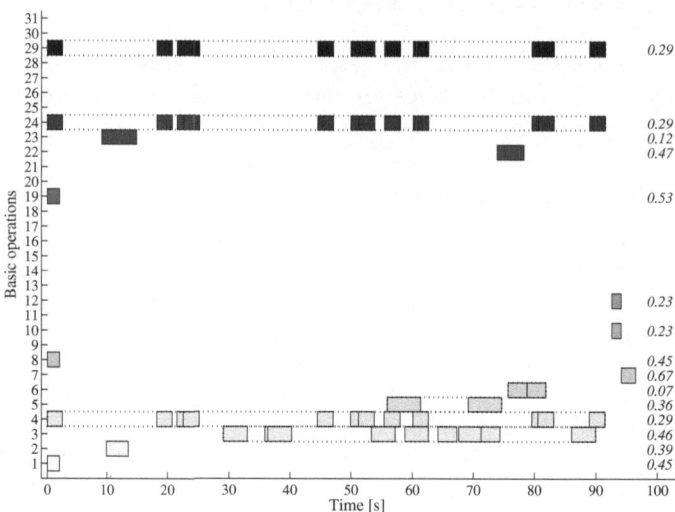

Fig. 4. Basic operations reported during activity $A_1 = \{O_1, \ldots, O_7\}$. The operations in the upper part are mistakenly inserted by the detectors. Multiple occurrences of operations are combined (dotted lines) to analyze the correct order in time.

up to 95% of the total inference time [9]. For the activity classification problem, however, Fact 3 suggests that the *largest-of-maximum* (LOM) defuzzification method is more appropriate because it favours the most probable activity to be selected at the output. In addition, as shown in Sec. 7, LOM defuzzification has a much faster execution time than the centroid method on sensor nodes.

5.2   Temporal Order of Basic Operations

In this section, we extend the recognition system for the case of sequences of ordered operations by incorporating time knowledge in the inference process. To illustrate the usefulness of this extension, we return to the example of activity A_1 from Table 2.

Fig. 4 shows one experimental data set collected when performing A_1 during the car assembly trial. The basic operations detected are represented as gray-tone rectangles. The position of the rectangles on the vertical axis gives the operation ID. The width of the rectangles gives the extent of time during which the operations were reported by the corresponding detectors. On the right side of the figure, we list the average confidence of each operation. We remember from Table 2 that activity A_1 consists of operations $O_1, O_2, ..., O_7$, in this order. Analyzing Fig. 4, we can make the following observations:

– The required operations $O_1, O_2, ..., O_7$ are all recognized. However, O_6 has very low confidence (0.07).
– There are recurring insertions, for example O_{29}, with confidence 0.29 on the average. Nevertheless, this does not generate a classification error because

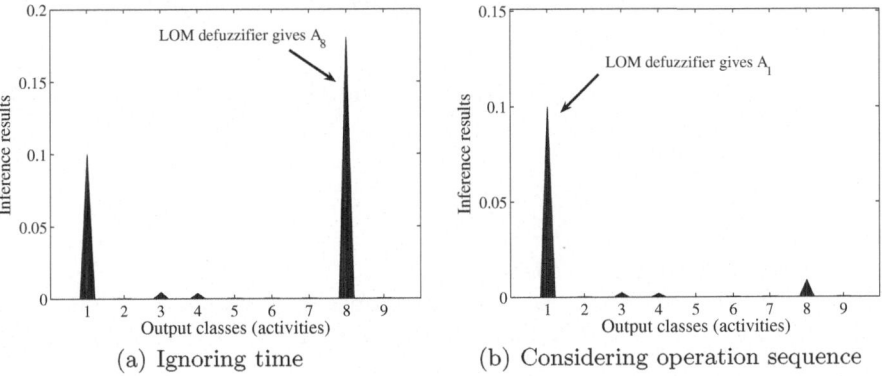

(a) Ignoring time   (b) Considering operation sequence

Fig. 5. (a) The insertions observed in Fig. 4 determine the incorrect result A_8 instead of A_1. (b) Using the penalty function T_A for operations appearing in the wrong time order, A_1 is correctly identified.

O_{29} is involved only in A_{11}, together with $O_{27}, O_{28}, O_{30}, O_{31}$, and none of these operations occurs during the experiment.

– There are also insertions of O_{19}, O_{22}, O_{23}, which form the activity A_8-"Mount hood rod". O_{19} and O_{22} have high confidences (0.47 and 0.53). Even though O_{23} has a lower confidence (0.12), this is fuzzified to a high value because O_{23} occurs with low confidence throughout the whole trial (so also in the training set). As shown in Fig. 5 (a), the result of the LOM defuzzification in this case is wrong; the FIS classifies the activity as A_8 instead of A_1.

Time knowledge can help overcome these problems. Let us analyze again Fig. 4. First, we build the unified time interval when a certain operation was recognized (represented as dotted line rectangles). Then, we define the strict temporal order relation $\prec$ between two operations O_i and O_j with their time intervals $T_i = [a_i; b_i]$ and $T_j = [a_j; b_j]$, respectively, as:

$$O_i \prec O_j \iff b_i < a_j \qquad (2)$$

The $\prec$ relation is used to identify the operations appearing in wrong order, i.e. the cases when O_i should be executed *after* O_j according to the activity description, but $O_i \prec O_j$ in the experimental data. However, when the unified time intervals T_i and T_j overlap, the operations are *not* considered in the wrong order, so $\prec$ is a strict relation. For example, in Fig. 4, operations $O_1, O_2, ..., O_7$ appear in the right order according to the definition of A_1. In contrast, $O_{19} \prec O_{23} \prec O_{22}$, while the correct sequence for A_8 would be O_{22}, O_{19}, O_{23}. If we filter out such incorrectly ordered insertions, we can prevent errors as in Fig. 5 (a). For this purpose, we need to add to the initial FIS an input variable for each activity class; these variables characterize how well the sequences of operations fit the activity descriptions.

Without the loss of generality, let us consider activity A defined as $A = \{O_1, O_2, ..., O_k\}$ and the maximal sequence of operations from experimental data

$S = \{O_{i_1}, O_{i_2}, ..., O_{i_l}\}$, with $S \subseteq A.$[2] We define the number of inversions in S with respect to A as:

$$inv_A(S) = \|\{(i_a; i_b) \mid a < b \text{ and } O_{i_b} \prec O_{i_a}\}\| \tag{3}$$

and the number of deletions:

$$del_A(S) = k - l \tag{4}$$

where $\|.\|$ is the cardinality measure.

In other words, $inv_A(S)$ measures the number of pairs of operations in the wrong order and $del_A(S)$ the number of deletions with respect to A. We use the measure:

$$T_A = \frac{inv_A(S) + del_A(S)}{k(k+1)/2} \tag{5}$$

as an input variable to the FIS for each activity A, where S is the maximal sequence included in A from the experimental data. By including both the inversions and deletions, T_A acts as a penalty function for matching the sequence of detected operations to the right activity description. For the example in Fig. 4 and activities A_1 and A_8, we have $T_1 = (0+0)/28 = 0$ and $T_8 = (2+0)/6 = 0.33$, respectively, so T_8 will induce a higher penalty.

The rule base is also extended to take the new inputs into account, so Rule 1 (see Sec. 5.1) becomes:

IF O_{19} is *High* AND O_{20} is *High* AND O_{21} is *High* AND T_7 is *Low*

THEN A_7 is *High* $\tag{6}$

The result of the fuzzy inference taking into account the new T_A variables is depicted in Fig. 5 (b). As we can see, the LOM defuzzification yields the correct classification result, A_1.

6   Results

As explained in Sec. 4, the car assembly trial consisted of 11 complex activities performed 10 times by each of the 2 subjects, so a total of 220 experiments. In this section, we present the performance results of both the normal FIS and the temporal order extension. As a basis for comparison, we refer to the previous work of Amft et al. [3], where the same car assembly trial was analyzed with two methods: *edit distance* and *event histogram*. The activity recognition is performed by processing the continuous flow of events from the detectors through an adaptive search window, sized according to the training data (see [3] for details). Fig. 6 gives a snapshot of the continuous recognition process, where the upper

---

[2] Insertions of operations not related to A are not taken into account in the input variable for A (T_A), but in the input variables for the other activities, according to their definitions.

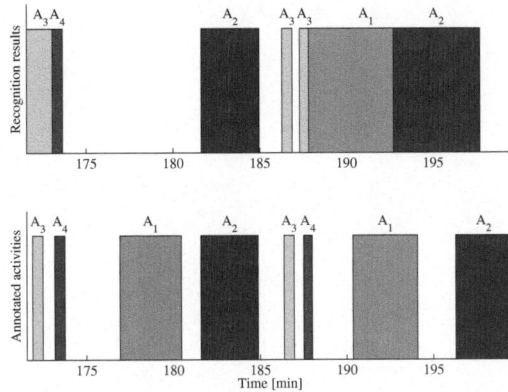

Fig. 6. Snapshot of the recognition results. The upper graph depicts the recognized activities and the lower graph shows the ground truth.

graph depicts the activities reported by the fuzzy inference and the lower graph represents the annotated data (ground truth). We notice that activity A_1 is not recognized at time 179 and A_4 is confused with A_3 at time 188. The remaining activities are correctly reported, within the approximate time intervals given by the search window.

In order to characterize thoroughly the recognition performance, we use the metrics *recall* and *precision*. Fig. 7 shows the recall and precision of the fuzzy-based system, using four-fold cross validation for the selection of training and validation data (the training of the membership functions follows the heuristic method described in Sec. 5.1). The normal fuzzy inference achieves an overall recall and precision of 0.81 and 0.79. The temporal order extension improves the recognition performance to 0.85 recall and 0.85 precision. Fig. 8(a) compares these results to the edit distance (0.77 recall and 0.26 precision) and event histogram (0.77 recall and 0.79 precision) methods from [3]. We can make the following observations:

1. Activities A_6 and A_7 display a recall and precision close to 1, while A_4 and A_{10} have a relatively low recognition performance. This behavior can be explained by analyzing again Fig. 3. We notice the clear separation in the input data for A_6 and A_7, which leads to the high recognition performance. Likewise, we notice the similitude between the input data for A_3 - A_4, which generates the confusion in the recognition results.
2. The normal FIS has the worst performance in the case of A_1 and A_2, which are the most complex activities, each comprising 7 basic operations. The temporal order extension improves considerably the recognition performance of these activities, increasing the recall and precision by 15% and 25% on average. Similarly, the recall and precision of A_{11} (composed of 5 operations) are increased by 10% and 9%, respectively. This shows that analyzing

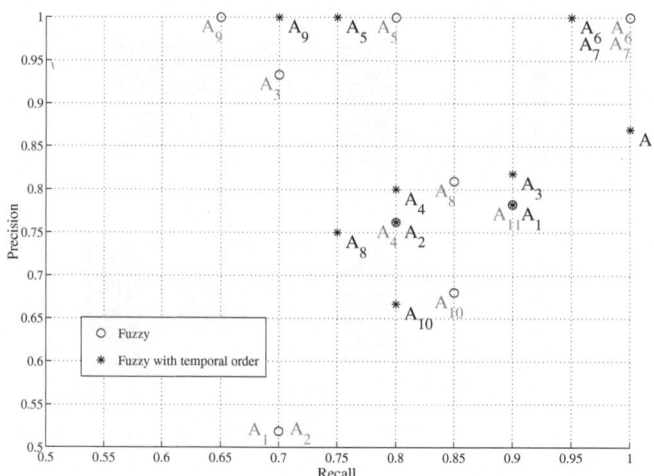

Fig. 7. Recognition performance of the normal FIS and the temporal order extension

the temporal order is particularly successful in the case of complex sequences of operations.

3. For less complex activities, such as A_8 (composed of only 3 operations), the performance of the temporal order extension may drop compared to the regular fuzzy inference. This is caused by the penalty for deletions introduced via the $del_A(S)$ factor (see Eq. 4), which becomes considerable in the case of a small k (number of operations).

4. The normal fuzzy method outperforms the event distance in terms of precision and has similar performance as the event histogram, with a slight improvement in the recall. The temporal order extension further increases the recognition performance with 4-6% on average. Therefore, the event histogram and normal fuzzy methods are appropriate for activities composed of few basic operations, while the temporal order extension is recommended for more complex activities.

As a final result, we present the trade-off between the overall recall and precision in Fig. 8(b). The trade-off is given by an activation threshold set on the FIS output: if the output is lower than the threshold, we consider the situation as "nothing happening" (the NULL class), otherwise we report the classified activity A_1-A_{11}. In this way, the recognition performance can be adapted according to what is more important from the application perspective: to have a low rate of false positives (good precision) or to minimize the false negatives (good recall).

7   Prototyping

In order to analyze the feasibility of our approach, we implemented the detectors algorithms and the fuzzy inference on a resource-constrained WSN platform.

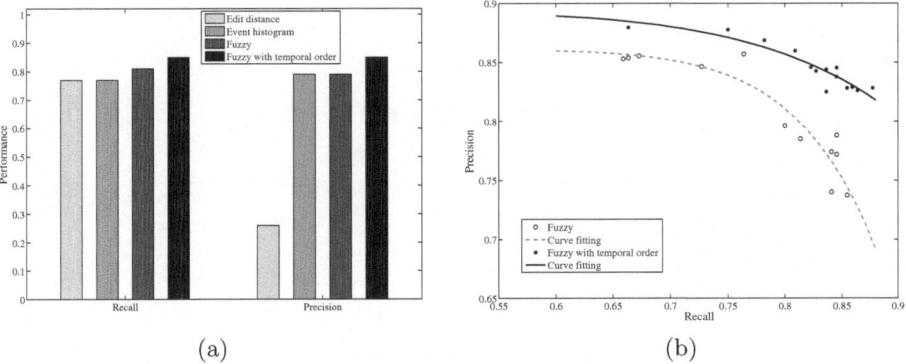

(a)   (b)

Fig. 8. (a) Comparison of the overall recall and precision achieved by four recognition methods: edit distance, event histogram [3], normal fuzzy and the temporal order extension. (b) Tuning the performance of the fuzzy-based methods.

We chose the Tmote Mini sensor node platform (see Fig. 9 (a)) for its compact, industry-standard miniSDIO form factor. The Tmote Mini integrates an MSP430 microcontroller and a TI/Chipcon CC2420 low-power radio in a single one square inch package.

The implementation of the detectors covers two main tasks: (1) to spot relevant sensor data segments in a continuous data flow and (2) to classify those segments as one of the detector's events. To this end, the detectors implement a feature similarity search (FSS) algorithm [2,3]. In the first phase, the data is segmented into windows of a detector-dependent size with a step size of 250ms. In the second phase, the algorithm extracts a detector-dependent number of features from the data segments, and forms the feature vector. The similarity (measured by the Euclidean distance) of this feature vector to a set learned from user-annotated data determines the classification decision.

Let us choose detector D_{10} (front door) for example. D_{10} implements 10 features (length, beginning, end, total difference from mean on 5 partial windows, variance) from acceleration data sampled at 50Hz using an ADXL330 3-axis accelerometer. The complexity of the FSS algorithm scales with the window length (200 samples), the number of features and the number of classes to be distinguished. The total execution time measured on the sensor node is 28.8ms, more precisely 26.8ms for the feature computation phase and 2.0 ms for the classification phase. The algorithm requires 1.3kB of RAM and 1.9kB of FLASH.

The implementation of the fuzzy inference covers the main components described in Sec. 5:

- *Fuzzification.* For computational simplicity, we use trapezoidal membership functions. The fuzzification is computational oriented due to the limited RAM available. To reduce the computational overhead, the maximum fuzzified value is scaled to a power of 2.
- *Inference.* The rules base is stored in a compact, binary form. We use *sum-product* inference as it proves more robust to deletions than max-min method.

FIS type	Inputs	T_{fuzz}	T_{inf}	T_{defuzz}	T_{total}	RAM
Normal	31	4.21	4.45	0.07	8.73	288
Time order	42	5.88	5.97	0.07	11.92	344

(a)   (b)

Fig. 9. (a) Tmote Mini sensor prototype platform. (b) FIS implementation details: execution time and memory footprint.

– *Defuzzification.* The result of the rule evaluation is defuzzified using the *largest-of-maximum* (LOM) method.

Fig. 9 (b) presents the performance parameters of our implementation, for both the normal FIS and the temporal order extension. We can conclude that a total execution time of less than 12ms is feasible on the MSP430 microcontroller. Moreover, the defuzzification time is almost negligible compared to fuzzification and inference. This result highlights the considerable benefit of using LOM instead of centroid defuzzification (for comparison, see [9]). The rightmost column of Fig. 9 (b) lists the amount of RAM required by the FIS, out of 10kB available. The code memory footprint amounts to 490 bytes, out of 48kB available. These figures confirm that a very lightweight implementation of the FIS is feasible.

A final important aspect is the numerical accuracy of the FIS running on the node. In order to quantify the impact of integer computation over the fuzzy inference, we input the experimental data sequentially to the sensor node via the serial port and collect the classification results. The overall error is 1.11%, which is a value within tolerance limits considering the impracticality of floating point computation on the node.

8   Discussion

The results presented so far give an evaluation of the activity recognition system for a particular experiment and the associated data set. In what follows, we provide a more general discussion of the relevant performance factors, with respect to the particularities of fuzzy-enabled WSN.

Distributed recognition architecture. A distributed architecture is required when implementing activity recognition on resource-constrained devices because (1) a powerful central node may not be available and (2) the WSN does not have the capacity of transporting the raw data to a central fusion point. However, as we showed in the previous sections, it is essential that the distributed recognition system considers the whole chain from sensor data to activity classification, with respect to both performance evaluation and implementation.

Recognition performance. Fuzzy logic proves to be robust to both unreliable sensor data and wireless link failures, which cause insertions and deletions. The analysis of the temporal order improves the recognition performance, especially for the activities with many operations executed in a clear sequence. However, the performance remains limited for the activities that are either very similar or have large overlapping areas in the input data. Two additional factors can further affect the overall performance. The first concerns filtering the NULL class, which makes a trade-off between precision and recall. The second is related to the segmentation problem: the recognition system running online cannot know the exact extent of the activities, so only the events within a certain time segment are detected. Consequently, the FIS may operate with incomplete inputs and yield an incorrect result.

Energy consumption. There are two main factors that influence the energy consumption. The first factor is the power dissipated on the actual sensors (e.g. accelerometers), which is independent of the activity recognition system. The second important factor is the wireless communication. The distributed recognition architecture keeps the network overhead to a minimum, by communicating only the spotted events. The number of messages sent by a detector node D_i is proportional to: (1) the amount of events D_i can detect, i.e. $\|\mathcal{E}_i\|$, (2) the frequency of occurring of each event E_{ij} and (3) the rate of insertions. The number of messages received by a node depends on its role in the data fusion process, which can be: (1) simple detector, (2) combining events into basic operations or (3) final activity classifier. It follows that the latter two roles should be assigned to nodes that have fewer (or none) events to detect.

Memory and computational overhead. The implementation on sensor nodes shows that the memory requirements are very low and both the event detection algorithms the fuzzy inference achieve execution time of approximately 40ms. Although the major concern in WSN is to keep the communication overhead to a minimum, fast execution times are equally important in our case, because three demanding tasks compete for CPU utilization: the sensor sampling, the medium access protocol and the activity recognition.

9   Conclusions

Wireless sensor nodes worn on the body and placed on the tools can recognize the activities of the users and assist them at work. We presented a distributed architecture that uses fuzzy logic for reliably detecting and classifying the user activities online. For performance evaluation we used the experimental data from 12 sensor nodes equipped with 3D accelerometers, obtained during a car assembly trial within an industrial setting. The fuzzy system achieved an overall recall and precision of 0.81 and 0.79. An interesting extension was to include temporal order knowledge about the sequences of operations into the fuzzy inference, which improved the recognition performance to 0.85 recall and 0.85 precision. In order to analyze the feasibility of our approach, we implemented both the detection algorithms and the fuzzy logic engine on the Tmote Mini sensor platform. The results showed that our method can run online, with execution times of approximately 40ms, memory overhead in the order of 1.5kB and an overal accuracy loss of 1.11%.

Acknowledgements. This paper describes work undertaken in the context of the SENSEI project, "Integrating the Physical with the Digital World of the Network of the Future" (www.sensei-project.eu), contract number: 215923.

References

1. WearIT@work Project, http://www.wearitatwork.com
2. Amft, O., Junker, H., Tröster, G.: Detection of eating and drinking arm gestures using inertial body-worn sensors. In: International Symposium on Wearable Computers (ISWC), pp. 160–163 (2005)
3. Amft, O., Lombriser, C., Stiefmeier, T., Tröster, G.: Recognition of user activity sequences using distributed event detection. In: European Conference on Smart Sensing and Context (EuroSSC), pp. 126–141 (2007)
4. Brand, M.: The "inverse hollywood problem": From video to scripts and storyboards via causal analysis. In: AAAI/IAAI, pp. 132–137 (1997)
5. Marin-Perianu, M., et al.: Decentralized enterprise systems: A multi-platform wireless sensor networks approach. IEEE Wireless Communications 14(6), 57–66 (2007)
6. Gemperle, F., Kasabach, C., Stivoric, J., Bauer, M., Martin, R.: Design for wearability. In: International Symposium on Wearable Computers (ISWC), pp. 116–123 (1998)
7. Ivanov, Y., Bobick, A.: Recognition of visual activities and interactions by stochastic parsing. IEEE Trans. Pattern Anal. Mach. Intell. 22(8), 852–872 (2000)
8. Li, S., Lin, Y., Son, S.H., Stankovic, J.A., Wei, Y.: Event detection services using data service middleware in distributed sensor networks. Telecommun Syst. 26(2), 351–368 (2004)
9. Marin-Perianu, M., Havinga, P.J.M.: D-FLER: A distributed fuzzy logic engine for rule-based wireless sensor networks. In: International Symposium on Ubiquitous Computing Systems (UCS), pp. 86–101 (2007)
10. Osmani, V., Balasubramaniam, S., Botvich, D.: Self-organising object networks using context zones for distributed activity recognition. In: International Conference on Body Area Networks (BodyNets) (2007)
11. Predd, J.B., Kulkarni, S.R., Poor, H.V.: Distributed learning in wireless sensor networks. IEEE Signal Processing Magazine 23(4), 56–69 (2006)
12. Ross, T.J.: Fuzzy Logic with Engineering Applications. Wiley, Chichester (2004)
13. Saligrama, V., Alanyali, M., Savas, O.: Distributed detection in sensor networks with packet losses and finite capacity links. IEEE T Signal Proces 54(11), 4118–4132 (2006)
14. Samarasooriya, V.N.S., Varshney, P.K.: A fuzzy modeling approach to decision fusion under uncertainty. Fuzzy Sets and Systems 114(1), 59–69 (2000)
15. Stäger, M., Lukowicz, P., Tröster, G.: Power and accuracy trade-offs in sound-based context recognition systems. Pervasive and Mobile Computing 3(3), 300–327 (2007)

16. Stiefmeier, T., Lombriser, C., Roggen, D., Junker, H., Ogris, G., Tröster, G.: Event-Based Activity Tracking in Work Environments. In: International Forum on Applied Wearable Computing (IFAWC) (March 2006)
17. Wang, T., Han, Y., Varshney, P., Chen, P.: Distributed fault-tolerant classification in wireless sensor networks. IEEE J. Sel. Area. Comm. 23(4), 724–734 (2005)
18. Wren, C.R., Minnen, D.C., Rao, S.G.: Similarity-based analysis for large networks of ultra-low resolution sensors. Pattern Recogn. 39(10), 1918–1931 (2006)

CaliBree: A Self-calibration System
for Mobile Sensor Networks

Emiliano Miluzzo[1], Nicholas D. Lane[1], Andrew T. Campbell[1], and Reza Olfati-Saber[2]

[1] Computer Science Department,
Dartmouth College, Hanover NH 03755, USA
{miluzzo,niclane,campbell}@cs.dartmouth.edu
[2] Thayer School of Engineering,
Dartmouth College, Hanover NH 03755, USA
olfati@dartmouth.edu

Abstract. We propose *CaliBree*, a self-calibration system for mobile wireless sensor networks. Sensors calibration is a fundamental problem in a sensor network. If sensor devices are not properly calibrated, their sensor readings are likely of little use to the application. Distributed calibration is challenging in a mobile sensor network, where sensor devices are carried by people or vehicles, mainly for three reasons: *i)* the sensing contact time, i.e., the amount of time nodes are within the sensing range of each other, can be very limited, requiring a quick and efficient calibration technique; *ii)* for scalability and ease of use, the calibration algorithm should not require manual intervention; *iii)* the computational burden must be low since some sensor platforms have limited capabilities. In this paper we propose CaliBree, a distributed, scalable, and lightweight calibration protocol that applies a discrete average consensus algorithm technique to calibrate sensor nodes. CaliBree is shown to be effective through experimental evaluation using embedded wireless sensor devices, achieving high calibration accuracy.

1   Introduction

Sensors calibration is a fundamental problem in sensor networks. Without proper calibration, sensor devices produce data that may not be useful or can even be misleading. There are many possible sources of error introduced into sensed data, including those caused by the sensing device hardware itself. Hardware error can be broken down into that caused by the sensor hardware component, the sensing device component, and the sensor drift (sensors' characteristics change by age or damage). The sensor hardware level error is corrected at the factory where a set of known stimuli is applied to the sensor to produce a map of the output. The sensing device error component is introduced when the sensor is mounted on the board itself that includes the microcontroller, the transceiver, and the circuitry that form a sensor node [1] (we call the sensor plus the supporting board a sensor device). To correct the sensing device error component calibration of the sensor device is required.

While device calibration must sometimes be done in the factory (e.g., for high precision medical sensors), a growing number of sensors are embedded in consumer devices [4] [5] and are currently used particularly in a number of popular recreational

S. Nikoletseas et al. (Eds.): DCOSS 2008, LNCS 5067, pp. 314–331, 2008.

domain [6], and emerging [7] [8] applications. This latter class of cheap sensors are generally shipped without any sensor device calibration and it is up to the user to perform the calibration procedure to make sure that the gathered sensed data is meaningful. Moreover, sensors drift from their initial calibration over time. This imparts a significant burden to the user of the sensor devices. Further, this manual method of calibration process does not scale when considering large scale people-centric deployments and applications.

We conjecture that in mobile sensor networks [3] [16] [17] there will be two classes of sensors: calibrated nodes that can be either static or mobile, and uncalibrated nodes. We refer to the nodes belonging to the former class as *ground truth* nodes. These ground truth nodes may exist as a result of factory calibration, or user manual calibration.

We propose CaliBree, a distributed, scalable, and lightweight protocol to automatically calibrate mobile sensor nodes in this environment. In CaliBree, uncalibrated nodes opportunistically interact with calibrated nodes to solve a discrete average consensus problem [9], leveraging cooperative control over their sensor readings. The average consensus algorithm measures the disagreement of sensor samples between the uncalibrated node and a series of calibrated neighbors. The algorithm eventually converges to a consensus among the nodes and leads to the discovery of the actual disagreement between the uncalibrated node's sensor and calibrated nodes' sensors. The disagreement is used by the uncalibrated node to generate (using a best fit line algorithm) the calibration curve of the newly calibrated sensor. The calibration curve is then used to locally adjust the newly calibrated node's sensor readings.

CaliBree relies on opportunistic rendezvous between uncalibrated nodes and ground truth devices because we want the calibration process to be transparent to the user. The convergence time of the algorithm depends on the density of the ground truth nodes. Still, if the density was low, the accuracy of the algorithm would not be impacted, only the convergence time would be extended. However, we expect urban sensor networks [3] [18] will have a high density of ground truth nodes. In the CitySense project, well calibrated sensor nodes are mounted on light poles in an urban area. Those sensors can be considered as ground truth nodes that could be used by mobile nodes running the CaliBree algorithm.

In order for the consensus algorithm to succeed, the uncalibrated sensor devices must compare their data when sensing the same environment as the ground truth nodes. Given the limited amount of time mobile nodes may experience the same sensing environment during a particular rendezvous, and the fact that even close proximity does not guarantee that the uncalibrated sensor and the ground truth sensor experience the same field of view, the consensus algorithm is run over time when uncalibrated nodes encounter different ground truth nodes. We experimentally determine that an uncalibrated node achieves calibration after running CaliBree with less than five different ground truth nodes.

The contribution of this paper is:

– It proposes, to the best of our knowledge, the first fully distributed approach to calibrating mobile sensor devices such as embedded sensor devices [6] [7] and sensor enabled cellphones [4] [5].
– It proves the existence of the *sensing factor* (see Section 2) which we believe is an important characteristic to be considered in the design of protocols and applications for mobile sensing systems.

– It presents a calibration technique which is efficient, scalable, and lightweight, therefore suitable to be applied to mobile sensing systems.
– It shows the experimental evaluation of the CaliBree protocol through validation using a testbed of static and mobile embedded sensor devices [1].

In the following sections, we describe the motivation, design, and evaluation of the CaliBree system. In Section 2 we motivate the need of an efficient and scalable calibration protocol for mobile sensor networks. In Section 3 we discuss the shortcomings of existing techniques proposed in the literature to achieve sensor networks calibration. The CaliBree design is illustrated in Section 4, and Section 5 describes the experimental approach we took to validate CaliBree. We summarize our work in Section 6 where we also discuss our future research direction.

2   Motivation

Curiously the issue of calibration of wireless sensor networks has received low attention in the literature despite it being recognized as a fundamental issue. Without calibration the data acquired from such networks is meaningless. This obvious fact and the difficulties in performing calibration in general is a repeated finding of real world sensor network deployments [19] [20]. Emerging mobile sensing architecture which are the focus of this paper will not be different.

To quantify the magnitude of the calibration problem we perform experiments with our own building sensor network testbed comprising both mobile and static sensors. We perform experiments to: *i)* show the variability of the individual calibration curves between multiple sensor nodes considering two different sensing modalities, and *ii)* quantify how the differences in these calibration curves would impact the actual reported sensor values from these nodes. For both of these experiments we used the Tmote Sky wireless sensor [1], a multimodal sensing platform commonly used by the experimental sensor network community.

Manual Sensor Calibration. For all the experiments performed unless otherwise noted we used a set of manually calibrated Tmote Sky sensor nodes as ground truth nodes. Given the linear response of the Tmote Sky sensors we took four calibration points for 21 different Tmote Sky on two of the available sensor suite on the node, namely the PAR (Photosynthetically Active Radiation) light sensor, which has a frequency response range approximately equivalent to that of a human eye, and the temperature Sensirion AG temperature/humidity sensor. Ground truth sensor readings were provided for temperature by the Extech SuperHeat Psychrometer RH350 [30] and for light by the Extech Dataloggin Light Meter 401036 [31]. As per the typical manual calibration process a calibration curve specific to the sensor in question was determined by taking a linear regression of measurements of the physical phenomena (provided by the ground truth sensors) and the raw output of the sensor in question. The value of the raw output used in the regression was the mean of 20 individual raw readings.

Figure 1(a) and Figure 1(b) present the larger and smaller bounds sensor specific calibration curves for, respectively, the PAR and temperature sensors. The larger bound curve is associated to a node to which we give the label "A" whereas the node associated to the lower bound curve is labeled with "B". The x-axis represents the raw output of the sensors while the y-axis provides actual light (in Lux) and temperature (in degrees Celsius) values that are expected to correspond to the Analog to Digital Converter (ADC) output of the sampled sensors.

We also derive the sensed data values from the factory Tmote Sky calibration curves and compare them to the values calculated with the manually derived calibration curve for same ADC outputs. The difference between the readings is plotted in Figure 1(c) and Figure 1(d) for the PAR and temperature sensors respectively and represents the calibration error of the sensor nodes coming off the factory. In Figures 1(c) and 1(d) the y-axis presents the calibration error of nodes A and B, relative to the ADC output of the sensor itself, which is reported in the x-axis. In Figure 1(d) errors of up to 55 degrees Celsius are shown depending on the temperature range. Similarly, in Figure 1(c) error as large as nearly 2,600 Lux are demonstrated.

Variations in the sensor specific calibration curves. Our experimentation demonstrated differences both in the gain and offset of the calibration curves for each sensor node and manufacturer provided generic calibration curve (see Figures 1(a) and 1(b)). Not only were there differences between the individual sensor nodes but there were patterns in the type of calibration error depending on the modality considered. The light sensor had very small difference in the offset of the calibration curves while it had substantial difference in the gain between each curve. For instance the error when the light sensor is exposed to bright environments, where the ADC's output is near 4000 units, is nearly 2600 Lux, whereas in darker environments (low ADC output) the error can become small. In contrast was temperature sensor for which the calibration curves had both gain and offset variation but with offset differences being more substantial. This suggests any general approach to calibration must be able to both determine the unique gain and offset values for each sensor device.

Existing calibration approaches within wireless sensor networks do not assume the presence of well calibrated sensor nodes in the network (i.e., [21] [23] [22]). This is in line with the typical assumption about the use of cheap, lower quality (i.e., radio interface and sensors), resource constrained sensor nodes that collectively form a dense network over the sensor field. However, these assumptions do not hold in an emerging class of sensor network architectures [16] [3] [14] [5] deployed in urban areas, where such restrictions are not longer motivated, and comprising new advanced forms of embedded sensor platforms such as sensor enabled cellphones [4]. In particular, this impacts how calibration should be performed within such networks. Specifically, the potential exists for a subset of the devices to be capable of acting as ground truth nodes, i.e., sources of reliable calibrated sensor data. Such nodes are able to support the rest of the network comprised by uncalibrated sensors. Due to the availability of ground truth sensor data a traditional approach to calibration becomes possible. This being the approach of determining the calibration curve for a particular sensor based upon a collection of sensor values that can be compared to the ground truth sensor data. This comparison is opportunistic in the sense that due to the uncontrolled mobility patterns of either or both the ground truth and uncalibrated nodes, the uncalibrated nodes probabilistically encounter ground truth nodes. We envision a scenario where urban-scale sensor network deployments provide a number of well calibrated sensors [3] [18] with which uncalibrated nodes could rendezvous in order to perform calibration.

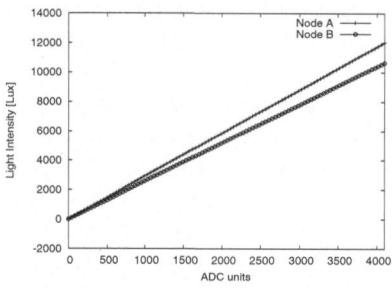

(a) Plot of the upper and lower bounds of the PAR calibration curves obtained by manually calibrating 21 Tmote Sky nodes.

(b) Plot of the upper and lower bounds of the temperature calibration curves obtained by manually calibrating 21 Tmote Sky nodes.

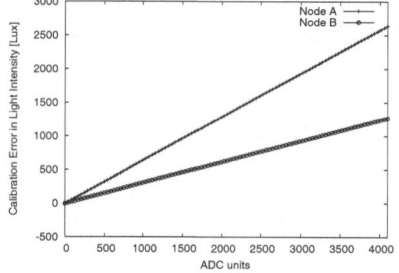
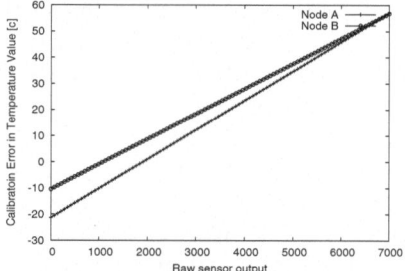

(c) Plot of the light calibration error measured by comparing the sensed data obtained from the factory calibration curves and the manually generated calibration curves.

(d) Plot of the temperature calibration error measured by comparing the sensed data obtained from the factory calibration curves and the manually generated calibration curves.

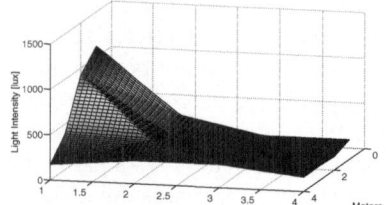
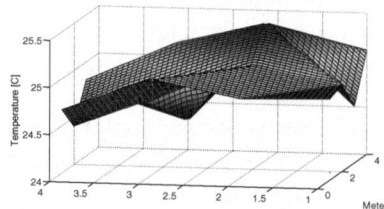

(e) Contour plot of the mean light intensity measured with a set of 21 Tmote Sky nodes arranged in a 3 by 7 grid with a 0.5 meters spacing within an indoor office space environment.

(f) Contour plot of the mean temperature gradient measured with a set of 21 Tmote Sky nodes arranged in a 3 by 7 grid with a 0.5 meters spacing within an indoor office space environment.

Fig. 1.

Superficial consideration of this approach to calibration suggests that performing calibration in such networks is trivial. However, several factors make the calibration challenging. Firstly, the calibration can only occur when the ground truth node and the uncalibrated node are experiencing identical sensing environment. This is necessary because the comparison between calibrated and uncalibrated data is only meaningful when the same input to the sensors is applied. Secondly, the calibration rendezvous is complicated by the existence of the *sensing factor*. The sensing factor is identified by the tendency of a physical phenomenon to be localized to a small region around the entity taking the measurement. If for example we consider the light sensing modality, given the high directional nature of light, the light readings reported by a light sensor are relative to the proximity region of the sensor. In contrast, for the temperature modality, the temperature gradient around a temperature sensor presents a much smaller variation. In general, the existence of the sensing factor is largely independent of the specific modality in question and can be related to a broad class of sensors (e.g, light, dust/pollen, CO_2, sound, etc.). The variability of the sensor data relative to an originating location of sampling increases rapidly as the distance from this origin increases (for example consider the exponential decay of various phenomena such as light and heat). We note that the sensing factor, since it is based upon invariant physical laws, will remain the same regardless of the components of the sensor devices themselves. Unlike discussions of short communication rendezvous durations found in DTNs (Delay Tolerant Networks) (such as discussed in [13]) which could be an artifact of the short range low power radios, the sensing factor will be present regardless the radio technology and more or less dependent on the sensing modality.

Characterizing the Sensing Factor. To quantify the sensing factor we performed an experiment where both light and temperature were sampled from a 3 by 7 static grid of 21 calibrated Tmote Skys separated by a 0.5 meters distance. The nodes were placed in an indoor environment during daylight hours. Figure 1(e) is a contour plot of light readings and Figure 1(f) is the contour plot for the temperature readings. Both Figures 1(e) and 1(f) clearly demonstrate the variability of light and temperature over relatively short distances. Gradients exist with the sampled phenomena and the variability of these gradients increases with distance. It is evident that the variation of the light intensity is larger than for the temperature (light drop is of about 500 Lux in just 0.5 meters whereas one Celsius degree variation in temperature is obtained over more than 2 meters). This implies that if for example a light ground truth node was positioned in the (x,y)=(1,0) location in Figure 1(e), an uncalibrated light sensor node needs to move close within 0.5 meters distance from the ground truth node in order to sample a sensing environment similar to the one of the ground truth node and perform accurate calibration. For the temperature sensor, the distance between ground truth node and uncalibrated nodes within which the calibration can be performed becomes larger.

In general, mobility combined with the sensing factor reduces the time interval in which nodes experience the same environment which is a requirement to perform accurate calibration. CaliBree is designed to operate quickly when uncalibrated nodes enter the same sensing environments of ground truth nodes. It also allows uncalibrated nodes to exploit distance information between themselves and ground truth nodes and make decisions about whether to rendezvous with them or not.

3   Related Work

A significant amount of work spanning many decades addresses the general problem of sensor calibration. However, relatively few solutions are developed for the more recently formed conception of wireless sensor networks [10] [12] [11]. The bulk of research in this more focused area deals primarily with energy-efficient networking and distributed computing, rather than with accurate sensing. In fact, the payload of packets in these networks is often treated as a black box that is ignored or abstracted away. The work that does exist in calibrating wireless sensor networks assumes a dense network of static and highly resource-constrained nodes [28], and is not directly applicable to sensor networks with uncontrolled mobility (e.g., [3]), the environment assumed in this paper. These networks comprise loose federations of heterogeneous nodes with variable mobility patterns (i.e., static and mobile) that lead to variable and often sparse nodes density.

Motivated by the unscalability of manual calibration techniques with a known standard input signal, the authors of [24] propose a technique called "macro-calibration" for use in networks of thousands of nodes. The technique builds an optimization problem from trends and relationships observed in the aggregate sensor data provided by the network to generate calibration equations. However, the design and evaluation of [24] focuses on the accuracy of range estimates between nodes to support localization. Others have also contributed solutions limited to the needs of localization [21] [22] [23] [29] and time synchronization rather than the sensor modality agnostic type of calibration that is our focus. More general and less modality and application specific calibration is considered in [25]. This work also adopts the aforementioned "macro-calibration" approach in densely deployed networks. More recent work [26] does not require the same levels of density, but assumes that sampled sensor data is band limited. By sampling this data above the Nyquist rate, the actual sensor values will exist in a lower dimensional subspace of the higher dimensional space comprised by the uncalibrated readings.

A calibration approach involving robotic network elements is presented in [27], whereby a robot with calibrated sensors gathers samples within the sensor field to allow already deployed static sensors to be calibrated/re-calibrated as required. Our work differs in that we do not depend on controlled robotic mobility but rather we exploit opportunistic rendezvous between mobile uncalibrated and calibrated sensor nodes carried by humans and their vehicles. Further, our solution does not require the introduction of costly and complex robotic hardware.

4   CaliBree Design

In this section we present the design of the CaliBree protocol. Recall the definition of sensing contact time as the time window in which mobile nodes experience approximately the same sensing environment. Similarly, we term the spatial region where nodes experience the same sensing environment as the *common sensing range*. The sensing contact time depends on several factors including mobile node's speed, sensor orientation, obstacles to a sensor's field of view, and the physical sensing range limit. The common sensing range varies with sensor type. For example, as shown in Section 2, the common sensing range for light tends to be small due to its highly directional nature, whereas for temperature the common sensing range is larger since the temperature gradient is typically small in the proximity of a human-carried sensor device. Given that an uncalibrated node should experience the same sensing context as the ground truth node during the calibration process, if the common sensing range is small the sensing contact time could be very short. Moreover, if either or both the uncalibrated or ground truth nodes are moving the sensing contact time may be further shortened. We show in Section 5 that in the case of a location-dependent sensing modality like the light sensor, the sensing contact time is in the order of few seconds under human mobility patterns. There is then a need to design a calibration protocol that is fast, completing during the short sensing contact time.

To this end, CaliBree is designed to solve a distributed average consensus problem. Equation (1) shows the formulation of the discrete consensus problem we use:

$$\bar{d}_i(k+1) = \begin{cases} (1-\varepsilon) \cdot \bar{d}_i(k) + \varepsilon \cdot \sum_{j=1}^{N_i} \frac{s_i(k)-s_j(k)}{N_i}, & k > 0, \\ d_i^{uncal}, & k = 0. \end{cases} \quad (1)$$

$\bar{d}_i(k)$ is the average disagreement measured by node i up to round k; $0 < \varepsilon < 1$ sets the weight given to the current round's disagreement consensus; N_i is the set of ground truth nodes in i's neighborhood at round k; $\frac{s_i(k)-s_j(k)}{N_i}$ is the average disagreement between i's sample and those of the ground truth nodes in range at round k; d_i^{uncal} is the difference between the uncalibrated data and one of the ground truth node's data when the calibration starts.

The consensus algorithm formulated in (1) works as follows. Each ground truth node periodically transmits a beacon advertising its availability to participate in the calibration routine for at least one sensor type. If an uncalibrated node wishes to calibrate its sensor of an advertised type, it replies to this advertisement, triggering the CaliBree protocol (a distance-based energy optimization to this trigger is described in Section 4.3). Upon receiving a reply to its advertisement, the ground truth node starts broadcasting a series of packets containing its instantaneous sensed data value for the sensor type under calibration. We call these broadcasts packets sent by the ground truth nodes *calibration beacons*. As the uncalibrated node starts receiving the calibration beacons it begins running the consensus algorithm. The uncalibrated node calculates the difference between its own sensed data and the sensed value from the ground truth node and feeds this difference into Equation 1 as $s_i - s_j$. Equation 1 outputs the current estimate of the average 'disagreement' between the uncalibrated and ground truth nodes. The average disagreement $\bar{d}$ from Equation 1 decreases as the uncalibrated node physically approaches the ground truth node(s) and their sensing ranges begin to overlap. Considering a particular node pair (i, j), the minimum $\bar{d}$ occurs at the time of maximum sensing range overlap between the uncalibrated node and the ground truth node (i.e., they both are sampling a similar environment). This $\bar{d}^{min}$ estimates the uncalibrated sensor device's true offset from the ground truth sensor.

In the following, we demonstrate the ability of the consensus algorithm to converge to the minimum disagreement. We investigate the calibration of the light sensor for this experiment and throughout the paper given the challenge implied by the highly directional nature of light. By showing that CaliBree is able to work for light, which potentially leads to small sensing contact times due to its sensitivity to sensor orientation and obstructions, we gain confidence that it works well for other sensor types, under less demanding constraints as well. We implement the consensus algorithm on Tmote Sky [1] wireless sensor nodes, and start by manually calibrating a single ground truth node. The ground truth node is placed on a shelf next to the window in our lab.

Figure 2(a) shows the evolution of the average disagreement $\bar{d}$ between the uncalibrated light sensor and the single ground truth node as a human carries the uncalibrated sensor periodically towards and then away from the ground truth sensor. The minima in Figure 2(a) represent the minimum disagreements and are achieved every time the uncalibrated node arrives within the common sensing range of the ground truth node. In this experiment the minimum disagreement is approximately 700 Lux. In Section 5 we show that the common sensing range, i.e., the spatial region where nodes experience the same sensing environment, depends on the relative context of the nodes (e.g., light sensor orientation).

At the moment the minimum disagreement is achieved, the uncalibrated sensed data plus the average minimum disagreement $\bar{d}^{min}$ gives the actual ground truth sensor readings. For uncalibrated node i and ground truth sensor j, we call the value of i's sensor at the moment of minimum disagreement s_i^{min} and the value of j's sensor at the same moment s_j^{min}. Then we have that $s_i^{min,j} + \bar{d}^{min,j} = s_j^{min}$, where $s_i^{min,j}$ is the value of i's sensor at the moment of minimum disagreement during the rendezvous with ground truth sensor j. Moreover, we use $ADC_i^{min,j}$ to refer to the output of the analog to digital converter (ADC) that samples the sensor on the sensing device i at the time of the minimum disagreement during the rendezvous with the ground truth node j. The $s_i^{min,j}$ value is a function of this ADC value, where the function is defined by the sensor manufacturer. The uncalibrated node stores the following bundle in an internal buffer for later reference when generating its calibration curve: $\{ADC_i^{min,j}, s_i^{min,j}, \bar{d}^{min,j}\}$.

In case ground truth and/or uncalibrated nodes follow a sleep schedule to reduce their power consumption, CaliBree is not triggered if either the ground truth nodes or the uncalibrated nodes are in sleep mode (use of a radio wake-up mechanism [33] is possible, but is outside the scope of this paper). If a rendezvous occurs when both ground truth and uncalibrated nodes are awake, the node(s) stay awake until the calibration rendezvous completes. Due to CaliBree's reliance on the uncalibrated node and the ground truth node sharing a common sensing context, which may quickly change due to human mobility, for best performance we recommend prioritizing the CaliBree service over other platform services during the ground truth rendezvous period. We conjecture this is not excessively disruptive for the current sensing device running application since as we show in Section 5.2 the calibration rendezvous lasts for two seconds at most.

4.1   Best Fit Line Algorithm

The best fit line algorithm takes as an input the collected bundles and generates a sensor device's calibration curve in the form $y = f(x)$. Here x represents the ADC input and y the calibrated sensed data output. The best fit line algorithm takes as input the two finite sets $\{x_i = ADC_i^{min}\}$ and $\{y_i = s_i^{min,j} + \bar{d}^{min,j}\}$, for $j \in [1,N]$ where N is the number of encountered ground truth nodes, and produces the calibration equation. In Section 5 we show that fewer than five ground truth nodes are needed by the best fit line algorithm to compute the calibration curve.

By running the calibration algorithm every time an uncalibrated node encounters ground truth nodes we reduce the calibration error (e.g., due to slightly differing

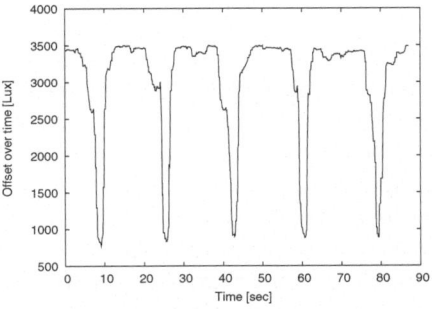

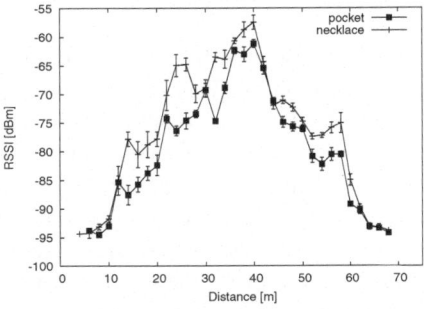

(a) Plot of the disagreement over time between an uncalibrated node and one ground truth node when the uncalibrated node rendezvous multiple times with the ground truth node.

(b) Plot of the RSSI measured at the mobile node from packets received by the ground truth node as a function of the distance from the ground truth node.

Fig. 2.

sensing context during the calibration rendezvous) that might be introduced by performing calibration by relying on one or few ground truth nodes.

4.2 Epsilon Adaptation

To make CaliBree more responsive to the dynamics of a mobile sensor network CaliBree adapts the ε value in Equation 1 according to the sensing context of the uncalibrated and ground truth nodes. Recall that as ε increases the weight given to the newly calculated disagreement value increases relative to the average historical value. CaliBree adaptively changes the value of ε when some ground truth readings might introduce large errors and increase the time to convergence of the minimum disagreement estimation. In particular, the value of ε is reduced (less weight to the current sample) when:

- the distance between the uncalibrated node and the ground truth node is large;
- the orientation of the uncalibrated node is different than the orientation of the ground truth node(s);
- the time since the ground truth node was last calibrated is large;
- there are hardware differences between the uncalibrated sensing device and the ground truth node(s) (e.g., different ADC scaling), such that comparison of the sensor readings is not possible.

To allow the uncalibrated node to make these determinations, the ground truth sensor includes its current orientation, calibration time stamp, and hardware specifications in each calibration beacon. Location is also stamped in the calibration beacon if a localization system is in place in the network. If this is not the case, we describe in the next section how well distance between nodes can be estimated using the RSSI values of exchanged packets. In our current implementation we do not yet make use of this information to adapt ε, but instead we experimentally find that a fixed value of ε is 0.05 balances the sensitivity and convergence time of the average consensus algorithm.

4.3 Distance Estimation

CaliBree is triggered by an uncalibrated node when it determines via advertisement packet reception that it is approaching a ground truth node. CaliBree is most efficient, in terms of number of calibration beacons sent, if the protocol is triggered as close as possible to a given ground truth node. This is true since the average consensus algorithm will not converge to the minimum disagreement until the uncalibrated node and the ground truth node share a common sensing range. Calculating incremental disagreements far outside this common sensing range provides no benefit and is therefore wasteful. To determine the relative distance between itself and a ground truth node, the uncalibrated node can leverage a localization system when available, or can perform estimates based on the Received Signal Strength Indicator (RSSI) measurements taken from advertisement packets transmitted by the ground truth node(s).

We run an experiment to verify whether the RSSI can be used as a satisfactory means to infer distance. The ground truth node is placed approximately in the middle of a long hallway. The ground truth node sends packets periodically from which a mobile node extracts and records RSSI information. The mobile mote is carried in one case in a necklace, and in the other case inside a pocket. Figure 2(b) shows a plot of RSSI at the single mobile node versus actual distance from the ground truth node. Error bars indicate the 95% confidence intervals. The ground truth node is placed about 38 meters from one end of an office building hallway ($x=0$). It can be seen that for both the necklace and pocket cases the measured RSSI increases as the mobile node approaches the ground truth node and decreases when the mobile node goes away from the ground truth node. While neither the rising edge or the falling edge of either curve is Figure 2(b) is monotonic (perhaps RSSI can not be used for accurate ranging) we conjecture that RSSI can be effective as a coarse proximity indicator. In future work we will report on determining an appropriate RSSI threshold to serve as a boundary between "close" and "not close", inasmuch as these labels pertain to triggering Calibree. In our current implementation of CaliBree, ground truth nodes send advertisement packets at the lowest possible transmit power setting, which on the Tmote Sky platform and in our experimental field gives a radio range of approximately 8 meters, to minimize the number of wasted calibration beacons. Approaches such as [15] can be used at close range to refine the ranging estimate within an error of few tens of centimeters.

Another way to determine the uncalibrated-ground truth nodes distance derives from more accurate forms of localization, e.g., on board GPS or hardware location engines [32], which will be largely used by future mobile sensing devices.

5 Experimental Evaluation

In this section we evaluate the performance of the CaliBree protocol using Tmote Sky [1] wireless embedded devices. We implement CaliBree in TinyOS [2] which is currently the de-facto open source standard operating system for embedded experimental wireless sensor systems. We use a testbed of 20 static Tmote Sky nodes, calibrated to provide ground truth sensed data, deployed across the three floors of an office building. The experiments characterize the performance of CaliBree in calibrating the light sensor of a single[1] Tmote Sky sensing device as it is carried by a human moving at walking speed around the building. The mobility of the human brings the uncalibrated sensor through the sensing ranges of various arrangements of ground truth nodes, as described in the following.

Although we evaluate CaliBree with Tmote Sky platforms, CaliBree could be used equally well with any sensing platform that requires post-factory calibration or calibrated nodes that experience calibration drift over time. We leave a survey of such devices to future work.

5.1   Sensing Contact Time

In the following, we quantify the sensing contact time, i.e., the amount of time two nodes experience the same sensing environment, between a mobile sensor and a static ground truth node. An uncalibrated sensor is carried at human walking speed through the sensing ranges of statically deployed ground truth nodes. In Figure 3(a) the output of the consensus algorithm, the average disagreement, is shown over time as the mobile uncalibrated node rendezvous with three ground truth nodes. The minima in the graph occur when the mobile node is within the common sensing range of each respective ground truth node, where the difference between uncalibrated and calibrated data is minimized. The plot inset in Figure 3(a) is a zoom in of the leftmost minimum. It shows that the amount of time the difference between the uncalibrated and ground truth data is minimum is on average about two seconds. This time interval, the sensing contact time, is relatively short even at human walking speeds when the ground truth node is static, underscoring the importance of efficient messaging and fast consensus convergence in the calibration protocol. Beyond this time interval the uncalibrated node no longer experiences the same sensing environment as the ground truth node and the accuracy of the calibration output decreases.

5.2   Ground Truth Nodes Beacon Rate

As described in Section 4, once an uncalibrated node triggers the calibration routine on the ground truth node, a series of calibration beacons are broadcast by the ground truth node. The broadcast nature of the calibration beacons allow other uncalibrated nodes to trigger their own calibration processes. These beacons allow the uncalibrated node to run the CaliBree consensus algorithm. In our implementation the calibration beacon packet is 18 bytes in size. In determining the best rate at which ground truth nodes should send calibration beacons, one must consider how the resultant consensus update rate (consensus round interval) of the average consensus algorithm impacts the ability of the mobile sensing device to detect when it has left the common sensing range of the ground truth node.

In our implementation, presence in the common sensing range is inferred by detecting the difference from the minimum in a moving window average of consecutive values of $\bar{d}$ from Equation 1. The output of Equation 1 is updated on the reception of every calibration beacon. Therefore, the speed at which a mobile node can detect it has

---

[1] CaliBree calibration accuracy is designed to be independent of the number of uncalibrated nodes that operate the calibration procedure concurrently.

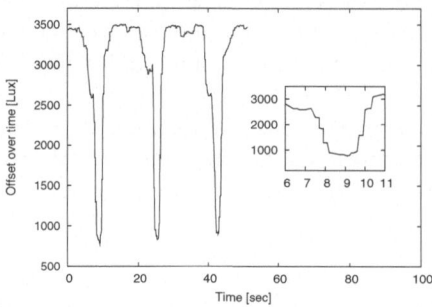
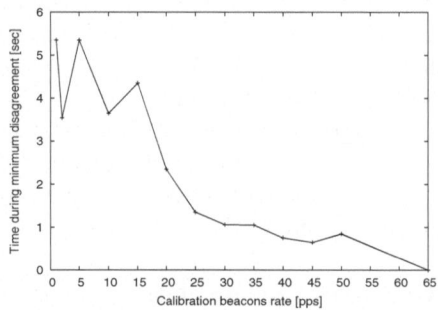

(a) Plot of the sensing contact time. An uncalibrated node experiences the same sensing environment as the ground truth node for a short time, even under favorable operating conditions (low speed).

(b) Plot of the amount of time an uncalibrated node overestimates its presence in the common sensing range of the ground truth node as a function of the calibration beacon transmission rate. The larger the transmission rate, the faster the consensus algorithm reacts to changed conditions and the lower the error.

Fig. 3.

left the common sensing range is proportional to the calibration beacon rate. Figure 3(b) shows a plot of the amount of time an uncalibrated node overestimates its presence in the common sensing range of the ground truth node as a function of the calibration beacon transmission rate. The overestimation presented here is calculated with respect to the common sensing range dwell time inferred when the beacon rate was 65 Hz, the maximum of our tested rates[2]. In order for the mobile node to immediately detect it has left the common sensing range, the calibration beacon rate should be infinite. Under more practical beacon rates and practical conditions (e.g., occasional packet loss), Figure 3(b) shows that when the calibration beacon rate is smaller than 25 beacons per second the consensus algorithm overestimates the dwell time in the common sensing area by 2.5 to 5.5 seconds, persisting in the minimum disagreement state even after leaving the common sensing range, and leading to inaccurate calibration results. For beacon rates larger than 25 pps the consensus algorithm updates the state faster and the estimated common sensing range dwell time is close to that given by the highest tested beacon rate. In our implementation we use a calibration beacon rate of 65 packets/sec. While this high data rate seems to be incurring high cost in terms of bandwidth and energy, we show that the calibration rendezvous lasts for few seconds (less than two) and it would be triggered only at the first time of usage of the sensing device and after long time scale (months or years) if needed due to sensor drift.

5.3   Node Calibration

In this section we show the performance of CaliBree when: *i)* the uncalibrated node comes across multiple co-located ground truth nodes, and *ii)* the uncalibrated node

---

[2] 65 packets/sec approaches the maximum possible packet transmission rate of the TinyOS networking stack on the Tmote Sky platform.

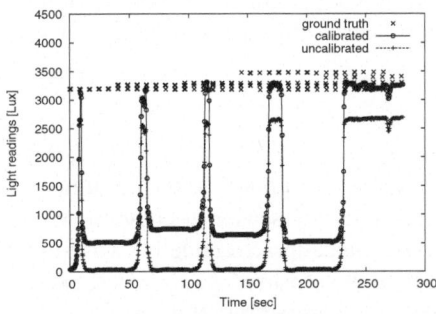 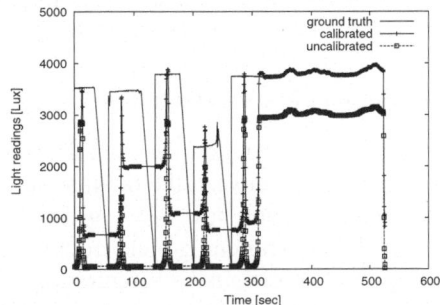

(a) Plot of the calibration performance when a set of five ground truth nodes sit within a $1m^2$ area.

(b) Plot of the calibration performance when the uncalibrated node sequentially encounters five different ground truth nodes.

Fig. 4.

encounters one ground truth node a time while being carried around the three floor office building.

Co-located ground truth nodes. In this experiment five ground truth nodes are co-located within a $1\ m^2$ area and are turned on sequentially at intervals of several seconds. The purpose of the experiment is to show the convergence of CaliBree when an uncalibrated node rendezvous with co-located ground truth nodes. The mobility of the uncalibrated sensor node carries it first towards the ground truth cluster, then away until five ground truth rendezvous have completed. After the fifth rendezvous the node remains co-located with the cluster for the duration of the experiment. The result is shown in Figure 4(a) where a comparison between the static ground truth data, the uncalibrated data, and calibrated sensor readings is reported. The values with the flat pattern on the upper part of the figure are the ground truth readings. As the number of active ground truth nodes increases, the calibration curve obtained with the best fitting line algorithm becomes more accurate and the final result when the fifth ground truth node is activated is that the calibrated data lays somewhere in between the highest and lowest ground truth data values. It is possible to see that precise calibration results can be obtained already after the first two rendezvous.

Sparse ground truth nodes. In the case of sparsely placed ground truth nodes across the building, the mobility of the uncalibrated node brings it through the sensing ranges of five different ground truth nodes. After the fifth rendezvous, the uncalibrated node remains near the fifth ground truth node to show that, after having computed the calibration curve from a data set of five data points from different locations, the uncalibrated node has achieved accurate calibration. In Figure 4(b), where again the flat segments in the upper part of the Figure represent the sensor readings of the ground truth nodes, we see that after the first few rendezvous the uncalibrated node starts producing accurate sensor readings. After the fifth rendezvous the CaliBree protocol is not run anymore and by placing the uncalibrated node near the last ground truth node for 200 seconds we observe that the uncalibrated node is accurately calibrated. In fact, as Figure 4(b) shows, the calibrated data curve overlaps the ground truth curve after the fifth rendezvous.

5.4 Sensor Nodes Orientation

In Section 4 we mention the need for adapting the ε value in Equation 1 according to the relationship between the orientation of the nodes. In support of this argument we show in this section how light sensing is impacted by the sensor orientation. We run an experiment with one ground truth node and one uncalibrated node in two different settings: *i)* a "horizontal" configuration where both the sensors face upwards, and *ii)* a "vertical" configuration where the uncalibrated node is tilted by 90 degrees and faces the ground truth node. In both the scenarios the distance between the uncalibrated and ground truth nodes in increased over time by 30 cm each measurement. The difference between the light sensed data of the two nodes for the horizontal and vertical configuration is plotted in Figure 5. It is shown that the vertical configuration doubles the common sensing range (from 30 cm to 60 cm). The experiment confirms the directional nature of the light sensor and shows that when an uncalibrated node is near a ground truth node the mutual orientation of the nodes matters. In the case where the sensors on the nodes have different orientations, a mechanism that reduces the weight of the ground truth sensed readings, like for example reducing the value of ε, could improve the performance of the calibration system. The mutual orientation between nodes could be inferred by compass/magnetometer readings for example. Assuming that the future generations of mobile sensor platforms will be equipped with compass sensors is reasonable considering the continuing advances in the embedded sensing technology and the increasing interest in providing smarter sensing devices for people-centric sensing applications [7].

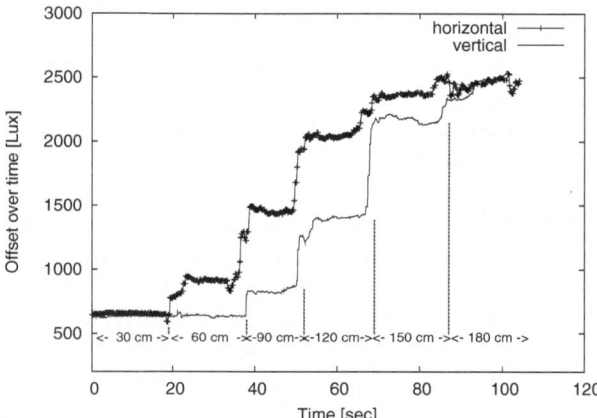

Fig. 5. Plot of the impact of the uncalibrated - ground truth nodes mutual orientation

6   Conclusions

We presented CaliBree, a distributed self-calibration protocol for mobile wireless sensor networks. CaliBree is a very promising technique and, to the best of our knowledge, it is an important first step towards the introduction of calibration algorithms for mobile sensing systems. CaliBree is scalable, robust, and self-adaptive to the dynamics of a mobile sensor network. We demonstrate through experimentation with real sensor devices the existence of the sensing factor which we believe will be one of the drivers in the design and implementations of protocols and applications in the mobile sensing systems domain. We also demonstrate that calibration can be achieved after rendezvous with less than five calibrated nodes. Thus, it can be considered a suitable technique to calibrate mobile sensor nodes in a scalable, lightweight and efficient way.

As part of the future research direction we intend to implement mechanisms to let the uncalibrated nodes infer the context of the ground truth nodes to make clear decisions about the suitability of the ground truth nodes to provide ground truth data. For context we mean those conditions, like for example mutual sensor orientations, that impact the overall calibration performance.

We also intend to implement and validate CaliBree on sensor enabled cellphones and finally, we plan to test the performance of CaliBree when the uncalibrated nodes move at a higher speed than pedestrian to verify the suitability of the protocol to vehicular mobility patterns as well.

Acknowledgments

This work is supported in part by Intel Corp., Nokia, NSF NCS-0631289, and the Institute for Security Technology Studies (ISTS) at Dartmouth College. ISTS support is provided by the U.S. Department of Homeland Security under award 2006-CS-001-000001, and by award 0NANB6D6130 from the U.S. Department of Commerce. The views and conclusions contained in this document are those of the authors and should not be interpreted as necessarily representing the official policies, either expressed or implied, of any funding body.

The authors would like to thank Shane Eisenman for his valuable technical feedback and great help in editing this manuscript.

References

1. Polastre, J., Szewczyk, R., Culler, D.: Telos: Enabling Ultra-Low Power Wireless Research. In: Proc. of IPSN/SPOT 2005, April 25-27 (2005)
2. TinyOS, http://tinyos.net
3. Campbell, A.T., Eisenman, S.B., Lane, N.D., Miluzzo, E., Peterson, R.A.: People-Centric Urban Sensing (Invited Paper). In: Proc. of 2nd ACM/IEEE Int'l Conf. on Wireless Internet, Boston (August 2006)
4. Nokia 5500 Sport Phone, http://www.nokia.com
5. SensorPlanet, http://www.sensorplanet.com
6. Nike + iPod, http://www.apple.com/ipod/nike

7. Eisenman, S.B., Miluzzo, E., Lane, N.D., Peterson, R.A., Ahn, G.A., Campbell, A.T.: The BikeNet Mobile Sensing System for Cyclist Experience Mapping. In: Proc. of SenSys 2007, Sydney, Australia, November 6-9 (2007)
8. Eisenman, S.B., Campbell, A.T.: SkiScape Sensing. In: Proc. of 4th ACM Conference on Embedded Networked Sensor Systems (SenSys 2006), Boulder, Colorado, November 1-3 (2006) (Poster abstract)
9. Olfati-Saber, R., Murray, R.: Consensus problems in networks of agents with switching topology and time-delays. IEEE Trans. Autom. Control 49(9), 1520–1533 (2004)
10. Culler, D., Estrin, D., Srivastava, M.: Guest Editors' Introduction: Overview of Sensor Networks. Computer 34 (2004)
11. Estrin, D., Girod, L., Pottie, G., Srivastava, M.: Instrumenting the world with wireless sensor networks. In: International Conference on Acoustics, Speech, and Signal Processing (ICASSP 2001), Salt Lake City, Utah (May 2001)
12. Estrin, D., Govindan, R., Heidemann, J., Kumar, S.: Next century challenges: scalable coordination in sensor networks. In: Proceedings of the 5th annual ACM/IEEE international conference on Mobile computing and networking, Seattle, USA (1999)
13. Chaintreau, A., Hui, P., Crowcroft, J., Diot, C., Gass, R., Scott, J.: Impact of Human Mobility on Opportunistic Forwarding Algorithms. IEEE Transactions on Mobile Computing 6(6), 606–620 (2007)
14. Abdelzaher, T., et al.: Mobiscopes for Human Spaces. IEEE Pervasive Computing - Mobile and Ubiquitous Systems 6(2) (April-June 2007)
15. Lowton, M., Brown, J., Finney, J.: Finding NEMO: On the Accuracy of Inferring Location in IEEE 802.15.4 Networks. In: Proc. of 2nd ACM Workshop on Real-World Wireless Sensor Networks (REALWSN 2006), Uppsala, Sweden, June 19 (2006)
16. Hull, B., et al.: CarTel: A Distributed Mobile Sensor Computing System. In: 4th ACM SenSys., Boulder, CO, USA (November 2006)
17. Ganti, R.K., Jayachandran, P., Abdelzaher, T.F., Stankovic, J.A.: SATIRE: a software architecture for smart AtTIRE. In: Proc. of MobiSys 2006, Uppsala, Sweden (2006)
18. CitySense. An Open, Urban-Scale Sensor Network Testbed, http://www.citysense.net
19. Tolle, G., et al.: A macroscope in the redwoods. In: Proc. of SenSys 2005, San Diego, California, USA (November 2005)
20. Szewczyk, R., Polastre, J., Mainwaring, A., Culler, D.: Lessons From A Sensor Network Expedition. In: Proc. of the First European Workshop on Sensor Networks (EWSN) (January 2004)
21. Ihler, A.T., et al.: Nonparametric belief propagation for self-calibration in sensor networks. In: Proc. of IPSN 2004, Berkeley, California, USA (2004)
22. Girod, L., Lukac, M., Trifa, V., Estrin, D.: The design and implementation of a self-calibrating distributed acoustic sensing platform. In: Proc. of SenSys 2006, Boulder, CO, USA (November 2006)
23. Taylor, C., et al.: Simultaneous localization, calibration, and tracking in an ad hoc sensor network. In: Proc. of IPSN 2006, Nashville, Tennessee, USA (2006)
24. Whitehouse, K., Culler, D.: Calibration as parameter estimation in sensor networks. In: WSNA 2002: Proc. of the 1st ACM international workshop on Wireless sensor networks and applications, Atlanta, Georgia, USA (2002)
25. Bychkovskiy, V., Megerian, S., Estrin, D., Potkonjak, M.: A Collaborative Approach to In-Place Sensor Calibration. In: Zhao, F., Guibas, L.J. (eds.) IPSN 2003. LNCS, vol. 2634, pp. 301–316. Springer, Heidelberg (2003)
26. Balzano, L., Nowak, R.: Blind calibration of sensor networks. In: In Proc. of IPSN 2007, Cambridge, Massachusetts, USA (April 2007)
27. LaMarca, A., et al.: Making Sensor Networks Practical with Robots. In: Pervasive 2002, Zurich, Switzerland, August 26-28 (2002)

28. Bychkovskiy, V., et al.: A Collaborative Approach to In-Place Sensor Calibration. In: Zhao, F., Guibas, L.J. (eds.) IPSN 2003. LNCS, vol. 2634, pp. 301–316. Springer, Heidelberg (2003)
29. Moses, R.L., Patterson, R.: Self-calibration of sensor networks. In: The Society of Photo-Optical Instrumentation Engineers (SPIE) Conference (2002)
30. Extech. SuperHeat Psychrometer RH350,
    http://www.extech.com/instrument/products/alpha/datasheets/RH350.pdf.
31. Extech. Dataloggin Light Meter 401036,
    http://www.extech.com/instrument/products/400_500/datasheets/401036.pdf.
32. Chipcon CC2431, http://www.ti.com/corp/docs/landing/cc2431/index.htm.
33. Shih, E., Bahl, P., Sinclair, M.J.: Wake on wireless: an event driven energy saving strategy for battery operated devices. In: Proc. of MobiCom 2002, Atlanta, Georgia, USA, September 23-28 (2002)

An Information Theoretic Framework for Field Monitoring Using Autonomously Mobile Sensors*

Hany Morcos[1], George Atia[2], Azer Bestavros[1], and Ibrahim Matta[1]

[1] Computer Science Department, Boston University, Boston, MA,
[2] Electrical and Computer Engineering Department, Boston University, Boston, MA

Abstract. We consider a mobile sensor network monitoring a spatio-temporal field. Given limited caches at the sensor nodes, the goal is to develop a distributed cache management algorithm to efficiently answer queries with a known probability distribution over the spatial dimension. First, we propose a novel distributed information theoretic approach assuming knowledge of the distribution of the monitored phenomenon. Under this scheme, nodes minimize an entropic utility function that captures the average amount of uncertainty in queries given the probability distribution of query locations. Second, we propose a correlation-based technique, which only requires knowledge of the second-order statistics, relaxing the stringent constraint of a priori knowledge of the query distribution, while significantly reducing the computational overhead. We show that the proposed approaches considerably improve the average field estimation error. Further, we show that the correlation-based technique is robust to model mismatch in case of imperfect knowledge of the underlying generative correlation structure.

1 Introduction

Early sensor network research assumed that sensors are static with very low computation and storage capabilities, and once deployed, these nodes are not likely to be recharged or moved. Hence, once separated from the network (*e.g.,* due to failure of nodes on the path to the rest of the network), nodes will remain disconnected until their batteries die. Sensor network technologies have matured to the degree that they are expected to be embedded in many platforms. Some of these platforms are mobile, *e.g.,* automobiles, handheld devices and wearable computers, giving rise to a rather new paradigm for sensor networks, which allows for the consideration of mobility, including the possibility of leveraging it for new classes of sensor network applications.

A paradigm, in which sensor networks are mobile, not only changes many traditional sensor network assumptions (*e.g.,* node isolation may be only temporary due to mobility), but also it gives rise to new applications, or to old applications under new settings. One such application is field monitoring. An extensive body of research studied this problem in the context of static sensor networks [8,21,17,13,23,7]. Dense node deployment is usually assumed. A unique party in the network (*i.e.,* the sink) is assumed to be

---

* This work has been partially supported by a number of National Science Foundation grants, including CISE/CSR Award #0720604, ENG/EFRI Award #0735974, CISE/CNS Award #0524477, CNS/NeTS Award #0520166, CNS/ITR Award #0205294, and CISE/EIA RI Award #0202067.

S. Nikoletseas et al. (Eds.): DCOSS 2008, LNCS 5067, pp. 332–345, 2008.

responsible for posing queries to the rest of the network. Flooding (whether network-wide or limited) is leveraged to discover the best forwarding paths to and from the sink. Lack of change in the network topology allows these paths to be useful for handling multiple queries, validating the cost of flooding.

Besides mobility, the field monitoring setting we consider in this paper is different from the above scenario. Specifically, sensor nodes are not viewed as reactive elements whose sole role is to sample a single location and respond to queries about this specific location. Rather, we view sensors as being embedded or attached to larger entities (*e.g.*, cars and handheld devices), which constitute points of interaction between the system and its users. As such, users may pose queries to the system, and get replies from the system through these points of interaction (nodes). As an example for this setup, consider a firefighter's backpack that contains a number of sensors (*e.g.*, temperature sensor, smoke sensor, carbon-monoxide sensor, *etc.*), along with a head-mounted display and a keyboard to allow interaction between each firefighter and the system [1]. In such a system, sensors could sample the environment in which firefighters work. Collected samples should be managed and stored in order to satisfy queries issued by firefighters to the system. A query can target any location in the scene, not only locations sampled by the inquirer. For example, if one firefighter needs to go to some location in the scene, then measurements of temperature, smoke levels, and concentration of carbon monoxide would prove valuable to this firefighter. Thus, the goal of the system is to provide an accurate estimation of the phenomenon of interest at the given query location. A defining characteristic of this system is the mobility pattern of the mobile hosts (firefighters in this example). This pattern is not governed by the need to optimize the system performance, rather it is governed by an overarching mission (*e.g.*, the need to save someone trapped in a room, or constraints due to how the fire progresses). This same setting applies equally well to a group of soldiers in a battlefield, or a group of researchers performing a study in some urban field.

Another important factor in the paradigm we consider in this paper is that, users may have specific preferences when posing queries to the system. Specifically, the spatial distribution of interest over the field might be skewed as opposed to uniform (*i.e.*, there might exist some zones in the field that users are likely to inquire about more frequently – *e.g.*, near exits). Also, different phenomena of interest (*e.g.*, temperature, and carbon monoxide) might have different interest distributions. Knowledge of such distributions can be leveraged to optimize the system performance.

We assume that in such systems the storage space of mobile nodes allotted to *each* phenomenon of interest is limited. This is a realistic assumption for two reasons: 1) considering the fact that data from different phenomena share the same storage space (or cache). Adding more sensor types increases the number of phenomena that the system is able to handle, but also increases contention over the limited memory available for storage. 2) As we alluded above the type of applications we target are *parasitic* applications; in the sense that these applications exploit mobility of the host and its resources (*e.g.*, storage of a firefighter wearable system) to provide some service. Hence, it is conceivable that, although the host might have plenty of storage, our target applications will be allowed access to a limited fraction of this storage. These two reasons motivate the need for a cache management algorithm. We assume that samples from different phenomena are independent, hence, solving the problem for one phenomenon is enough.[1]

To this end, in this paper we propose two cache management algorithms for tackling this problem. Our techniques aim to minimize some utility function that captures the average amount of uncertainty in queries given the distributional characteristics of query locations. Our contributions are as follows:

- Assuming knowledge of the entire spatio-temporal distribution of the target phenomenon, we develop an information-theoretic framework to optimize the cache content, and provide accurate answers to queries (Section 3).
- We propose a different approach based on optimizing a correlation-based function relaxing the stringent constraint of full distribution knowledge. We develop a strategy that only requires knowledge of the second order statistics of the phenomenon of interest. Furthermore, this technique lowers the required computational complexity (Section 4).
- We provide extensive performance evaluation of our techniques, showing (and quantifying the impact of) the various factors and parameters that affect performance (Section 6). We, also, study the robustness of the technique developed in Section 4 to model mismatch in case of imperfect knowledge of the correlation structure.

The rest of the paper is organized as follows. In Section 2 the setup and problem definition are provided. Details of the proposed techniques are presented in Sections 3 and 4 together with an analysis of their corresponding computational complexity. Based on these two cache management algorithms, we show how to design a cooperative scheme in Section 5, where nodes benefit from samples cached at their neighbors to obtain more accurate query estimates. We then present in Section 6 an evaluation of the cache management strategies for two phenomena generated using different processes. We provide a summary of related work in Section 7, discuss future work and conclude the paper in Section 8.

2 Problem Definition

We start with the problem definition along with a description of the system goal. The setup, system parameters, and notation we use are as follows:

- The system consists of n autonomously mobile nodes (*i.e.*, node mobility is not controlled by the system).
- Each node has a cache of size c.
- The nodes move in a field $\mathcal{F}$ with area $A = L \times L$.
- While roaming the field, sensor nodes sample a target phenomenon and this process continues for T time units.
- Location information is accessible to the sensor nodes, such that they can associate each sample with the location where it was collected.

---

[1] We leave the relaxation of this assumption to future work on this problem.

- We use capital letters to represent random variables and small letters to represent realizations of these random variables.
- $V_{\ell,t}$ is a random variable that represents the value of the field phenomenon at location ℓ and time t. $v_{\ell,t}$ denotes a realization of this random variable.
- We use the boldfaced letter $\mathbf{s}_t^i = [s_1, s_2, .., s_c] \in \mathbb{R}^c$ to denote the c-dimensional cache content vector of node i at time t. To simplify notation and since we would be generally referring to any arbitrary node i, we will drop the superscript i, unless it is not clear from the context. Note that any cached sample s_j corresponds to a field value v_{ℓ_j, t_j}, where ℓ_j is the location from which this sample was collected and t_j its corresponding time stamp.
- It is assumed that a query posed at any time instant τ inquiring about location ℓ targets the value of the field phenomenon $v_{\ell,\tau}$.
- The field phenomenon is fully characterized by a space-time multivariate probability distribution $p(\{v_{\ell,t}\}; \ell \in \mathcal{F}, 0 \leq t \leq T)$ with a $L^2 \times T \times L^2 \times T$ correlation matrix R, such that $R(v_{\ell_1,t_1}, v_{\ell_2,t_2})$ represents the correlation between two values of the phenomenon with space-time coordinates (ℓ_1, t_1) and (ℓ_2, t_2), respectively.
- Define the random variable $L(q)$ as the location which query q targets (called the *query target*). We assume that $L(q)$ follows some spatial distribution Q, where $Q(\ell(q))$ is the probability of querying field location $\ell(q)$. Q is assumed to be stationary. Similarly, we use $t(q)$ to denote the time at which query q was posed. Obviously, the best answer to q would be $v_{\ell(q),t(q)}$.

System Goal: After some warm-up time, each node in the system is expected to answer queries about the target phenomenon in the field. The query specifies some field location, the node is expected to provide an estimate of the phenomenon at the query target and the goal is to minimize the mean square estimation error (MSE) of the system's response. Hence, the nodes are required to maintain an efficient cache content to be able to answer queries reliably. In the next sections, we develop different strategies for cache management at the sensor nodes.

3   Information Theoretic Cache Management

In this section we develop an information theoretic strategy via which nodes locally update their caches based on knowledge of the space-time distribution of the phenomenon of interest.

3.1   DEBT Cache Maintenance Strategy

At each time instant, local decisions are made at the mobile nodes concerning which samples to keep, and whether or not a new sample should be acquired at the current location. These decisions are made so as to minimize an entropic utility function that captures the average amount of uncertainty in queries given the probabilistic query target distribution — hence the name of the strategy: Distributed Entropy Based Technique (DEBT). Specifically, at each time instant t, a node i greedily decides in favor of the cache content that minimizes the conditional differential entropy averaged over the query distribution Q, *i.e.*,

$$\mathbf{s_t} = \arg\min h(V_{L(q),t}/\mathbf{s_t}, L(q))$$

$$= \arg\min_{\mathbf{s_t} \in \mathcal{S}_t} \int_{\ell(q) \sim Q} Q(\ell(q)) h(V_{\ell(q),t}/\mathbf{s_t}, \ell(q)) \tag{1}$$

where, $\mathbf{s_t} \in \mathbb{R}^c$ is the cache content selected by node i at time t, and $h(V_{L(q),t}/\mathbf{s_t}, L(q))$ is the differential entropy of the values of the phenomenon, conditioned on a given cache content,[2] at the possible query locations $\ell(q)$ which follow a spatial distribution Q. $\mathcal{S}_t$ is the set of all possible decisions leading to all possible cache contents at node i at time t which is given by:

$$\mathcal{S}_t = \{\mathbf{s_t} : \mathbf{s_t} \in \mathcal{C}_{c,c+1}(\mathbf{s_{t-1}} \bigcup \{v_{\ell_t,t}\})\} \tag{2}$$

where $\mathcal{C}_{c,c+1}(\mathcal{A})$ denotes all the ($c+1$ choose c) possible combinations of the elements of a set $\mathcal{A}$ and $v_{\ell_t,t}$ denotes the value of the phenomenon at the current location of the i-th node, ℓ_t.

The expression above simply enumerates all the possible cache contents at time t; the options being to drop any of the samples from time $t-1$ and acquiring the new sample at the current location of node i, or just keep the old set of samples.

The intuition behind DEBT is that a node always keeps a cache content that minimizes the uncertainty in the values of the phenomenon (captured by the conditional entropy) given the knowledge of the spatial distribution of the query targets over the field of interest. It might well be true that an old sample taken at a specific location is more valuable, and hence is worth caching than a newer sample taken at a different location given the aggregate effect of the spatial query distribution and the spatio-temporal distribution of the phenomenon.

It is worth mentioning that the computation of $h(V_{\ell(q),t}/\mathbf{s_t})$ (Eq.3 [5]) requires knowledge of the posterior density $p(v_{\ell(q),t}/\mathbf{s_t})$, which can be generally obtained by proper marginalization of the full space-time distribution. For the Gaussian case, this simplifies to a computation of the conditional mean and variance $\mu_{v_{\ell(q),t}/\mathbf{s_t}}$ and $\lambda_{v_{\ell(q),t}/\mathbf{s_t}}$.

$$h(V_{\ell(q),t}/\mathbf{s}) = - \int_{v_{\ell(q),t}} p(v_{\ell(q),t}/\mathbf{s}) \ln p(v_{\ell(q),t}/\mathbf{s}) dv_{\ell(q),t} \tag{3}$$

3.2 Least Square Error (LSE) Query Response Strategy

To answer a posed query q, a node computes an estimate of the phenomenon at the query target given its cache content. Given the knowledge of the space-time distribution, it would be natural to resort to a Bayesian Least Square Estimate (BLSE), which is given by the conditional expectation of the posterior density, to minimize the mean

---

[2] Note that the differential entropy $h(V_{L(q),t}/\mathbf{s})$ that we use in the minimization of Equation(1) is conditioned on a given realization of the cache content. That is to say, no averaging is taken over the conditioning random vector since we are dealing with real-time selection of the samples. This is clearly different from the standard quantity $h(V_{L(q),t}/\mathbf{S})$ with $\mathbf{S}$ being a random variable.

square estimation error. Hence each node's task is to compute the expected value of the phenomenon at q given its cache content $\mathbf{s}$, that is:

$$\hat{V}_{\ell(q),t(q)} = E[V_{\ell(q),t(q)}/\mathbf{s}] \tag{4}$$

where $\hat{V}_{\ell(q),t(q)}$ is the node estimate. Again we point out that this generally requires the computation of the posterior density $p(v_{\ell(q),t(q)}/\mathbf{s_t})$. Under Gaussian assumptions, the BLSE estimate in Eq.(4) is always linear in the cache content, that is the BLSE is equal to the Linear Least Square Estimate (LLSE). For general distributions, the computational complexity could be reduced if we only restrict ourselves to linear functions of the cache content, $i.e.$ LLSE, which would only require knowledge of the second-order statistics of the phenomenon. Note that the LLSE, $\hat{X}_{LLSE}$, of a random variable X with mean μ_X, given a random vector $Y = y$, with mean vector μ_Y is given by [22]:

$$\hat{X}_{LLSE} = \mu_X + \Lambda_{XY}\Lambda_Y^{-1}(y - \mu_Y) \tag{5}$$

where Λ_{XY} denotes the cross-covariance between X and Y, and, Λ_Y is the covariance matrix of the observation vector Y. While the DEBT/LSE techniques outlined in this section are expected to yield accurate performance, they are not practical. Specifically, we note the following two types of limitations on DEBT practicality:

- *Informational Limitations:* DEBT assumes knowledge of the entire distribution of the target phenomenon. Such information may not be always available, or if available (*e.g.*, through historical monitoring of the phenomenon of interest), it may not be accurate.
- *Computational Limitations:* In order to provide optimized decisions about whether or not to sample visited field locations, and how to manage the cache, DEBT calculates the conditional differential entropy of the query distribution Q given any cache setting. This requires performing multiple numerical integration operations, which might not be always suitable due to the limited computational capabilities at the sensor nodes.

This motivates taking a different approach that is less-demanding in terms of knowledge about the spatio-temporal field. In the next section, we propose a more practical (yet quite competitive) strategy that only requires knowledge of the correlation structure, *i.e.*, second-order statistics.

4 Correlation-Based Cache Management

In this section, we propose a Correlation-Based Technique (CBT) as a practical alternative to the DEBT approach presented before.

CBT averts the limitations of DEBT by only assuming knowledge of the space-time correlation structure of the field phenomenon R. Namely, instead of calculating the conditional entropy to make caching decisions, CBT decides which samples to cache using only the correlation structure of the target phenomenon R. Notice that defining R implies only knowledge of the second-order statistics of the target phenomenon, as opposed to knowledge of the entire distribution in case of DEBT. Like DEBT, the crux of the CBT technique is to be able to assign a measure of utility capturing knowledge about the field to any given set of samples $\mathbf{s} = \{s_1, s_2, .., s_c\}$ with respect to the query distribution Q. Then, it retains the set of samples that maximizes the utility. First, we need to assign a measure of utility $u(q, \mathbf{s})$ to a set of samples $\mathbf{s}$ with respect to a specific query q with location $\ell(q)$, and time $t(q)$. Then by averaging $u(q, \mathbf{s})$ over the spatial distribution Q, we get a weighted information metric over the entire field, $M(Q, \mathbf{s})$. More specifically, for a query q, we gauge the utility of $\mathbf{s}$ with respect to q as follows:

$$u(q, \mathbf{s}) = \frac{Q(\ell(q))}{\Lambda_{q|\mathbf{s}}} \tag{6}$$

Averaging $u(q, \mathbf{s})$ over Q, we get

$$M(Q, \mathbf{s}) = \int_Q u(q, \mathbf{s}) = \int_{\ell \sim Q} \frac{Q(\ell)}{\Lambda_{q|\mathbf{s}}} \, d\ell \tag{7}$$

where $Q(\ell(q))$ is the probability of querying field location $\ell(q)$, and $\Lambda_{q|\mathbf{s}}$ is the conditional covariance of $q|\mathbf{s}$, given by

$$\Lambda_{q|\mathbf{s}} = \Lambda_q - \Lambda_{q,\mathbf{s}} \Lambda_{\mathbf{s}}^{-1} \Lambda_{q,\mathbf{s}}^T \tag{8}$$

where Λ_q is the variance of the stationary process, $\Lambda_{q,\mathbf{s}}$ is the cross-covariance between q and $\mathbf{s}$, and $\Lambda_{\mathbf{s}}$ is the covariance matrix of the cache content $\mathbf{s}$. Notice that calculation of $\Lambda_{q|\mathbf{s}}$ only requires knowledge of the correlation matrix R. Then, CBT makes its caching decisions by maximizing the total utility over the choice of possible cache content $\mathbf{s}$ (*i.e.*, $\max_{\mathbf{s}} M(Q, \mathbf{s})$).

5   Nodes Cooperation

So far we have described operation of a single node. However, in a mobile network of numerous nodes, cooperation between nodes could be engineered to yield a better performance. In this paper, we limit our attention to cooperation concerning query response. This is done as follows. Whenever a node i gets a query q, i broadcasts q to its direct neighbors. Upon receiving the query, each neighbor j of i estimates its answer based on its local cache content, then, submits the estimate back to i along with a measure of confidence in this answer. Node i performs the same task, and receives query replies from its neighbors. The answer with the highest confidence is used as the query response. In our setting we use the conditional covariance $\Lambda_{q|\mathbf{s}}$ (Equation 8) as the measure of confidence in the estimated answer. The intuition is that a lower conditional covariance corresponds to less uncertainty about the query. Notice that, the radius of flooding the query could be increased to values larger than one (*i.e.*, consult nodes beyond direct neighbors), however, we choose not to do this in order to avoid query flooding and its associated communication overhead.

Also, notice that, while we chose to limit nodes cooperation to the query handling (*i.e.*, estimation) plan, cooperation between nodes could be done on different plans, for example, the sample caching (*i.e.*, decision-making) process. In this case, nodes would take decisions as to which samples to cache and which ones to evict based not only on the contents of local cache, but on the contents of neighboring caches as well. This would require broadcasting the cache content (or a summary of it thereof) to neighbors, which is a costly process in terms of power. Also, performing cooperation on the decision making plan requires more coordination in presence of mobility, since the set of neighbors changes with time. In this paper we evaluate the first option, and leave investigation of the second to future work.

6 Performance Evaluation

In this section we evaluate the performance of the different proposed cache management techniques. We start in Subsection 6.1 with a description of the data generation models we used to generate the input data. In Subsection 6.2, we provide the details of our evaluation methodology. Next, in Subsection 6.3, we introduce the performance metrics we use in our evaluation. Finally, we present the results of our experiments in Subsections 6.4, and 6.5.

6.1 Data Generation Model

In this subsection, we describe the two data generation models we used in this study.

Model 1: A Gaussian Phenomenon: In the first model, the underlying space-time distribution of the phenomenon is a multivariate Gaussian. Thus, the field distribution is fully captured by the mean vector and the joint spatio-temporal correlation (STC) matrix R, $L^2 \times T \times L^2 \times T$. To generate the field, we first generate the data to satisfy the spatial correlation using the standard Cholesky decomposition transformation by pre-multiplying a matrix of independent Gaussian random variables by the square root of the desired spatial covariance [18]. Each individual temporal signal associated with a given location is then filtered using a temporal filter to provide the correct spectral shape. This approach results in an STC covariance structure where the off-diagonal blocks are scalings of the diagonal blocks with a scaling factor that depends on the corresponding time lag. Here we note that other methods based on techniques described in [6] could also be used for generation of fields with arbitrary joint space-time correlation.

Model 2: A Random Phenomenon: In the second model, the generated data does not follow a Gaussian distribution. The purpose of this experiment is to study the performance of the CBT technique proposed in Section 4, which only requires knowledge of the second-order statistics, when the underlying field follows an arbitrary distribution. We generated data that satisfies a desired STC by first applying a spatial transformation to a vector V of uniformly distributed random variables, and then by filtering the resulting vector through an autoregressive (AR) digital filter to introduce the desired temporal correlation. The coefficients of the autoregressive filter were obtained using the standard Levinson-Durbin algorithm which takes as input the targeted correlation for the different time lags, and outputs the filter coefficients for the specified order [9]. Since the driving noise (V) we used in the first place is non-Gaussian, the resulting process is also non-Gaussian, and only matches the second-order statistics requirements.

6.2 Simulation Model and Methodology

We assume that n nodes, each with a cache of size c, perform a random walk in a 2-D field of dimensions $L \times L$. At every time unit, each node decides whether or not to sample its current location. This decision is made based on the utility that this new sample provides compared to utility of the original cache content. If the new sample does not increase the utility of the cache, it is not kept in the cache. Otherwise, one of the old samples that provides the least utility is evicted in favor of the newly acquired one. After allowing a warmup period of w time units, each node is required to answer a query every time unit. The query specifies a location in the field, referred to as *query target*. A query answer is an estimate of the value of the phenomenon at the query target given each node's locally cached field samples. Notice that each node is asked an independent query whose target is drawn from the spatial query distribution Q. This distribution is assumed to be a bivariate normal distribution whose mean is the center of the field, and variance is $\sigma_Q^2 \times I$, where I is the identity matrix of size 2×2. The answer to any query is calculated using Eq. (5), where Y in Eq. (5) is the vector of samples cached by the queried node.

In the experiment with the Gaussian phenomenon, evaluation of the posterior densities by the mobile nodes only required evaluation of a mean vector and a covariance matrix which capture the entire distribution. However, in the non-Gaussian scenario, the computational complexity of DEBT becomes prohibitively expensive, especially for large cache sizes. The reason is that the evaluation of the posteriors requires marginalization of the space-time distribution over the range of the variables of interest for the entire duration of the evaluation (*i.e.*, length of the simulation in time units). Hence, in the experiment with the Random phenomenon, we only evaluate CBT.

In order to assess the robustness of CBT to model mismatch, we also conducted another experiment in which noise is added to the second-order statistics knowledge used by the nodes for managing their caches (to reflect uncertainty in correlation knowledge). We then evaluate the performance for different signal-to-noise ratios (SNR), where SNR is defined as:

$$SNR = 10\, log_{10} \frac{\sigma^2}{\sigma_{noise}^2} \tag{9}$$

where σ^2 is the variance of the phenomenon, and the added noise is Gaussian with mean $\mu = 0$, and variance σ_{noise}^2. We experimented with SNR's = 2db, and 15db.

To quantify the gains achieved by the proposed techniques, we compare them to random caching, which provides us with a lower bound on performance. With random caching, at every time unit, each node randomly decides whether or not to sample its current location. If a node decides to sample its current location, and its cache is full, it randomly chooses one of its local samples to be evicted to accommodate the newly acquired sample.

In the following evaluation, we set the default value of the parameters of our simulation and data models as follows. $L = 8$, $c = 10$, $n = 5$, simulation time = 100 time units, warmup time $w = 50$ time units, variance of the Gaussian phenomenon σ_G^2 = 50, variance of the random phenomenon σ_R^2 = 50, and variance of the spatial query distribution σ_Q^2 = 4. The default mobility model is a random walk on a 2D discrete field, under which, each node is initially placed at random location in the field. Then at every time unit, each node moves to one of its four neighboring locations with the same probability (*i.e.,* 0.25 for each location).

6.3   Performance Metrics

The main performance metric we used in our evaluation is the Mean Squared Error (MSE): Given a specific query, a node returns an estimate of the value of the phenomenon at the query location. We then measure the mean squared error associated with this estimate. Thus, given a query q at time t whose target is $\ell(q)$, the MSE in the estimation of q is:

$$MSE = E[(V_{\ell(q),t} - \hat{V}_{\ell(q),t/\mathbf{s_t}})^2] \tag{10}$$

We calculate the MSE for each query received by each node after the warmup period, then we report the average of 20 independent simulation runs.

We start by showing results of a single node as a function of the cache size c, and the variance of the query distribution σ_Q^2. Then we show results of cooperation between a number of nodes. More results can be found in the extended version of this paper [15].

6.4   Single-Node Results

Effect of Cache Size: Figure 1 (left) shows the effect of cache size on the MSE of the different considered strategies for a Gaussian and non-Gaussian phenomena. Intuitively, as the cache size increases, the better the MSE performance of CBT and DEBT since a larger cache size implies a better reconstruction of the phenomenon by the queried nodes. DEBT has a lower MSE compared to CBT, however, CBT's performance is very competitive at a much lower computational cost.

Similar effects could also be observed for the non-Gaussian phenomenon (Figure 1 right), regarding the efficiency of CBT. CBT outperforms random caching by a factor of two orders of magnitude. As expected, adding noise to the correlation structure of the phenomenon (*i.e.,* decreasing SNR), degrades the CBT performance. However, even with SNR of as low as 2db, CBT still outperforms random caching with a significant gain.

Query Spatial Distribution Variance: Figure 2 quantifies the effect of a larger variance, σ_Q^2, for the query distribution on the MSE for both Gaussian and non-Gaussian phenomena. Intuitively, a larger variance implies more uncertainty in the target query locations for a fixed cache size and a fixed number of nodes, which explains the decrease in estimation quality for the various schemes.

In case of a Gaussian phenomenon (Figure 2 left) both DEBT and CBT have MSE that is an order of magnitude lower than that of random caching. While in case of a non-Gaussian phenomenon (Figure 2 right), CBT achieves a huge improvement over random caching, with respect to the MSE. Adding noise to the correlation information decreases the performance of CBT, but is still much better than random caching.

6.5   Multi-node Results

In the following experiments, we gauge the performance improvement due to cooperation between multiple nodes, as we explained it in Section 5 for a non-Gaussian

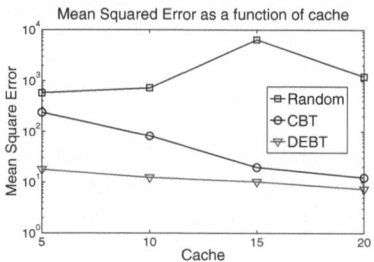 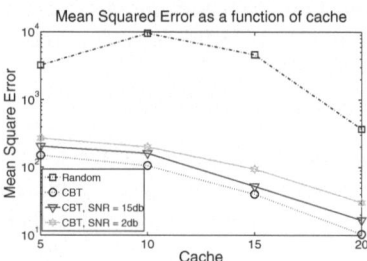

Fig. 1. Performance as a function of the cache size for a Gaussian phenomenon (left), and a non-Gaussian phenomenon (right)

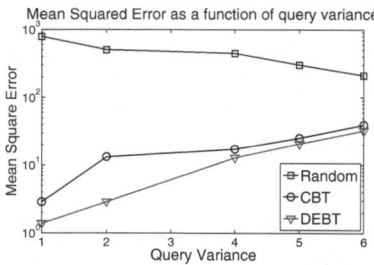 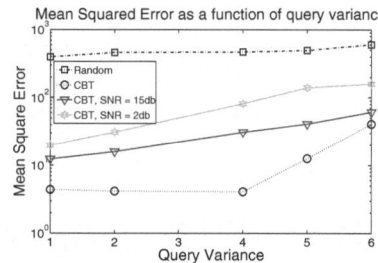

Fig. 2. Performance as a function of the variance of the query distribution for a Gaussian phenomenon (left), and a non-Gaussian phenomenon (right)

phenomenon. Intuitively, we expect cooperation between nodes to improve the performance of all techniques, where the degree of improvement depends on the density of the nodes. We study this effect by varying the cache size and the number of nodes in the field. We also plot the cooperation gain, which is defined as the ratio between MSE from experiments with one node to MSE of the same node when there are n cooperating nodes in the network. In [15], we show results of varying the variance of the distribution of query targets. In the following experiments, $n = 5$, and communication range = 8.

Effect of Cache Size: Figure 3 shows the effect of cache size on the MSE of the different considered strategies for a non-Gaussian phenomenon. The improvement of MSE due to cooperation is evident. It is clear that, after increasing the cache size to a certain point, cooperation causes the gap between random and CBT to shrink. The reason is that, at this point, there is enough storage capacity in the system, such that the performance of a smart algorithm and that of a naive algorithm seem to be close. However, the improvement of performance comes at a cost of added communication overhead. This is an important factor in system design. It implies that, in dense systems where nodes are not power-limited, a smart caching algorithm is not the only option to consider. However, in sparse systems, or in systems where nodes are power-constrained, applying a smart caching algorithm makes a noticeable difference in performance.

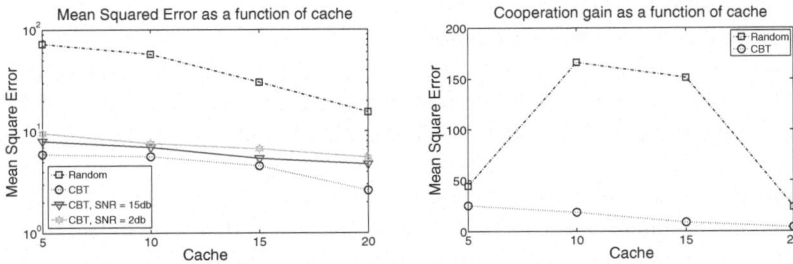

Fig. 3. Performance of multiple nodes as a function of the cache size for a random phenomenon (right), cooperation gain (ratio of MSE with a single node and with n nodes)

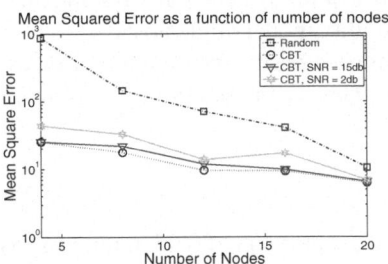

Fig. 4. Performance of multiple cooperative nodes as a function of the number of nodes n

Effect of Number of Nodes: Figure 4 shows the effect of varying the number of nodes, n, on the MSE of CBT and random caching for a non-Gaussian phenomenon. Increasing the number of nodes increases the amount of cooperation between nodes, and the storage capacity of the entire system. This improves the estimation by all nodes. Random caching has noticeable improvement as we increase the number of nodes. This trend matches the expectation that when storage is abundant, the caching algorithms make a minor difference. However, for all the parameter ranges we experimented with, CBT, even with noisy versions, performs better than random caching.

7 Related Work

The main goal of data placement in sensor networks is to minimize the access cost [16,20], where cost is quantified in terms of communication energy.

In order to save energy in the context of caching, Kotidis [10] tries to optimize energy consumption by trying to put some sensor nodes to sleep mode, without affecting the query ability of the network. This is done by building a correlation model for the samples of sleeping nodes in neighboring active nodes. However, the built model is only local and can not be used to answer general queries about the entire network. It also involves packet exchange and fitting neighbors' data to a linear model. In this paper, given knowledge of the spatio-temporal correlation model, we use it to locally (with no packet exchange) answer queries about the entire network.

In all of the above efforts, the entire network is assumed to be static, while our work considers mobility, which is a harder problem.

Spatio-temporal queries have been studied in static networks with both static [4] and mobile [12] sinks. Our model is different in that, queries are handled only locally. Moreover, the temporal dimension to the problem is manifested in the correlation structure of the phenomenon.

Caching and replication have been considered in ad hoc networks [24,11,19]. Nodes are assumed to be interested in a fixed set of objects such that each object has a well-defined source. In our case, queries may target field locations that may not have been sampled by any node.

Leveraging mobile sensor networks to perform field monitoring has been studied [2,3,25]. While these efforts assume control over the mobility pattern and optimize it in order to maximize the utility of the system, our work maximizes the utility of the cache given the uncontrolled mobility model of the hosts.

Finally, we utilized information theory to assign a measure of merit to any set of samples. Information theory has been used in similar problems [14].

8   Conclusion

In this paper we focused on the problem of field monitoring using autonomously mobile sensor nodes. Nodes make local decisions about whether or not to sample their current location and how to manage their limited storage. We proposed a distributed entropic based technique (DEBT) to solve this problem. DEBT assumes knowledge of the entire distribution of the target phenomenon, and leverages this knowledge to make decisions about the cache management. DEBT has two major limitations: 1) high computational complexity, and 2) knowledge of the entire distribution of the target phenomenon is not always feasible. We then proposed CBT, a more practical approach, which assumes knowledge of only second-order statistics of the target phenomenon. CBT has a much lower computational complexity, and very competitive performance. We evaluated both techniques and showed that the resulting gains in MSE are substantial for both Gaussian and random phenomena. Furthermore, CBT still delivers very good estimation of the field, even when its knowledge about the correlation structure is not perfect.

We intend to extend the model we presented here to incorporate node cooperation on the caching (*e.g.,* decision making) plan, such that nodes can benefit from the knowledge attained by their neighbors in sample management. We also intend to study the effect of different mobility models on the performance of different cache management techniques.

References

1. Wireless sensor system guides urban firefighters,
   http://mrtmag.com/mag/radio_wireless_sensor_system/
2. Wang, G., Cao, G., La Porta, T.F.: Movement-assisted sensor deployment. IEEE Transactions on Mobile Computing 5(6), 640–652 (2006)

3. Batalin, M.A., Rahimi, M., Yu, Y., Liu, D., Kansal, A., Sukhatme, G.S., Kaiser, W.J., Hansen, M., Pottie, G.J., Srivastava, M., Estrin, D.: Call and response: experiments in sampling the environment. In: SenSys (2004)
4. Coman, A., Nascimento, M.A., Sander, J.: A framework for spatio-temporal query processing over wireless sensor networks. In: DMSN (2004)
5. Cover, T.M., Thomas, J.A.: Elements of Information Theory. Wiley Series in Telecommunications and Signal Processing. Wiley-Interscience, Chichester (2006)
6. Hatke, G.F., Yegulalp, A.F.: A novel technique for simulating space-time array data, Pacific Grove, CA, USA, vol. 1, pp. 542–546 (2000)
7. Heinzelman, W.R., Chandrakasan, A., Balakrishnan, H.: Energy-efficient communication protocol for wireless microsensor networks. In: HICSS (2000)
8. Intanagonwiwat, C., Govindan, R., Estrin, D.: Directed diffusion: a scalable and robust communication paradigm for sensor networks. In: MobiCom 2000, pp. 56–67 (2000)
9. Kay, S.: Modern Spectral Estimation: Theory and Application. Prentice-Hall, Englewood Cliffs (1988)
10. Kotidis, Y.: Snapshot queries: Towards data-centric sensor networks. In: The 21st International Conference on Data Engineering (ICDE), Tokyo, Japan (April 2005)
11. Lau, W.H.O., Kumar, M., Venkatesh, S.: A cooperative cache architecture in support of caching multimedia objects in manets. In: WOWMOM, pp. 56–63 (2002)
12. Lu, C., Xing, G., Chipara, O., Fok, C., Bhattacharya, S.: A spatiotemporal query service for mobile users in sensor networks. In: ICDCS 2005, pp. 381–390 (2005)
13. Madden, S., Franklin, M.J., Hellerstein, J.M., Hong, W.: The design of an acquisitional query processor for sensor networks. In: SIGMOD 2003, pp. 491–502 (2003)
14. Marco, D., Duarte-Melo, E.J., Liu, M., Neuhoff, D.: On the many-to-one transport capacity of a dense wireless sensor network and the compressibility of its data (2003)
15. Morcos, H., Bestavros, A., Matta, I.: An information theoretic framework for field monitoring using autonomously mobile sensors. Technical Report BUCS-TR-2008-003, Computer Science Department, Boston University, 111 Cummington Street, Boston, MA 02135 (January 2008)
16. Prabh, K.S., Abdelzaher, T.F.: Energy-conserving data cache placement in sensor networks. ACM Trans. Sen. Netw. 1(2), 178–203 (2005)
17. Ratnasamy, S., Karp, B., Yin, L., Yu, F., Estrin, D., Govindan, R., Shenker, S.: Ght: a geographic hash table for data-centric storage. In: WSNA, pp. 78–87 (2002)
18. Rubinstein, R.Y.: Simulation and the Monte Carlo Method. John Wiley & Sons, Chichester (1981)
19. Sailhan, F., Issarny, V.: Cooperative caching in ad hoc networks, pp. 13–28 (2003)
20. Sheng, B., Li, Q., Mao, W.: Data storage placement in sensor networks. In: MobiHoc 2006, pp. 344–355 (2006)
21. Shenker, S., Ratnasamy, S., Karp, B., Govindan, R., Estrin, D.: Data-centric storage in sensornets. SIGCOMM Comput. Commun. Rev. 33(1), 137–142 (2003)
22. Trees, H.L.V.: Detection, Estimation, and Modulation Theory, Part I. John Wiley and Sons, Chichester (2001)
23. Ye, F., Luo, H., Cheng, J., Lu, S., Zhang, L.: A two-tier data dissemination model for large-scale wireless sensor networks. In: MobiCom 2002, pp. 148–159 (2002)
24. Yin, L., Cao, G.: Supporting cooperative caching in ad hoc networks. In: Infocom, Hong Kong, March 2004, IEEE Infocom (2004)
25. Zou, Y., Chakrabarty, K.: Sensor deployment and target localization based on virtual forces, San Francisco, CA, USA (2003)

Coverage Estimation in the Presence of Occlusions for Visual Sensor Networks

Cheng Qian[1] and Hairong Qi[2]

[1] Viatronix Inc, Stony Brook, New York, 11790
cqian@Viatronix.com
[2] University of Tennessee, Knoxville, TN, 37996
hqi@utk.edu

Abstract. Visual coverage is an essential issue in the research on visual sensor networks. However, because of the presence of visual occlusions, the statistics of visual coverage blend the statistics of nodes and targets and are extremely difficult to derive. By assuming the deployment of nodes as a stationary Poisson point process and ignoring boundary effects, this paper presents the first attempt to estimate the probability that an arbitrary target in the field is visually k-covered. The major challenge for the estimation is how to formulate the probability (q) that a node captures a target in its visual range. To tackle this challenge, we first assume a visual detection model that takes visual occlusions into account and then derive several significant statistical parameters of q based on this model. According to these parameters, we can finally reconstruct the probability density function of q as a combination of a Binomial function and an impulse function. With the estimated coverage statistics, we further propose an estimate of the minimum node density that suffices to ensure a K-coverage across the field.

1 Introduction

With recent advances in CMOS imaging devices, system-on-chip (SOC) technologies, and wireless communications, the concept of visual sensor networks (VSNs) surfaced and has been attracting increasing attention. In a VSN, each node carries sensing, computing, and wireless communication capacities. Once deployed in a certain field, these nodes are able to automatically form an ad-hoc network and collaboratively execute various vision tasks such as environmental surveillance, object tracking, and remote videography [1] [2] [3].

Coverage is always an essential issue for sensor networks. Having a proper analysis of sensing coverage paves the way to solving problems like deciding the minimum node density, designing efficient sleep schedules, and optimizing software performances. However, this issue is very difficult to tackle for VSNs. It is because the occurrence that a node visually senses a target not only depends on whether the target stands inside the field of view(FOV) of the node but also on whether nearby targets do not block the view of the target of interest. Therefore, for those applications where targets are crowded and occlusions ubiquitously exist, an accurate estimation of visual coverage should not only deal with the deployment of nodes but also the spatial distribution of targets.

S. Nikoletseas et al. (Eds.): DCOSS 2008, LNCS 5067, pp. 346–356, 2008.

The discussion in this paper is set in a context where a large population of targets is crowdedly located across a widespread geographical area where a large population of nodes has been deployed. For instance, an ad-hoc VSN is deployed across the Times Square to track pedestrians, or numerous nodes are air-dropped into a desert to detect adversary vehicles. Within this context, it is reasonable to ignore the boundary effect and assume both targets and nodes are uniformly distributed across the field. Thus, we can evaluate the coverage by the probability of the number of the nodes that an arbitrary target in the field is visually captured by. To the best of our knowledge, this paper is the first that derives an explicit estimate of visual coverage for a randomly deployed visual sensor network. Based on this estimate, we can further obtain an estimate of the minimum node density that is sufficient to ensure a K-coverage across the field (i.e. each target is simultaneously captured by no less than K nodes). For simplicity, we assume all the visual sensors in the network are mounted at a same height and pointed horizontally. Therefore we can ignore the vertical dimension and discuss the coverage problem on a 2D plane spanned by focal points of visual sensors.

1.1 Related Works

This coverage issue has been extensively discussed for sensor networks formed by scalar sensor nodes such as acoustic nodes, where a simple sensing model is assumed. In this sensing model, the sensing region of each acoustic node is an isotropic disc with radius ρ, and any target standing within the disc can be sensed with the amplitude of the receiving signal decreasing as the distance to the node increases. If we further assume the deployment of the acoustic nodes as a stationary Poisson point process with density λ and ignore the boundary effect, the coverage of the network follows a Poisson distribution with mean $\pi\rho^2\lambda$. Wan et al. [4] [5] reevaluated the probability by taking the boundary effect into account. Ca et al. [6] [7] extended the discussion to the situation when nodes and targets are mobile and when a sleep scheduling is employed. Wang [8] upgraded "being sensed" to "being accurately localized" and provided a corresponding conservative estimate of the necessary node density. Wang [9] also found that the covered region is virtually increased and the necessary node density can be reduced if neighboring nodes can cooperate with each other to fulfill the localization task. For some applications, it is more appropriate to evaluate the coverage as the maximal or minimal exposure of a path that traverses through the field. That way, the coverage is not only related to local node densities in different regions but also the node density coherence across the field. Relevant research can be found in [10] [11] [12].

Compared to acoustic sensors, the sensing model of visual nodes is much more complicated. First of all, the sensing region becomes directional. The field of view (FOV) of most visual sensors is less than $180°$. If we model the deployment of nodes as a stationary Poisson point process and ignore the boundary effect, the coverage of a directional sensor network will still follow a Poisson distribution but with mean equal to $\pi\rho^2\theta\lambda$, where ρ and θ respectively denote the range and angle of the FOV. Note that Ai [13] and Adriaens [14] also examined the coverage problem for directional sensor networks respectively from the perspective of linear programming and Voronoi geometry. Meanwhile, Isler [16] and Yang [15] applied their coverage analysis to derive the number of relevant directional-sensing nodes that suffice to guarantee accurate estimation respectively for the purpose of target localization and occupancy reasoning. Secondly, visual occlusions between targets come into existence, and a target of interest in the FOV of a visual node may be blocked by other targets standing between the target and the node. This means that "being visually captured by a node" depends not only on whether the target stands in the FOV of the node but also on the locations of neighboring targets. As a consequence, the visual coverage blends the statistics of the nodes and the targets and is extremely difficult to estimate.

As a matter of fact, if we assume the locations of the nodes conform to a stationary Poisson point process, the statistics of visual coverage can be easily obtained once the probability of the event that a node captures an arbitrary target within its visual range, say q, is derived. As discussed above, the probability q is a random value with respect to the random locations of targets. Therefore, our challenging work comes down to deriving the probability density function of q, i.e. $f(q)$. In this paper, we first assume a detection model in which we consider targets as 2D discs on the plane and extend an occlusion zone from a node to each target within its visual range. Obviously, q is equal to the probability that the node is oriented towards the arbitrary target and their in-between occlusion zone is clear of other targets. Then, instead of running into the complication of pursuing an exact form of $f(q)$, we try to derive several significant parameters of q such as its maximum, minimum, and expectation and their corresponding density function values. Finally, we construct an approximate probability density function based on these parameters, i.e. $\tilde{f}(q)$. By using $\tilde{f}(q)$, we can easily obtain the probability that an arbitrary target in the field is covered by k nodes and furthermore an estimate of the minimum node density that ensures a K-coverage across the field.

1.2   Paper Organization

Section 2 presents a detection model that takes visual occlusions into account. Section 3 presents the visual coverage estimate in the absence of visual occlusions. Section 4 presents the entire derivation of the coverage estimate in the presence of visual occlusions. Based on the coverage analysis, Section 5 provides an estimate of the minimum node density that suffices to ensure certain visual coverage. The paper is concluded in Section 6.

2   Visual Detection Model

We consider the 2D plane (sensing field) is very large and ignore the boundary effect. We model each node as an infinitesimal point on the plane and ignore the displacement between the focal point and the location of a node. We assume the locations of nodes on the plane conform to a Poisson point process with node density λ_s and the orientations of nodes are uniformly distributed over $[0°, 360°)$. We also assume all the nodes have a uniform FOV and denote the radius and angle of the FOV by ρ and θ.

Let us model targets as isotropic discs on the 2D plane with uniform radius r and define that a node captures a target only if the front arc of the disc bounded by two tangent viewing rays is completely visible to the node, as shown in Figure 1. More specifically, the arc of the disc bounded by the tangent rays must be inside the FOV of the node, and the centers of all other discs should be outside a corresponding occlusion zone between the disc and the node, which is illustrated as the bold-boundary region in Figure 1. This target model can be justified by real crowds of objects with similar heights and horizontal scales, such as human bodies and vehicles. Each object has a vertical axis, and most of its texture and shape signatures are contained in a cylinder space around the axis, as shown in Figure 2. Once the cylinder facade is fully captured by a node, sufficient silhouette contour and texture information will be projected onto the image to be used for identifying this target from background and other nearby targets.

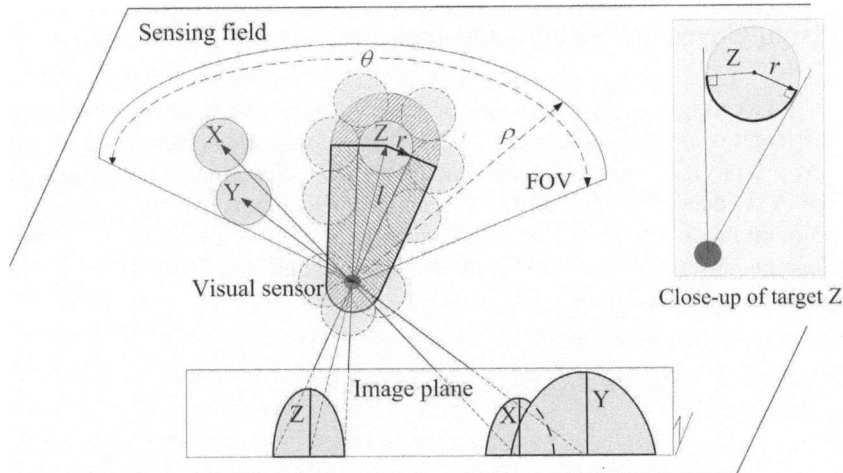

Fig. 1. Target model. Each target is represented by a disc. A node captures target Z, because the target facade (the entire bold arc in the close-up window) is visible to the node. At that time, all other targets must stay outside the hatched region in which the sub-region with bold boundary is the occlusion zone for target Z. Target X is occluded by target Y, because target Y steps into the occlusion zone of target X.

Fig. 2. A cylinder model for human bodies

We further assume that the centers of targets (discs) are uniformly distributed across the 2D plane on the condition that these discs do not overlap with each other. We also define target density λ_t as the ratio of the overall number of targets on the plane to the area of the entire plane. When the overall area occupied by targets is still a relatively small portion of the 2D plane, we can approximately estimate the probability that the centers of all the targets on the plane stay outside a finite area A as

$$\lim_{\Omega \to \infty} (\frac{\Omega - A}{\Omega})^{\lambda_t \Omega} = e^{-\lambda_t A} \tag{1}$$

where Ω denotes the area of the entire 2D plane. This formula will be frequently used in the derivations below.

3   Visual Coverage without Occlusions

If $r \to 0$, we can ignore the occlusion zone and consider that a node captures a target once the target is located inside the FOV. In other words, nodes that capture the target must stay in a circle of radius ρ centered at the target and be oriented towards the target. Let $\mathcal{P}(k, A, \lambda)$ denote the probability that an area A contains exactly k points from a Poisson point process with density λ, i.e. $e^{-\lambda A}(\lambda A)^k / k!$. Let C_i^k denote the number of k-element subsets from an i-element set. The probability that an arbitrary target is captured by exactly k nodes is

$$
\begin{aligned}
p(k) &= \sum_{i=k}^{\infty} \mathcal{P}(i, \pi\rho^2, \lambda_s) C_i^k (\frac{\theta}{2\pi})^k (1 - \frac{\theta}{2\pi})^{i-k} \\
&= e^{-\lambda_s \pi\rho^2} (\lambda_s \pi\rho^2)^k \frac{1}{k!} (\frac{\theta}{2\pi})^k \sum_{i=k}^{\infty} (\lambda_s \pi\rho)^{i-k} \frac{1}{(i-k)!} (1 - \frac{\theta}{2\pi})^{i-k} \\
&= \frac{1}{k!} e^{-\lambda_s \pi\rho^2} (\lambda_s \pi\rho^2 \times \frac{\theta}{2\pi})^k e^{\lambda_s \pi\rho^2 \times (1 - \frac{\theta}{2\pi})} \\
&= \mathcal{P}(k, \pi\rho^2 \times \frac{\theta}{2\pi}, \lambda_s)
\end{aligned}
\tag{2}
$$

4   Visual Coverage with Occlusions

Different from the case above, when target radius r is a finite value, the event that a node captures a target not only depends on whether the target stands in its FOV but also depends on whether other targets stand outside of the occlusion zone between the node and the target of interest.

As illustrated by Figure 3, to be able to capture a target, the maximum distance from a node to the disc center l_m is equal to $\sqrt{\rho^2 + r^2}$, which is reached when two ends of the disc arc touch the distant edge of the FOV. The minimum distance l_o is equal to $r/sin(\theta/2)$, which is reached when the disc blocks the entire FOV. Let A_u denote the area of the annulus with outer radius l_m and inner radius l_o. Only nodes located inside the annulus have the chance to capture the target. However, for each node inside the annulus, the probability of capturing the target, which we denote by q, becomes a random value with respect to the randomness of the locations of other targets. Let $f(q)$ denote the probability density function of q, which is parameterized by λ_t. The probability that an arbitrary target on the plane is captured by exactly k nodes can be expressed as

$$p(k) = \int \sum_{i=k}^{\infty} \mathcal{P}(i, A_u, \lambda_s) C_i^k q^k (1-q)^{i-k} f(q) dq \tag{3}$$

Following a similar derivation for Eq 2, we have

$$p(k) = \int \mathcal{P}(k, q A_u, \lambda_s) f(q) dq \tag{4}$$

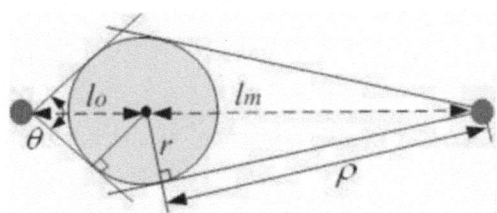

Fig. 3. Maximum and minimum distance from a node to a "visible" target

Pursuing an explicit form of $f(q)$ is very difficult, but it is feasible to derive some informative statistical parameters of q. With these parameters, we can construct an approximate function $\tilde{f}(q)$ for use in Eq 4.

Let q_m and q_o respectively denote the maximum and minimum of q. q reaches the maximum q_m when the centers of all other targets stand outside of a circle centered at the arbitrary target with radius $l_m + r$. See Figure 4.a. Then, for each node inside the annulus and facing the arbitrary target, there are no any occlusions ahead, so we have

$$q_m = \frac{1}{A_u} \int_{l_o}^{l_m} \frac{\theta - 2\sin^{-1}(r/l)}{2\pi} 2\pi l dl$$

$$= \frac{1}{A_u} \int_{l_o}^{l_m} [\theta - 2\sin^{-1}(r/l)] l dl \tag{5}$$

Based on Eq 1, the probability that $q = q_m$, which we denote by F_m, is equal to $e^{-\pi(l_m+r)^2 \lambda_t}$. Since F_m is a finite value, $f(q)$ must be an impulse function at q_m with amplitude F_m. To derive the boundary condition of $f(q)$ at the left side of q_m, let us imagine a situation when $q = q_m - \Delta q$, where Δq is an infinitesimal value. This situation happens only if a target encroaches on an infinitesimal area ΔA in the big empty circle from outside as illustrated in Figure 4.a. Since nodes in ΔA would have $\sin^{-1}(r/l_m)/\pi$ chance to capture the target, $\Delta q = \Delta A \sin^{-1}(r/l_m)/\pi$. Meanwhile, based on Eq 1, the probability that $q \in [q_m - \Delta q, q_m)$ is equal to $\Delta F = e^{-[\pi(l_m+r)^2 - \Delta A]\lambda_t} - e^{-\pi(l_m+r)^2 \lambda_t} = \lambda_t e^{-\pi(l_m+r)^2 \lambda_t} \Delta A$. Hence, the left-hand limit of $f(q)$ at q_m is

$$f(q_m^-) = \lim_{\Delta A \to 0} \frac{\Delta F}{\Delta q} = \frac{\pi \lambda_t e^{-\pi(l_m+r)^2 \lambda_t}}{\sin^{-1}(r/l_m)} \tag{6}$$

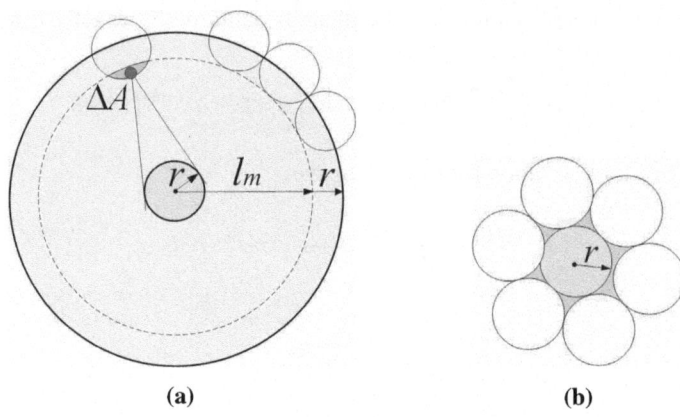

Fig. 4. (a) When q is maximum (b) When q is minimum

q reaches the minimum q_o when the arbitrary target is tightly surrounded by six other targets. See Figure 4.b. Then, only nodes falling into the tiny gap area between the seven discs(the small shaded area in Figure 4.b) have the chance to capture the target. Notice that, when $\theta > 120^o$, even those nodes in the shaded area have no chance to capture the target. We ignore this tiny area and consider

$$q_o = 0 \tag{7}$$

Let $E(q)$ denote the expectation of q, and we estimate $E(q)$ as follows. Suppose there is a node located at a distance l to the arbitrary target. If the node intends to capture the target, its FOV should face the target, and all other targets should stand outside the occlusion zone illustrated in Figure 1. Combined with the area within r distance to the center of the arbitrary target, the entire forbidden zone that the centers of other targets can not step into is illustrated as the hatched region in Figure 1, whose area can be expressed as

$$A_{FB}(l) = 3r\sqrt{l^2 - r^2} + r^2[\pi - 2sin^{-1}(r/l)] + \\ 4r^2[\pi + 2sin^{-1}(r/l)]$$

Based on Eq 1, the probability that $A_{FB}(l)$ is empty is $e^{-A_{FB}(l)\lambda_t}$. Combined with the probability that the node faces the target, $E(q)$ can be expressed as

$$E(q) = \frac{1}{A_u} \int_{l_o}^{l_m} [\theta - 2sin^{-1}(r/l)]e^{-A_{FB}(l)\lambda_t} l\, dl \tag{8}$$

Based on the parameters derived above, we can approximate $f(q)$ by an impulse function with amplitude F_m at $q = q_m$ and a scaled Binominal distribution function over $[0, q_m)$ as

$$\tilde{f}(q) = \begin{cases} \dfrac{(1 - F_m)N}{q_m}C_N^i\gamma^i(1 - \gamma)^{N-i}, & i = int(Nq/q_m) \text{ if } 0 \le q \le q_m^- \\ F_m & \text{if } q = q_m \end{cases}$$

where $int(x)$ returns the maximum integer not greater than x, and N and γ can be decided by two constraints respectively imposed by Eq 6 and 8. By assuming N is large enough, the two constraints can be expressed as:

$$\frac{(1 - F_m)N}{q_m}\gamma^N = f(q_m^-)$$

$$(1 - F_m)\gamma q_m + F_m q_m = E(q)$$

Trivially, q with its probability density function equal to $\tilde{f}(q)$ also satisfies the two constraints imposed by Eq 5 and 7.

Therefore, $p(k)$ can be estimated as:

$$\begin{aligned}
p(k) &= \int \mathcal{P}(k, qA_u, \lambda_s)\tilde{f}(q)dq \\
&\approx F_m\mathcal{P}(k, q_m A_u, \lambda_s) + (1 - F_m) \times \\
&\quad \sum_{i=0}^{N} \mathcal{P}(k, \frac{q_m(i + 0.5)}{N}A_u, \lambda_s)C_N^i\gamma^i(1 - \gamma)^{N-i}
\end{aligned} \tag{9}$$

As $r \to 0$, $A_u \to \pi\rho^2$, $q_m \to E(q) \to \theta/2\pi$, and $f(q_m^-) \to \infty$, which means $\gamma \to 1$ and $N \to \infty$. Hence, $\tilde{f}(q)$ converges to a unit impulse function at q_m, and Eq 9 converges to Eq 2.

5   Node Density and Visual Coverage

Suppose we try to find the minimum node density sufficient to ensure the probability that each target is captured by less than K nodes is smaller than a tolerance value ε. Trivially, if nodes are randomly deployed, this probability decreases as node density λ_s increases. Therefore the minimum node density is the smallest positive root $\hat{\lambda}_s$ of the following equation

$$\sum_{k=0}^{K-1} p(k) = \varepsilon \tag{10}$$

where $p(k)$ is parameterized by λ_s as shown in Eq 9. As a special case, the minimum node density that ensures 1-coverge is the smallest positive root of the equation $p(0) = \varepsilon$.

Results. The effect of visual occlusions on visual coverage can be studied by examining the results of the minimum node density estimation based on Eq 10. The amount of visual occlusions in the field increases as the target density increases. By setting the FOV of nodes as $\rho = 6m$ and $\theta = 120°$, target radius $r = 0.2m$, and the tolerance value $\epsilon = 0.05$, we change the node density λ_t from 0.1 to 1. Figure 5.a shows the minimum node density $\hat{\lambda}_s$ corresponding to different target densities λ_t under different coverage requirements K. We observe that $\hat{\lambda}_s$ increases as either K or λ_t increases.

Figure 5.b shows the ratio of each $\hat{\lambda}_s$ to the minimum node density that corresponds to a same λ_t but only ensures 1-coverage. This ratio indicates the economic cost for upgrading the coverage performance of the network. Note that the ratio is always smaller

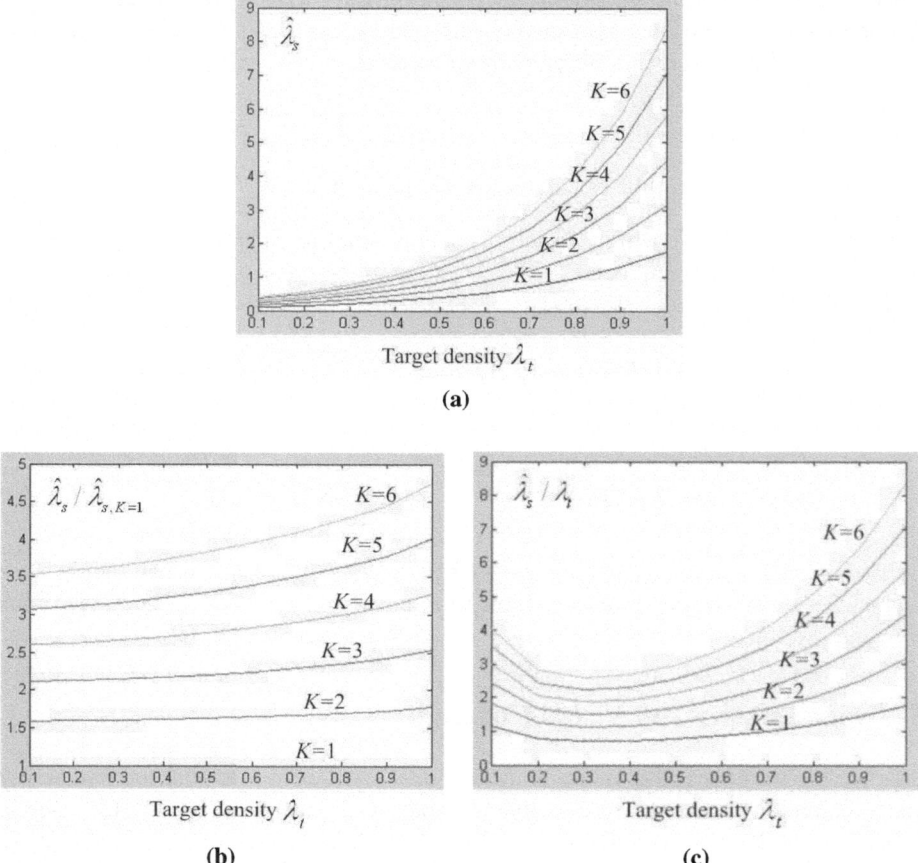

Fig. 5. (a) Minimum node density $\hat{\lambda}_s$ vs target density λ_t. (b) Ratio of $\hat{\lambda}_s$ to the minimum node density corresponding to 1-coverage but a same target density λ_t vs target density λ_t. (c) Density ratio $\hat{\lambda}_s/\lambda_t$ vs target density λ_t.

than the corresponding K value over the range of λ_t in the figure, which indicates that each node is capable of capturing multiple targets within its wide FOV. However, this ratio also grows as λ_t increases, which means that the capability of capturing multiple targets is declining with the FOV becoming crowded with targets. In other words, for the purpose of target detection, the increase of visions in the field created by the increase of node density is counteracting the increase of visual occlusions created by the increase of target density.

The ratio of $\hat{\lambda}_s$ to λ_t indicates the economic cost of the network with respect to the scale of the target problem to solve. The lower the ratio, the more economic the network will be. Figure 5.c shows this ratio corresponding to different λ_t and different K. In the figure, each curve has a corner, and the ratio decreases before λ_t hits the corner and increases afterwards. As mentioned above, when the target size is relatively small compared to the wide vision of nodes, each node is able to capture multiple targets.

Therefore, when λ_t is small and existing targets are still sparsely distributed in front of nodes, instead of simply deploying more nodes, the wide vision of nodes can be exploited to maintain existing coverage as new targets entering the field. However, as λ_t increases to the corner point, the vision of nodes will be packed with existing targets, and the only solution to ensure both the existing and new targets are covered by enough nodes is to deploy more nodes into the field.

The amount of visual occlusions in the field also increases if the target size increases. By setting the FOV as $\rho = 6m$ and $\theta = 120°$, target density $\lambda_t = 0.2$, and the tolerance value $\epsilon = 0.05$, we change the target radius from $0.1m$ to $0.5m$. Figure 6.a shows the minimum node density $\hat{\lambda}_s$ corresponding to different target radiuses r under different coverage requirements K. Figure 6.b shows the ratio of each $\hat{\lambda}_s$ to the minimum node density that corresponds to a same r but only ensures 1-coverage. As we can see, these two figures tell the same story about the effect of visual occlusions on the necessary node density as Figure 5.a and Figure 5.b.

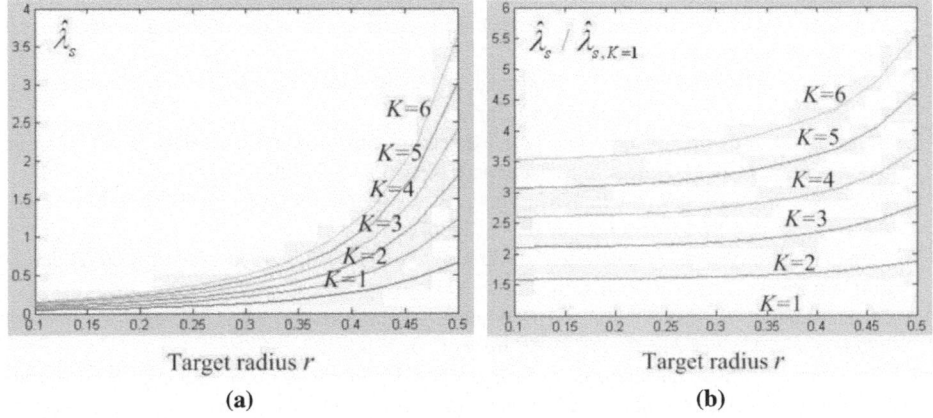

Fig. 6. (a) Minimum node density $\hat{\lambda}_s$ vs target radius r. (b) Ratio of $\hat{\lambda}_s$ to the minimum node density corresponding to 1-coverage but a same target radius r vs target radius r.

6   Conclusions

Confronted with the difficulties caused by the presence of visual occlusions, we derive an explicit coverage estimate for visual sensor networks. The derivation features a detection model that takes visual occlusions into account and the methods of estimating several significant statistical parameters related to visual converge. In the future, our research can be improved at least from three aspects. Firstly, in order to improve the estimation accuracy, we will try to derive higher order moments of q and even modify the Binominal approximate function. Secondly, instead of assuming all the targets have a uniform size or shape, we will formulate the statistics of their sizes and shapes and incorporate these statistics into the estimation. Thirdly, besides the coverage, we will use more factors, such as energy efficiency and networking performances, to constrain the minimum node density.

Acknowledgement

This research is sponsored in part by NSF ECS-0449309 (CAREER). The authors are also indebted to Prof. Lynne Parker for her valuable review.

References

1. Akyildiz, I.F., Su, W., Cayirci, E.: Wireless sensor networks: A survey. Computer Networks 38(4), 393–422 (2002)
2. Obraczka, K., Manduchi, R., Garcia-Luna-Aveces, J.: Managing the information flow in visual sensor networks. In: Proceedings of the Fifth International Symposium on Wireless Personal Multimedia Communications (2002)
3. Akdere, M., Cetintemel, U., Crispell, D., Jannotti, J., Mao, J., Taubin, G.: Data-centric visual sensor networks for 3d sensing. In: Second Geosensor Networks Conference (2006)
4. Wan, P.J., Yi, C.W.: Coverage by randomly deployed wireless sensor networks. IEEE Transactions on Information Theory 52(6), 2658–2669 (2006)
5. Yen, L.H., Yu, C.W., Cheng, Y.M.: Expected k-coverage in wireless sensor networks. Ad Hoc Networks 4(5), 636–650 (2006)
6. Cao, Q., Yan, T., Abdelzaher, T., Stankovic, J.: Analysis of target detection performance for wireless sensor networks. In: DCOSS (2005)
7. Brass, P.: Bounds on coverage and target detection capabilities for models of networks of mobile sensors. ACM Transactions on Sensor Networks 3(2) (2007)
8. Wang, W., Srinivasan, V., Chua, K.C., Wang, B.: Coverage for target localization in wireless sensor networks. In: Proceedings of IPSN (2006)
9. Wang, W., Srinivasan, V., Chua, K.C., Wang, B.: Energy-efficient coverage for target detection in wireless sensor networks. In: IPSN (2007)
10. Meguerdichian, S., Koushanfar, F., Potkonjak, M., Srivastava, M.: Coverage problems in wireless ad-hoc sensor networks. IEEE Infocom 3, 1380–1387 (2001)
11. Clouqueur, T., Phipatanasuphorn, V., Ramanathan, P., Saluja, K.K.: Sensor deployment strategy for target detection. In: 1st ACM International Workshop on Wireless Sensor Networks and Applications (2006)
12. Veltri, G., Huang, Q., Qu, G., Potkonjak, M.: Minimal and maximal exposure path algorithms for wireless embedded sensor networks. In: ACM SenSys (2003)
13. Ai, J., Abouzeid, A.: Coverage by directional sensors in randomly deployed wireless sensor networks. Journal of Combinatorial Optimization 11(1), 21–41 (2006)
14. Adriaens, J., Megerian, S., Potkonjak, M.: The third annual IEEE Communications Society Conference on Sensor, Mesh and Ad Hoc Communications and Networks (2006)
15. Yang, D.B., Shin, J., Ercan, A., Guibas, L.J.: Sensor tasking for occupancy reasoning in a camera network. In: IEEE/ICST 1st Workshop on Broadband Advanced Sensor Networks (BASENETS) (2004)
16. Isler, V., Bajcsy, R.: The sensor selection problem for. bounded uncertainty sensing models. In: Proceedings of IPSN (2005)

Time-Bounded and Space-Bounded Sensing
in Wireless Sensor Networks

Olga Saukh, Robert Sauter, and Pedro José Marrón

Universität Bonn, Bonn, Germany and Fraunhofer IAIS, St. Augustin, Germany
{saukh,sauter,pjmarron}@cs.uni-bonn.de

Abstract. Most papers on sensing in wireless sensor networks use only very simple sensors, e.g. humidity or temperature, to illustrate their concepts. However, in a large number of scenarios including structural health monitoring, more complex sensors that usually employ medium to high frequency sampling and post-processing are required. Additionally, to capture an event completely several sensors of different types are needed which have to be in range of the event and used in a timely manner. We study the problem of time-bounded and space-bounded sensing where parallel use of different sensors on the same node is impossible and not all nodes possess all required sensors. We provide a model formalizing the requirements and present algorithms for spatial grouping and temporal scheduling to tackle these problems.

1   Introduction

Many wireless sensor network (WSN) applications focus on monitoring scenarios like structural health monitoring of bridges, tunnels and buildings, environmental monitoring and tracking of mobile objects. Complex events, which are common in these scenarios, can only be detected by a combination of several environmental characteristics captured at one point in time and space. Therefore, these applications require sampling and in-network processing of several kinds of complex data which together provide a complete description of a complex event. However, it is practically impossible to install a sensor node with all required sensors to cover every point of the target area. Therefore, distributed wireless sensor nodes usually provide approximations of the description of events.

Although it is often theoretically possible to equip the sensor nodes with all required complex sensors, this might cause several problems. First, each additional complex sensor requires energy, which degrades the lifetime of the individual sensor node and of the whole network. Second, it is often difficult or impossible to trigger and sample several complex sensors at the same time. Moreover, the triggers of two separate sensors activated by an event will not happen simultaneously, which makes simultaneous sampling quite complex. Third, each sensor has its own area of regard. Therefore, some sensors might register abnormality of one environmental characteristic when capturing events, with no confirmation from other attached sensors with a smaller sensing range. Fourth, attaching all sensors to every sensor node still requires much cable to place the sensors at

S. Nikoletseas et al. (Eds.): DCOSS 2008, LNCS 5067, pp. 357–371, 2008.

meaningful locations. For example in bridge monitoring scenario, an acceleration sensor used to measure cable forces must be attached to the corresponding cable whereas an acoustic sensor must be embedded into the bridge floor to be able to acquire acoustic waves which might indicate cracks in the construction. Therefore, in many cases there is a strong need for the distribution of sensors among several sensor nodes and for further in-network cooperation of space-bounded sensors and grouping of time-bounded sensor values in order to detect and characterize an event. Fifth, sensing takes time and for highly dynamic events, like the occurrence of a crack in a bridge, it is impossible to inquire several sensors sequentially within the event duration.

In this paper we assume that every sensor node has one or several sensors attached. These sensors try to capture events that can only be detected within a limited range in space and within a limited period in time. On the one hand, we present algorithms that establish space-bounded non-disjoint groups of sensor nodes which can be seen as one logical sensor node for recognizing an event. On the other hand, we present a scheduling algorithm that allows the distribution of sensing tasks in every group and creates a local task schedule for every sensor on every individual sensor node.

The rest of this paper is structured as follows. In Section 2 we present related approaches. We provide the definition of the problem of time-bounded and space-bounded sensing and discuss the model and assumptions in Section 3. In Section 4 we present algorithms to build groups and generate schedules followed by a thorough evaluation of these algorithms in Section 5. Conclusions and an outlook to future work in Section 6 conclude this paper.

2   Related Work

Event detection is a popular research area in WSN [1,2,3,4]. The event detection system presented in [5] allows the detection of composite events in case nodes have heterogeneous sensing capabilities. The results described in [1,2,6] provide algorithms for the detection of k-watched composite events, where each event occurrence can be detected by at least k sensors. In [3], the authors concentrate on state transitions of the environment rather than on states only and discuss a generalized view on event detection in WSNs. They model state transitions with finite automata. However, this model is impractical due to its complexity. The authors of [4] consider the problem of describing events or states and state transitions of the environment with an event description language. The main difference of the mentioned works to our approach is that only the spatial characteristic of event detection has been considered. Since an event happens at some point in space and *time*, we also consider its temporal characteristic. Moreover, most papers in this group consider spatial node grouping to increase the confidence of the sensing results. In this paper we group nodes in order to be able to process complex queries and detect complex events.

A number of research projects including TinyDB [7] and Cougar [8] have considered a query-based database abstraction of the WSN. However, these works assume that the sensor nodes possess the same unordered set of sensors and that the actual access of these sensors is not time limited. Therefore, the sequential execution of the sensing tasks that compose the query on every node is always possible. In this paper we motivate and provide a solution for the case when sequential execution of sensing tasks on every node is impossible.

There are a number of papers that consider the problem of spatial node grouping or clustering [9, 10] in WSN. The usual reason for this grouping is to allow for efficient data aggregation in sensor networks and, therefore, save energy of individual nodes. In this paper we present node grouping algorithms that try to construct the maximum number of complete groups – groups that have all required sensing capabilities to fulfil the query (or detect an event)

The problem of job scheduling is closely related to our work. This problem is usually formulated as follows: Given a directed acyclic graph, where vertices represent jobs and an edge (u, v) indicates that task u must be completed before task v. The goal is to find a schedule of tasks which requires the minimum amount of time or machines. Additional resource constraints on every machine and resource requirements for every task may exist. There are also a number of solutions to this problem in different formulations [11]. However, in this paper we also consider concurrency constraints between tasks, which reduces the applicability of existing solutions to our problem. Therefore, we present a new scheduling algorithm for sensor networks which allows the scheduling of sensing tasks between different sensor nodes taking concurrency constraints between individual tasks into account.

3 Distributed Sensing

3.1 Terminology, Assumptions and Problem Statement

The *sensor network* is usually modeled as an undirected graph $G(V, E)$ embedded into the plane, where V is a set of nodes and E is the set of edges between nodes that can communicate. Every node $v_i \in V$ in this *embedding* $p : V \to \mathbb{R}^2$ has coordinates $(x_i, y_i) \in \mathbb{R}^2$ on the plane. However, as explained later, it is not necessary for our approach that the nodes are aware of their coordinates. We consider that all sensor nodes are embedded within some *target area* $A \subset \mathbb{R}^2$ on the plane. Additionally, the function $v : V \to 2^{\mathbb{S}}$ defines the sensor types each node possesses where $\mathbb{S}$ is the domain of sensor types used in the scenario. This implies that each node contains at most one sensor of each type.

A *query* or *event description* $Q = (S, R, D, C, Pred)$ is defined as a 5-tuple. The set $S \subseteq \mathbb{S}$ describes the sensor types used in the query. The functions $R : S \to \mathbb{R}$ and $D : S \to \mathbb{D}$ describe the sensing range and sensing time respectively for each sensor type in S. We assume a sensing area to be a disc with radius $r \in \mathbb{R}$. The sensing time only describes the actual time needed to access the sensor readings. This duration does not include any data processing, filtering or aggregation actions which can be arbitrarily postponed. Additionally, the concurrency constraints between different sensor types are captured by $C : S \times S \to R$. For each pair of sensor types, an element of C defines the maximum duration between starting the sensing of the corresponding sensors. *Pred* is a predicate which maps the sensor values of each sensor type and the combination of these values to {true,false}. We do not further discuss possible definitions of this predicate since we do not extend the large amount of prior work regarding this topic. The combination of these characteristics is extremely important because, for example, a break in a concrete structure event can be captured by an acoustic sensor only within several microseconds in a range of several meters, whereas a fire event results in a temperature increase that can be measured by a temperature sensor within several minutes within a range dependent on the fire event itself (centimetres for a fire of a candle till hundreds of meters for a fire in a forest). We refer to the sensing duration and concurrency constraints and to the sensing range of a query or event description as *event time and space constraints*.

The sensing range of an event with respect to a specific sensor coincides with a sensing range the sensor has and depends on the sensitivity of the sensor. The event duration is, however, usually longer than the sensing duration of the sensor. For example, an acoustic event results in a series of acoustic emissions or fire event lasts longer than needed by a temperature sensor to read the value. The event duration as well as the additional dependencies between the sensing with different sensor types have to be captured by the concurrency constraints.

Our main assumption is that if an event happens at some location $P \in \mathbb{R}^2$ and at some point in time $T_0 \in \mathbb{R}$ and is sensed by a group of sensors which comprises all required sensor types S and satisfies time and space constraints D, C, R with respect to the starting point of the event in time and the point in space respectively, then the sensor readings obtained are a good approximation of the event characteristics. This assumption is very natural for sensor networks.

We define a group of sensors $G_i \subseteq S \times V$, if $\forall (s_j, v_k^i) \in G_i, s_j \in v(v_k^i)$ and $\forall s_j \in S, \exists (s_j, v_k^i) \in G_i$. This ensures, that only sensor types of a node are used if this node actually possesses the type and that all sensor types required by the query are contained at least once in the group. Let dist denote the Euclidean distance, we say that a group G_i is *space-bounded* with respect to a certain event, if $\forall (s_j, v_k^i) \in G_i, \text{dist}(P, p(v_k^i)) \leq R(s_j)$. This requires that all sensors of each type are in range of the event. However, the definition of a group does not require that for every node all sensor types it possesses are used in the group. This allows the creation of groups where for some nodes only the long-range sensors (e.g. temperature) are used and the short range sensors are ignored. The missing sensor types have to be supplied by nodes closer to the event.

We assume, that parallel sampling of two sensors on the same sensor node is impossible. Two sensors can be accessed sequentially only. Every sensor s_j has its sensing duration $D(s_j)$ – the time the sensor node requires to obtain the sensor readings from this sensor.

We refer to a *local schedule* J_i^{local} for a sensor node v_i as a sequence of sensing tasks that need to be executed for the complete or a fraction of an event description Q. The schedule contains the start times for the access of every sensor type. As described above, the sensing tasks may not overlap; however, it is possible to insert idle times between two sensing tasks. A local schedule is *time-bounded* if the sequence of sensing tasks fulfils the concurrency constraints C. A *group schedule* J_i for a group G_i is defined correspondingly to the local schedule but includes information for all sensing tasks of all group members. A group schedule is time-bounded if the local schedules it includes are time-bounded and additionally the concurrency constraints C between sensor types on different nodes are met, which can require the inclusion of idle times on individual nodes.

The problem of time-bounded and space-bounded sensing is: *Define a group of nodes and a schedule that enable the execution of the query Q under the given time, concurrency and space constraints.*

In the following subsections we consider the problem of time-bounded and space-bounded sensing separately as well as in combination with distributed coordination.

3.2 Time-Bounded Sensing

Consider the case when an individual sensor node is equipped with all required sensors to process the query Q. It is easy to construct a schedule that sequentially fetches sensor readings from each of these sensors in order to evaluate Q and send the aggregated result to the data sink. Here a simple algorithm can be used to try every possible permutation in order to find a schedule that fulfils the concurrency constraints. Since the number of sensors on a node is usually quite low (≤ 5), it is easy even on resource constrained sensor network devices to enumerate all possible permutations.

Additional concurrency dependencies between the individual nodes must also be satisfied when distributing sensing tasks including the distribution delays. We model the scheduling problem as a graph in which the vertices represent sensing tasks with their corresponding task durations and the edges represent the concurrency constraints between each pair of tasks. The concurrency constraints express the maximum difference in time between the start times of each pair of tasks.

We define the duration $\mathrm{Dur}(J)$ of a schedule as the time from starting to sense the first task to the completion of the last task. Additionally, we define a general metric $\varphi(J)$ indicating the badness of a schedule. We use a badness metric instead of a quality metric since for most intuitive approaches, e.g. the duration of a schedule, higher values mean worse results. Using this metric, we describe the challenge of time-bounded sensing as an optimization problem:

$$\varphi(J_i) \rightarrow \mathbf{min}, \ J_i \text{ is a valid schedule as defined above} \tag{1}$$

In particular, J_i must fulfil the concurrency constraints between the execution start times of different tasks. Note that classical scheduling algorithms from the literature do not assume any concurrency dependencies between the tasks.

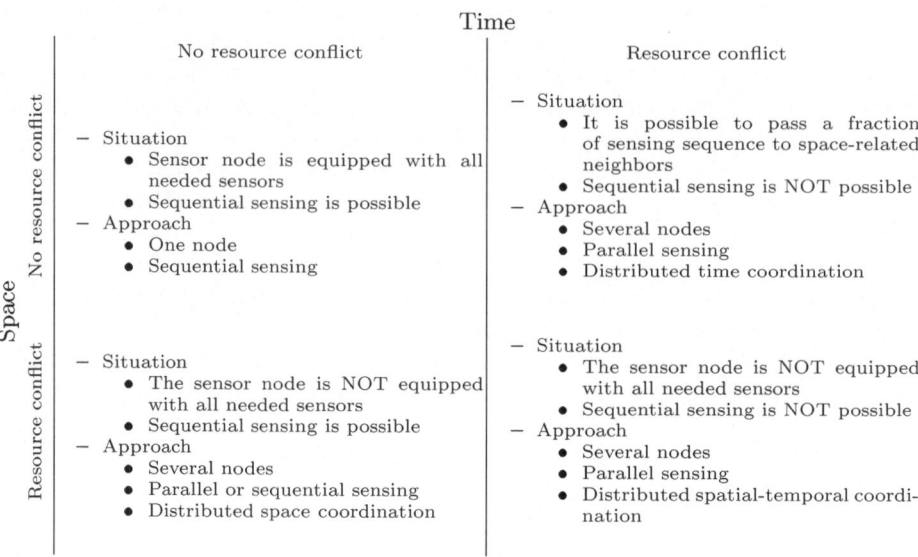

Fig. 1. Problem Space

3.3 Space-Bounded Sensing

If an individual sensor node does not possess all required sensors to fulfil the query or check the event description, popular approaches, e.g. TinyDB [7], fail to consider this node. However, it might be possible to combine two nodes, each with an incomplete set of sensors, located close to each other. From the viewpoint of the application, a space-bounded group plays the role of *one logical sensor node* which possesses all sensors that the group has and allows to perform operations over such space-bounded sensor readings, e.g. calculate if the sensor readings confirm to a certain event description.

Since the location of an event is not known, we introduce a dispersion metric $\psi : S \times V \to \mathbb{R}$. Our approach can work with different metrics. Besides metrics based on actual node coordinates, e.g. the maximum distance from the centroid of the group, metrics based only on some proximity information such as RSSI or the number of shared neighbors are possible. Therefore, we formulate the optimization problem of space-bounded sensing as follows:

$$\psi(G_i) \to \mathbf{min}, \ G_i \text{ is a valid group as defined above} \qquad (2)$$

3.4 Distributed Coordination

The formulated time-bounded and space-bounded sensing problem considers time and space as limited resources in the context of event detection or phenomena monitoring.

Consider the case when all nodes are equipped with a complete set of sensors. A *time resource conflict* occurs, when it is impossible to acquire an event with a single sensor node via sequential scheduling of sensing tasks due to concurrency constraints. This problem can be solved by intelligent relocation of a fraction of the local schedule to the nodes close-by.

A *spatial resource conflict* occurs if a node does not possess the required set of sensors to process the query. In this case, the event must be captured by in-network cooperation of a group of sensor nodes. If the sequential scheduling on every individual node does not result in any problems, the solution involves group coordination mechanism which partitions the sensor network in space-bounded groups equipped with all required sensors to process the query. Then every group operates as a single logical sensor node.

In real world deployments both problems might occur. In this case, the distribution of sensing tasks in space by space-bounded group coordination and intelligent distributed scheduling of sensing tasks is the only possible solution. Fig. 1 summarizes the described problem space.

4 Algorithms

4.1 Ordering of Sensors

As described in the problem statement, both algorithms have to solve an optimization problem. We introduce a total order among sensor types to greatly reduce the complexity of the algorithms. This order does not pose any unduly restrictions for the query definition, since the priorities can often be derived quite naturally from the characteristics of the employed sensor types. For example, humidity and temperature sensors are not very time sensitive and can be sampled with low priority and are thus among the last of the order. Very complex sensors, e.g. acoustic emission sensors, that may even rely on hardware triggers and thus initiate the complete event detection process, should be placed first in the order.

To express the total order, we redefine a query for our algorithms in terms of vectors and matrices. A query $\tilde{Q} = (\boldsymbol{S}, \boldsymbol{R}, \boldsymbol{D}, C, Pred)$ on n sensor types comprises the sensor types $\boldsymbol{S} = (s_1, \ldots, s_n), s_i \in \mathbb{S}$, the associated sensing ranges $\boldsymbol{R} = (r_1, \ldots, r_n), r_i \in \mathbb{R}$ and sensing durations $\boldsymbol{D} = (d_1, \ldots, d_n), d_i \in \mathbb{R}$ as well as a symmetric $n \times n$ matrix C where each element $c_{ij} \in \mathbb{R}$ defines the maximum duration between the start times of sensor type i and sensor type j.

Additionally, we reduce the number of allowed sensors of the same type in a group to one. Therefore, a group $\boldsymbol{G} = (v_1, \ldots, v_n), v_i \in V$ simply indicates for each sensor type the corresponding node. Accordingly, a schedule $\boldsymbol{J} = (t_1, \ldots, t_n), t_i \in \mathbb{R}$ defines the start times of each sensor type.

4.2 Space-Bounded Group Establishment

As stated above, the goal of space-bounded group establishment is to build *tight* groups, where tight is defined in terms of a dispersion metric $\psi : S \times V \to \mathbb{R}$.

We distinguish between three parts of our algorithm, which we describe in detail below. For each part, we have developed two alternatives. Since all combinations are possible, we have evaluated eight different algorithm combinations.

For each group, one node is distinguished as the leader of the group. This is always the node of which the first sensor type is used. The groups are built iteratively on the basis of the order of sensor types. At initialization, all nodes that possess the first sensor start forming groups by looking for nodes with the second sensor type. Which nodes are available for addition to a group is determined by the *selection rule*. Of course only nodes which have the correct sensor type are considered. The best node is then selected based on the *dispersion metric*. This step is performed for the following sensor types until the group is complete or as long as a suitable node is found. Each sensor of each node can only be used in one group. We now describe each step in more detail and show the implementations we used.

Dispersion metric. The dispersion metric has to be computed at each step and with each candidate node. Let G_p denote a partially complete group consisting of the already selected sensor types and a candidate node for the sensor type that is currently required. Let $k = |G_p|$ denote the number of sensor types in this partial group. Then $\text{centroid}(G_p) = (\frac{1}{k}\sum_{v_i \in G_p} x_i, \frac{1}{k}\sum_{v_i \in G_p} y_i)$ defines the centroid of the partial group. We define the standard deviation $SDev = (\frac{1}{k}\sum_{v_i \in G_p} dist(p(v_i), \text{centroid}_x(G_p))^2)^{\frac{1}{2}}$ and the maximum of the distances to the centroids $Rad = \max_{v_i \in G_p} dist(p(v_i), \text{centroid}_x(G_p))$ as our two metrics.

Selection rule. The selection rule defines which nodes to consider for the next sensor type. We define the *Group* selection rule where all neighbors of all group members are considered. The *Leader* selection rule allows only the neighbors of the leader of a group to be added. These rules differ obviously in communication cost, since the collection of candidates as well as the actual selection requires message transmissions. Therefore, the *Group* selection rule allows choosing from a larger number of candidates at the expense of increased communication costs.

Grouping algorithm. The grouping algorithm defines the overall process of group building. We implemented a rather simple *FirstChoice* algorithm. A leader simply selects the next node based on the selection rule and the dispersion metric. The first leader that chooses the sensor on another node wins. Our second approach *BestChoice* is more complicated. A leader informs a selected node not only with its ID but also with the value of the dispersion metric, which the new node saves. If another leader later selects the same sensor on the same node and the dispersion metric of this group is better than the saved one, the node changes the group and informs the former leader about its decision. The former leader then discards all selected sensors of a lower priority and begins rebuilding its group. This algorithm is detailed in Fig. 2[1].

---

[1] The *BestChoice* algorithm converges if the dispersion metric has a lower bound. Proof by induction on the number of sensor nodes in the group. The transitivity of the min relation ensures that no cycles occur.

Procedure bestChoice
for $j = 2$ to n do
   | leader$_j \leftarrow$ inf
   | leaderMetric$_j \leftarrow$ inf
endfor
if *isLeader* then
   | search(*1*)
endif

Procedure search(*sensor*)
$(v_\mathrm{cand}, \mathrm{metric}) \leftarrow$
findCand(sensor, G_i, $V_\mathrm{ignore}^\mathrm{sensor}$)
if *metric* > 0 then
   | send(v_{cand}, *SELECT*, *sensor*,
   | *metric*)
endif

Procedure   onReceive(*what*,
sender, sensor)

if *what* = *ACCEPT* then
   | $G_k \leftarrow$ sender
   | if *sensor* < *n* then
   | | search(*sensor+1*)
   | endif
endif
if *what* = *REJECT* then
   | $V_\mathrm{ignore}^\mathrm{sensor} \leftarrow V_\mathrm{ignore}^\mathrm{sensor} \cup \{v_\mathrm{sender}\}$
   | search(*sensor*)
endif
if *what* = *SELECT* then
   | if *leaderMetric$_{sensor}$* > *metric*
   | then
   | | if *leader$_{sensor}$* $\neq$ inf then
   | | | send(*leader$_{sensor}$*,
   | | | *REJECT, sensor*)
   | | endif
   | | leaderMetric$_\mathrm{sensor} \leftarrow$ metric
   | | leader$_\mathrm{sensor} \leftarrow$ sender
   | | send(*sender, ACCEPT,*
   | | *sensor*)
   | else
   | | send(*sender, REJECT,*
   | | *sensor*)
   | endif
endif

Procedure schedule
/* returns start times t of all
   sensor tasks   */
$t \leftarrow 0$
minStart $\leftarrow 0$
current $\leftarrow 2$
for $1 = 1$ to n do
   | nodeBusy$_{G_i} \leftarrow 0$
endfor
nodeBusy$_{G_1} \leftarrow d_1$
while *current* $\leq n$ do
   minTime $\leftarrow$ max(nodeBusy$_{G_{current}}$,
   minStart)
   good $\leftarrow$ **TRUE**
   for $i = current\text{-}1$ to n **step** -1
   do
      minTime $\leftarrow$ max(*minTime*,
      $t_i - C_{currenti}$))
      if *minTime* > $t_i - C_{currenti}$
      then
      | if $i = 1$ then
      | // impossible
      | | **return** -1
      | endif
      | minStart $\leftarrow$ minTime
      | $-C_{currenti}$
      | nodeBusy $\leftarrow 0$
      | for $j \leftarrow 1 \ldots i - 1$ do
      | | nodeBusy$_{G_j} \leftarrow t_j + d_j$
      | endfor
      | current $\leftarrow i$
      | good $\leftarrow$ **FALSE**
      | break
      endif
   endfor
   if *good* then
   | $t_\mathrm{current} \leftarrow$ minTime
   | nodeBusy$_{G_\mathrm{current}} \leftarrow$
   | minTime + d_current
   | current $\leftarrow$ current+1
   | minStart $\leftarrow 0$
   endif
endw
return t

endif

Fig. 2. The BestChoice algorithm and the scheduling algorithm

4.3 Time-Bounded Scheduling

We have developed an algorithm that generates a group schedule fulfilling the concurrency constraints and minimizing the duration $Dur(J)$ of the whole schedule. The total order of sensor types allows a very efficient non-recursive algorithm, which is detailed in Fig. 2. The algorithm is based on backtracking. However, due to the priorities, the start time of a task may only monotonically increase, which limits the number of backtrackings. The algorithm always finds the schedule with the minimal duration. As a slight simplification, we do not include the time that is necessary to notify all members of the group. However, this interval can simply be inserted at the beginning of the local schedules. The problem of inter-group conflicts when an event can be sensed by several groups and the sensors of one node are used in different groups is left for future work. However, priorities provide a deterministic method to select which sensor on a node is used first in case of a conflict.

5 Evaluation

The evaluation of our approach is based on thorough simulations with different node densities, different numbers of sensor types required to detect an event or process the query and two settings of concurrency constraints. The sensor nodes are distributed in an area of 900×530 m^2. In all figures we distributed between 150 and 450 sensor nodes in the area which results in average node degrees between 9 and 26. We constructed topologies using perturbed grids to create the scenarios: the nodes are placed randomly inside of circles arranged in a regular grid (the radius is equal to the grid spacing). This type of topologies is a good match for real-world deployments where the goal usually is a more or less regular coverage of the sensing area. We assume that two nodes can communicate with each other, if they are within a distance of at most $R_c = 100$m (communication radius). Therefore, the communication graph is a *unit disk graph*. Every sensor node possesses several different sensor types. Every type of sensor is uniformly distributed over the sensing area so that a certain percentage of nodes possess this sensor. Typically, the communication radius is larger than the sensing radius R_s [12], therefore, we chose $R_s = \frac{1}{2}R_c$ as the sensing radius for all sensor types.

5.1 Space-Bounded Quality of Sensing

Fig. 3 shows the performance of spatial node grouping algorithms under the setting, that every sensor node can be equipped with more than one sensor. In this scenario, 50% of the sensor nodes possess the same sensor type. In Fig. 3a) we present the number of complete groups constructed by different algorithms. The number of complete groups increases with higher node densities and a greater number of deployed sensor nodes. Here, the *FirstChoice* algorithm dominates by constructing a slightly higher number of groups. As expected, both grouping algorithms combined with the *Group* selection rule require a higher number of sent packets per node (Fig. 3b)). Interestingly, the algorithms combined with

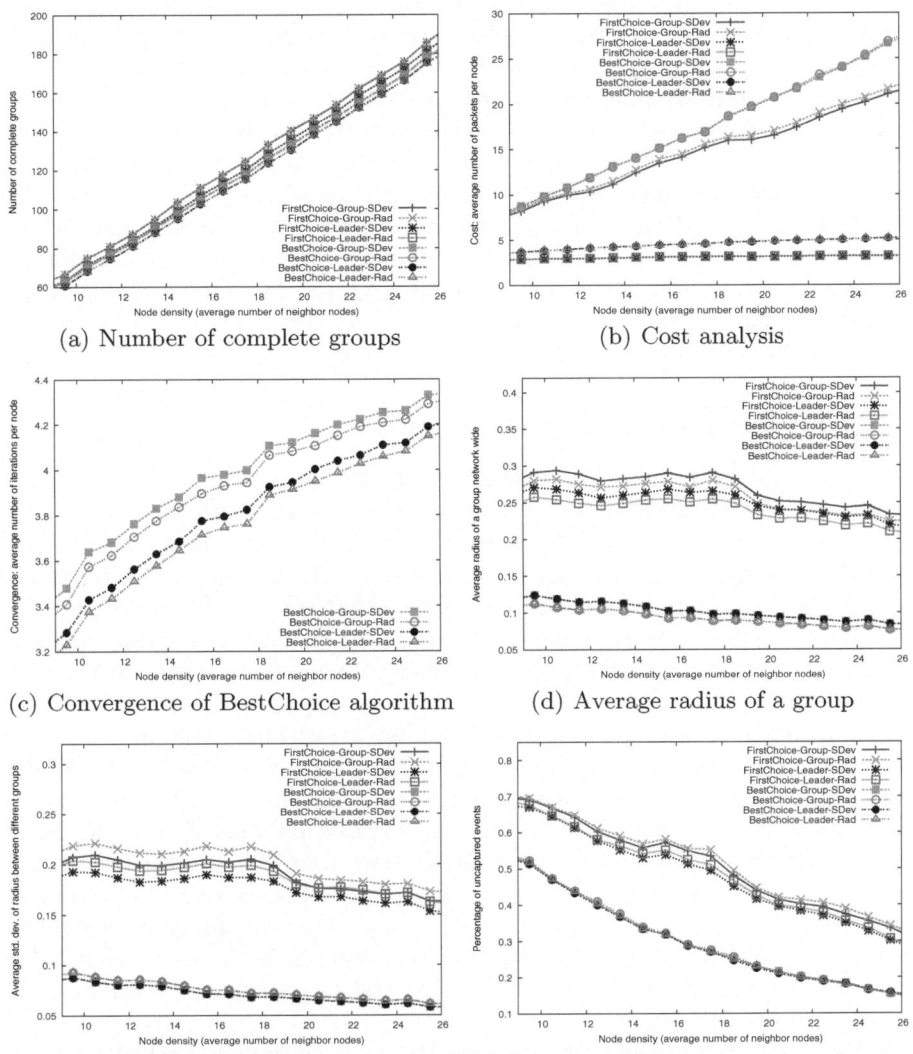

(a) Number of complete groups

(b) Cost analysis

(c) Convergence of BestChoice algorithm

(d) Average radius of a group

(e) Average standard deviation in a group (f) Uncaptured events or uncovered region

Fig. 3. Evaluation of the spatial grouping algorithms: 4 sensor types (50% of nodes possess the same sensor type, each node might possess more than 1 sensor types)

the *Leader* metric produce almost constant message overhead irrespective to the node density in the network, which makes this selection rule more applicable to dense scenarios than *Group*. Moreover, although the *BestChoice* algorithm requires several steps to converge, in combination with *Single* it does not produce a large increase in communication cost (Fig. 3c). Fig. 3d,e) present the average radius and its standard deviation of constructed groups relative to the communication range of the sensor nodes R_c. In all combinations, the *BestChoice* algorithm generates the best solution. Finally, we evaluate the coverage of the monitored area by the constructed groups and plot in Fig. 3f) the percentage of events that remain uncaptured. For this test, we generated 1000 events at random points within the monitoring area. We consider that an event is captured if there is at least one spatial group nearby such that the event is in the sensing range of all sensors of this group. Obviously, the complement to the percentage of uncaptured events represents the percentage of the covered area by the constructed groups. A small group radius and a high number of successfully constructed spatial groups result in only 10% of uncaptured events and, therefore, to 90% of area coverage as presented in Fig. 3f). Again, the *BestChoice* algorithm outperforms the *FirstChoice* algorithm, as it tends to construct tighter groups and, therefore, increases the common area covered by all sensors of the group.

5.2   Time-Bounded Quality of Sensing

In this section we evaluate the presented scheduling algorithm. There are five types of sensors involved in the evaluation. The first three types of sensors are distributed uniformly among the sensor nodes in the network so that 50% of the sensor nodes possess the same type of sensor. The last two types of sensors are distributed randomly over the network so that 60% of the nodes have the same sensor. These sensors might be more common ones like temperature or humidity. The first four sensors have a sensing range of $R_s = \frac{1}{2} R_c$ and the last sensor $R_s = R_c$.

The *BestChoice* and *FirstChoice* grouping algorithms combined with the *Group* selection rule and the *SDev* dispersion metric were used to build spatial groups. Moreover, we constructed two sets of concurrency constraints to evaluate the performance of the scheduling algorithms: "hard" and "medium" presented in Fig. 5. We assume that the sensing tasks are ordered from left to right in the presented graphs. Notice, that neither of these sets of concurrency constraints can be fulfilled when considering sequential scheduling on one sensor node equipped with all required sensors. Therefore, in both cases the schedule must consider the relocation of some sensing tasks to other sensor nodes in the spatial group.

In Fig. 4 we present the evaluation results of the constructed schedules in every complete group in the network. Fig. 4a) plots the percentage of groups able to fulfil the schedule compared to the overall number of complete groups constructed by the spatial grouping algorithms. This graph shows a big difference in the difficulty of both schedules: up to 90% of groups were able to compute a valid "medium" schedule, whereas less than 30% of groups succeeded to execute the "hard" one. However, for both schedules the *BestChoice* algorithm performs worse than *FirstChoice*. Moreover, in Fig. 4b,c) we evaluate changes to the average spatial radius of the groups able to compute a valid schedule compared to the average spatial radius of all constructed groups. Only the worst groups constructed by the *BestChoice* algorithms computed a valid schedule. The problem is, that the *BestChoice* algorithm constructs tighter groups than *FirstChoice* and, therefore, tends to group fewer nodes where more sensor types on each node

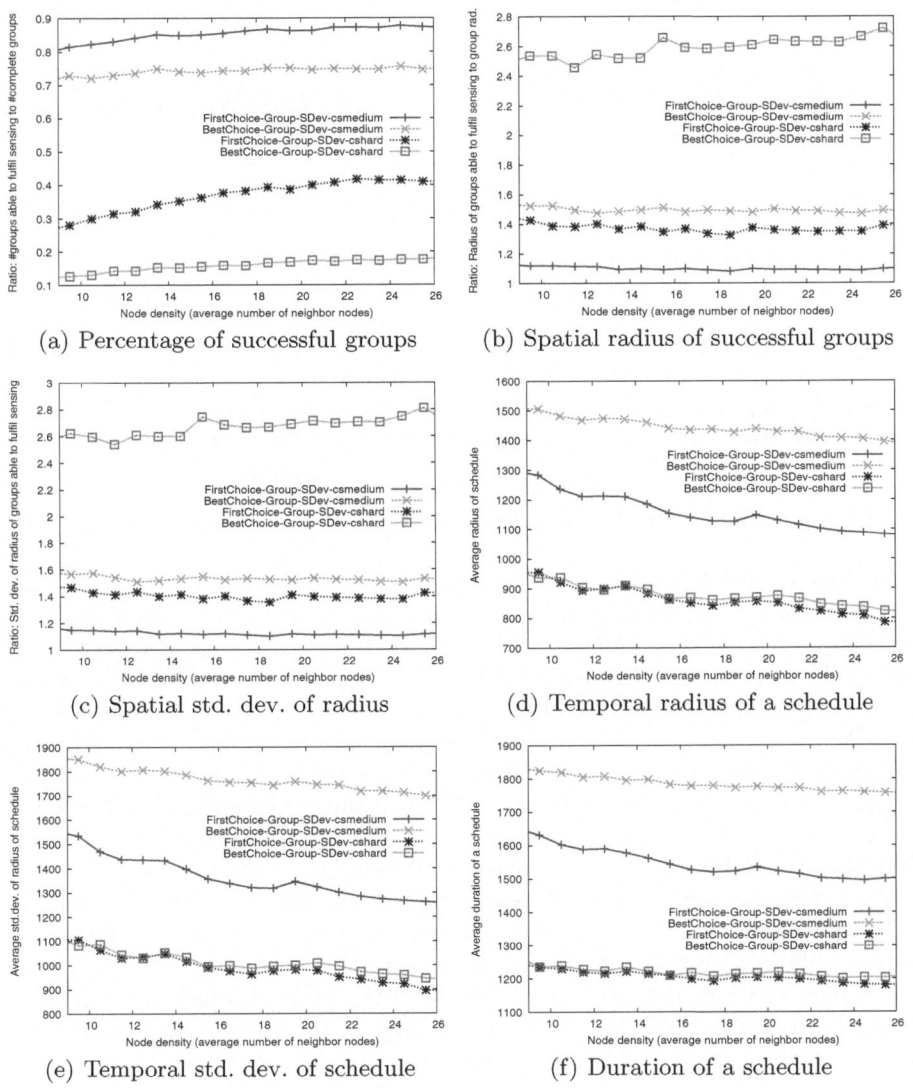

(a) Percentage of successful groups

(b) Spatial radius of successful groups

(c) Spatial std. dev. of radius

(d) Temporal radius of a schedule

(e) Temporal std. dev. of schedule

(f) Duration of a schedule

Fig. 4. Spatial and temporal evaluation of the scheduling algorithm

are used. This shows that the *BestChoice* algorithm is better suited for events that require distributed space coordination whereas the *FirstChoice* algorithm is better suited for "hard" concurrency constraints and, therefore, for distributed time coordination.

In Fig. 4d-f) we evaluate the temporal characteristics of the constructed schedules: its average radius, standard deviation of the average radius in every group and average duration of a schedule. We define *radius of a schedule* as the maximum difference between the start execution times of the earliest and the latest

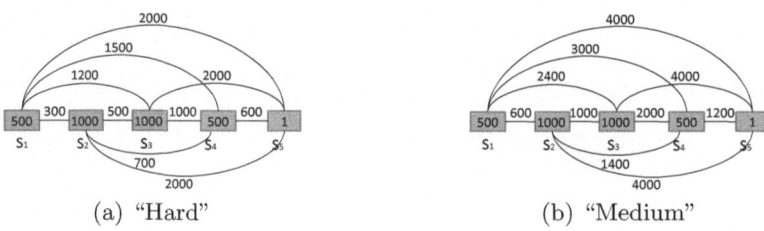

Fig. 5. Two sets of concurrency constraints used for evaluation: "hard" and "medium"

task in the schedule. If the number of sensors in the group is equal to the number of sensor nodes in this group, the radius of a valid schedule equals 0. The *FirstChoice* algorithm outperforms the *BestChoice* algorithm also in the temporal characteristics of constructed schedules due to the reasons explained above.

6   Conclusions and Future Work

In this paper we formulate and provide a solution for the problem of time-bounded and space-bounded sensing in wireless sensor networks in order to enable complex event detection or query execution. Two cases are considered that require the distribution of sensing tasks among several nodes in vincinity: when a node is not equipped with all required sensors and when the event duration and the concurrency dependencies between sensing tasks preclude sequential sensing on one node. We analyse several spatial grouping algorithms for constructing groups of sensor nodes that can act together as one logical node equipped with all needed sensors to recognize an event. Additionally, we provide a scheduling algorithm to enable the efficient relocation of sensing tasks to group members. Our evaluation results show that time-bounded and space-bounded sensing provides good results for complex event detection even in case when no single sensor node is able to accomplish this task on its own.

We plan to implement the described concepts for real sensor nodes as part of our future work. We also want to look into integrating these concepts into existing query and event detection systems for wireless sensor networks.

References

1. Vu, C.T., Beyah, R.A., Li, Y.: Composite event detection in wireless sensor networks. In: Proc. of the IEEE International Performance, Computing, and Communications Conference (2007)
2. Ould-Ahmed-Vall, E., Riley, G.F., Heck, B.S.: Distributed fault-tolerance for event detection using heterogeneous wireless sensor networks. Technical report, Georgia Institute of Technology (2006)
3. Römer, K., Mattern, F.: Event-based systems for detecting real-world states with sensor networks: A critical analysis. In: DEST Workshop on Signal Processing in Sensor Networks at ISSNIP, pp. 389–395 (2004)

4. Mansouri-Samani, M., Sloman, M.: GEM: a generalized event monitoring language for distributed systems. Distributed Systems Engineering 4(2), 96–108 (1997)
5. Janakiram, D., Phani Kumar, A.V.U., Adi Mallikarjuna Reddy, V.: Component oriented middleware for distributed collaboration event detection in wireless sensor networks. In: Proc. of the 3rd International Workshop on Middleware for Pervasive and Ad-Hoc Computing (MPAC 2005) (2005)
6. Krishnamachari, B., Iyengar, S.: Distributed bayesian algorithms for fault-tolerant event region detection in wireless sensor networks. IEEE Trans. Comput. 53(3), 241–250 (2004)
7. Madden, S.R., Franklin, M.J., Hellerstein, J.M., Hong, W.: TinyDB: an acquisitional query processing system for sensor networks. ACM Trans. Database Syst. 30(1), 122–173 (2005)
8. Fung, W.F., Sun, D., Gehrke, J.: Cougar: the network is the database. In: Proc. of the 2002 ACM SIGMOD international conference on Management of data (SIGMOD 2002), pp. 621–621. ACM, New York (2002)
9. Yoon, S., Shahabi, C.: The clustered aggregation (CAG) technique leveraging spatial and temporal correlations in wireless sensor networks. ACM Trans. on Sensor Networks 3(1), 3 (2007)
10. Handy, M., Haase, M., Timmermann, D.: Low energy adaptive clustering hierarchy with deterministic cluster-head selection. In: 4th International Workshop on Mobile and Wireless Communications Networks, pp. 368–372 (2002)
11. Brucker, P.: Scheduling Algorithms, 5th edn. Springer, Heidelberg (2007)
12. Funke, S., Klein, C.: Hole detection or: How much geometry hides in connectivity?. In: Proc. of the 22nd Symp. on Computational Geometry (2006)

SAKE: Software Attestation for Key Establishment in Sensor Networks

Arvind Seshadri, Mark Luk, and Adrian Perrig

Carnegie Mellon University, Pittsburgh PA 15213, USA
arvinds@cs.cmu.edu, mark.luk@gmail.com, perrig@cmu.edu

Abstract. This paper presents a protocol called SAKE (Software Attestation for Key Establishment), for establishing a shared key between any two neighboring nodes of a sensor network. SAKE guarantees the secrecy and authenticity of the key that is established, without requiring any prior authentic or secret information in either node. In other words, the attacker can read and modify the entire memory contents of both nodes before SAKE executes. Further, to the best of our knowledge, SAKE is the only protocol that can perform key re-establishment after sensor nodes are compromised, because the presence of the attacker's code in the memory of either protocol participant does not compromise the security of SAKE. Also, the attacker can perform any active or passive attack using an arbitrary number of malicious, colluding nodes. SAKE does not require any hardware modification to the sensor nodes, human mediation, or secure side channels. However, we do assume the setting of a computationally-limited attacker that does not introduce its own computationally powerful nodes into the sensor network.

SAKE is based on ICE (Indisputable Code Execution), a primitive we introduce in previous work to dynamically establish a trusted execution environment on a remote, untrusted sensor node.

Keywords: Key establishment, software attestation, sensor networks.

1 Introduction

Sensor networks are expected to be deployed in the near future in many safety-critical applications such as critical infrastructure protection and surveillance, fire and burglar alarm systems, home and office automation, inventory control systems, and medical applications such as patient health monitoring. Therefore, securing sensor networks is of paramount importance.

Many security mechanisms proposed for sensor networks rely on cryptography, which makes the problem of key establishment one of the central issues in sensor network security. The problem of key establishment can be described as instantiating secret keys in the nodes of a sensor network in a manner that the authenticity and confidentiality of keys can be guaranteed. Even though key establishment in sensor networks is a well-studied problem, a major body of research that addresses the problem assumes the presence of some secret information in the nodes [1,2,3,4,5,6]. The techniques that do not assume prior secret

S. Nikoletseas et al. (Eds.): DCOSS 2008, LNCS 5067, pp. 372–385, 2008.

information in the nodes namely Integrity Codes [7], Smart-Its Friends [8], Are You with Me [9], Shake Them Up [10], Key Infection [11], and Message in a Bottle (MiB) [12] all either require human mediation, extra hardware, or are insecure.

In this paper, we present SAKE, a key establishment protocol for establishing shared keys between any two neighboring nodes in a sensor network. SAKE is immune to all active and passive attacks without requiring any prior secrets or authentic information in either of the nodes. In order words, the attacker is free to read and modify the entire memory contents of both participants before the protocol executes. In addition, SAKE does not require any secure side channels between the principals. The secrecy and authenticity of the key that is established are guaranteed against an attacker that can perform any active or passive attack using a arbitrary number of compromised, colluding nodes. One consequence of this is that SAKE guarantees that a good node will never establish a key with a malicious node i.e., SAKE is fail-safe.

Another interesting property of SAKE is that the attacker's code can exist in the memory of the participating nodes during the execution of SAKE. In other words, the protocol participants can be compromised. Yet the attacker will not be able to compromise the confidentiality of the key that is established. To the best of our knowledge, this property is unique to SAKE. All other research in the area of sensor network key establishment assumes a network-based adversary, whereby the protocol participants are assumed to be uncompromised. This property of the SAKE makes it suitable for key re-establishment after node compromise since the presence of the attacker's code in either (or both) participants does not compromise the security of SAKE.

The results we present in this paper appear even more surprising since we assume commodity sensor nodes, i.e., no hardware modification is required to the sensor nodes, and do not require human mediation or secure side channels. However, we do assume that the attacker does not introduce its own computationally powerful nodes into the network. Even with this assumption, we have dramatically increased the level of difficulty for an attacker. In order to circumvent our protocol, the attacker's hardware must always be present and must participate actively in the SAKE protocol, which increases the chance of detection. For example, researchers have recently shown that it is possible to use radio fingerprinting to uniquely identify the origin of messages in a sensor network [13]. Such techniques can be used to create a list of nodes that are authorized members of the network, thereby detecting the presence of the attacker's devices. Compared to today's security scenario, where an attack could be launched remotely, SAKE presents a substantial hurdle for the attacker by requiring physical presence. As we discuss in the related work section, the SAKE protocol covers a new point in the design space, offering an appealing trade-off between ease-of-use and security.

SAKE is based on a primitive called ICE (Indisputable Code Execution) that we introduced in our earlier work [14]. Using ICE, a challenger can verify if the execution of an arbitrary piece of code on a remote sensor node will be untampered by any malicious code that might be present. In SAKE, either principal can use ICE to verify if the execution of the SAKE protocol code will be untampered, thereby obtaining the guarantee that the key that is established will not be revealed to malicious code. We implement ICE and SAKE on Telos sensor nodes.

2   Problem Definition, Attacker Model, and Assumptions

This section presents our problem definition, attacker model, and sensor network architecture and assumptions, in that order.

2.1   Problem Definition

When nodes are newly deployed or when they are compromised, they need a mechanism to establish pairwise shared keys with their neighbors. Once pairwise shared keys between neighbors are established, it is trivial to establish shared keys between non-neighboring nodes. Therefore, establishing pairwise shared keys is the fundamental problem in key establishment.

The protocol used to establish shared keys must guarantee the authenticity and confidentiality of the keys that are established. This requires the protocol to be resilient to passive and active attacks. Also, the protocol cannot assume the existence of any secret or authentic information in the participants, since such information may not exist in newly deployed nodes or might have been compromised in the case of compromised nodes.

2.2   Attacker Model

We assume an attacker that can compromise an arbitrary number of sensor nodes and introduce its own sensor nodes into the network. Upon compromising a sensor node, the attacker can read and modify its entire memory contents. Also, the attacker's code can exist in the memory of the SAKE protocol participants during the execution of the SAKE protocol.

The malicious nodes controlled by the attacker can arbitrarily collude. However, the attacker does not introduce computationally powerful devices, such as laptop computers, into the network. Note that introducing new hardware into the network increases the attacker's chance of being detected. The attacker's hardware must always be present in the network and must actively participate in the SAKE protocol. For example, researchers have recently shown that it is possible to use radio fingerprinting to uniquely identify the origin of messages in a sensor network [13]. Such techniques can be used to create a list of nodes that are authorized members of the network, thereby detecting the presence of the attacker's devices. In our future work, we will consider an attacker who is present in the sensor network.

An example attacker that fulfills these requirements would be a remote attacker. Such an attacker exploits software vulnerabilities, such as a buffer overflow vulnerability, to compromise a sensor node. Once a node is compromised, the attacker can gain full control over the node. For example, the attacker can inject and run arbitrary code, and steal cryptographic keys.

2.3   Sensor Network Architecture and Assumptions

We consider a wireless sensor network consisting of one or more base stations and several sensor nodes. All sensor nodes have identical hardware configurations. The sensor nodes communicate among themselves and the base station using wireless communication links. Every sensor node has an unique, immutable identifier, hereafter referred to as its *Silicon ID*. Such Silicon IDs are available on commodity motes. Examples include the 64-bit DS2401 silicon serial number chip present on the Mica2 motes [15], and the 48-bit serial ID chip present on the Telos motes [16]. The immutability of the Silicon ID implies that an attacker cannot modify it, even when it compromises the sensor node. The sensor network is assumed to have a reliable transport layer protocol like PSFQ [17]. The SAKE protocols require that all protocol messages be reliably delivered between the participants. Finally, the nodes are assumed to have a hardware source of randomness. This source of randomness is used to generate the shared key that is established and for other cryptographic operations required by SAKE. A hardware source of randomness ensures that the attacker cannot know the output of the random source during the SAKE's execution, even if it knows the entire memory contents of the sensor node before SAKE beings executing.

The base stations are the gateway between the sensor network and the outside world. Other sensor networks or computers on the Internet can send network packets directly to the sensor nodes but these packets will always pass through at least one of the base stations. This assumption is justified since the nodes have short-range wireless radios. We also assume that the base stations are trusted. This assumption is commonly made by most sensor network security researchers since compromise of the base stations in most cases means the compromise of the network.

3   The SAKE Protocol

In this section we describe the SAKE protocol. Section 3.1 presents an overview of SAKE. Section 3.2 describes the construction of the SAKE in detail. In Section 3.3, we discuss how we select cryptographic primitives for SAKE and present our evaluation.

3.1   Overview

SAKE uses the well-known Diffie-Hellman (DH) key exchange protocol to establish a shared key between the two participating neighboring nodes. The DH protocol is immune to passive attacks. However, it is vulnerable to the man-in-the-middle (MitM) attack, unless the two participating nodes have some means of authenticating each other's DH protocol messages. Then, the key question we face is: how to authenticate the DH protocol messages without relying on the existence of prior secret or authentic information in both participating nodes?

We use the Guy Fawkes protocol by Anderson et al. [18] to authenticate the DH protocol messages. This protocol uses one-way hash chains. We choose the

Guy Fawkes protocol rather than constructs based on asymmetric cryptography, such as digital signatures. Hash chains, which are based on symmetric cryptography, are computationally less expensive to generate. This is an advantage since SAKE has to execute on resource-constrained sensor nodes.

To use the Guy Fawkes protocol, each of the two participants has to "commit" to the other participant's hash chain. That is, each participant needs to obtain at least one authentic value of the other participant's hash chain. The Guy Fawkes protocol does not mention how the participants commit to each other's hash chains. Instead it assumes that an out-of-band mechanism exists for this purpose. In SAKE, we use the ICE primitive to build a mechanism that allows the two participating nodes to commit to each other's hash chains, thereby addressing this shortcoming of the Guy Fawkes protocol.

All the above constructions ensure that a network-based active or passive attacker cannot compromise the confidentiality of the key that is established. The final issue that needs to be addressed is: how can we prevent the attacker that might be present on one of the participants from learning the key? This question is particularly relevant in the case of key re-establishment after node compromise. Even in the case of newly introduced nodes, we cannot be sure that there is no malware present, unless we can trust the distribution channels through which the new nodes are obtained. What we need is a mechanism that prevents the attacker's code, if present on either participating node, from compromising the confidentiality of the key. The ICE primitive provides this mechanism.

3.2   Constructing SAKE

In this section, we progressively construct SAKE by first describing how we use the ICE primitive to achieve two goals: one, prevent an attacker that may be present on one of the participating nodes from compromising the confidentiality of the key and two, build a hash chain commitment mechanism for the Guy Fawkes protocol. Then, we describe how we use the Guy Fawkes protocol to authenticate the DH protocol messages, thereby completing our construction of SAKE.

Using ICE for SAKE ("SAKE on the Rocks")
We start off with a short overview of ICE. Then, we describe how ICE can be used to prevent the attacker's code, that might be present on either participating node, from compromising the confidentiality of the shared key. After that, we discuss the asymmetry in the ICE primitive that we use as the basis of SAKE's hash chain commitment mechanism. Finally, we describe SAKE's hash chain commitment mechanism.

ICE overview. ICE is a challenge-response protocol through which the challenger can verify if an arbitrary piece of code (*target code*), that it invokes on the responder, will execute untampered by any malicious code, that might be present on the responder. In other words, when the ICE protocol succeeds, the challenger obtains an indisputable guarantee that the expected target code executes unmolested by any potential malware present on the responder.

The verification in ICE is done using two checks. First, check that the integrity of the target code is intact before it is invoked for execution. This check ensures that the correct target code image is invoked for execution. Second, check that the execution environment for the target code is set up so that no malicious code will execute concurrently with the target code. Given that the correct target code is invoked for execution, the second check verifies that malicious code will never execute as long as the target code is executing. This prevents the malicious code from attacking the target code during the latter's execution, thereby rendering the malicious code harmless.

ICE is based on a special checksum function that sets up the execution environment for the target code, and then computes a checksum over the memory region containing itself, the target code and the CPU state corresponding to the execution environment. The checksum constitutes a proof of integrity of the target code and the correct instantiation of its execution environment. By verifying the correctness of the checksum it obtains, the challenger can satisfy itself of the indisputable execution of the target code. However, the attacker can try to tamper with the checksum code to forge the correct checksum. To detect such tampering, we construct the checksum function so that its execution time increases on any attempt to tamper with its execution. Therefore, if the challenger obtains the correct checksum within the expected period of time, it obtains the guarantee of indisputable execution of the target code. This overview skips several details of ICE, which can be found in our earlier work [14].

Achieving key confidentiality using ICE. The indisputable execution property of ICE offers a direct mechanism to prevent any malware, that may be present on either SAKE participant, from compromising the confidentiality of the key that is established. Each of the two participants acts as the ICE challenger in turn and verifies that the SAKE protocol code will execute untampered by malware on the other participant. If the verification fails, the challenging node refuses to establish a key with the responding node and terminates the SAKE protocol. This ensures that the SAKE protocol is fail-safe. The confidentiality of the shared key is thus ensured, as long as the SAKE protocol is executing.

Further, as we discuss in our earlier work on ICE [14], we can use ICE to undo the attacker's modifications to the sensor node software. By undoing the attacker's modifications after the shared key is established, we can guarantee the confidentiality of the shared key even after the SAKE protocol terminates. This enables SAKE to re-establish shared keys after node compromise.

Asymmetry in ICE. We use the asymmetry in the checksum value and the computing time of the ICE checksum, between the genuine checksum function and a modified checksum function, as the basis of the mechanism for hash chain commitment. The key observation here is that if we include the Silicon ID of a node as part of the input used to compute the checksum, only the node with the Silicon ID will be able to return the the correct ICE checksum within the expected time. In other words, we can use the ICE checksum as a short-lived secret. No other node in the network will be able to compute the correct checksum until sometime after the node with the correct Silicon ID finishes its computation.

The above conclusion assumes that all nodes in the network have the same computing power and that a malicious node does not ask a remote colluding device, with greater computing power, to compute the checksum on its behalf. The first assumption is valid since the attacker does not introduce its own computationally powerful nodes into the network. To justify the second assumption, first note that one of the base stations has to act as the gateway between any malicious node and a remote collaborator. Since the expected time taken to compute the ICE checksum is around 3 seconds [14], by delaying all outgoing packets by 3 seconds, the base stations can ensure that the checksum response computed by a remote collaborator arrives too late at the malicious node.

If the ICE checksum function directly reads the Silicon ID during checksum computation, to incorporate it into the checksum, then there is possibility of the following attack. Since the Silicon ID is a constant, the attacker could place the correct Silicon ID in the memory of a node with a different Silicon ID and use this in-memory copy during checksum computation. Since reading the in-memory copy of the Silicon ID is faster than reading the Silicon ID off the chip, the attacker could actually compute the checksum faster, despite using a node with an incorrect Silicon ID to compute the checksum!

To detect this attack, we design the ICE checksum function so that it reads the Silicon ID into memory before starting checksum computation and refers to the in-memory copy during the checksum computation. Now, we need a mechanism to check whether the correct Silicon ID was read into memory before the checksum computation started. We construct a function, called the *ID check function*, that performs this check. After the checksum function sends the checksum to the challenger, it invokes the ID check function. This function reads the Silicon ID of the node off the chip and compares it against its in-memory copy. If the comparison does not succeed, the ID check function sends an error message to the challenger.

The above approach works because the ICE checksum function computes the checksum over itself, the ID check function, the SAKE protocol code, and the execution environment. Therefore, as long as the challenger obtains the correct ICE checksum within the expected time, it knows that the attacker has not modified the ID check function and cannot tamper with its execution. Then, if the challenger does not receive an error from the ID check function on the responder, it obtains the guarantee that the responder used the correct Silicon ID to compute the ICE checksum.

Using ICE for hash chain commitment. Now that we have a mechanism that guarantees that the ICE checksum, computed by the node with the correct Silicon ID will be a short-lived secret, we can use ICE to build the hash chain commitment mechanism for SAKE. Figure 1 shows SAKE's hash chain commitment mechanism based on ICE.

Each of the two nodes participating in the SAKE protocol acts as a ICE challenger, and requests the other node to compute the ICE checksum using the responder's Silicon ID as one of the inputs. After computing the checksum, the responding node uses the computed checksum and a random value to generate its hash chain. Using a random value for generating the hash chain ensures that the attacker will not know the hash chain even if it knows the entire memory contents of the sensor node before the SAKE protocol executes.

The responding node then sends the first element of its hash chain along with a Message Authentication Code (MAC) of this element, computed using the checksum as the key, to the challenger. After sending this packet to the challenger, the responding node invokes the ID check function. If the ID check function finds that the in-memory copy of the responding node's Silicon ID does not match the actual value of its Silicon ID, it sends an error message to the challenger and terminates the SAKE protocol.

If the MAC verifies correctly, the challenger obtains the guarantee that the hash chain element it received was computed by a node with the correct in-memory copy of the Silicon ID. This is the case since a node with an incorrect in-memory copy of the Silicon ID will either return an incorrect checksum or will return the correct checksum outside the expected time window. Further, if the ID check function does not return an error, the challenging node knows that the responding node has the correct hardware Silicon ID as well. Together, these two checks assure the challenger that it has obtained an authentic value of the responder's hash chain. In this manner, the challenger commits to the responder's hash chain.

Authenticating DH Protocol Messages

In the previous section, we described how the two nodes participating in SAKE, commit to each other's hash chains. We now describe how we use these hash chains in the Guy Fawkes protocol to authenticate the DH protocol messages. We start off with a brief overview of the Guy Fawkes protocol.

The Guy Fawkes protocol. The Guy Fawkes protocol enables two participants that have previously committed to each other's hash chains to exchange authenticated messages. Each message exchanged between the participants is accompanied by the MAC of the message. The sending participant computes the MAC using the next undisclosed element of its hash chain, as the key. Once the receiving participant acknowledges the receipt of the packet containing the message and its MAC value in an authentic manner, the sending participant discloses element of its hash chain that it used to compute the MAC. The receiving participant then authenticates this hash chain element using the a previously obtained authentic hash chain element. The receiving participant then verifies the authenticity of the message by verifying its MAC using the newly received (now authentic) hash element as the key. In this manner, both principals can use their respective hash chains to exchange authenticated messages with each other, provided they can acknowledge each others messages in an authentic manner.

The Guy Fawkes protocol can be used for sending authenticated acknowledgements as well. The receiving participant simply sends the next undisclosed element of its hash chain to acknowledge the receipt of a message from sending participant. The sending participant verifies the authenticity of the receiving participant's acknowledgement by authenticating the hash chain element it receives, using a previously received authentic hash chain element. If the authentication

$A \rightarrow B :$ $\langle$ICE Challenge$\rangle$

$A :$ $T_1 =$ Current time

$B :$ Compute ICE checksum over memory region
containing the SAKE protocol code,
Check ID function, ICE code, and
in-memory copy of Silicon ID
$C_B =$ ICE checksum
$r_B \overset{R}{\leftarrow} \{0,1\}^{128}$
Generate one-way hash chain,
$b_3 = H(C_B \parallel r_B)$,
$b_2 = H(b_3), b_1 = H(b_2), b_0 = H(b_1)$

$B \rightarrow A :$ $\langle b_0, MAC_{C_B}(b_0) \rangle$

$A :$ $T_2 =$ Current time
Verify $(T_2 - T_1) \leq$ Time allowed to compute
ICE checksum
Compute MAC of b_0 with
ICE checksum computed by A
If MAC of b_0 computed by self equals MAC
of b_0 sent by B
then B's ICE checksum is correct

$B :$ Invoke ID check function

$B \rightarrow A :$ $\langle$Result of ID check function$\rangle$

$A :$ If ID check function on B sends error, abort

Now B acts as the challenger and obtains the first element
of A's hash chain in a similar manner

Fig. 1. Using ICE to construct the hash chain commitment mechanism for SAKE. A and B are two neighboring nodes. This commitment mechanism enables A and B to commit to each other's hash chains without requiring any prior secret or authentic information, and without the existence of secure side channels. A and B can then use their respective hash chains in the Guy Fawkes protocol to authenticate the DH protocol messages.

is successful, then the sending participant knows that the receiving participant has correctly received the message, and therefore, it is now safe to release the hash chain element that the sending participant used to compute the MAC of the message. If the sending participant were to disclose the hash element without requiring an authenticated acknowledgement, then the attacker could suppress the sending participant's packet before it reaches the receiving participant, and later on use the disclosed hash chain element to generate fake messages with the correct MAC value.

As can be seen from the above description, the Guy Fawkes protocol proceeds in a lock-step manner between the two participants, with each of them successively disclosing the next element of their respective hash chains. The sending participant first sends the message along with the MAC. The receiving participant then discloses the next undisclosed element of its hash chain by way of acknowledging the message. Finally, the sending participant discloses the element of its hash chain that it used to compute the MAC of the message.

Using the Guy Fawkes protocol. We now describe how SAKE uses the Guy Fawkes protocol to authenticate the DH protocol messages.

Figure 1 shows how the two nodes participating in the SAKE protocol commit to each other's hash chain, which is a requirement for using the Guy Fawkes protocol. Figure 2 shows how the two nodes now use the Guy Fawkes protocol to authenticate the DH protocol messages. Once both nodes have the authentic first element of each other's hash chain, they send acknowledgements to each other indicating that they have successfully committed to each other's hash chains and are now ready to perform a DH key exchange. After verifying the other participant's acknowledgement, each of node generates its DH half-key. The nodes then exchange their respective half keys and authenticate this exchange using the Guy Fawkes protocol. Finally both nodes compute their shared secret using the half keys they receive.

$$A \rightarrow B : \langle a_1 \rangle$$
$$B \rightarrow A : \langle b_1 \rangle$$

$A :$ Verify $b_0 = H(b_1)$

$B :$ Verify $a_0 = H(a_1)$

With the above acknowledgements A and B indicate that they are now ready to perform the DH key exchange protocol

$A :$ $x \xleftarrow{R} \{0,1\}^{112}$

Compute $(g^x) \bmod p$

$A \rightarrow B : \langle (g^x) \bmod p, MAC_{a_2}((g^x) \bmod p) \rangle$

$B \rightarrow A : \langle b_2 \rangle$

$A :$ Verify $b_1 = H(b_2)$

$A \rightarrow B : \langle a_2 \rangle$

$B :$ Verify MAC of $(g^x) \bmod p$ using a_2

$B \rightarrow A : \langle b_3 \rangle$

$A :$ Verify $b_2 = H(b_3)$

At this point A knows that B has obtained A's authentic DH half-key

Now B sends its DH half key $(g^y) \bmod p$ to A in a similar manner

$A :$ Compute $(g^y \bmod p)^x \bmod p$

$B :$ Compute $(g^x \bmod p)^y \bmod p$

Fig. 2. Using the Guy Fawkes protocol to authenticate DH protocol messages. A and B are two neighboring nodes. This protocol uses the hash chains that A and B committed to in the protocol shown in Figure 1.

3.3 Cryptographic Primitive Selection and Evaluation

In this section, we first discuss how we select the cryptographic primitives for use in SAKE, that are suitable for resource-constrained sensor nodes. Then, we present our evaluation that shows the time taken to execute SAKE and energy consumed.

Selection of cryptographic primitives. We have designed SAKE for sensor nodes with limited computing resources. Hence, the choice of the cryptographic primitives requires careful thought. To save program memory, we suggest code reuse by using a block cipher to implement all required cryptographic primitives. We suggest the use of RC5 [19] as the block cipher because of its small code size and efficiency. In a prior work in sensor network security [20], Perrig et al. stated that an optimized RC5 algorithm can encrypt an eight byte block in 120 cycles on an 8-bit microcontroller. A MAC function can be obtained using the CBC-MAC construction. A hash function can also be constructed with a block cipher as follows: $h(x) = C(x) \oplus x$, where $C(x)$ stands for the encryption of x using the block cipher. The encryption uses a publically known value as the key. Both the MAC function and the hash function are constructed out of the RC5 block cipher with 128 bit blocks, 8 rounds, and 128 bit keys.

Diffie-Hellman parameters. Generally, it is considered impractical to perform expensive asymmetric cryptographic operations on sensor nodes because of resource constraints. In our work, by carefully picking the parameters, it is possible to carry out the DH key exchange protocol on the Telos motes.

The DH key exchange protocol uses an exponentiation operation given by g^x mod p. The security of the DH key exchange protocol is based on the length of the exponent x and the length of the prime p. For SAKE, we use 512 bit modulus p and 112 bit exponent x. According to Lenstra and Verheul, these parameters are deemed to be secure in the year 1990 considering the state of the art technology at that time [21]. Since we are dealing with low cost, mass produced sensor nodes, 1990 levels of security are sufficient. Of course, the attacker can break the keys of the sensor nodes with considerable effort by employing technology that is state-of-the-art. Greater security can be obtained by increasing the sizes of x and p. The cost, though, is the computation time and energy. Since g is unimportant for the security of the DH key exchange protocol, we set g to be two in order to speed up computation. Using these parameters, the Telos motes were able to perform g^x mod p in 6.19 seconds.

4   Related Work

In this section, we review related work in the area of sensor network key establishment.

Many researchers have considered key establishment protocols, however, all these efforts assume the presence of prior secret information to prevent man-in-the-middle attacks [3,5,2,4,20,22].

There are very few key establishment protocols that prevent man-in-the-middle attacks without assuming the presence of authentic or secret information.

Some of these protocols require human-mediation or special hardware [8,9,10,12]. SAKE requires neither but does assume that the attacker does not introduce its own computationally-powerful nodes into the network. In this manner, SAKE is an interesting point the key establishment space, offering a unique trade-off between security and ease-of-use.

Integrity-code [7] attempts to perform key establishment without prior secrets. In this scheme, the presence of an RF signal represents a binary '1', while its absence represents a '0'. Assuming an attacker cannot remove radio energy, an attacker would only be able to modify messages by changing a '0' into a '1', but not the other way around. A carefully selected encoding scheme renders it impossible for an attacker to modify the encoded messages. This ensures message authenticity. However, the drawback to this approach is that the threshold energy value that differentiates a '0' from a '1' needs to be known to the protocol participants in an authentic manner. When posed with this question, the authors declined to specify such a method.

Like SAKE, Key Infection [11] also performs key establishment without assuming prior secret or authentic information. However, since the key is initially transmitted in the clear, the scheme is insecure.

SAKE is the only protocol we are aware of that can also perform key re-establishment after node compromise. This is the case since the presence of the attacker's code in the memory of the protocol participants does not affect the security of SAKE.

5   Conclusion

We present SAKE, a new protocol for key establishment in sensor networks. SAKE's unique feature is that it can establish new keys without requiring any secret values in any node, yet, active or passive attacking sensor nodes cannot perform man-in-the-middle attacks, as long as the attacker is remote and does not insert its own computationally more powerful nodes into the network. The key idea behind our approach is to set up a short-lived shared secret between the protocol participants through the use of the ICE primitive [14].

The main application domains we envision for SAKE are sensor networks that are deployed in physically secure environment, such as in nuclear power plants, financial institutions, military operations, or other critical infrastructures. In these domains, SAKE provides a powerful primitive to establish keys upon network setup and to re-establish secret keys after a remote attacker has compromised nodes, without requiring any human intervention. Hence, SAKE represents a useful new primitive in the key establishment design space, offering a unique trade-off between security and ease-of-use.

Acknowledgments

We thank the anonymous reviewers for their comments and suggestions.

References

1. Perrig, A., Szewczyk, R., Wen, V., Culler, D., Tygar, J.D.: SPINS: Security protocols for sensor networks. Wireless Networks 8(5), 521–534 (2002)
2. Eschenauer, L., Gligor, V.: A key-management scheme for distributed sensor networks. In: Proceedings of Conference on Computer and Communication Security, November 2002, pp. 41–47 (2002)
3. Chan, H., Perrig, A., Song, D.: Random key predistribution schemes for sensor networks. In: IEEE Symposium on Security and Privacy (May 2003)
4. Liu, D., Ning, P.: Establishing pairwise keys in distributed sensor networks. In: Proceedings of ACM Conference on Computer and Communications Security (CCS), October 2003, pp. 52–61 (2003)
5. Du, W., Deng, J., Han, Y., Varshney, P.: A pairwise key pre-distribution scheme for wireless sensor networks. In: Proceedings of ACM Conference on Computer and Communications Security (CCS), October 2003, pp. 42–51 (2003)
6. Karlof, C., Sastry, N., Wagner, D.: TinySec: A link layer security architecture for wireless sensor networks. In: Proceedings of ACM Conference on Embedded Networked Sensor Systems (SenSys) (November 2004)
7. Cagalj, M., Capkun, S., Rengaswamy, R., Tsigkogiannis, I., Srivastava, M., Hubaux, J.P.: Integrity (I) codes: Message integrity protection and authentication over insecure channels. In: IEEE Symposium on Security and Privacy (May 2006)
8. Holmquist, L.E., Mattern, F., Schiele, B., Alahuhta, P., Beigl, M., Gellersen, H.W.: Smart-its friends: A technique for users to easily establish connections between smart artefacts. In: Proceedings of Ubicomp (2001)
9. Lester, J., Hannaford, B., Borriello, G.: Are you with me? Using accelerometers to determine if two devices are carried by the same person. In: Proceedings of Pervasive (2004)
10. Castelluccia, C., Mutaf, P.: Shake them up! a movement-based pairing protocol for cpu-constrained devices. In: Proceedings of ACM/Usenix Mobisys (2005)
11. Anderson, R., Chan, H., Perrig, A.: Key infection: Smart trust for smart dust. In: Proceedings of IEEE Conference on Network Protocols (ICNP) (October 2004)
12. Kuo, C., Luk, M., Negi, R., Perrig, A.: Message-in-a-bottle: User-friendly and secure key deployment for sensor nodes. In: Proceedings of the ACM Conference on Embedded Networked Sensor System (SenSys) 2007 (2007)
13. Rasmussen, K., Capkun, S.: Implications of radio fingerprinting on the security of sensor networks. In: Proceedings of the Third International Conference on Security and Privacy for Communication Networks (SecureComm) (September 2007)
14. Seshadri, A., Luk, M., Perrig, A., van Doorn, L., Khosla, P.: SCUBA: Secure code update by attestation in sensor networks. In: ACM Workshop on Wireless Security (WiSe) (September 2006)
15. Hill, J., Szewczyk, R., Woo, A., Hollar, S., Culler, D., Pister, K.: System architecture directions for networked sensors. In: Architectural Support for Programming Languages and Operating Systems, pp. 93–104 (2000)
16. Polastre, J., Szewczyk, R., Culler, D.: Telos: Enabling ultra-low power wireless research. In: Proceedings of International Conference on Information Processing in Sensor Networks: Special track on Platform Tools and Design Methods for Network Embedded Sensors (IPSN/SPOTS) (April 2005)
17. Wan, C.Y., Campbell, A.T., Krishnamurthy, L.: PSFQ: A reliable transport protocol for wireless sensor networks. In: Proceedings of ACM Workshop on Wireless Sensor Networks and Applications (WSNA) (September 2002)

18. Anderson, R., Bergadano, F., Crispo, B., Lee, J., Manifavas, C., Needham, R.: A new family of authentication protocols. ACM Operating Systems Review 32(4), 9–20 (1998)
19. Rivest, R.: The RC5 encryption algorithm. In: Proceedings of Workshop on Fast Software Encryption, pp. 86–96 (1994)
20. Perrig, A., Szewczyk, R., Wen, V., Culler, D., Tygar, J.D.: SPINS: Security protocols for sensor networks. In: Proceedings of Conference on Mobile Computing and Networks (Mobicom) (July 2001)
21. Lenstra, A., Verheul, E.: Selecting cryptographic key sizes. Journal of Cryptology: The Journal of the International Association for Cryptologic Research (1999)
22. Zhu, S., Setia, S., Jajodia, S.: LEAP: Efficient security mechanisms for large-scale distributed sensor networks. In: Proceedings of ACM Conference on Computer and Communications Security (CCS) (October 2003)

Improving the Data Delivery Latency in Sensor Networks with Controlled Mobility

Ryo Sugihara* and Rajesh K. Gupta

Computer Science and Engineering Department, University of California, San Diego
{ryo,rgupta}@ucsd.edu

Abstract. Unlike traditional multihop forwarding among homogeneous static sensor nodes, use of mobile devices for data collection in wireless sensor networks has recently been gathering more attention. It is known that the use of mobility significantly reduces the energy consumption at each sensor, elongating the functional lifetime of the network, in exchange for increased data delivery latency. However, in previous work, mobility and communication capabilities are often underutilized, resulting in suboptimal solutions incurring unnecessarily large latency. In this paper, we focus on the problem of finding an optimal path of a mobile device, which we call "data mule," to achieve the smallest data delivery latency in the case of minimum energy consumption at each sensor, i.e., each sensor only sends its data directly to the data mule. We formally define the path selection problem and show the problem is $\mathcal{NP}$-hard. Then we present an approximation algorithm and analyze its approximation factor. Numerical experiments demonstrate that our approximation algorithm successfully finds the paths that result in 10%-50% shorter latency compared to previously proposed methods, suggesting that controlled mobility can be exploited much more effectively.

1 Introduction

Exploiting mobility is gaining popularity as a means to solve several issues in traditional multihop forwarding approach in wireless sensor networks and mobile ad-hoc networks. Studies have shown that mobility significantly reduces energy consumption at each node, thus prolongs the network lifetime [1][2][3][4]. Controlled mobility, as opposed to random or predictable mobility by the classification in [5], refers to the case that the observers have the control on the motion of mobile devices, and has the biggest potential for improving the performance of the network. We focus on data collection application in sensor networks and use the term "data mules" to refer to such mobile devices from now on. There are some recent applications that employ data mules for data collection in sensor networks, e.g., a robot in underwater environmental monitoring [6] and a UAV (unmanned aerial vehicle) in structural health monitoring [7].

---

* R. Sugihara was supported by IBM Tokyo Research Laboratory.

S. Nikoletseas et al. (Eds.): DCOSS 2008, LNCS 5067, pp. 386–399, 2008.

There are some studies that analyze the use of data mules in terms of energy efficiency [2][5][8][9]. However, in these studies, mobility and communication capabilities are often underestimated, leading to suboptimal solutions that incur unnecessarily large latency. Some of the underestimations are: the data mule can only move at a constant speed, the data mule needs to go to each node's exact location to collect data from it, the data mule needs to stop during communication with each node, etc.

In this paper, we are interested in improving the data delivery latency in data collection using a data mule. We achieve this through a better formulation of the problem and an efficient approximation algorithm that finds near-optimal solutions. To capture the mobility capability precisely, we assume that the data mule can select the path to traverse the sensor field and also can change its speed under a predefined acceleration constraint. As for the communication capability, we assume that each node can send the data to the data mule when it is within its communication range, regardless of whether the data mule is stopped or moving. With all these assumptions together, we can formulate the problem as a scheduling problem that has both time and location constraints. We focus on the path selection problem in this paper. A heuristic algorithm for optimal speed control and job scheduling can be found in [10].

Our contributions in this paper are:

- Formulating the path selection problem for data collection in sensor networks with a data mule such that the mobility and communication capabilities are precisely captured,
- An efficient approximation algorithm that produces near-optimal paths that enable faster data delivery, and
- Demonstrating the validity and effectiveness of the formulation and the approximation algorithm through numerical experiments by comparing the data delivery latency with previous approaches.

This paper is structured as follows. In Section 2 we introduce the data mule scheduling problem and the related work. In Section 3 we give a formal definition of the path selection problem and describe the preliminary experiments to choose an appropriate cost metric. In Section 4 we present an approximation algorithm and analyze its computational complexity and approximation factor. Section 5 shows some results from numerical experiments and Section 6 concludes the paper.

2   Data Mule Scheduling

The Data Mule Scheduling (DMS) problem is how to control a data mule so that it can collect data from the sensors in a sensor field in the shortest amount of time. As shown in Figure 1, we can decompose the DMS problem into the following three subproblems:

1. Path selection: determines the trajectory of the data mule; produces a set of location jobs

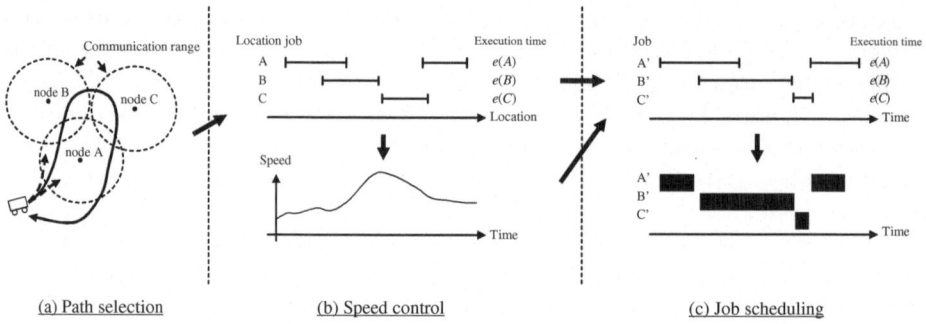

Fig. 1. Subproblems of the data mule scheduling (DMS) problem

2. Speed control: determines how the data mule changes the speed; produces a
   set of jobs
3. Job scheduling: determines the schedule of data collection jobs from individual sensors

The focus of this paper is on the path selection problem. The other two problems
have been formulated as the 1-D DMS problem and we have presented an efficient
heuristic algorithm that yields near-optimal solutions [10].

Path selection is to determine the trajectory of the data mule in the sensor
field. To collect data from each sensor node, the data mule needs to go within the
node's communication range at least once. Depending on the mobility capability
of data mule, there can be some constraints on the path, such as the minimum
turning radius.

2.1   Why Do We Minimize the Latency?

As mentioned earlier, data mules can be used as an alternative to multihop forwarding in sensor networks. The use of data mules in collecting data introduces
the trade-off between energy consumption and data delivery latency. Our objective is to optimize this trade-off, so that energy consumption is minimized under
some latency constraint or vice versa.

Protocol designers have tried to optimize the multihop forwarding in both
energy and latency through sophisticated MAC protocols [11][12][13]. Data mule,
or its combination with multihop forwarding, is a relatively nascent area. In this
paper, we focus on the pure data mule approach, in which each node uses only
direct communication with the data mule and no multihop forwarding. Energy
consumption related to communication is already minimized in this case, since
each node only sends its own data and does not forward others' data. Naturally,
our objective is to minimize the data delivery latency by minimizing the travel
time of the data mule.

2.2   Example Application: SHM with UAV

Our problem formulation is based on our experience with the example application described in [7]. It is a structural health monitoring (SHM) application to do post-event (e.g., earthquakes) assessments for large-scale civil infrastructure such as bridges. Automated damage assessment using sensor systems is much more efficient and reliable than human visual inspections.

In this application, the sensor nodes operate completely passively and do not carry batteries, for the sake of long-term measurement and higher maintainability. Upon data collection, an external mobile element provides energy to each node via microwave transmission, wakes it up, and collects data from it. The prototype system uses a radio-controlled helicopter as the mobile element that is either remotely-piloted or GPS-programmed. Each sensor node is equipped with ATmega128L microcontroller, a 2.4GHz XBee radio, antennas for transmission/reception, and a supercapacitor to store the energy. Each node has two types of sensors. One is a piezoelectric sensing element integrated with nuts and washers to check if the bolt has loosened. The other is capacitive-based sensors for measuring peak displacement and bolt preload. Since the size of data from these sensors are small, communication time is almost negligible; however, it takes a few minutes to charge a supercapacitor through microwave transmission in the current prototype. The team is currently investigating a new design to improve the charging time down to tens of seconds.

The data collected by the UAV is brought back to the base station and analyzed by researchers using statistical techniques for damage existence and its location/type. Since the primary purpose of this application is to assess the safety of large civil structures after a disaster such as an earthquake, every process including data collection and analysis needs to be as quick as possible for prompt recovery. Furthermore, shorter travel time is required in view of the limited fuel on the helicopter.

Thus the goal of our formulation is to achieve data collection from spatially distributed wireless sensors in the minimum amount of time. It also provides another reason for using controlled mobility instead of multihop forwarding approach: simply because the SHM sensors are not capable of doing multihop communication. Furthermore, use of UAVs implies the need for more precise mobility model that takes acceleration constraint into consideration, as opposed to the simple "move or stop" model used in majority of the related work.

2.3   Related Work

The term "data mule" was coined by Shah et al. in their paper in 2003 [3]. They proposed a three-tier architecture having mobile entities called Data MULEs (Mobile Ubiquitous LAN Extensions) in the middle tier on top of stationary sensors under wired access points. As we have also assumed, Data MULEs collect data from sensor nodes when they are in close proximity and deposit it at the wired access points. The difference is that they assumed Data MULEs are not controllable and move randomly, so their routing scheme is rather optimistic.

The use of controlled mobility in sensor networks has been studied in several papers. Kansal et al. [5] analyzed the case in which a data mule (which is called "mobile router" in the paper) periodically travels across the sensor field along a fixed path. Jea et al. [14] used similar assumptions but further assumed multiple data mules are simultaneously on the sensor field. In their models, they can only change the speed of data mule and path selection is out of scope. Path selection problem has been considered in several different problem settings. Somasundara et al. [8] studied path selection problem, assuming that each sensor generates data at a certain rate and that the data mule needs to collect data before the buffer of each sensor overflows. Gu et al. [15] presented an improved algorithm for the same problem settings. Xing et al. [9] presented path selection algorithms for rendezvous based approach. In these work, it is assumed that data mule needs to go to the sensor node's exact location to collect data (i.e., no remote communication)[1]. Although this assumption facilitates TSP-like formulation of the problem, the communication capability is underutilized, since the data mule can actually collect data from nodes without visiting their exact locations via wireless communications. Ma and Yang [2] also discussed the path selection problem but under different assumptions. They consider remote wireless communication and also multihop communication among nodes. However, they assumed the data transmission time is negligible. In a recent paper from the authors [16], they consider constant bit rate case, but they also assume the data mule stops during the communication, whereas we allow communication while in motion.

There are a number of studies on exploiting mobility also in mobile ad-hoc networks (MANETs) area. Among these, our work is most analogous to Message Ferrying [17]. They assume a controllable mobile node (called "ferry") that mediates communications between sparsely deployed stationary nodes. The speed of ferry is basically constant but can be reduced when it is necessary to communicate more data with a node. Further, they consider the extent of wireless communication range to optimize the movement. In our work, we employ a more precise mobility model with acceleration constraint and also realize a more optimized path selection where the data mule only needs to visit subset of nodes as long as it travels inside the communication ranges of all nodes.

3   Path Selection Problem

In this section we give a formal definition of the path selection problem. To make the problem tractable, we first simplify the problem, where the path consists of the line segments between the nodes. The problem is to find a minimum-cost path that intersects with the communication ranges of all nodes. We prove the problem is still $\mathcal{NP}$-hard. Then we do a preliminary experiment to choose an appropriate cost metric before proceeding to solve the problem.

---

[1] In [9], remote communication is used for gathering data at rendezvous points via multihop forwarding, but it is assumed that data mule needs to go to the exact location of these rendezvous points.

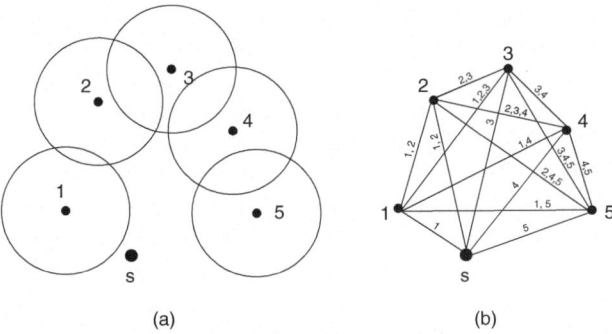

Fig. 2. Simplifying the path selection problem using a labeled graph representation: (a) Instance of path selection problem. (b) Corresponding labeled graph.

3.1 Problem Description

As we have discussed in the previous section, the ultimate objective of the path selection problem is to find a path such that the shortest travel time (= latency) can be realized in the corresponding 1-D DMS problem. However, it is not clear which path results in shorter travel time when there is an acceleration constraint. For example, even if the path length is short, the travel time would be long if the intersections of the path and communication range of each node are short, because the data mule needs to slow down to collect all the data and then accelerate again. Moreover, it is also difficult to search an optimal path in a brute-force manner when the data mule can freely move around within the space.

To deal with these issues, we simplify the path selection problem. To reduce the solution space, we consider a complete graph having vertices at sensor nodes' locations and assume the data mule moves between vertices along a straight line. Each edge is associated with a cost and a set of labels, where the latter represents the set of nodes whose communication ranges intersect with this edge. In other words, the data mule can collect data from these nodes while traveling along this edge. We want to find a minimum-cost tour that the data mule can collect data from all the nodes. We discuss later how we assign the cost to each edge so that a tour with smaller cost results in shorter travel time.

Figure 2 is an example that depicts the basic idea of the formulation. Figure 2(a) shows five nodes and their communication ranges, in addition to the starting point (shown as "s"). From this input, we construct a labeled undirected complete graph as shown in Figure 2(b). Each edge e has a set of labels $L(e) \subseteq L$ and cost $c(e)$, where $L = \{l_1, ..., l_n\}$ is the set of all labels and n is the number of sensor nodes. We determine $L(e)$ as follows: $l_i \in L(e)$ if node i's communication range intersects edge e. Intuitively, by moving along edge e, the data mule can collect data from the nodes whose labels are in $L(e)$.

Now we define the problem formally as follows:

LABEL-COVERING TOUR. Given an undirected complete graph $G = (V, E)$ where each vertex in $V = \{x_0, x_1, ..., x_n\}$ is a point in $\mathbf{R}^2$, a cost function on edges $c : E \rightarrow \mathbf{Q_0^+}$, a set $L = \{l_1, ..., l_n\}$ of labels, and a constant r. Each edge $e_{ij} \in E$ is associated with subset $L_{ij} \subseteq L$. For $k = 1, ..., n$, $l_k \in L_{ij}$ iff the Euclidean distance between x_k and an edge e_{ij} is equal to or less than r. A tour T is a list of points that starts and ends with x_0, allowing multiple visits to each point. A tour T is "label-covering" when it satisfies at least one of the followings for $k = 1, ..., n$: 1) $\exists e_{ij} \in T(E), l_k \in L_{ij}$, where $T(E)$ is the set of edges traversed by T, or 2) $dist(x_0, x_k) \leq r$, where $dist(x_i, x_j)$ is the Euclidean distance between x_i and x_j. Find a label-covering tour T that minimizes the total cost $\sum_{e_{ij} \in T(E)} c_{ij}$.

Unfortunately, this simplified problem is still $\mathcal{NP}$-hard.

Theorem 1. *LABEL-COVERING TOUR is $\mathcal{NP}$-hard.*

Proof. We show metric TSP is a special case. First we choose the cost function c to satisfy the triangle inequality (e.g., Euclidean distance). For a given set of points $V = \{x_0, ..., x_n\}$, by choosing a small r, we can make $dist(x_0, x_i) > r$ for all $i > 0$, $L_{ij} = \{l_i, l_j\}$ for all $i, j > 0$, and $L_{0j} = \{l_j\}$ for all $j > 0$ For such r, any label-covering tour must visit all the points. An optimal label-covering tour does not visit any point multiple times except x_0 at the start and the end of the tour, since in such cases, we can construct another label-covering tour with smaller total cost by "shortcutting". Therefore, an optimal label-covering tour is an optimal TSP tour for V. □

3.2   Choice of Cost Metric

In the definition of LABEL-COVERING TOUR, the cost c_{ij} is a critical parameter. In a restricted scenario, in which the data mule can either move at a constant speed or stop and no remote communication is used, Euclidean distance is the optimal cost metric in the sense that the shortest travel time is realized when the total path length is minimum. However, it is not clear for the general case in which the speed is variable under an acceleration constraint.

Since we minimize the total cost, and also we want to choose a tour that we can achieve the shortest travel time, a good cost metric should be strongly correlated with the travel time. Therefore we measure the goodness of cost metric by the correlation coefficient between cost and total travel time in the corresponding 1-D DMS problem. When the correlation is high, smaller cost implies shorter total travel time, and thus finding a minimum cost tour makes more sense.

We compare three different cost metrics that seem reasonable:

- Number of edges: $c_{ij} = 1$
- Euclidean distance: $c_{ij} = dist(x_i, x_j)$
- Uncovered distance: $c_{ij} = \sum_{s \subseteq e_{ij}, \forall k, dist(x_k, s) > r} |s|$, i.e., total length of intervals in edge e_{ij} that are not within the communication ranges of any nodes.

Uncovered distance is apparently a reasonable cost metric because it represents the total distance that the data mule "wastes", i.e., travels without communicating with any nodes.

Table 1. Correlation coefficients between total cost and total travel time for different cost metrics: 20 nodes, $a_{max} = 1$, $v_{max} = 10$

Radius (d)	150				500			
Comm. range (r)	10		100		10		100	
Exec. time (e)	2	20	2	20	2	20	2	20
Num. edge	0.992	0.987	0.982	0.850	0.984	0.982	0.988	0.988
Euclidean dist.	0.997	0.996	0.990	0.835	0.999	0.999	0.999	0.999
Uncovered dist.	0.992	0.993	—	—	0.999	0.999	0.935	0.935

Experimental Methods. We assume nodes are deployed in the circular area of radius d that has a start (i.e., point x_0) in the center. We randomly place other nodes within the circle so that they are uniformly distributed. For each edge connecting a pair of nodes, we assign a set of labels by calculating the distance from the line segment and each node.

For each of the node deployments, we randomly generate label-covering tours. A tour is generated by random walk which, at each point, chooses next point randomly, repeats this until all the labels are covered, and goes back to x_0. We measure the cost of the tour in three different cost metrics as listed above.

Using the tour, we transform the original problem to 1-D DMS problem and find a near-optimal latency by using the heuristic algorithm presented in [10]. We assume that each node has the same execution time e and the communication range r, and also that the speed of data mule needs to be zero at each point where it changes the direction[2]. For each node deployment, the cost and the total travel time are normalized among different random tours so that the mean is zero and standard deviation is one. For the collective set of data for same d, r, and e, we calculate the correlation coefficient between the normalized cost and the normalized total travel time for each cost metric.

For each (d, r, e), we generate 1000 examples, which consist of 20 random tours for each of 50 node deployments. We use $d = 150, 500$ and $r = 10, 100$ for LABEL-COVERING TOUR. For 1-D DMS, we use the execution time $e = 2, 20$, the maximum absolute acceleration $a_{max} = 1$, and the maximum speed $v_{max} = 10$. These parameter values roughly simulate data collection by a helicopter as in [7] by assigning units as follows: meters for d and r, seconds for e, m/s^2 for a_{max}, and m/s for v_{max}. Also the values of r are chosen to simulate the communication ranges of IEEE 802.15.4 and 802.11, respectively.

Results. Table 1 shows the correlation coefficients. Except one case, both number of edges and Euclidean distance had correlation coefficient above 0.98, and Euclidean distance had higher correlation than number of edges. In the exceptional case $(d, r, e) = (150, 100, 20)$, the correlation was weaker than other cases. This is most likely because the travel time is more influenced by the execution time rather than the moving time, since the deployment area is small relative to

---

[2] Otherwise, it would require infinite acceleration, since we assume a path consists of only line segments and not curves.

- Make a TSP tour T using an exact or approximation algorithm for metric TSP
- Initialize $d[0] \leftarrow 0$, $d[1...n] \leftarrow +\infty$, $tour[0] \leftarrow \{T(0)\}$, and $tour[1...n] \leftarrow \emptyset$.
- For $i = 0$ to $n - 1$ do
  - For $j = i + 1$ to n do
    * Check if the line segment $T(i)T(j)$ is within the distance r from each of the nodes $T(i + 1), ..., T(j - 1)$.
    * If yes and $d[i] + |T(i)T(j)| < d[j]$, update the tables by $d[j] \leftarrow d[i] + |T(i)T(j)|$ and $tour[j] \leftarrow append(tour[i], T(j))$.
- Return $tour[n]$.

Fig. 3. Approximation algorithm for LABEL-COVERING TOUR: $T(i)$ is the i-th vertex that the tour T visits. $T(0)$ is the starting vertex.

the size of communication range and also the execution time is long. Uncovered distance had similar results but had no data when $(d, r, e) = (150, 100, 2)$ and $(d, r, e) = (150, 100, 20)$. These are the cases when the communication range is so broad that the total cost measured by uncovered distance is always zero.

These results suggest that number of edges and Euclidean distance are both appropriate metrics that precisely measure the goodness of paths. In the rest of the paper, we use Euclidean distance as the cost metric.

4 Approximation Algorithm for the Path Selection Problem

In the previous section we have formulated the path selection problem as LABEL-COVERING TOUR and shown its $\mathcal{NP}$-hardness. In this section, we design an approximation algorithm for this problem.

As discussed earlier, Euclidean distance is an appropriate cost metric. This enables us to design an approximation algorithm by using known algorithms for metric TSP where the triangle inequality holds. Figure 3 shows the approximation algorithm for LABEL-COVERING TOUR. It first finds a TSP tour T by using any algorithm (exact or approximate) for TSP. Then, using dynamic programming, it finds the shortest label-covering tour that can be obtained by applying shortcutting to T. For the dynamic programming, we use two tables $d[i]$ and $tour[i]$, where $tour[i]$ is the shortest path that is obtained by shortcutting T and covers the labels $T(0), ..., T(i)$, and $d[i]$ is the length of $tour[i]$.

4.1 Analysis

Computation time of the algorithm is $\mathcal{C}_{TSP} + O(n^3)$, where $\mathcal{C}_{TSP}$ denotes the computation time of the algorithm used for solving TSP.

Next we analyze the approximation factor of the algorithm. Let T_{OPT}, T_{APP} denote the optimal label-covering tour and the approximate label-covering tour,

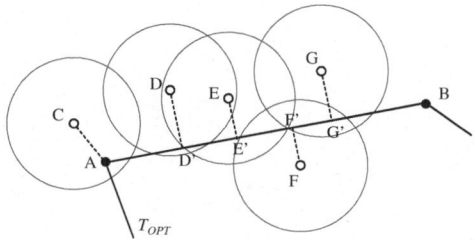

Fig. 4. Constructing a TSP tour from the optimal label-covering tour T_{OPT}: every non-visited point is within distance r from T_{OPT}

respectively. Total length of tour T is denoted as $|T|$. Also let α be the approximation factor of the TSP algorithm used in the first step of the approximation algorithm. Then we have the following theorem:

Theorem 2. $|T_{APP}| \leq \alpha(|T_{OPT}| + 2nr)$

Proof. Clearly $|T_{APP}| \leq \alpha|T_{TSP}|$, where T_{TSP} is the optimal TSP tour. We give a lower bound to T_{OPT} by constructing another TSP tour by modifying T_{OPT}. Figure 4 shows the idea of construction. The points A and B (shown in filled circles) are visited by T_{OPT} and other points in the figure (shown in non-filled circles) are not. We call the former "visited points" and the latter "non-visited points". By the definition of label-covering tour, any non-visited points are within distance r from either a traversed edge or a visited point of a label-covering tour. For example in the figure, all of AC and DD', ..., GG' have the length less than r. Then we can construct a "tour"[3] that is identical to T_{OPT} but takes a detour to visit each non-visited point (e.g., ACAD'DD'...B). Since there are at most n non-visited points, total length of detour is at most $2nr$. This "tour" is easily converted to a shorter TSP tour by skipping all additional points (e.g., D', E', ...) and apply shortcutting so that each point is visited exactly once. Therefore, we have $|T_{OPT}| + 2nr \geq |T_{TSP}|$. The theorem follows by combining this and $|T_{APP}| \leq \alpha|T_{TSP}|$.   □

5   Performance Evaluation

We evaluate the performance of the approximation algorithm by numerical experiments. We have implemented the algorithm in MATLAB. We use the same method and parameters as in Section 3.2 for test case generation, and use Concorde TSP solver[4] to find an optimal TSP tour. For each node deployment, we obtain a label-covering tour by running the approximation algorithm. Using the node deployment and the tour, we get a set of "location jobs", which is the

---

[3] This is not a tour in our definition because it does not consist of edges between the nodes.

[4] http://www.tsp.gatech.edu/concorde/index.html

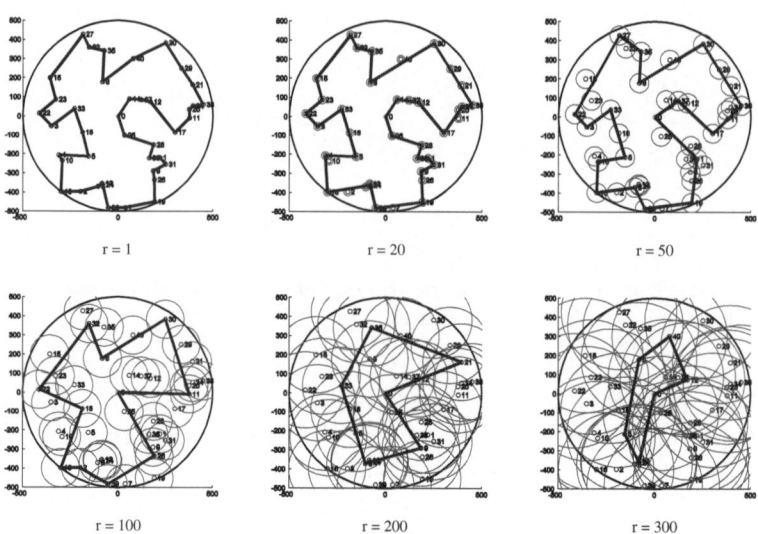

Fig. 5. Label-covering tours for different communication ranges: 40 nodes, $d = 500$; Path of data mule is shown in bold line

input to 1-D DMS problem. A location job is the notion defined in [10], and is intuitively understood as a real time job whose feasible interval is defined on 1-D location axis instead of time axis. Then we run the heuristic algorithm for 1-D DMS problem [10] and obtain the total travel time, which is near-optimal for the given tour. We use the total travel time as the evaluation metric.

Figure 5 shows some examples of label-covering tours for a node deployment with different communication ranges. As the communication range grows, the number of visited points becomes less and the path length becomes shorter.

5.1 Effect of Node Density and Network Size

Figure 6(a) shows the relation between the communication range and the total travel time for different node density. To see how broader communication range affects the travel time, we have normalized the total travel time by the one when the communication range is zero. The graph shows that the total travel time is reduced in all cases by up to 60% for this parameter set, suggesting the proposed problem formulation and algorithm altogether successfully exploit the breadth of communication range. The amount of reduction is bigger when the density is higher (i.e., smaller d), except the case of $d = 150$ for large communication ranges. This is because the total travel time is already very close to the lower bound, which is the product of the execution time and the number of nodes.

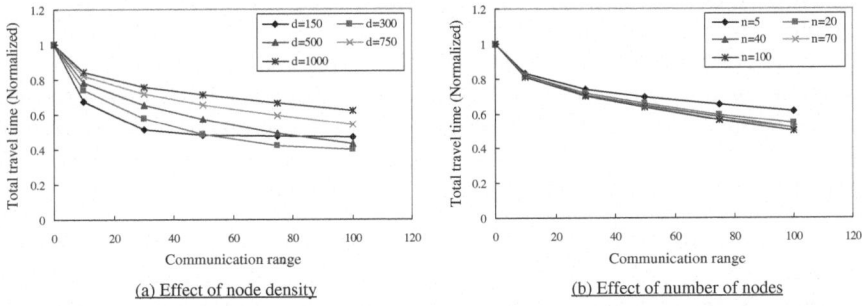

Fig. 6. Comparison of total travel time for (a) different node density (40 nodes) and (b) different number of nodes ($d = 500$ for 20 nodes): $e = 10$, $a_{max} = 1$, $v_{max} = 10$

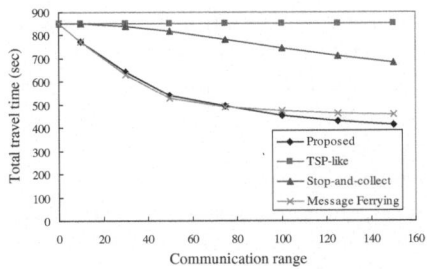

Fig. 7. Comparison of total travel time for different path selection algorithms: 40 nodes, $d = 500$, $e = 10$, $a_{max} = +\infty$, $v_{max} = 10$

Figure 6(b) shows the effect of number of nodes, varied from $n = 5$ to $n = 100$. We set d to 500 when $n = 20$, and changed d in proportion to $\sqrt{n}$ so that the density remains constant. The results show the reduction of total travel time for large communication ranges, but no big difference for different number of nodes.

5.2   Comparison with Other Strategies

Next we compare the travel time of our approximation algorithm with those of other algorithms as listed below.

- TSP-like: Based on the model used in [8]. Data mule visits all nodes. It stops at each node location to collect data and moves to the next node. While moving, the speed is constant at v_{max}. We use optimal TSP tours.
- Stop-and-collect: Based on the model used in [16]. Data mule takes a label-covering tour, as in our approximation algorithm. However, it stops to collect data when it is in the communication range of each node. While moving,

speed is constant at v_{max}. We find tours by using our approximation algorithm with optimal TSP tours[5].

- Message Ferrying: Based on the algorithm proposed in [17]. Data mule visits all nodes as in the TSP-like algorithm, but communication is also done while moving. Speed is variable between 0 and v_{max}. Speed and data collection schedule are determined by solving a linear program such that the total travel time is minimized. We use optimal TSP tours.

To allow direct comparison, we set $a_{max} = +\infty$ for our proposed approximation algorithm, since all other algorithms assume data mule can change its speed instantly. Note that when $a_{max} = +\infty$, we can obtain an exact solution for 1-D DMS problem by solving a linear program (see [10] for details).

Figure 7 shows the results for a representative case for 40 nodes. When the communication range is small, the travel time does not differ among the algorithms. As the communication range grows, Message Ferrying and the proposed algorithm show larger improvements than other two methods, and the proposed algorithm gets gradually better than Message Ferrying. When the communication range is 150, the proposed algorithm is nearly 10% better than Message Ferrying, 40% better than Stop-and-collect, and more than 50% better than TSP-like method.

When there is an acceleration constraint (i.e., $a_{max} \neq +\infty$), which none of these studies has addressed, the gaps between the proposed algorithm and others are expected to be larger. This is because all of these methods require the data mule to stop more frequently than the proposed algorithm does.

These results suggest that the proposed algorithm effectively exploits broader communication range for planning the path of the data mule.

6   Conclusions and Future Work

Controlled mobility for data collection in wireless sensor network provides flexibility in the trade-off between energy consumption and data delivery latency. With the goal of optimizing this trade-off, in this paper we focused on improving the latency in the pure data mule approach with the minimum energy consumption. The formulation of path selection problem, together with speed control and job scheduling problems, enables us to capture two-dimensional data mule scheduling problem under precise mobility and communication models. We have designed an approximation algorithm for the problem and experimentally demonstrated that it finds near-optimal paths and achieves much smaller latency compared to previously proposed algorithms.

Our ongoing work includes design of a hybrid approach that optimizes the energy-latency trade-off. The idea is to extend the problem framework shown in Figure 1 by adding "forwarding" subproblem before the path selection. In

---

[5] We could not use the path selection algorithm proposed in [16], since it has a restriction on the configuration of data mule and deployment area. Specifically, it assumes the data mule starts from the left end of the deployment area, travels toward the right end, and comes back to the initial position.

forwarding problem, we determine how each node forwards the data to other nodes within given energy consumption limit. Then the data mule collects data only from the nodes that have data after the forwarding. Since the results in this paper suggest the problem formulation along with the approximation algorithm allows significant reduction in latency, the hybrid approach is likely to achieve better trade-off between energy and latency compared to previous work.

References

1. Chakrabarti, A., Sabharwal, A., Aazhang, B.: Using predictable observer mobility for power efficient design of sensor networks. In: IPSN, pp. 129–145 (2003)
2. Ma, M., Yang, Y.: SenCar: An energy efficient data gathering mechanism for large scale multihop sensor networks. In: DCOSS, pp. 498–513 (2006)
3. Shah, R.C., Roy, S., Jain, S., Brunette, W.: Data MULEs: modeling a three-tier architecture for sparse sensor networks. In: SNPA, pp. 30–41 (2003)
4. Wang, W., Srinivasan, V., Chua, K.C.: Using mobile relays to prolong the lifetime of wireless sensor networks. In: MobiCom., pp. 270–283 (2005)
5. Kansal, A., Somasundara, A.A., Jea, D.D., Srivastava, M.B., Estrin, D.: Intelligent fluid infrastructure for embedded networks. In: MobiSys., pp. 111–124 (2004)
6. Vasilescu, I., Kotay, K., Rus, D., Dunbabin, M., Corke, P.: Data collection, storage, and retrieval with an underwater sensor network. In: SenSys., pp. 154–165 (2005)
7. Todd, M., Mascarenas, D., Flynn, E., Rosing, T., Lee, B., Musiani, D., Dasgupta, S., Kpotufe, S., Hsu, D., Gupta, R., Park, G., Overly, T., Nothnagel, M., Farrar, C.: A different approach to sensor networking for SHM: Remote powering and interrogation with unmanned aerial vehicles. In: Proceedings of the 6th International workshop on Structural Health Monitoring (2007)
8. Somasundara, A.A., Ramamoorthy, A., Srivastava, M.B.: Mobile element scheduling for efficient data collection in wireless sensor networks with dynamic deadlines. In: RTSS, pp. 296–305 (2004)
9. Xing, G., Wang, T., Xie, Z., Jia, W.: Rendezvous planning in mobility-assisted wireless sensor networks. In: RTSS, pp. 311–320 (2007)
10. Sugihara, R., Gupta, R.K.: Data mule scheduling in sensor networks: Scheduling under location and time constraints. UCSD Tech. Rep. CS2007-0911 (2007)
11. Polastre, J., Hill, J., Culler, D.: Versatile low power media access for wireless sensor networks. In: SenSys, pp. 95–107 (2004)
12. Rhee, I., Warrier, A., Aia, M., Min, J.: Z-MAC: a hybrid MAC for wireless sensor networks. In: SenSys, pp. 90–101 (2005)
13. Ye, W., Heidemann, J., Estrin, D.: An energy-efficient MAC protocol for wireless sensor networks. In: INFOCOM, pp. 1567–1576 (2002)
14. Jea, D., Somasundara, A.A., Srivastava, M.B.: Multiple Controlled Mobile Elements (Data Mules) for Data Collection in Sensor Networks. In: Prasanna, V.K., Iyengar, S.S., Spirakis, P.G., Welsh, M. (eds.) DCOSS 2005. LNCS, vol. 3560, pp. 244–257. Springer, Heidelberg (2005)
15. Gu, Y., Bozdağ, D., Brewer, R.W., Ekici, E.: Data harvesting with mobile elements in wireless sensor networks. Computer Networks 50(17), 3449–3465 (2006)
16. Ma, M., Yang, Y.: SenCar: An energy efficient data gathering mechanism for large-scale multihop sensor networks. IEEE Trans. Parallel and Distributed System 18(10), 1476–1488 (2007)
17. Zhao, W., Ammar, M.: Message ferrying: Proactive routing in highly-partitioned wireless ad hoc networks. In: FTDCS, pp. 308–314 (2003)

Decoding Code on a Sensor Node

Pascal von Rickenbach and Roger Wattenhofer

Computer Engineering and Networks Laboratory, ETH Zurich, Switzerland
{pascalv,wattenhofer}@tik.ee.ethz.ch

Abstract. Wireless sensor networks come of age and start moving out
of the laboratory into the field. As the number of deployments is increasing the need for an efficient and reliable code update mechanism
becomes pressing. Reasons for updates are manifold ranging from fixing
software bugs to retasking the whole sensor network. The scale of deployments and the potential physical inaccessibility of individual nodes
asks for a wireless software management scheme. In this paper we present
an efficient code update strategy which utilizes the knowledge of former
program versions to distribute mere incremental changes. Using a small
set of instructions, a delta of minimal size is generated. This delta is
then disseminated throughout the network allowing nodes to rebuild the
new application based on their currently running code. The asymmetry
of computational power available during the process of encoding (PC)
and decoding (sensor node) necessitates a careful balancing of the decoder complexity to respect the limitations of today's sensor network
hardware. We provide a seamless integration of our work into Deluge,
the standard TinyOS code dissemination protocol. The efficiency of our
approach is evaluated by means of testbed experiments showing a significant reduction in message complexity and thus faster updates.

1   Introduction

Recent advances in wireless networking and microelectronics have led to the
vision of sensor networks consisting of hundreds or even thousands of cheap
wireless nodes covering a wide range of application domains. When performing
the shift from purely theoretical investigations to physical deployments the need
for additional network management services arises. Among other facilities this
includes the ability to reprogram the sensor network [1,2]. Software updates are
necessary for a variety of reasons. Iterative code updates on a real-world testbed
during application development is critical to fix software bugs or for parameter
tuning. Once a network is deployed the application may need to be reconfigured
or even replaced in order to adapt to changing demands.

Once deployed, sensor nodes are expected to operate for an extended period
of time. Direct intervention at individual nodes to install new software is at best
cumbersome but may even be impossible if they are deployed in remote or hostile
environments. Thus, network reprogramming must be realized by exploiting the
network's own ability to disseminate information via wireless communication.
Program code injected at a base station is required to be delivered to all nodes

S. Nikoletseas et al. (Eds.): DCOSS 2008, LNCS 5067, pp. 400–414, 2008.
© Springer-Verlag Berlin Heidelberg 2008

in its entirety. Intermediate nodes thereby act as relays to spread the software within the network. Given the comparatively small bandwidth of the wireless channel and the considerable amount of data to be distributed, classical flooding is prone to result in serious redundancy, contention, and collisions [3]. These problems prolong the update completion time, i.e. the time until all nodes in the network fully received the new software. Even worse, sensor nodes waste parts of their already tight energy budgets on superfluous communication. Current code distribution protocols for sensor networks try to mitigate the broadcast storm problem by incorporating transmission suppression mechanisms or clever sender selection [4, 5, 6].

The radio subsystem is one of the major cost drivers in terms of energy consumption on current hardware platforms. Therefore, communication should be limited to a minimum during reprogramming in order not to reduce the lifetime of the network too much. Orthogonal to the above mentioned efforts the amount of data that is actually disseminated throughout the network should be minimized. Data compression seems to be an adequate answer to this problem. As knowledge about the application currently executed in the sensor network is present[1] differential compression, also known as delta compression, can be applied. Delta algorithms compress data by encoding one file in terms of another; in our case encoding the new application in terms of the one currently running on the nodes. Consequently, only the resulting delta file has to be transferred to the nodes which are then able to reconstruct the new application by means of their current version and the received delta. There exists a rich literature proposing a plethora of different algorithms for delta compression, e.g. [7,8,9,10,11,12]. These algorithms shine on very large files. However, neither time nor space complexity is crucial considering the small code size of today's sensor network applications. It is much more important to account for the asymmetry of disposable computational power at the encoder and the decoder. While almost unlimited resources are available to generate the delta file on the host machine special care must be taken to meet the stringent hardware requirements when decoding on the nodes.

In this paper we present an efficient code update mechanism for sensor networks based on differential compression. The delta algorithm is pursuing a greedy strategy resulting in minimal delta file sizes. The algorithm operates on binary data without any prior knowledge of the program code structure. This guarantees a generic solution independent of the applied hardware platform. We refrain from compressing the delta any further as this would exceed the resources available at the decoder. Furthermore, in contrast to other existing work we directly read from program memory to rebuild new code images instead of accessing flash memory which is slow and costly. The delta file is also structured to allow sequential access to persistent storage. All this leads to a lean decoder that allows fast and efficient program reconstruction at the sensor nodes.

Our work is tightly integrated into Deluge [5], the standard code dissemination protocol for the TinyOS platform. Deluge has proven to reliably propagate large

---

[1] This assumption is based on the fact that sensor networks are normally operated by a central authority.

objects in multi-hop sensor networks. Furthermore, it offers the possibility to store multiple program images and switch between them without continuous download. We support code updates for all program images even if they are not currently executed. Performance evaluations show that update size reductions in the range of 30% for major upgrades to 99% for small changes are achieved. This translates to a reprogramming speedup by a factor of about 1.4 and 100, respectively.

The remainder of the paper is organized as follows: After discussing related work in the next section, we give an overview of the code update mechanism in Section 3. Section 4 describes the update creation process as well as the decoder. In the subsequent section we present an experimental evaluation of our reprogramming service. Section 6 concludes the paper.

2   Related Work

The earliest reprogramming systems in the domain of wireless sensor networks, e.g. XNP [13], did not spread the code within the network but required the nodes to be in transmission range of the base station in order to get the update. This drawback was eliminated by the appearance of MOAP [4] which provides a multi-hop code dissemination protocol. It uses a publish-subscribe based mechanism to prevent saturation of the wireless channel and a sliding window protocol to keep track of missing information.

Deluge [5] and MNP [6] share many ideas as they propagate program code in an epidemic fashion while regulating excess traffic. Both divide a code image into equally sized pages, pipelining the transfer of pages and thus making use of spatial multiplexing. A bit vector is used to detect packet loss within a page. Data is transmitted using an advertise-request-data handshake. Deluge uses techniques such as a sender suppression mechanism borrowed from SRM [14] to be scalable even in high-density networks. In contrast, MNP aims at choosing senders that cover the maximum number of nodes requesting data. There have been various proposals based on the above mentioned protocols, e.g. [15,16], that try to speed up program dissemination. However, all these approaches share the fact that the application image is transmitted in its entirety. This potentially induces a large amount of overhead in terms of sent messages but also in terms of incurred latency.

There have been efforts to update applications using software patches outside the sensor network community in the context of differential compression that arose as part of the string-to-string correlation problem [18]. Delta compression is concerned with compressing one data set, referred to as the *target image*, in terms of another one, called the *source image*, by computing a *delta*. The main idea is to represent the target image as a combination of copies from the source image and the part of the target image that is already compressed. Sections that cannot be reconstructed by copying are simply added to the delta file. Examples of such delta encoders include *vdelta* [7], *xdelta* [9], and *zdelta* [10]. They incorporate sophisticated heuristics to narrow down the solution space at the prize of decreased memory and time complexity as it is important to perform well on very large input files. However, these heuristics result in suboptimal compression. The *zdelta* algorithm further encodes the delta file using Huffman coding. This raises the decoder complexity to a level which does not match the constraints of current sensor network hardware. In [17], the *xdelta* algorithm is used to demonstrate the efficiency of incremental linking in the domain of sensor networks. However, the authors do not give a fully functional network reprogramming implementation but use the freely available *xdelta* encoder to evaluate the fitness of their solution. There exists other work in the domain of compilers and linkers trying to generate and layout the code such that the new image is as similar as possible to a previous image. Update-conscious compilation is addressed in [19] where careful register and data allocation strategies lead to significantly smaller difference files. In [20] incremental linkers are presented that optimize the object code layout to minimize image differences. All these approaches are orthogonal to our work and can be integrated to further increase the overall system performance. We refrain from including one of them since they are processor specific and thus do not allow a generic solution.

Similar to the above mentioned delta algorithms, *bsdiff* [11] does also encode the target image by means of copy and insert operations. The algorithm does not search for perfect matches to copy from but rather generates approximate matches where more than a given percentage of bytes are identical. The differences inside a match are then corrected using small insert instructions. The idea is that these matches roughly correspond to sections of unmodified source code and the small discrepancies are caused by address shifts and different register allocation. Delta files produced by *bsdiff* can be larger than the target image but are highly compressible. Therefore, a secondary compression algorithm is used (in the current version *bzip2*) which makes the algorithm hardly applicable for sensor networks. The authors of [21] propose an approach similar to *bsdiff*. Their algorithm also produces non-perfect matches which are corrected using repair and patch operations. The patch operations work at the instruction level[2] to recognize opcodes having addresses as arguments which must be moved by an offset given by the patch operation. The algorithm shows promising results but depends on the instruction set of a specific processor.

In [8] *rsync* is presented that efficiently synchronizes binary files in a network with low-bandwidth communication links. It addresses the problem by using two-stage fingerprint comparison of fixed blocks based on hashing. An adaptation to *rsync* in the realm of sensor networks is shown in [22]. As both the source image and the target image reside on the same machine various improvements were introduced. The protocol was integrated into XNP. Besides the fact that XNP does only allow single-hop updates, the protocol does not overcome the limitations of *rsync* and performs well only if the differences in the input files are small.

FlexCup [23] exploits the component-based programming abstraction of TinyOS to shift from a monolithic to a modular execution environment.

---

[2] Their work is based on the MSP430 instruction set.

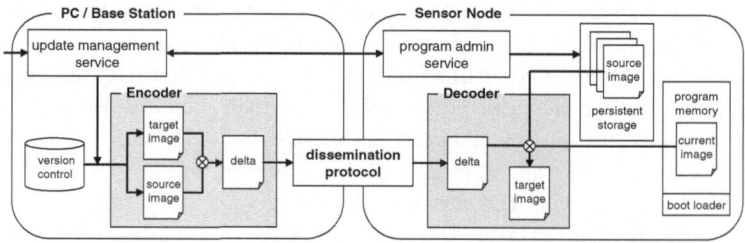

Fig. 1. Components involved in the process of wireless reprogramming

Instead of building one single application image FlexCup produces separate object files which are then linked to an executable on the sensor node itself. Thus, code changes are handled at the granularity of TinyOS components. This solution does no longer allow global optimizations. Furthermore, since the linking process requires all memory available at the sensor node, FlexCup is not able to run in parallel to the actual application. In [24], dynamic runtime linking for the Contiki operating system [25] is presented.

Besides TinyOS, there exist other operating systems for sensor networks which are inherently designed to provide a modular environment [25, 1]. They provide support for dynamic loading of applications and system services as a core functionality of the system. However, this flexibility implies additional levels of indirection for function calls which add considerable runtime overhead. The update process for changed components is limited to these components as they are relocatable and address independent.

Virtual machine architectures for sensor networks [26, 27, 28] push the level of indirection one step further. They conceal the underlying hardware to offer high-level operations to applications through an instruction interpreter. Updates are no longer native code but normally considerable smaller application scripts. This renders reprogramming highly efficient. However, the execution overhead of a virtual machine is considerable and outweighs this advantage for long-running applications [26, 24].

A temporary alternative to supply in-network programmability inside the sensor network itself is to provide a parallel maintenance network [29]. This is particularly useful during the development process as one does not have to rely on the network being operational to update it. Furthermore, new protocols can be tested and evaluated without the reprogramming service distorting the results.

3   Overview

Updating code in wireless sensor network is a non-trivial task and requires the interaction of multiple system components. In general, application reprogramming can be broken down into three steps: image encoding, image distribution, and image decoding. Figure 1 shows a schematic view of all involved components and how they are interrelated in our code update mechanism. On the left-hand side, all services are consolidated that run on the host machine or base station, respectively. On the right, the required components on a sensor node are depicted. The dissemination protocol is responsible to reliably distribute the encoded update in the entire sensor network. We make use of Deluge as it is widely accepted as the standard dissemination protocol and has shown its robustness in various real-world deployments. We give a brief overview of Deluge's data management as it has direct implications on all other system components.[3]

Deluge enables a sensor node to store multiple application images. It divides a junk of the external flash memory (EEPROM) into slots, each of them large enough to hold one image. In conjunction with a bootloader Deluge is then able switch between these images. To manage the program image upload, Deluge divides images into pages of fixed size.[4] So far, Deluge transmits images at the page granularity. That is, all packets of the last page in use were distributed no matter how many of them actually containing data of the new application image. The residual space of the last page is thereby filled with zero bytes. This overhead of up to one kilobyte might be of minor concern if the application image is transmitted in its entirety. However, it becomes unacceptable in the context of small changes leading to delta files of only a few bytes. Deluge was therefore adapted to just transmit packets containing vital information about the new image. The remaining bytes of the last page are then padded with zeros on the sensor node itself to enable a 16-bit cyclic redundancy check on the pages. By requiring a node to dedicate itself to receiving a single page at a time, it is able to keep track of missing packets using a fixed-size bit vector. Packets also include CRC checksums. Redundant data integrity checks at both packet and page level is critical as erroneous data is otherwise propagated throughout the whole network due to the epidemic nature of Deluge.

The protocol also incorporates an administration service that allows the base station to retrieve information about all stored images including which one is currently running. On the host machine, Deluge offers an update management service to inject new images into the network. To allow differential updates a version control system is required at the host machine in order to know all application images currently residing on the sensor nodes. In the current version a file-system-based image repository is used to archive the latest program versions stored in each slot on the sensor nodes. If a new target image is supposed to be injected to a given slot, the update manager first queries the nodes to retrieve metadata about all loaded images. Based on this information a crosscheck in the version control system is performed to ensure that the latest image version for the requested slot is present in the repository. Once the validity of the source image in the repository is verified it is used as input for the delta encoder along with the target image. The encoder processes both images and generates the corresponding delta file. The delta is then disseminated using Deluge as if it was a normal application image. However, it is not stored in the designated

---

[3] The interested reader is referred to [5] for a detailed description of Deluge.

[4] In the current version of Deluge one page sums up to 1104 bytes. In turn, this results in 48 data packets per page.

slot of the target image but in an additional EEPROM slot reserved for delta files. Upon complete delta reception, a node starts the decoding process using additional information from external flash memory and program memory. The target image is thereby directly reconstructed in its intended EEPROM slot. In the meantime, the delta is further disseminated within the network. We now give a detailed description of the encoding algorithm employed on the host machine as well as of the decoder that resides on the sensor nodes.

4 Update Mechanism

All delta algorithms introduced in Section 2 use some kind of heuristic to speed up the generation of copy commands and consequently to reduce the overall execution time of the encoder. In [30] a greedy algorithm is presented that optimally solves the string-to-string correction problem which lies at the heart of differential updating. While its time complexity is undesirable for very large input files it poses no problem in the context of sensor networks where program size is limited to a few hundred kilobytes.[5] Hence, the design of our delta encoder is based on the findings in [30]. Before we give a detailed description of the encoder itself we specify the employed instruction set and how instructions are arranged in the delta file.

4.1 Delta Instructions and Delta File Organization

We adopt the set of delta instructions specified in VCDIFF [31] which is a portable data format for encoding differential data. It is proposed to decouple encoder and decoder implementations to enable interoperability between different protocol implementations. It distinguishes three types of instructions: add, copy and run. The first two instructions are straightforward; add appends a number of given bytes to the target image and copy points to a section in the source image to be copied to the target image. The run instruction is used to encode consecutive occurrences of the same byte efficiently. It has two arguments, the value of the byte and the number of times it is repeated. Making use of the fact that the target image is decoded into the same slot in external memory where the source image already resides, we introduce a fourth instruction. The shift instruction is used to encode sections of the image that have not changed at all from one version to the next. It is used to prevent unnecessary EEPROM writes. The only effect of a shift instruction is the adjustment of the target image pointer at the decoder.

These instructions are designed to minimize the overhead of the delta file. Each instruction code has a size of one to three bytes dependent on the number of bytes in the target image the corresponding instruction encodes. Table 1 comprises the

---

[5] The maximal application memory footprint of state-of-the-art sensor network hardware is limited to 48kB for nodes equipped with MSP430 microcontrollers or 128kB for ATmega128 platforms, respectively.

Table 1. Instruction codes, arguments, and overall costs in bytes if the instructions reconstruct less than 32 bytes. The length of the instruction is encoded in the first five bits of the instruction code.

Instruction	Code	Arguments	Cost [bytes]
shift	xxxxx100	none	1
run	xxxxx101	byte to be repeated	2
copy	xxxxx110	start address	3
add	xxxxx111	data to be added	1+#bytes

arguments and costs of all four instruction types if they reconstruct less than 32 bytes of the target image. The actual length is directly encoded in the first 5 bits of the instruction code in this case. The cost of an instruction increases by one if the encoded fragment spans up to 255 bytes, or by two if it is larger than that, as the instruction length occupies one or two additional bytes, respectively.

We refrained from using the delta file organization as proposed in VCDIFF. It splits the file in three sections, one for data to be added, one for the actual instructions, and one for the addresses of the copy instructions. This enables better secondary compression of the delta file. As we try to keep the decoder complexity to a minimum to meet the nodes' hardware limitations no such compression is applied. We could still use the VCDIFF format without secondary compression. However, the fact that an instruction has to gather its arguments from different places within the delta file results in unfavorable EEPROM access patterns. Random access to external memory—as it would be the case if the VCDIFF format was employed—results in increased overhead during the decoding process. This is caused by the discrepancy between the small average delta instructions and the rather coarse-grained EEPROM organization. On average, the delta instruction length is below four bytes for all experiments described in Section 5. In contrast, external memory access is granted at a page granularity with page sizes of 256 bytes for flash chips of modern sensor network hardware. As EEPROM writes are expensive (see Table 2) current flash storage incorporates a limited amount of cached memory pages to mitigate the impact of costly write operations.[6] However, it is important to notice that even though a read itself is cheap, it may force a dirty cache page to be written back to EEPROM which renders a read operation as expensive as a write.

To allow for the above mentioned EEPROM characteristics the delta file is organized by appending instructions in the order of their generation. That is, each instruction code is directly followed by its corresponding arguments. Furthermore, all instructions are ordered from left to right according to the sections they are encoding in the target file. This permits a continuous memory access during the execution of the delta instructions.

---

[6] The Atmel AT45DB041B flash chip on the TinyNode platform has a cache size of two pages.

Table 2. Relative energy consumption of different operations in comparison to a packet reception for the TinyNode platform

Operation	Current Draw	Time	Rel. Power Drain
Receive a packet	14 mA	5 ms	1
Send a packet	33 mA	5 ms	2.36
Read EEPROM page	4 mA	0.3 ms	0.017
Write EEPROM page	15 mA	20 ms	4.29

4.2   Delta Encoder

The severe hardware constraints of wireless sensor networks let our delta encoder differ in various points from common delta compression algorithms to optimize the decoding process. Besides the objective to minimize the delta file size one also has to consider the energy spent on reconstructing the new image at the nodes. In particular, special care has to be taken to optimize external flash memory access.

The only instruction that requires additional information from the source image to reconstruct its section of the target image is copy. To avoid alternating read and write requests between source image and delta file potentially causing the above discussed EEPROM cache thrashing problem we derive the data required by copy instructions directly from program memory. This decision has several implications. Most important, copies must be generated based on the currently executed image even if it is not identical to the image we would like to update. That is, the decoder is actually using a third input file, namely the currently executed image, to reconstruct the target image. Second, decoding is sped up since reading from program memory is fast in comparison to accessing external flash memory. Third, we are able to directly overwrite the source image in external memory without wasting an additional slot during the reconstruction process. This renders shift instructions possible. As a drawback, it is no longer allowed to use an already decoded section of the target image as origin for later copies. This potentially results in larger delta files. However, the aforementioned positive effects compensate this restriction.

In a first phase, the encoder analyzes both source and target image. The algorithm runs simultaneous over both input files and generates shift instructions for each byte sequence that remains unchanged. Then, the target image is inspected and run instructions are produced for consecutive bytes with identical values. In a third pass, for each byte in the target image a search for the longest common subsequence in the source image is performed. A copy is then generated for each byte with a matching sequence of size at least three as copy instructions of length three or larger start to pay off compared to an add.

In a second phase a sweep line algorithm is employed to determine the optimal instruction set for the target image minimizing the size of the resulting delta. All instructions produced in the first phase reconstruct a certain section of the target image determined by their start and end address. The algorithm processes the image from left to right and greedily picks the leftmost instruction based on its

Table 3. Target and delta sizes for the different settings. Additionally, times required to encode the images are given.

Setting	Target Size [bytes]	Delta Size[bytes]	Size Reduction	Encoding Time
Case 1	28684	322	98.88%	8453 ms
Case 2	27833	5543	80.08%	5812 ms
Case 3	28109	7383	73.73%	6797 ms
Case 4	34733	17618	49.28%	7563 ms
Case 5	21508	14904	30.70%	4781 ms

start addresses. Then, the next instruction is recursively chosen according to the following rules. First, the instruction must either overlap with or be adjacent to the current instruction. Second, we choose the instruction among those fulfilling the previous requirement whose endpoint is farthest to the right. The instruction costs are used for tie breaking. To avoid redundancy in the delta file the new instruction is pruned to start right after the end of the current one if they overlap. If no instruction satisfies these demands an add is generated. These add instructions span the sections not covered by any of the other three instruction types. Once the algorithm reaches the end of the target image the delta file is generated according to the rules stated in the previous section.

4.3   Delta Decoder

The delta decoder is mapped as a simple state machine executing delta instruction in rotation. Prior to the actual decoding metadata is read from the head of the delta file. This information is appended by the encoder and contains the delta file length and additional metadata for the target image required by Deluge. The length is used to determine completion of the decoding. The metadata comprises version and slot information of the target image. This information is used to verify the applicability of the delta to the image in the given slot. In case of failure the decoding process is aborted and the image is updated traditionally without the help of differential reprogramming. If the delta is valid, the instructions are consecutively executed to rebuild the target image.

Once the decoder has fully reconstructed the image it signals Deluge to pause the advertisement process for the newly built image. This has the effect that the delta is disseminated faster within the network than the actual image enabling all nodes to reprogram themselves by means of the delta.

5   Experimental Evaluation

In this section we analyze the performance of our differential update mechanism on real sensor network hardware. The experiments were run on TinyNode 584 [32] sensor nodes operating TinyOS. The TinyNode platform comprises a MSP430 microcontroller featuring 10 kB of RAM and 48kB of program memory. Furthermore 512 kB external flash memory are available.

Table 4. The number of occurrences of each instruction type for all scenarios. Furthermore, the average number of bytes encoded by one instruction is given.

Setting	#shift	avg. size	#run	avg. size	#copy	avg. size	#add	avg. size
Case 1	5	57334	0	0	0	0	5	2.80
Case 2	16	79.19	3	244	884	27.43	859	1.83
Case 3	71	26.01	5	91	1183	20.13	1153	1.72
Case 4	30	37.13	12	40.32	2757	9.68	2472	2.64
Case 5	25	7.78	20	54.70	2219	6.44	1858	3.19

To prove the fitness of the proposed approach in a wide range of application scenarios five different test cases are consulted ranging from small code updates to complete application exchanges. Except one, all applications are part of the standard TinyOS distribution. Before evaluating the performance of our reprogramming approach the different test settings are discussed in the following.

Case 1: This case mimics micro updates as they occur during parameter tuning. We increase the rate at which the LED of the Blink application is toggled. This change of a constant has only local impact and should therefore result in a small delta.

Case 2: The Blink application is modified to facilitate concurrent program executions. The application logic is therefore encapsulated in an independent task (see BlinkTask). This leads to additional calls to the TinyOS scheduler and a deferred function invocation.

Case 3: The CntToLeds application, which shows a binary countdown on the LED's, is extended to simultaneously broadcast the displayed value over the radio
(CntToLedsAndRfm). This is a typical example of a software upgrade that integrates additional functionality.

Case 4: In this setting Blink is replaced with the Oscilloscope application. The latter incorporates multi-hop routing to convey sensor readings towards a base station. Both application share a common set of system components. This scenario highlights the ability of our protocol to cope with major software changes.

Case 5: Here we switch from Blink to Dozer [33], an energy-efficient data gathering system. Among other features, Dozer employs a customized network stack such that the commonalities between the two applications are minimal. Furthermore, Dozer is the only application in this evaluation that has no built-in Deluge support.

The Blink application produces a memory footprint of 24.8 kB in program memory and 824 bytes RAM using the original Deluge. In comparison, the enhanced Deluge version including delta decoder sums up to 27.6 kB ROM and 958 bytes of RAM. Consequently, our modifications increase the memory footprint of an application by 2.8 kB in program memory and 134 bytes of RAM.

The performance of our delta encoder for all five cases is shown in Table 3. For Case 1, the encoder achieves a size reduction by a factor of 100. Actually, the mere delta is only 32 bytes long. The other 290 bytes consist of metadata overhead introduced by Deluge such as 256 bytes of CRC checksums. As already mentioned in the case descriptions, the increasing delta sizes indicate that the similarity between source and target image decreases from Case 1 to 5. The delta file produced due to major software changes is still only about half the size of the original image. Moreover, even if we replace an application with one that has hardly anything in common with the former, such as in Case 5, the encoder achieves a size reduction of about 30%. Table 3 also contains the execution times of the encoder for the five different scenarios. Note that the encoding was computed on a customary personal computer. The encoding process for the considered settings takes up to nine seconds. Compared to the code distribution speedup achieved by smaller delta files this execution time is negligible.

Table 4 shows the number of occurrences of all four instruction types in all five settings. It also contains the average number of bytes covered by one instruction. One can see that the modifications in Case 1 are purely local as only 14 bytes have to be overwritten and the rest of the image stays untouched. For the other four scenarios, copy and add instructions constitute the dominating part of the delta files. It is interesting to see that the average size of a copy is larger if the source and target images have a higher similarity. The opposite is true for the add instruction. In the case of minor application updates many code blocks are only shifted to a different position within the code image but not changed at all. This fact is exploited by the copy instructions enabling a relocation of these sections with constant overhead. However, if the new application is completely unrelated to the one to be replaced as in Case 5, the image exhibits less opportunities for copies. Consequently, more add instructions are necessary to rebuild the target image.

To evaluate the decoder we measure the reconstruction time of the target image on a sensor node. Table 5 shows the decoding time for all cases dependent on the available buffer size at the decoder. If no input buffer is available, each delta instruction is read separately from external memory before it is processed. If an input buffer is allocated, the decoder consecutively loads data blocks of the delta file from EEPROM into this buffer. Before a new block of data is fetched, all instructions currently located in the buffer are executed. Similar to the input buffer handling, the decoder writes the result of a decode instruction directly to external memory if no output buffer is present. In contrast, if an output buffer is available, it is filled with the outcomes of the processed delta instructions and only written back to EEPROM if it is full. We limit the maximum buffer size to 256 bytes thereby matching the EEPROM page size of the TinyNode platform.

For Case 1 the decoding process takes approximately 1.2 seconds no matter which buffer strategy is applied. This can be explained by the fact that only 10 delta instructions are involved (see Table 4) where five of them are shift instructions which do not lead to EEPROM writes. In contrast, the decoder takes about 15 seconds to update from Blink to the Oscilloscope application if neither input nor output buffers are used. However, decoding time decreases to 8.68

Table 5. Time to reconstruct the target image on a sensor node as a function of the available input and output buffer sizes in bytes at the decoder (input | output)

Setting	none | none	256 | none	256 | 256	128 | 128	64 | 64	32 | 32	16 | 16
Case 1	1.20 s	1.24 s	1.24 s	1.24 s	1.24 s	1.23 s	1.23 s
Case 2	7.03 s	5.05 s	4.11 s	4.20 s	4.38 s	4.74 s	5.40 s
Case 3	8.51 s	5.52 s	4.25 s	4.36 s	4.54 s	5.02 s	5.74 s
Case 4	15.14 s	8.68 s	5.63 s	5.79 s	6.06 s	6.82 s	7.84 s
Case 5	11.27 s	6.47 s	4.08 s	4.15 s	4.35 s	4.74 s	5.45 s

or 5.63 seconds if an input buffer of 256 bytes or both, input and output buffers of size 256 bytes are employed, respectively. That is, decoding with maximum input and output buffer reduces the execution time by a factor of 2.7 in Case 4.

Due to the promising results with large buffer sizes, we also study the impact of varying buffer sizes on the decoding speed. The reconstruction times for all scenarios were evaluated for buffer sizes of 16, 32, 64, 128, and 256 bytes, respectively. Table 5 shows that execution times increase with decreasing buffer sizes. However, the increases are moderate: Reducing the buffers from 256 to 16 bytes—and thus saving 480 bytes of RAM—results in an at most 40% longer decoding time. Furthermore, the execution times with buffers of 16 bytes are roughly the same as if only an input buffer of 256 bytes is used.

6   Conclusions

Sensor networks are envisioned to operate over a long period of time without the need of human interaction. Once deployed, remote reprogramming of the sensor network is therefore crucial to react to changing demands. This paper introduces an efficient code update strategy which is aware of the limitations and requirements of sensor network hardware. We exploit the fact that the currently running application on the sensor nodes is know at the time a new program is supposed to be installed. Differential compression is employed to minimize the amount of data that has to be propagated throughout the network. The proposed delta encoder achieves data size reductions from 30% to 99% for a representative collection of update scenarios. The delta decoder is integrated into the Deluge framework which guarantees reliable data dissemination within the sensor network. The energy spent for the image distribution is directly proportional to the transmitted amount of data. Thus, our system reduces the power consumption of the code dissemination by the same percentage as it compresses the input data.

References

1. Han, C.-C., Kumar, R., Shea, R., Kohler, E., Srivastava, M.: A Dynamic Operating System for Sensor Nodes. In: Int. Conference on Mobile Systems, Applications, and Services (MobiSys), Seattle, Washington, USA (2005)

2. Wang, Q., Zhu, Y., Cheng, L.: Reprogramming Wireless Sensor Networks: Challenges and Approaches. IEEE Network 20(3), 48–55 (2006)
3. Ni, S.-Y., Tseng, Y.-C., Chen, Y.-S., Sheu, J.-P.: The Broadcast Storm Problem in a Mobile Ad Hoc Network. In: Int. Conference on Mobile Computing and Networking (MobiCom), Seattle, Washington, USA (1999)
4. Stathopoulos, T., Heidemann, J., Estrin, D.: A Remote Code Update Mechanism for Wireless Sensor Networks, UCLA, Tech. Rep. CENS-TR-30, (2003)
5. Hui, J.W., Culler, D.: The dynamic behavior of a data dissemination protocol for network programming at scale. In: Int. Conference on Embedded Networked Sensor Systems (SENSYS), Baltimore, Maryland, USA (2004)
6. Kulkarni, S.S., Wang, L.: MNP: Multihop Network Reprogramming Service for Sensor Networks. In: Int. Conference on Distributed Computing Systems (ICDCS), Columbus, Ohio, USA (2005)
7. Hunt, J.J., Vo, K.-P., Tichy, W.F.: Delta Algorithms: An Empirical Analysis. ACM Trans. Software Engineering and Methodology 7(2), 192–214 (1998)
8. Tridgell, A.: Efficient Algorithms for Sorting and Synchronization, Ph.D. dissertation, Astralian National University (1999)
9. MacDonald, J.: File System Support for Delta Compression, Masters thesis. Department of Electrical Engineering and Computer Science, University of California at Berkeley (2000)
10. Trendafilov, D., Memon, N., Suel, T.: zdelta: An Efficient Delta Compression Tool, Polytechnic University, Tech. Rep. TR-CIS-2002-02 (2002)
11. Percival, C.: Naive Differences of Executable Code (2003), http://www.daemonology.net/papers/bsdiff.pdf
12. Agarwal, R.C., Gupta, K., Jain, S., Amalapurapu, S.: An approximation to the greedy algorithm for differential compression. IBM J. of Research and Development 50(1), 149–166 (2006)
13. C.T.Inc., Mote In-Network Programming User Reference (2003), http://www.xbow.com
14. Floyd, S., Jacobson, V., Liu, C.-G., McCanne, S., Zhang, L.: A Reliable Multicast Framework for Light-Weight Sessions and Application Level Framing. IEEE/ACM Trans. Networking 5(6), 784–803 (1997)
15. Simon, R., Huang, L., Farrugia, E., Setia, S.: Using multiple communication channels for efficient data dissemination in wireless sensor networks. In: Int. Conference on Mobile Adhoc and Sensor Systems (MASS), Washington, DC, USA (2005)
16. Panta, R.K., Bagchi, I.K.S.: Stream: Low Overhead Wireless Reprogramming for Sensor Networks. In: Int. Conference on Computer Communications (INFOCOM), Anchorage, Alaska, USA (2007)
17. Koshy, J., Pandey, R.: Remote Incremental Linking for Energy-Efficient Reprogramming of Sensor Networks. In: European Workshop on Wireless Sensor Networks (EWSN), Istanbul, Turkey (2005)
18. Wagner, R.A., Fischer, M.J.: The String-to-String Correlation Problem. J. ACM 21, 168–173 (1974)
19. Li, W., Zhang, Y., Yang, J., Zheng, J.: UCC: Update-Conscious Compilation for Energy Efficiency in Wireless Sensor Networks. In: ACM SIGPLAN Conference on Programming Language Design and Implementation (PLDI), San Diego, California, USA (2007)
20. von Platen, C., Eker, J.: Feedback Linking: Optimizing Object Code Layout for Updates. In: ACM SIGPLAN/SIGBED Conference on Language, Compilers, and Tool Support for Embedded Systems (LCTES), Ontario, Canada (2006)

21. Reijers, N., Langendoen, K.: Efficient Code Distribution in Wireless Sensor Networks. In: Int. Conference on Wireless Sensor Networks and Applications (WSNA), San Diego, California, USA (2003)
22. Jeong, J., Culler, D.: Incremental Network Programming for Wireless Sensors. In: Int. Conference on Sensor and Ad Hoc Communications and Networks (SECON), Santa Clara, California, USA (2004)
23. Marrón, P.J., Gauger, M., Lachenmann, A., Minder, D., Saukh, O., Rothermel, K.: FlexCup: A Flexible and Efficient Code Update Mechanism for Sensor Networks. In: European Conference on Wireless Sensor Networks (EWSN), Zurich, Switzerland (2006)
24. Dunkels, A., Finne, N., Eriksson, J., Voigt, T.: Run-Time Dynamic Linking for Reprogramming Wireless Sensor Networks. In: Int. Conference on Embedded Networked Sensor Systems (SENSYS), Boulder, Colorado, USA (2006)
25. Dunkels, A., Gronvall, B., Voigt, T.: Contiki - A Lightweight and Flexible Operating System for Tiny Networked Sensors. In: Int. Conference on Local Computer Networks (LCN), Orlando, Florida, USA (2004)
26. Levis, P., Culler, D.: Maté: A Tiny Virtual Machine for Sensor Networks. ACM SIGOPS Operating System Review 36(5), 85–95 (2002)
27. Levis, P., Gay, D., Culler, D.: Active Sensor Networks. In: Int. Conference on Networked Systems Design and Implementation (NSDI), Boston, Massachusetts, USA (2005)
28. Balani, R., Han, C.-C., Rengaswamy, R.K., Tsigkogiannis, I., Srivastava, M.: Multi-Level Software Reconfiguration for Sensor Networks. In: Int. Conference on Embedded Software (EMSOFT), Seoul, Korea (2006)
29. Dyer, M., Beutel, J., Thiele, L., Kalt, T., Oehen, P., Martin, K., Blum, P.: Deployment Support Network - A Toolkit for the Development of WSNs. In: European Conference on Wireless Sensor Networks (EWSN), Delft, Netherlands (2007)
30. Reichenberger, C.: Delta Storage for Arbitrary Non-Text Files. In: Int. Workshop on Software Configuration Management (SCM), Trondheim, Norway (1991)
31. Korn, D., MacDonald, J., Mogul, J., Vo, K.: The VCDIFF Generic Differencing and Compression Data Format, RFC 3284 (Proposed Standard) (June 2002), http://www.ietf.org/rfc/rfc3284.txt
32. Dubois-Ferrier, H., Meier, R., Fabre, L., Metrailler, P.: TinyNode: a comprehensive platform for wireless sensor network applications. In: Int. Conference on Information Processing in Sensor Networks (IPSN), Nashville, Tennessee, USA (2006)
33. Burri, N., von Rickenbach, P., Wattenhofer, R.: Dozer: Ultra-Low Power Data Gathering in Sensor Networks. In: Int. Conference on Information Processing in Sensor Networks (IPSN), Cambridge, Massachusetts, USA (April 2007)

Local PTAS for Independent Set and Vertex Cover in Location Aware Unit Disk Graphs
(Extended Abstract)

Andreas Wiese[1],[*],[**] and Evangelos Kranakis[2],[***]

[1] Technische Universität Berlin, Institut für Mathematik, Germany
[2] School of Computer Science, Carleton University, 1125 Colonel By Drive, Ottawa, Ontario, Canada K1S 5B6

Abstract. We present the first local approximation schemes for maximum independent set and minimum vertex cover in unit disk graphs. In the graph model we assume that each node knows its geographic coordinates in the plane (location aware nodes). Our algorithms are local in the sense that the status of each node v (whether or not v is in the computed set) depends only on the vertices which are a constant number of hops away from v. This constant is independent of the size of the network. We give upper bounds for the constant depending on the desired approximation ratio. We show that the processing time which is necessary in order to compute the status of a single vertex is bounded by a polynomial in the number of vertices which are at most a constant number of vertices away from it. Our algorithms give the best possible approximation ratios for this setting.

The technique which we use to obtain the algorithm for vertex cover can also be employed for constructing the first global PTAS for this problem in unit disk graph which does not need the embedding of the graph as part of the input.

1 Introduction

Locality plays an important role in wireless and ad-hoc-networks. In such networks, there is often no global entity to organize the network traffic. So the nodes have to negotiate their coordination in a distributed manner. As in most cases the entire network is much too large to be explored by a single node, we are interested in local algorithms. These are algorithms, in which the status of a node v (e.g., whether or not it is in the independent set or vertex cover) depends only

---

[*] Research conducted while the authors were visiting the School of Computing Science at Simon Fraser University, Vancouver.
[**] Research supported by a scholarship from DAAD (German Academic Exchange Service).
[***] Research supported in part by NSERC (Natural Science and Engineering Research Council of Canada) and MITACS (Mathematics of Information Technology and Complex Systems).

S. Nikoletseas et al. (Eds.): DCOSS 2008, LNCS 5067, pp. 415–431, 2008.

on the nodes which are a constant number of hops away from v. This constant has to be independent of the size of the overall network. The locality concept is also advantageous for disaster recovery and dynamically changing network. If only parts of the network are changed or lost, using a local algorithm only fractions of the solution need to be recomputed and so we do not need to redo the entire calculation.

We model the wireless network with location aware Unit Disk Graphs (UDGs). This represents the situation where all nodes have an identical transmission range and know about their geographic position in the plane. Whether communication between two nodes is possible depends only on their Euclidean distance. This is due to the fact that wireless devices naturally have a limited transmission range. Since positioning devices, like GPS, become more and more common the concept of location awareness becomes relevant.

In wireless networks, clustering is an important aspect of organizing network traffic. A maximal independent set can be used for clustering by defining its nodes to be clusterheads. A cluster consists of the nodes which are adjacent to a single clusterhead. The latter are responsible for the communication of nodes within their cluster with other nodes. Vertex covers and independent sets are widely used for providing a good initial virtual backbone (e.g., see [2]) and for enabling efficient routing in wireless networks (e.g., see [16]). Both problems are closely related since the complement set of an independent set is a vertex cover. However, approximation ratios for the above problems are not preserved by this operation, so we need to design approximation algorithms for both problems.

1.1   Related Work

Maximum independent set and minimum vertex cover are both NP-hard in general graphs [7]. For independent set it is even impossible to approximate the problem in polynomial time with a better factor than $|V|^{1/2-\epsilon}$, unless $P = NP$ [8]. (Observe that finding a maximum independent set in a graph is the same as finding a maximum clique in its complementary graph.) However, for vertex cover there are several polynomial time approximation algorithms which achieve an approximation factor of 2, e.g., in [1]. But it is NP-hard to find an approximation better than $10\sqrt{5} - 21 \approx 1,3607$ [6], so there can be no PTAS, unless $P = NP$.

When restricting the problem to unit disk graphs, both problems are still NP-hard [4]. But there are constant ratio algorithms known [12] (in the case of vertex cover with a ratio of $3/2$ which is better than in the general case). If the embedding of the graph is part of the input there are PTASs known due to Hunt III et al. [9]. Note however that finding an embedding for a unit disk graph is NP-hard [3] (since the recognition of a unit disk graph is NP-hard). Even finding an approximation for the embedding is NP-hard [10]. For the case that the embedding of the graph is unknown, Nieberg et al. found a PTAS for the weighted independent set problem in UDGs [14]. Kuhn et al. proposed a local approximation scheme for independent set [11] for growth-bounded graphs (this class includes UDGs). However, their definition of locality is not the same as assumed in this paper. In their algorithm, whether a vertex v is in the independent set depends on the vertices which are up to $O(\log^* n)$ hops away from v which is not constant as it depends on the size of the graph.

1.2   Main Result

We present local approximation algorithms for maximum independent set and minimum vertex cover with approximation ratios $1 - \epsilon$ and $1 + \epsilon$ respectively. They are the first local approximation schemes for these problems in unit disk graphs. Our technique is very similar to the one of the local dominating set algorithm presented in [15]. We give upper bounds for the locality distance (the maximum k such that whether or not a vertex is in the computed set depends only on the vertices which are at most k hops away from it) depending on the desired approximation ratio. In order to be able to guarantee the approximation factor for vertex cover, we prove that the set of all vertices of a unit disk graph forms a factor 12 approximation for minimum vertex cover (assuming that there is at least one edge to cover in the graph). The technique used in this proof can be expanded to show that in every $K_{1,m}$–free graph the set of all vertices forms a factor $2m$ approximation for vertex cover.

The technique to derive the PTAS for minimum vertex cover has applications in other settings as well. We explain how it can be used to construct the first global PTAS for vertex cover in unit disk graph which does not need the embedding of the graph as part of the input.

1.3   Organization of the Paper

The remainder of this paper is organized as follows: In Section 2 we introduce some basic concepts and definitions including a tiling of the plane in hexagons which our algorithms are using. Our $1 - \epsilon$ approximation algorithm for independent set is presented in Section 3. After that in Section 4 we prove our bound for vertex cover and present our $1 + \epsilon$ approximation algorithm for this problem. There we also discuss how this technique can be used to derive a global PTAS for vertex cover which does not rely on the embedding of the graph. Finally in Section 5 we summarize our results and discuss open problems.

2   Preliminaries

We give some basic definitions and explain a tiling of the plain which our algorithms are using. Then we introduce the concept of 1-separated collections which will enable us to establish a lower bound for the maximum independent set problem.

2.1   Definitions

An undirected graph $G = (V, E)$ is a *unit disk graph* if there is an embedding in the plane for G such that two vertices u and v are connected by an edge if and only if the Euclidean distance between them is at most 1. The graph G we consider for our algorithms is a unit disk graph.

A set $I \subseteq V$ is an *independent set* if for every pair of vertices v, v' with $v \in I$ and $v' \in I$ it holds that $\{v, v'\} \notin E$. A set $VC \subseteq V$ is called a *vertex cover* if for every edge $e = \{u, v\}$ it holds that either $u \in VC$ or $v \in VC$. Equivalently, a set VC is a vertex cover if and only if $V \setminus VC$ is an independent set.

Definition 1. *For two vertices u and v let $d(u, v)$ be the hop-distance between u and v, that is the number of edges on a shortest path between these two vertices.*

The hop-distance is not necessarily the geometric distance between two vertices. Denote by $N^r(v) = \{u \in V \mid d(u, v) \leq r\}$ the r-neighborhood of a vertex v. For ease of notation we set $N^0(v) := \{v\}$, $N(v) := N^1(v)$ and for a set $V' \subseteq V$ we define $N(V') = \bigcup_{v' \in V'} N(v')$. Note that $v \in N(v)$. We define the diameter of a set of vertices $V' \subseteq V$ as $diam(V') := \max_{u, v \in V'} d(u, v)$.

Denote by the *locality distance* (or short the *locality*) of an algorithm the minimum α such that the status of a vertex v (e.g., whether or not v is in an independent set or vertex cover) depends only on the vertices in $N^\alpha(v)$. In all algorithms presented in this paper we will prove that α depends only on the desired approximation factor for the respective problem.

2.2   Tiling of the Plane

This method of tiling the plane is taken from [5] and [15]. The plane is split into hexagons and a class number is assigned to each hexagon. The tiling has the following properties:

- Each vertex is in exactly one hexagon.
- Two vertices in the same hexagon are connected by an edge.
- Each hexagon has a class number.
- The distance between two vertices in different hexagons with the same class number is at least a certain constant positive integer.
- The number of hexagonal classes is bounded by a constant.

We achieve these properties as follows: First we define the constant c to be the smallest even integer such that $(2c + 1)^2 < \left(\frac{1}{1-\epsilon}\right)^c$. We consider a tiling of the plane with tiles. Each tile consists of hexagons of diameter one that are being assigned different class numbers (see Figures 1). Denote by H the set of all hexagons containing vertices of G (only these hexagons are relevant for us) and by b the number of hexagons in one tile. Ambiguities caused by vertices at the border of hexagons are resolved as shown in Figure 1(b): The right borders excluding the upper and lower apexes belong to a hexagon, the rest of the border does not. We assume that the tiling starts with the coordinates $(0, 0)$ being in the center of a tile of class 1. We choose the number of hexagons per tile in such a way that two hexagons of the same class have an Euclidean distance of at least $2c + 1$. Note that this implies that two vertices in different hexagons of

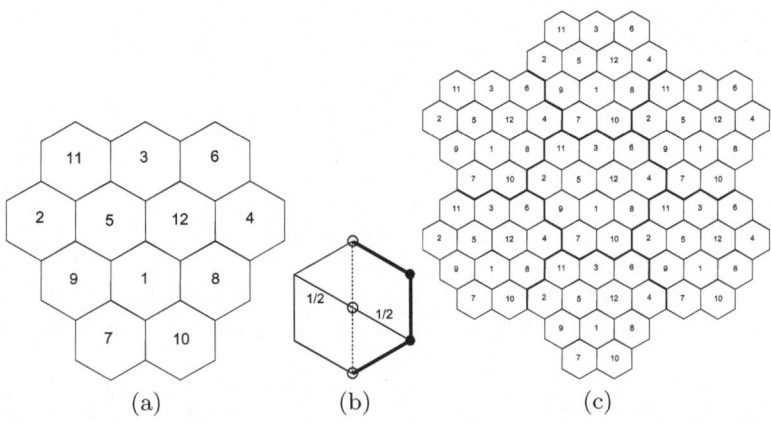

Fig. 1. (a) A tile divided into 12 hexagons. Having 12 hexagons in one tile achieves a minimum Euclidean distance between to hexagons of the same class of 2. (b) One hexagon of the tiling. The bold lines indicate the parts of its border that belong to this hexagon. (c) Several tiles glued together.

the same class number are at least $2c + 1$ hops away from each other. Later we will show that we need at most $12c^2 + 18c + 7$ hexagons per tile to ensure this, i.e., $12c^2 + 18c + 7 \leq b$. Let $class(h)$ be the class number of a hexagon h.

2.3   1-Separated Collections

We present the concept of 1-separated collections. We use it in order to establish an upper bound for an optimal independent set. Let $G = (V, E)$ be a unit disk graph.

Definition 2. *Let H be an index set and let the sets S_h with $h \in H$ be subsets of V. The sets S_h are called a 1-separated collection if for any two vertices $s \in S_h$ and $s' \in S_{h'}$ with $h \neq h'$ it holds $d(s, s') > 1$.*

Let $I : \mathcal{P}(V) \to \mathcal{P}(V)$ be an operation returning an *independent set of maximum cardinality* for the subset of vertices given as argument to it. Later in the algorithm we will construct sets S_h and will surround them by sets T_h such that $S_h \subseteq T_h \subseteq N(S_h)$. Now we establish our upper bound for a maximum independent set for G.

Lemma 1. *Let the sets T_h, $h \in H$ be sets such that $V = \bigcup_{h \in H} T_h$. We have that $|I(V)| \leq \sum_{h \in H} |I(T_h)|$.*

Proof. Since $I(T_h)$ is a maximum independent set for T_h it holds that $|I(V) \cap T_h| \leq |I(T_h)|$. As $V = \bigcup_{h \in H} T_h$ every vertex $v \in I(V)$ is contained in at least one set T_h. So it follows that $|I(V)| \leq \sum_{h \in H} |I(T_h)|$.

So we see that the cardinality of the union $\bigcup_{h \in H} I\,(T_h)$ is an upper bound for the cardinality of a maximum independent set. The idea is now to construct a 1-separated collection with sets S_h and a surrounding set T_h for each set S_h such that $T_h \subseteq N\,(S_h)$. We want to do it in such a way that $|I\,(T_h)|$ is not much larger than $|I\,(S_h)|$. If then $\bigcup_{h \in H} T_h = V$ we can show that $\bigcup_{h \in H} I\,(S_h)$ is not much smaller than an optimal independent set for G.

Lemma 2. *Let $\mathcal{S} = \bigcup_{h \in H} S_h$ be a 1-separated collection in G and let $T_h, h \in H$ be subsets of V with $S_h \subseteq T_h$ for all $h \in H$. If there exists an $\epsilon > 0$ such that $(1 - \epsilon) \cdot |I\,(T_h)| \leq |I\,(S_h)|$ holds for all $h \in H$ and if $V = \bigcup_{h \in H} T_h$ then the set $\bigcup_{h \in H} I\,(S_h)$ is a $(1 - \epsilon)$-approximation of a maximum independent set in G.*

Proof. Let I_{OPT} be an optimal independent set for G. It holds that $\left|\bigcup_{h \in H} I\,(S_h)\right| = \sum_{h \in H} |I\,(S_h)| \geq (1 - \epsilon) \cdot \sum_{h \in H} |I\,(T_h)| \geq (1 - \epsilon) \cdot I_{OPT}$.

3   Independent Set

We present a local $1 - \epsilon$ approximation algorithm for maximum independent set. Its technique is very similar to the one of the local algorithm for minimum dominating set presented in [15].

3.1   The Algorithm

Before giving the formal algorithm we present the main structure and provide an intuitive description of the algorithm. For some hexagons h we construct a set S_h and a set T_h with $S_h \subseteq T_h \subseteq N\,(S_h)$. The sets T_h contain all vertices in their respective hexagon h and the vertices in a certain surrounding area. They are disjoint and have certain properties that ensure the desired approximation ratio of $1 - \epsilon$. We call vertices contained in a set T_h *covered*. The construction of the sets T_h is done by iterating over the class numbers of the hexagons. First we cover hexagons of class 1 by computing sets T_h for all hexagons h of class 1. Assume that all hexagons of class i have already been covered. We proceed to cover all hexagons of class $i + 1$ whose vertices have not been completely covered so far by computing sets T_h for those hexagons. We stop when all vertices in all hexagons have been covered. Moreover, the number of iterations does not exceed the total number of classes. Finally we compute for all sets S_h a maximum independent set $I\,(S_h)$. We output $I := \bigcup_h I\,(S_h)$.

Now we present the algorithm in detail. Fix $\epsilon > 0$ and let b be the number of hexagonal classes. For all $h \in H$ we initialize the sets $S_h = T_h = \emptyset$. If all vertices of a hexagon have been covered, call this hexagon *covered*. For $i = 1, ..., b$ do the following: Consider a hexagon $h \in H$ of class i which is not covered. Define the vertex v_h which is closest to the center of h and which is not covered yet to be the *coordinator vertex* of h. Ambiguities are resolved by choosing the vertex with the smallest x-coordinate among vertices with the least distance to the center of h. Denote by $C(i)$ all vertices which are covered in previous iterations where hexagons of classes $i' < i$ were considered. Compute for all $r \leq c$ the r-neighborhoods $N^r (v_h)$ and compute the maximum independent sets $I(N^r(v_h))$. We determine the smallest value of r with $r \leq c - 1$ such that

$$(1 - \epsilon) \cdot \left| I\left(N^{r+1}(v_h) \setminus C(i)\right) \right| \leq \left| I\left(N^r(v_h) \setminus C(i)\right) \right| \tag{1}$$

holds and denote it by $\bar{r}$. Later we will prove that there is at least one value for r with $r \leq c - 1$ such that Inequality 1 does indeed hold (see Lemma 3). Now mark all vertices in $T_h := N^{\bar{r}+1}(v_h) \setminus C(i)$ as *covered*. We define $S_h := N^{\bar{r}}(v_h) \setminus C(i)$. In Lemma 4 we will prove that the sets S_h (for various hexagons h) form a 1-separated collection. We assign all vertices in $I(S_h)$ to the independent set. We do this procedure for all hexagons of class i which are not covered yet. As two vertices in different hexagons of the same class number are at least $2c + 1$ hops away from each other the order in which the hexagons are processed does not matter. We output $I := \bigcup_{h \in H} I(S_h)$.

The previous discussion is presented in Algorithm 1.

Algorithm 1. Local algorithm for finding an independent set in a unit disk graph

1 // Algorithm is executed independently by each node v;
2 *inSet*:=false;
3 **if** $\exists$ *hexagon h with $v \in S_h$* **then**
4 | compute $I(S_h)$;
5 | **if** $v \in I(S_h)$ **then** *inSet*:=true
6 **end**
7 **if** *inSet=true* **then** Become part of the independent set I **else** Do not become part of I

3.2 Proof of Correctness

We prove the correctness of Algorithm 1, its approximation factor, its locality and its processing time in Theorem 1.

Theorem 1. *Let G be a unit disk graph and let $\epsilon > 0$. Algorithm 1 has the following properties:*

1. *The computed set I is an independent set for G.*
2. *Let I_{OPT} be an optimal independent set. It holds that $|I| \geq (1 - \epsilon) \cdot |I_{OPT}|$.*
3. *Whether or not a vertex v is in I depends only on the vertices at most $O\left(\frac{1}{\epsilon^6}\right)$ hops away from v, i.e. Algorithm 1 is local.*
4. *The processing time for a vertex v is bounded by a polynomial in the number of vertices at most $O\left(\frac{1}{\epsilon^6}\right)$ hops away from v.*

We will prove the four parts of this theorem in four steps. In each step we first give some lemmas which are required to understand the proof of the theorem. It is very similar to the proof given in [15] for the correctness of the minimum dominating set algorithm presented there.

Correctness. We want to prove that the set I is indeed an independent set for G. As mentioned above we first prove that it is sufficient to examine values for r with $r \leq c - 1$ while computing $I(N^r(v))$, employing an argument used in [14].

Lemma 3. *Let v be a coordinator vertex. While computing its neighborhood $N^r(v)$ the values of r that need to be considered to find a value $\bar{r}$ such that*

$$(1 - \epsilon) \cdot \left| I\left(N^{\bar{r}+1}(v) \setminus C(i)\right)\right| \leq \left| I\left(N^{\bar{r}}(v) \setminus C(i)\right)\right| \tag{2}$$

are bounded from above by $c - 1$.

Proof. Assume on the contrary that Inequality 2 is false for all $r \in \{0, 1, ..., c-1\}$, i.e. for these values of r it holds that $\left| I\left(N^{r+1}(v) \setminus C(i)\right)\right| > \left(\frac{1}{1-\epsilon}\right) \cdot \left| I\left(N^r(v) \setminus C(i)\right)\right|$. By Corollary 3 in [13] the number of vertices in a maximum independent set for a neighborhood $N^c(v)$ is bounded by $(2c+1)^2$. It holds that $\left| I\left(N^0(v)\right)\right| = |I(\{v\})| = 1$. So we have that

$$(2c+1)^2 \geq |I(N^c(v) \setminus C(i))| > \left(\frac{1}{1-\epsilon}\right) \cdot \left| I\left(N^{c-1}(v) \setminus C(i)\right)\right| > \left(\frac{1}{1-\epsilon}\right)^2 \cdot \left| I\left(N^{c-2}(v) \setminus C(i)\right)\right|$$

$$> ...$$

$$> \left(\frac{1}{1-\epsilon}\right)^c \cdot \left| I\left(N^0(v) \setminus C(i)\right)\right|$$

$$\geq \left(\frac{1}{1-\epsilon}\right)^c.$$

But from the definition of c we know that $(2c+1)^2 < \left(\frac{1}{1-\epsilon}\right)^c$ which is a contradiction. So at least for one value of $r \in \{0, 1, ..., c-1\}$ it holds that $\left| I\left(N^{r+1}(v)\right)\right| \leq \left(\frac{1}{1-\epsilon}\right) \cdot \left| I\left(N^r(v)\right)\right|$.

Lemma 4. *The sets S_h, $h \in H$ form a 1-separated collection.*

Proof. Let $S_1, S_2, ..., S_m$ be the sets in the order in which they were computed by the algorithm (as mentioned above, the order in which the hexagons of one class are being processed does not matter). We prove the claim by induction on k.

We begin with $k = 1$. As $N(S_1) = T_1$ and by construction $S_j \cap T_1 = \emptyset$ for all sets S_j it follows that the distance between S_1 and any set S_j is strictly larger than 1.

Now assume the claim is true for all sets S_k with $k \leq i - 1$. From the construction of S_i it follows that $T_i = N(S_i) \setminus \bigcup_{j=1}^{j<i} T_j$. By construction it follows that $S_j \cap T_i = \emptyset$ for all sets S_j with $j > i - 1$. This completes the proof.

Proof. (of part 1 of Theorem 1): Each set $I(S_h)$ is an independent set. From Lemma 4 it follows that the sets S_h form a 1-separated collection. So no two vertices $v \in S_h$, $v' \in S_{h'}$ with $h \neq h'$ are connected by an edge. So the union $\bigcup_{h \in H} I(S_h) = I$ is an independent set.

Approximation Ratio. We prove that the size of I is by at most a factor $1 - \epsilon$ smaller than the size of a maximum independent set.

Lemma 5. *The sets T_h cover all vertices of the graph.*

Proof. Assume on the contrary that there is a vertex v which is not covered by any $T_h, h \in H$. Let h be the hexagon to which v belongs and let i be its class number. At some point in the algorithm, the hexagons of class i were considered. Then there were vertices in h which were not covered yet (at least v). So the coordinator vertex of h must have marked a set T_h as covered. However, as the hexagons have a diameter of 1 (and $\bar{r} + 1 \geq 1$) it follows that v is contained in T_h and therefore covered by T_h which is a contradiction.

Proof. (of part 2 of theorem 1): From the construction we can see that for every pair S_h, T_h it holds that $(1 - \epsilon) \cdot |I(T_h)| \leq |I(S_h)|$. From Lemma 5 it follows that $\bigcup_{h \in H} T_h = V$. So the conditions of Lemma 2 are satisfied and it holds that
$$|I| = \left| \bigcup_{h \in H} I(S_h) \right| \geq (1 - \epsilon) \cdot I_{OPT}$$

Locality. Now we want to prove that Algorithm 1 is local (part 3 of Theorem 1). We prove that whether or not a vertex v belongs to the computed set I depends only on the vertices at most a constant α hops away from v. This constant depends only on ϵ. We give an upper bound for α in terms of ϵ.

First we introduce three technical lemmas before we can prove part 3 of Theorem 1. Due to the lack of space their proofs are moved to the appendix.

Lemma 6. *Algorithm 1 satisfies the following three locality properties:*

1. *Let v_h be the coordinator vertex of a hexagon h of class k. What vertices are in T_h and S_h depends only on the vertices which are at most $2c \cdot (k - 1) + c$ hops away from v_h.*
2. *Let v' be any vertex. Whether v' is contained in a set $T_{h'}$ or a set $S_{h'}$ with $class(h') \leq k$ depends only on the vertices which are at most $2c \cdot k$ hops away from v'. If v' is contained in a set $T_{h'}$ (or $S_{h'}$) then what other vertices are in $T_{h'}$ (or $S_{h'}$ respectively) depends only on the vertices which are at most $2c \cdot k$ hops away from v'.*
3. *Let v'' be any vertex in a hexagon h'' of class k. Whether or not v'' is the coordinator vertex of h'' depends only on the vertices which are at most $1 + 2c \cdot (k - 1)$ hops away from v''.*

Proof. For ease of notation we introduce the sequences a_k, b_k and c_k. Let a_k be the smallest integer such that what vertices are in T_h and S_h depends only on the vertices which are at most a_k hops away from v_h. So in order to prove property 1 we want to show that $a_k \leq 2c \cdot (k - 1) + c$. Let b_k be the smallest integer such that whether v' is contained in a set $T_{h'}$ or a set S_h with $class(h') \leq k$ depends only on the vertices which are at most b_k hops away from v' and if v' is contained in such a set $T_{h'}$ (or S_h) then what vertices are in $T_{h'}$ (or S_h respectively) depends only on the vertices which are at most b_k hops away from

v'. For proving property 2 we need to show that $b_k \leq 2c \cdot k$. Let c_k be the smallest integer such that whether or not v'' is the coordinator vertex of h'' depends only on the vertices which are at most c_k hops away from v''. So for proving property 3 we need to show that $c_k \leq 1 + 2c \cdot (k-1)$.

Proof by induction. We begin with $k = 1$. As we need to explore the vertices at most c hops away from v_h in order to compute T_h and S_h, we conclude that $a_1 \leq c$.

Let v'' be a vertex in a class 1 hexagon h''. To find out whether v'' is the coordinator vertex of h'', we need to explore the vertices which are at most 1 hop away from v''. So $c_1 \leq 1$.

Let v' be a vertex. We want to find out whether there is a hexagon h' with $class(h') \leq 1$ such that v' is contained in the set $T_{h'}$ or $S_{h'}$. If yes, the coordinator vertex $v_{h'}$ of h' can be at most c hops away from v'. So we need to explore all vertices which are at most c hops away from v' to find all vertices in class 1 hexagons because only they could possibly be coordinator vertices for their hexagon h' such that $v' \in T_{h'}$ or $v' \in S_{h'}$. To find out if any of them is the coordinator vertex of their respective hexagon h' we need to explore the area $c_1 \leq 1$ hop around them. If one of them is a coordinator vertex, we need to explore the vertices at most $a_1 \leq c$ from it in order to compute $T_{h'}$ and $S_{h'}$ and to find out whether $v' \in T_{h'}$ or $v' \in S_{h'}$. If this is the case, we immediately know the sets $T_{h'}$ and $S_{h'}$ as well. So we only need to explore the vertices which are at most $b_1 \leq c + \max(a_1, c_1) \leq 2c$ hops away from v' in order to compute this task.

Assume that the claims in the lemma hold for all $k \leq i - 1$. Let v_h be the coordinator vertex of a hexagon h of class i. In order to compute T_h and S_h we need to explore the vertices which are at most c hops away from v_h and therefore need to find out for each vertex in $N^c(v_h)$ whether it has been covered by a set $T_{h'}$ with $class(h') < i$. So for computing T_h and S_h we need to explore the vertices which are $a_i \leq c + b_{i-1}$ hops away from v_h.

Let v'' be a vertex in a hexagon h'' of class i. To find out whether v'' is the coordinator vertex for h'' we need to explore all other vertices in h'' and find out if they have been covered by a set $T_{h'}$ with $class(h') < i$. For this we need to explore the vertices which are at most $c_i \leq 1 + b_{i-1}$ hops away from v''.

Now let v' be a vertex. We want to find out whether v' is covered by a set $T_{h'}$ or $S_{h'}$ with $class(h') \leq i$. So first we need to explore all vertices at most c hops away from v'. This is the set $N^c(v')$. Only vertices in this set can possibly be coordinator vertices for a hexagon h' such that $T_{h'}$ or $S_{h'}$ contains v'. To check if a vertex in $N^c(v')$ is a coordinator vertex, we need to explore all vertices which are at most c_i hops away from it. If a vertex $v_{h'}$ in $N^c(v')$ is the coordinator vertex for its hexagon h', we need to explore the vertices which are at most a_i hops away from $v_{h'}$ in order to compute $T_{h'}$ and $S_{h'}$. Then we can check if $v' \in T_{h'}$ or $v' \in S_{h'}$. If this is the case, we immediately know the sets $T_{h'}$ and $S_{h'}$ as well. This gives us $b_i \leq c + \max(a_i, c_i) \leq c + \max(c + b_{i-1}, 1 + b_{i-1}) \leq c + a_i$.

So we have shown that $a_1 \leq c$, $b_1 \leq 2c$, $c_1 \leq 1$, $c_i \leq 1 + b_{i-1}$, $b_i \leq c + a_i$ and $a_i \leq c + b_{i-1}$. This implies $a_i \leq c + b_{i-1} \leq c + c + a_{i-1} \Rightarrow a_i \leq 2c \cdot (i-1) + c$, $b_i \leq 2c \cdot i$ and $c_i \leq 1 + 2c \cdot (i-1)$.

Lemma 7. *Let v be vertex. Whether or not v is in I depends only on the vertices which are at most $1 + 2c \cdot (b - 1) + c$ hops away from v.*

Proof. First we prove the following claim: Let v be a vertex in a hexagon h with $class(h) = k$ and let v be contained in a set $T_{h'}$. Let $S_{h'}$ be the set of the 1-separated collection which is contained in $T_{h'}$. We claim that what vertices (other than v) are in $T_{h'}$ and what vertices are in $S_{h'}$ depends only on the vertices which are at most $1 + 2c \cdot (k - 1) + c$ hops away from v.

Proof of the claim: We use the notation a_k, b_k and c_k as introduced in the proof of Lemma 6. For computing the set $T_{h'}$ we need to check whether v is covered by a set $T_{h'}$ with $class(h') < class(h)$. This depends only on the vertices at most $b_{k-1} \leq 2c \cdot (k - 1)$ hops away from v (see Lemma 6). If there is such a set $T_{h'}$ with $v \in T_{h'}$ then $T_{h'}$, $S_{h'}$ and whether $v \in S_{h'}$ depends only on the vertices which are at most b_{k-1} hops away from v' as well (see Lemma 6).

If there is no set $T_{h'}$ such that $class(h') < class(h)$ and $v \in T_{h'}$ the algorithm has to find out whether there is another vertex $v' \neq v$ in h that is the coordinator vertex for h. In order to do this, we need to explore the vertices at most $c_k \leq 1 + 2c \cdot (k - 1)$ hops away from v. If there is such a vertex v' then we need to explore the vertices which are at most $1 + a_k \leq 1 + 2c \cdot (k - 1) + c$ hops away from v in order to compute the sets $T_{h'}$ and $S_{h'}$ (which are then in fact T_h and S_h). If not, then v is the coordinator vertex for h. Then we need to explore the vertices which are at most $a_k \leq 2c \cdot (k - 1) + c$ hops away from v in order to compute $T_{h'}$ and $S_{h'}$. Altogether we conclude that for computing $T_{h'}$ and $S_{h'}$ we need to know only about the vertices at most $1 + 2c \cdot (k - 1) + c$ hops away from v. This proves the claim.

When a vertex v computes whether it is in the set I it determines whether it is included in a set S_h. If this is the case, it computes the maximum independent set $I(S_h)$. Then v is part of the independent set if it is contained in $I(S_h)$. So we need only to explore the the vertices which are at most $1 + 2c \cdot (k - 1) + c$ hops away from v.

Lemma 8. *For ensuring a minimum Euclidean distance of d between two hexagons of the same class number, we need at most $3d^2 + 3d + 1$ hexagons per tile. So for a minimum distance of $2c + 1$ we need at most $12c^2 + 18c + 7$ hexagons per tile.*

Proof. This proof is straightforward, we give only a sketch here. We construct a tile by placing a hexagon h in the center and surrounding it by several circles of hexagons as shown in Figure 2. We can show that it is sufficient to add d circles of hexagons around h to ensure a minimum distance of $d/2$ between any point in h and the border of the tile. So in a tiling with such tiles there will be a minimum Euclidean distance of d between two hexagons of the same class as h. By symmetry this follows for the hexagons of the other class numbers as well. The other term in the claim follows by substituting $2c + 1$ for d.

Proof. (of part 3 of Theorem 1): We want to show that whether or not a vertex v is in I depends only on the vertices at most $O\left(\frac{1}{\epsilon^6}\right)$ away from v. With some

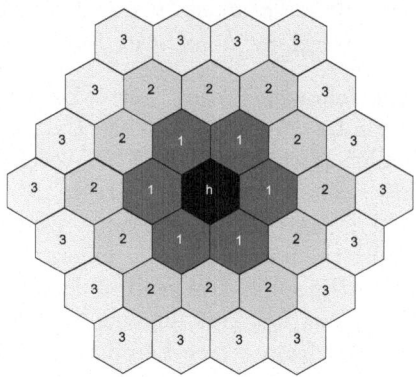

Fig. 2. Left: A hexagon with three circles of hexagons around it. Hexagons with the same number belong to the same circle.

basic calculations it can be shown that $c \in O\left(\frac{1}{\epsilon^2}\right)$.Denote by $\alpha(\epsilon)$ the locality distance of Algorithm 1 when run with a performance guarantee of $1 - \epsilon$. From Lemma 7 we know that we need to explore the vertices at most $1 + 2c \cdot (b - 1) + c$ hops away from v. From the definition of b, the above lemmas and Lemma 8 we get $\alpha(\epsilon) \leq 1 + 2c \cdot (b - 1) + c \leq 1 + 2c \cdot (12c^2 + 18c + 7 - 1) + c$ and therefore $\alpha(\epsilon) \in O\left(\frac{1}{\epsilon^6}\right)$.

Processing Time. The processing time is the time that a single vertex needs in order to compute whether or not it is part of the independent set. We measure it with respect to the number of vertices which are at most α hops away from a vertex v since these are all vertices that a vertex v needs to explore when computing its status. We denote this number by $n_\alpha(v)$ (i.e. $n_\alpha(v) = |N^\alpha(v)|$). We show that the processing time is bounded by a polynomial in $n_\alpha(v)$.

Proof. (of part 4 of Theorem 1). When executing the algorithm for a single vertex v, maximum independent sets for the sets $N^r(v')$ with $r \in \{0, 1, ..., c\}$ and $v' \in N^\alpha(v)$ must be computed. First we show that this can be done in polynomial time. By Corollary 3 in [13] the number of vertices in a maximum independent set for a neighborhood $N^r(v')$ is bounded by $(2r + 1)^2$. So the computation of such a set can be done in $O\left(n_\alpha(v)^{(2c+1)^2}\right)$, e.g. by enumeration. For each vertex $v' \in N^\alpha(v)$ we might have to compute maximum independent sets $I(N^r(v'))$ for each $r \in \{0, 1, ..., c\}$. As this dominates the processing time of the algorithm, we find that it is in $O\left(n_\alpha(v)^{(2c+1)^2} \cdot n_\alpha(v) \cdot c\right)$ and therefore bounded by $n_\alpha(v)^{O(1/\epsilon^4)}$.

4   Vertex Cover

In this section we show how the local $1 - \epsilon$ approximation algorithm for independent set presented in Section 3 can be used for locally computing a $1 + \epsilon$

approximation of the minimum vertex cover problem. First we show that taking all vertices of a unit disk graph leads to a 12 approximation for vertex cover (if the graph contains edges which are to be covered by the vertex cover). Then we present our algorithm and prove its correctness.

4.1   Factor 12 Upper Bound

We consider a connected unit disk graph with at least two vertices. We prove an upper bound of 12 for the number of all vertices in comparison with the number of vertices in a minimum vertex cover.

Theorem 2. *Let $G = (V, E)$ be a connected unit disk graph with $|V| \geq 2$ and let VC_{OPT} be a minimum vertex cover. It holds that $|V| \leq 12 \cdot |VC_{OPT}|$.*

Proof. We partition the vertices V into two sets V_1 and V_2 such that $V_1 \cap V_2 = \emptyset$ and $V = V_1 \cup V_2$. We prove that $|V_1| \leq 2 \cdot |V_{OPT}|$ and $|V_2| \leq 10 \cdot |VC_{OPT}|$. As $|V| = |V_1| + |V_2|$ it follows then that $|V| \leq 12 \cdot |VC_{OPT}|$.

First we define the set V_1. Let $M \subseteq E$ be a maximal matching (i.e. a matching which cannot be extended by adding another edge to M). We define $V_1 := \{u, v | \{u, v\} \in M\}$. As M is a matching it follows that $|V_1| \leq 2 \cdot |VC_{OPT}|$.

Now we define $V_2 := V \setminus V_1$. Since M is a maximal matching it follows that V_2 does not contain any adjacent vertices. Since G is a unit disk graph and therefore does not contain a $K_{1,6}$ it follows that every vertex $v \in V_1$ is adjacent to at most 5 vertices in V_2. Since $|V| \geq 2$ and G is connected it follows that $|V_1| \geq 1$ and $|V_2| \leq 5 \cdot |V_1| \leq 10 \cdot |VC_{OPT}|$. So we conclude $|V| = |V_1| + |V_2| \leq 2 \cdot |VC_{OPT}| + 10 \cdot |VC_{OPT}| \leq 12 \cdot |VC_{OPT}|$.

Corollary 1. *For every $K_{1,m}$ free graph $G = (V, E)$ it holds that $|V| \leq 2m \cdot |VC_{OPT}|$.*

4.2   Local $1 + \epsilon$ Approximation Algorithm for Minimum Vertex Cover

Let $G = (V, E)$ be a connected unit disk graph. The main idea of our algorithm is the following: We compute an approximate solution I for maximum independent set. Then we define $VC := V \setminus I$ as our vertex cover.

Let $1 + \epsilon$ be the desired approximation factor for minimum vertex cover. We define $\epsilon' := \min\left(\frac{1}{11}\epsilon, \frac{1}{2}\right)$. Using Algorithm 1 we locally compute a $1 - \epsilon'$ approximation for maximum independent set (note that $1 - \epsilon' > 0$ since $\epsilon' \leq 1/2$). Denote by I the computed set. We define the vertex cover VC by $VC := V \setminus I$. We output VC. We refer to the above as Algorithm 2.

4.3   Proof of Correctness

We prove the correctness of Algorithm 2, its approximation factor, its locality and its processing time in Theorem 3.

Algorithm 2. Local algorithm for finding a vertex cover in a unit disk graph

1 // Algorithm is executed independently by each node v;
2 define $\epsilon' := \min\left(\frac{1}{11}\epsilon, \frac{1}{2}\right)$;
3 Run Algorithm 1 with approximation ratio $1 - \epsilon'$;
4 // Denote by I the computed independent set;
5 **if** $v \in I$ **then** do NOT become part of the vertex cover VC **else** become part of VC

Theorem 3. *Let G be a unit disk graph and let $\epsilon > 0$. Algorithm 2 has the following properties:*

1. *The computed set VC is a vertex cover for G.*
2. *Let VC_{OPT} be an optimal vertex cover. It holds that $|VC| \leq (1 + \epsilon) \cdot |VC_{OPT}|$.*
3. *Whether or not a vertex v is in VC depends only on the vertices at most $O\left(\frac{1}{\epsilon^6}\right)$ hops away from v, i.e. Algorithm 2 is local.*
4. *The processing time for a vertex v is bounded by a polynomial in the number of vertices at most $O\left(\frac{1}{\epsilon^6}\right)$ hops away from v.*

We will prove the four parts of this theorem in four steps.

Correctness. We prove that the set VC is a vertex cover for G.

Proof. (of part 1 of Theorem 3): The set I is an independent set for G (see Theorem 1). Assume on the contrary that there is an edge $e = (u, v)$ with $u \notin VC$ and $v \notin VC$. As $VC = V \setminus I$ it follows that $u \in I$ and $v \in I$. This is a contradiction since I is an independent set. So VC is a vertex cover for G.

Approximation Ratio. We prove that for an optimal vertex cover VC_{OPT} it holds that $|VC| \leq (1 + \epsilon) \cdot |VC_{OPT}|$.

Proof. (of part 2 of Theorem 3): Let I_{OPT} be an optimal independent set. First we discuss the case where G has no edges and therefore consist of a single vertex v (since we assume that G is connected). In this case the maximum independent set is $\{v\}$. Since $0 < (1 - \epsilon')$ we have that $0 < (1 - \epsilon') \cdot |I_{OPT}| \leq |I|$ and so for the computed independent set I it holds that $I = \{v\}$. So then $V \setminus I = VC = \emptyset$ which is the optimal vertex cover.

Now assume that $|V| \geq 2$. So we can apply Theorem 2 and conclude that $|V| \leq 12 \cdot |VC_{OPT}|$. We know that $(1 - \epsilon') \cdot |I_{OPT}| \leq |I|$. From $|V| - |I_{OPT}| = |VC_{OPT}|$ we have $\frac{|I_{OPT}|}{|VC_{OPT}|} = \frac{|V|}{|VC_{OPT}|} - 1$. We compute that

$$\frac{|VC|}{|VC_{OPT}|} = \frac{|V|}{|VC_{OPT}|} - \frac{|I|}{|VC_{OPT}|} = \frac{|V|}{|VC_{OPT}|} - (1 - \epsilon')\left(\frac{|V|}{|VC_{OPT}|} - 1\right)$$

$$= 1 + \epsilon'\left(\frac{|V|}{|VC_{OPT}|} - 1\right)$$

$$\leq 1 + \epsilon'(12 - 1)$$

$$\leq 1 + \epsilon$$

So it follows that $|VC| \leq (1 + \epsilon) \cdot |VC_{OPT}|$.

Locality. Now we want to prove that Algorithm 2 is local (part 3 of Theorem 3).

Proof. (of part 3 of Theorem 3): According to Theorem 1 whether or not a vertex v belongs to I depends only on the vertices which are at most $O\left(\frac{1}{\epsilon'^6}\right) = O\left(\frac{1}{\epsilon^6}\right)$ hops away from v. Computing the set $VC = V \setminus I$ does not affect the locality of our algorithms. So whether a vertex v belongs to VC depends only on the vertex which are at most $O\left(\frac{1}{\epsilon^6}\right)$ hops away from v.

Processing Time. We prove that the processing time of Algorithm 2 is bounded by a polynomial. Again, we measure the processing time of a vertex v in $n_\alpha(v)$, that is the number of vertices which are at most α hops away from v, where α denotes the locality distance of Algorithm 2.

Proof. The processing time of Algorithm 2 is dominated by the processing time of Algorithm 1. So it is bounded by $n_\alpha(v)^{O\left(1/\epsilon'^4\right)}$ and therefore bounded by $n_\alpha(v)^{O\left(1/\epsilon^4\right)}$.

4.4 Global PTAS for Vertex Cover without Embedding of the Graph

Combining the PTAS for independent set and the upper bound for vertex cover in unit disk graphs (the set of all vertices giving a factor 12 approximation) we derived a PTAS for the vertex cover problem. This technique can be used in every setting where a PTAS for independent set is known and there is a constant upper bound for the number of vertices in a graph in comparison with the size of a minimum vertex cover.

In particular, Nieberg et al. [14] presented a global PTAS for maximum independent set in unit disk graphs which does not need the embedding of the graph as part of the input. Using our result for the upper bound for vertex cover in unit disk graphs their algorithm can be extended to the first global PTAS for vertex cover in unit disk graphs which does not rely on the embedding of the graph. This can be done in the same way as when employing our local PTAS for independent set in order to obtain our local $1 + \epsilon$ approximation algorithm for vertex cover: We run the algorithm for independent set with an approximation ratio of $1 - \epsilon'$ (with $\epsilon' = \min\left(\frac{1}{11}\epsilon, \frac{1}{2}\right)$) and output the inverse of the computed independent set. The proof that this gives an approximation ratio of $1 + \epsilon$ is the same as in Theorem 3.

5   Conclusion

We presented the first local $1 - \epsilon$ and $1 + \epsilon$ approximation algorithms for maximum independent set and minimum vertex cover, respectively. Local algorithms cannot compute optimal solutions for these problems (note that this holds no matter if $P = NP$ or $P \neq NP$). So our algorithms give the best possible approximation ratios for these problems in our setting. Despite the locality constraint, our algorithms achieve the same approximation factors which can be guaranteed by the best known global polynomial time algorithms for the discussed problems.

We estimated the locality distances of our algorithms depending on ϵ. It is evident that further improvements are desirable. It remains open to design local PTASs for the considered problems with lower locality distances. In order to guarantee the $1 + \epsilon$ approximation factor for minimum vertex cover we proved that all vertices of a unit disk graph form a factor 12 approximation for vertex cover (assuming that there are edges in the graph which are supposed to be covered). With the same method it can be proven for any $K_{1,m}$-free graph that the set of all vertices forms a factor $2m$ approximation for vertex cover. It remains open to improve this bound or to show that it is tight. If the bound for unit disk graphs could be improved, this would immediately prove a lower locality distance of our local PTAS for vertex cover. Additionally, due to the design of our algorithms a better locality distance for minimum independent set would imply an improved locality for vertex cover as well.

Also of interest would be to generalize the concepts used in our algorithms to related types of graphs like quasi-Unit Disk Graphs or graphs defined by other geometric shapes.

References

1. Bar-Yehuda, R., Even, S.: A local-ratio theorem for approximating the weighted vertex cover problem. Annals of Discrete Mathematics 25, 27–45 (1985)
2. Basagni, S.: Finding a maximal weighted independent set in wireless networks. Telecommunication Systems 18(1-3), 155–168 (2001)
3. Breu, H., Kirkpatrick, D.G.: Unit disk graph recognition is NP-hard. Computational Geometry. Theory and Applications 9(1-2), 3–24 (1998)
4. Clark, B.N., Colbourn, C.J., Johnson, D.S.: Unit disk graphs. Discrete Math 86(1-3), 165–177 (1990)
5. Czyzowicz, J., Dobrev, S., Fevens, T., González-Aguilar, H., Kranakis, E., Opatrny, J., Urrutia, J.: Local algorithms for dominating and connected dominating sets of unit disc graphs with location aware nodes. In: Proceedings of LATIN 2008. LNCS, vol. 4957 (2008)
6. Dinur, I., Safra, S.: The importance of being biased. In: Proceedings of the 34th Annual ACM Symposium on Theory of Computing (STOC 2002), May 19–21, 2001, pp. 33–42. ACM Press, New York (2002)
7. Garey, M.R., Johnson, D.S.: Computers and intractability: A guide to the theory of NP-completeness (1979)
8. Håstad, J.: Clique is hard to approximate within $n^{1-\epsilon}$. Electronic Colloquium on Computational Complexity (ECCC) 4(38) (1997)
9. Hunt III, H.B., Marathe, M.V., Radhakrishnan, V., Ravi, S.S., Rosenkrantz, D.J., Stearns, R.E.: NC-approximation schemes for NP- and PSPACE-hard problems for geometric graphs. J. Algorithms 26(2), 238–274 (1998)
10. Kuhn, F., Moscibroda, T., Wattenhofer, R.: Unit disk graph approximation. In: DIALM-POMC 2004: Proceedings of the 2004 joint workshop on Foundations of mobile computing, pp. 17–23. ACM Press, New York (2004)

11. Kuhn, F., Nieberg, T., Moscibroda, T., Wattenhofer, R.: Local approximation schemes for ad hoc and sensor networks. In: DIALM-POMC 2005: Proceedings of the 2005 joint workshop on Foundations of mobile computing, pp. 97–103. ACM Press, New York (2005)
12. Marathe, M.V., Breu, H., Hunt III, H.B., Ravi, S.S., Rosenkrantz, D.J.: Simple heuristics for unit disk graphs. Networks 25(1), 59–68 (1995)
13. Nieberg, T., Hurink, J.L.: A PTAS for the minimum dominating set problem in unit disk graphs. In: Erlebach, T., Persinao, G. (eds.) WAOA 2005. LNCS, vol. 3879, pp. 296–306. Springer, Heidelberg (2006)
14. Nieberg, T., Hurink, J.L., Kern, W.: A robust PTAS for maximum weight independent sets in unit disk graphs. In: Hromkovič, J., Nagl, M., Westfechtel, B. (eds.) WG 2004. LNCS, vol. 3353, pp. 214–221. Springer, Heidelberg (2004)
15. Wiese, A., Kranakis, E.: Local PTAS for Dominating and Connected Dominating Set in Location Aware UDGs (to appear, 2007)
16. Wu, J., Li, H.: On calculating connected dominating set for efficient routing in ad hoc wireless networks. In: DIAL-M, pp. 7–14. ACM, New York (1999)

Multi-root, Multi-Query Processing in Sensor Networks*

Zhiguo Zhang, Ajay Kshemkalyani, and Sol M. Shatz

Department of Computer Science, University of Illinois at Chicago,
Chicago, Illinois 60607

Abstract. Sensor networks can be viewed as large distributed databases, and SQL-like high-level declarative languages can be used for data and information retrieval. Energy constraints make optimizing query processing particularly important. This paper addresses for the first time, multi-root, multi-query optimization for long duration aggregation queries. The paper formulates three algorithms - naive algorithm (NMQ), which does not exploit any query result sharing, and two proposed new algorithms: an optimal algorithm (OMQ) and a heuristic (zone-based) algorithm (ZMQ). The heuristic algorithm is based on sharing the partially aggregated results of pre-configured geographic regions and exploits the novel idea of applying a grouping technique by using the location attribute of sensor nodes as the grouping criterion. Extensive simulations indicate that the proposed algorithms provide significant energy savings under a wide range of sensor network deployments and query region options.

1 Introduction

One way to extract sensor data from a distributed sensor network is by using mobile agents that selectively visit the sensors and incrementally fuse appropriate measurement data [12]. Another technique, which is the subject of this paper, is to inject queries into the network, treating the sensors as a distributed database [4]. To reduce energy use associated with communication while gathering data, in-network aggregation [8] can be used, in addition to special network routing to minimize messages needed for query processing [1,6]. For example, we previously proposed a grouping technique based on query-informed routing to make in-network aggregation more energy efficient [13]. The sensors are programmed through declarative queries in a variant of SQL. The following is an example query for monitoring the radiation in a nuclear power plant:

SELECT room, AVG(radiation) FROM sensordb WHERE building = ERF GROUP BY room HAVING AVG(radiation) > 100 DURATION 30 days EVERY 1 minute

---

* This material is based upon work supported by the U.S. Army Research Office under grant number W911NF-05-1-0573.

S. Nikoletseas et al. (Eds.): DCOSS 2008, LNCS 5067, pp. 432–450, 2008.

Most previous research on query processing in sensor networks has focused on the processing of a single long-running aggregation query (see, for example [8,13]). As an extension to this line of research, Trigoni et al. [11] and Emekci et al. [3] considered the case of reducing message transmission by sharing sensor readings for multiple queries, where queries are represented by particular query regions. A query region is the geographical region a query is interested in retrieving information from. For example, for the query SELECT AVG(temperature) FROM sensorDB where $position.X \geq 25$ and $position.X \leq 75$ and $position.Y \geq 25$ and $position.Y \leq 75$ DURATION 1 day EVERY 1 minute, the region (25, 25)(75, 75) is the query region.

These existing multi-query processing techniques work for centralized environments only, and they require that the queries arrive at a common root node. In this case, since all query regions are known by one root node, intersection regions (the intersection areas among query regions) can be computed at the root node, making it possible to share partially aggregated results of intersection regions. Since these methods use centralized computation of intersection regions at the root node, they are not directly suitable for multiple queries injected from different root nodes. In addition, although existing methods attempt to share the sensor value readings of intersection regions, they do not account for the effect of the routing structure on the efficiency of aggregation, and do not address the problem of how to group sensor nodes according to query regions. As a result, many intermediate nodes need to unnecessarily wake up and transfer messages for sensor value readings of nodes that lie in the same query region but belong to different queries. As pointed out in [13], the routing tree structure of sensor networks can have significant impact on the aggregation efficiency of data retrieval in sensor networks. Therefore, using query information in the construction of a routing tree can provide improvement by reducing message transmission.

In this paper, we formulate and address for the first time the problem of multi-root, multi-query processing for long duration aggregation queries. This problem arises in many applications where loosely-coupled, or independent, stakeholders want to gather information from a common (shared) sensor network. As a specific example, consider a case of environmental monitoring, where scientists studying wildlife migration and climatologists studying pollution patterns are operating from different locations, but both need to monitor average rainfall volumes associated with different regions in a forest-based sensor field. Another example arises in a battlefield situation, where two remotely located battalions want to monitor enemy troop movements in different but partially overlapping battlefield sectors.

We consider the most general case, where multiple queries are injected asynchronously into the network at different root nodes. Since there is no global knowledge of the different queries, completely distributed solutions are required. We formulate, and compare, three algorithms: a naive algorithm (NMQ), an optimal algorithm (OMQ), and a heuristic algorithm (ZMQ). ZMQ is based on sharing partially aggregated results of pre-configured geographic regions (called zones [7]), and exploits the novel idea of applying a grouping technique for optimization of multi-root, multi-query processing, by using the location attribute of sensor nodes as the grouping criterion. This optimization aims to maximally share the reading and transmission of sensor node values belonging to multiple queries. Once sensors are deployed, the sensor field can be viewed as being divided into zones, and a logical data aggregation tree is established to hierarchically represent the zones. The idea of using such a recursive tree for query dissemination and data retrieval is not a new idea; authors in [7] use a similar so-called quad-tree structure for handling spatial aggregation queries in sensor networks. In [5], the authors use this kind of quad-tree structure for optimizing queries that have frequently changing aggregation groups. Since the goal of our heuristic approach is to share the sensor readings and data transmission among different queries if their query regions intersect, we group together sensor nodes in the same zone, so that sensor nodes in the intersection region of multiple queries only need to send their sensed value once, independent of the number of queries. In a distributed and asynchronous manner, a query taps into the data aggregation tree at the lowest possible tree node such that the zone represented by that node's sub-tree contains the geographic area of its query coverage. Our approach becomes more effective as the regions associated with multiple queries increasingly overlap.

We performed extensive simulations on the proposed OMQ, NMQ, and ZMQ algorithms. The NMQ algorithm treats queries independently and does not do any sharing of the aggregated data for sharing of message transmission. Our simulation studies indicate that the OMQ and ZMQ algorithms provide significant reduction in messages (and thus energy saving) under a wide range of network conditions and query region options.

2   Background

2.1   Assumptions and Challenges

We make the following assumptions regarding the sensor network system model and solution framework:

1. All queries of interest are querying the same type of sensor data, like the temperature of the environment.
2. Each node in the network knows its geographical position and the scale of the sensor field. Since the use of GPS in each sensor node incurs a high cost and high power consumption, it may not be practical to have GPS on all sensor nodes. However, GPS-free localization techniques [9,10] make our assumption still reasonable.
3. A query is characterized by a rectangular query region, like in the example in Section 1. An arbitrarily shaped query region can be split into rectangular regions to get approximate results since there is typically a high degree of redundancy in sensor networks.
4. Either all queries use the same sampling rate, or the effective sampling rate is set as the highest of the individual sampling rates. As noted in [3], this way

of handling different sampling rates is based on the observation that those sensor readings with higher sampling rate can give a better approximation of the environment.

5. Queries are first injected into the sensor network via root nodes, which are regular sensor nodes of the network. Users can inject different queries into the sensor network from different root nodes of their choice.

Our approach to multi-root, multi-query processing in sensor networks is motivated by the goal of sharing sensor readings and data transmission among different queries if the query regions of different queries intersect. In designing a distributed solution for optimization of multi-root, multi-query processing, we identify two challenges:

Challenge One (C1): How to determine the intersection regions of multiple queries, especially if those queries are injected at different sensor nodes.

Challenge Two (C2): How to make nodes in different query regions group together for aggregation efficiency.

To address challenge **C1** we use the notion of zones [7] to represent query regions. A zone is a subdivision of the geographical extent of a sensor field, and each sensor node can compute the zones according to the scale of the sensor field independently. We will give more detail about the definition of zones in Section 2.3. Since zones are predefined when the network is deployed, intersection zones are easy to decide even though the queries are input at different root nodes. To address challenge **C2**, we apply a grouping technique to group sensor nodes in the same zone to form an aggregation efficient tree topology for multiple queries. Grouping sensor nodes in the same zone together in a sub-tree not only increases the aggregation efficiency, but also makes possible the sharing of partially aggregated results of zones. The grouping technique is reviewed next.

2.2 A Grouping Technique for Query-Informed Routing

Query processing in sensor networks typically proceeds in three phases: (i) disseminating queries into the network, (ii) sensing data, and (iii) retrieving data from the network. For phases (i) and (iii), a tree topology is formed using some variants of the broadcast and convergecast techniques (e.g., [2] presents an optimized broadcast protocol using an adaptive-geometric approach). Here, each node performs two actions: 1) according to the messages it receives, each node decides its own level and selects a parent node with respect to the tree topology being created, and 2) the node broadcasts its own id and tree level. Once all nodes in the network have established their tree levels and parent nodes, the tree topology is defined.

In previous methods [8], sensor nodes select their parents using only tree levels. The grouping technique used in [13] is motivated by the fact that it is common for queries in a sensor network to be aggregation queries (such as COUNT, MAX, MIN, AVERAGE, etc.) using GROUP BY or WHERE. Such queries can form aggregation groups according to a specific attribute of the sensor nodes, and there are situations where such queries must remain active over long durations. The basic idea of the grouping technique is to try to force those sensor nodes with the same specific attribute, used in the GROUP BY or WHERE clause, to be logically close to each other when forming the tree topology.

Consider the query: SELECT SUM(value) FROM sensordb WHERE color =blue. Figure 1 illustrates the main idea. B represents blue node. The aggregation is completed at node $S_{3,3}$ if the tree topology, as shown in Fig 1a, is formed by using the grouping technique. However, if sensor nodes select parent nodes non-deterministically, we may end up with a tree topology as in Fig 1b.

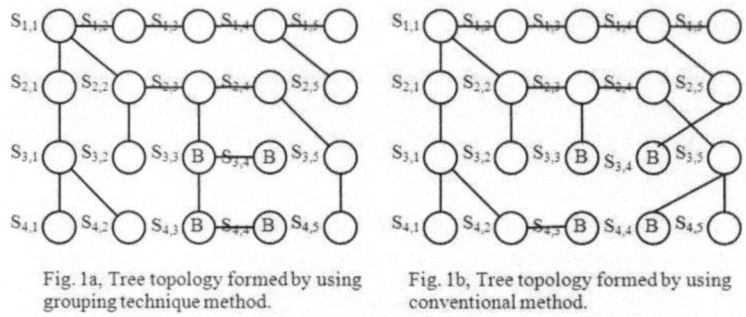

Fig. 1a, Tree topology formed by using grouping technique method.

Fig. 1b, Tree topology formed by using conventional method.

Fig. 1. An example for queries using WHERE clause

2.3   Sensor Field Division Using Zones

Zones for sensor networks have been used in the context of range queries [7]. Zones are a subdivision of the geographic extent of a sensor field. A zone is defined by the following constructive procedure. Consider a rectangle R on the x-y plane. Intuitively, R is the bounding rectangle that contains all sensors within the network. We call a sub-rectangle Z of R a zone, if Z is obtained by dividing R k times, $k \geq 0$, using a procedure that satisfies the following property: After the i-th division, $0 \leq i \leq k$, R is partitioned into $2i$ equal sized rectangles. If i is odd (even), the i-th division is along the values of the y-axis (x-axis). Thus, the bounding rectangle R is first sub-divided into two zones at level 1 by a vertical line that splits R into two equal pieces. Each of these sub-zones is split into two zones at level 2 by a horizontal line, and so on. The integer k is the level of zone Z, i.e., level(Z) = k.

A zone can be identified either by a zone code code(Z) or by an address addr(Z). The code code(Z) is a bit string of length level(Z), and is defined as follows. If Z lies in the left half of R, the first (from the left) bit of code(Z) is 0, else 1. If Z lies in the bottom half of R, the second bit of code(Z) is 0, else 1. The remaining bits of code(Z) are recursively defined on each of the four quadrants of R. This definition of the zone code matches the definition of zones given above, encoding divisions of the sensor field geography by bit strings.

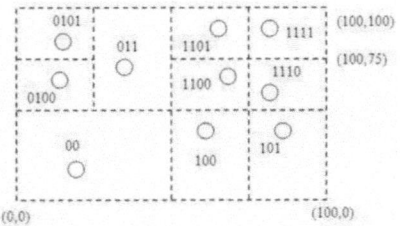

Fig. 2. Zone codes and boundaries

Figure 2 shows a deployed sensor network, and the zone code for each zone. The zone in the top right corner of R has a zone code of 1111, and its level is 4. The address of a zone Z, addr(Z), is defined to be the rectangle defined by Z. Each representation of a zone (its code and its address) can be computed from the other.

The zone with code 1111 represents the region $[75, 100] \times [75, 100]$ in sensor network space $[0, 100] \times [0, 100]$, where space $[xmin, xmax] \times [ymin, ymax]$ represents the rectangular region with the left bottom at (xmin, ymin) and right top at (xmax, ymax). Similarly, given a region $[25, 75] \times [50, 100]$, we can know that it contains zones 011 and 110. We use the same prefix of zones to represent a bigger zone that contains those zones. For example, zone with code 11 includes zones 1100, 1101, 1110, and 1111. Let *Prefix(codea, codeb)* be the longest common prefix of *codea* and *codeb*. For e.g., Prefix(1110, 11) equals 11.

As each node knows its position and the scale of the sensor field, these geographic zones are predefined once the network is deployed. Each sensor node knows its own zone code and the scale of the sensor field; hence it knows the geographical region that any zone code represents. Given a query represented by a query region, all sensor nodes identify the same set of zones that represent the region. This allows the distributed computing of intersection regions.

3   Multi-Query Modeling

3.1   The Naive Method for Multi-root, Multi-Query Processing (NMQ)

In Figure 3, Q1 and Q2 are injected from two different nodes R1 and R2. The different rectangles represent the different query regions. The NMQ algorithm sets up different tree structures for Q1 and Q2 separately. This naive algorithm does not share any sensor readings. The grey nodes, which are the nodes in the intersection region for query regions Q1 and Q2, need to send the same readings to different parent nodes twice, once for Q1 and once for Q2. Figure 3 shows the tree structures for Q1 and Q2 separately. A better algorithm would allow Q1 and Q2 to share the readings and messages of the grey nodes.

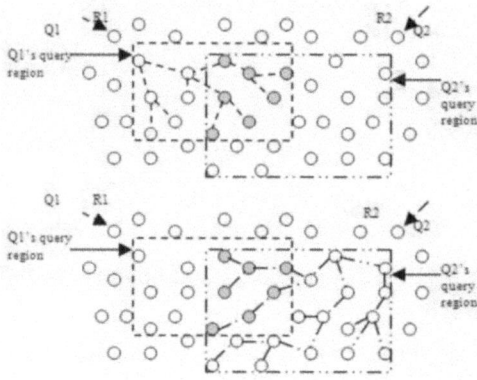

Fig. 3. NMQ for multi-root multiple query

3.2   Optimal Multi-root, Multi-Query Processing (OMQ)

To lower the cost of multi-query processing, we can share the readings of sensor nodes in the intersection region. Here, the two challenges **C1** and **C2** need to be solved.

To solve **C1**, reading-sharing methods need to know the identity of the nodes in the intersection regions. Consider Figure 3. Since Q1 is known only to R1 and Q2 is known only to R2, one option is to let the network first construct the tree structure for Q1 and then adjust the tree structure in the intersection region when Q2 is propagated in the network. This readjustment in the middle of processing Q1 would be problem-prone and energy consuming. The optimal algorithm (OMQ) pre-constructs the sub-trees for the intersection regions and the other regions of queries. Here, one and only one sub-tree is constructed for each region, and then the paths from R1 and R2 to those sub-trees are pre-setup, as in Figure 4. However, since the regions are known only after all queries are injected, this method is of limited use even for common-root multi-query processing. It is applicable to multi-root, multi-query processing only in the static case, where the regions of Q1 and Q2 are known in advance and do not change. Still, it can serve as a benchmark.

If we can know the regions, i.e., if **C1** is solved, we can use the grouping technique to solve **C2** -by grouping nodes in the same region into one sub-tree.

3.3   Zone-Based Multi-root, Multi-Query Processing (ZMQ)

To increase the sharing of data provided by sensor nodes in the presence of dynamically arriving queries, a practical way is to predefine a set of globally known regions in the network, and then represent query regions as pre-defined regions. This gives different nodes the same view of the field and lets them use the same known regions to represent the same query region.

Figure 5 shows an example of such pre-defined globally known regions, represented as a tree. Each node in the tree represents a globally known region. For

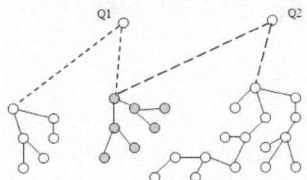

Fig. 4. OMQ method for multi-query processing

example, the network $[0, X] \times [0, Y]$ is one globally known region, as is the region represented by $[0, X/2] \times [0, Y/2]$. In the tree structure, the globally known region represented by a parent node consists of the globally known regions represented by the children nodes. Since a node cannot know ahead of time what the intersection regions might be, it is not practical to pre-setup exactly one region for each intersection region. So, we use pre-defined globally known regions to represent all possible query regions, and an intersection region is represented by one or more such globally known regions. Each time a query is injected into the network, the root node computes the globally known regions that can be used to represent the query region, and sets up paths from itself to the root nodes of those globally known regions. Queries can share the partially aggregated results of globally known regions if they have globally known regions in common. Although the roots cannot know the intersection regions ahead of knowing the queries, all roots can use identical views of the globally known regions to represent query regions. The intersection regions for any set of queries would then be a set of globally known regions.

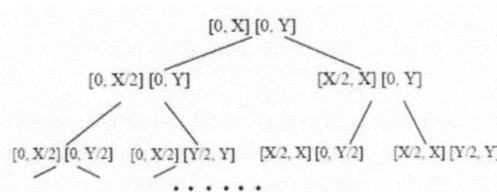

Fig. 5. Examples of predefined globally known regions

The framework of globally known regions solves challenge (**C1**). This framework is implemented using zones (Section 2.3). To solve challenge (**C2**), the ZMQ algorithm uses the grouping technique, reviewed in Section 2.3, to group nodes in each zone into one sub-tree. All those zone sub-trees form a globally pre-setup tree. Furthermore, the representing globally known regions are easy to compute given a query region. Given the size of the sensor field and a zone code, a node can easily compute the region represented by this zone. Therefore, intersection regions are very easy to determine even in a distributed environment.

4   Algorithm ZMQ (Zone-Based Multi-root, Multi-Query Processing)

4.1   Zone Setup

The system model assumes that each sensor node knows its location and the scale of the network region (see Section 2). Each node learns the locations of neighbors within radio range through direct-broadcast communication. Upon hearing any neighbor node, the node, say node A, calls an algorithm BUILD_ZONE to build its zone code and boundaries accordingly. The first split of the whole sensor field creates two sub-zones, 0 and 1.

On finding a new neighbor, a sensor node uses algorithm BUILD_ZONE to split its current zone either vertically or horizontally into two sub-zones, and then adjusts its zone code. Using the BUILD_ZONE algorithm, each node knows its own zone code. These zones form the zone-tree structure, See Figure 6. The parent-child relations in Figure 6 represent containment, where the parent zone is comprised of child zones. Each node in the zone tree is a zone; the path from the root node to the current node is the zone code of the zone represented by the current node.

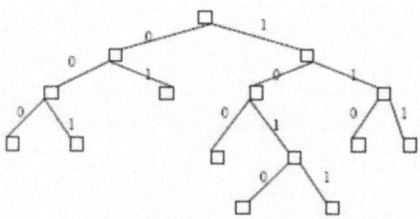

Fig. 6. The zone tree for all zones

The rectangular region represented by each zone is decided once the construction of zone codes is complete. For example, zone 1011 represents the region $[3X/4, X] \times [Y/4, Y/2]$. After computing their zone codes, sensor nodes in the same zone automatically form groups based on the computed zone codes.

4.2   Sensor Node Grouping in Zones

In the algorithm GROUPING_ZONE, A is the sensor node executing the algorithm, Z_A is the zone represented by A, and code(Z_A) is the zone code of Z_A. B is any neighbor node of A. Zone code ES (empty string) represents the whole sensor field zone. Operator "$\gg n$" eliminates the last n characters of a string; e.g., $110011 \gg 1$ is 11001.

Each node executes this algorithm after it detects that all its neighbor nodes have decided their zones. The idea is that a node A first searches neighbor nodes in its immediate parent zone (the smallest zone contains Z_A), to find a node with minimum zone code. If such a node B exists, and its zone code is smaller than

GROUPING_ZONE
1. $code = code\_old = code(Z_A)$ // initialize zone code variables
2. While $(code \neq ES)$ // Iterate until code represents the whole network
3. $code = code\_old \gg 1$ //code rep. the zone containing the zone rep. by code_old
4. $T = \varnothing$ // initialize the temporary set T
5.   For each B where B is neighbor of A
6.   If $(Prefix(code, code(Z_B)) = code$ and $Prefix(code\_old, code(Z_B)) \neq code\_old)$
7. $T = T \cup \{B\}$ // put any neighbor B into set T, if B is not inside
8.   // zone of *code_old*, but is in the zone of *code*
9.   If $(T = \varnothing)$
10.   *code_old* = *code*
11.   Continue   //back to line 2 to try the upper level zone
12.   Let C be the node in T with smallest code
13.   If $(code(Z_C) < code(Z_A))$
14.   Select C as A's grouping parent node
15.   Return   // parent node of A is C
16.   *code_old* = *code*
17.   Return   // A is the root node

A's zone code, then A selects B as its grouping parent. Otherwise, A searches neighbor nodes in the parent zone of its immediate parent zone, and so on, until it finds a parent. Consider Figure 7. Node 1111 has three neighbors, 1100, 1101, and 1110. This node first searches the direct parent zone of zone 1111, which is zone 111, and finds node 1110, which has smaller zone code than itself. Therefore, node 1111 selects node 1110 as its grouping parent node. The root node of zone 111 is node 1110. Similarly, node 1110 also has three neighbors. This node first searches its direct parent zone 111, and finds node 1111, which has bigger zone code than itself. So node 1110 searches zone 111's parent zone which is zone 11, and it finds that node 1100's zone code is the smallest one. Node 1110 would select 1100 as its grouping parent node.

Definition 1. *Zone links for the grouping-tree network are the parent-child links formed using the GROUPING_ZONE algorithm. A node's zone link is the link from the node to its parent.*

All the links in Figure 7 are zone links. Zone links are pre-constructed once a network is deployed. Each node has a unique zone link. The zone-link tree formed by zone links has the property that nodes in the same zone are in one sub-tree. This achieves the method of grouping by zone. When queries are received, the routing network only needs to be adjusted to include paths from the root nodes of zones to the root nodes of queries. These paths are established by "forward links"; see Section 4.3.

Note that for unevenly distributed networks, or if nodes fail, the algorithm GROUPING_ZONE may form several tree structures in a network. Consider Figure 7. If node 100 is absent, then node 101 cannot communicate with node 00, and two tree structures get formed - one rooted at 00 and the other at 101.

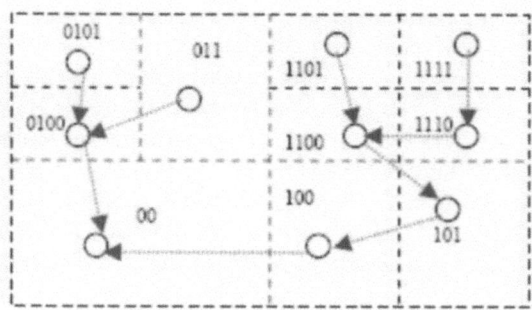

Fig. 7. A Grouping-tree example

However, the routing-tree construction algorithm of ZMQ would still form one routing tree for each query.

4.3   Query Handling

Definition 2. *Forward links are the parent-child links that connect root nodes of zones to root nodes of queries. While a node only has one zone link, it can have multiple forward links, one for each different query.*

Forward links are created in the grouping-tree in response to queries being injected to root nodes (see Section 4.3.2). These links transfer partially aggregated value of zones to the root nodes of queries during query processing. During query processing, forward links are all active, while zone links may be in an active or inactive state (i.e., some zone links may not be used to process some queries and thus they are not a part of the final routing topology).

Figure 8 shows an example of active zone links, inactive zone links and forward links. Two queries, Q1, with query region $[0,75] \times [50, 100]$, and Q2, with query region $[25, 100] \times [50, 100]$, are injected at different root nodes. After the queries have been propagated into the network, a routing topology is created. The dashed arcs illustrate the forward links, which form the paths from roots of zones to root nodes of queries. The dotted arcs illustrate inactive zone links, while the solid arcs illustrate active zone links, which form the sub-trees for queried zones.

4.3.1   Region Representation

Observation 1. *A region R (of a query Q) can be uniquely represented by a set of zones $S = \{Z1, Z2, \ldots\}$, where (i) zones in S do not overlap with each other (i.e., no node in $Zi \in S$ is in $Zj \in S$, for $Zi \neq Zj$), and (ii) no two zones in S can be siblings.*

Observation 1 is the basis for data sharing of intersection regions in our multi-root, multi-query processing. Once sensor nodes are deployed, they first use

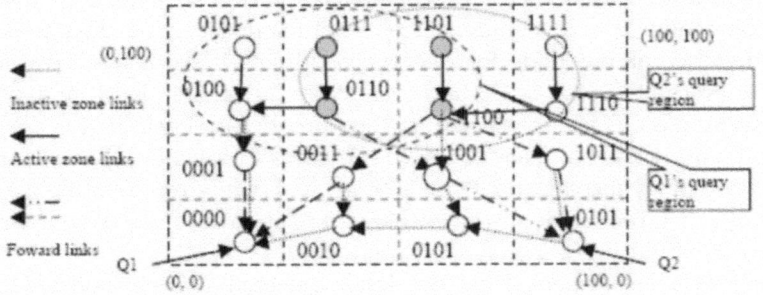

Fig. 8. Example routing structure for query processing

BUILD_ZONE to compute individual zone codes, and then use GROUPING_ZONE to group nodes in zones into sub-trees and thus form a complete zone-link tree. This was the pre-setup process. Now we address the process of handling queries, which requires an algorithm for setting up paths from root nodes of queries to root nodes of zones for queries, and algorithms for data retrieval using the paths.

4.3.2 Routing and Data Retrieval

In the tree built by the algorithm GROUPING_ZONE, each zone can compute its partial aggregation result in the zones root node in every sampling epoch. The processing for a query Q needs to set up paths from the root node of Q to each of the root nodes of the zones that belong to Q's zone representation. This is handled by an algorithm called BUILD_ROUTING_TOPOLOGY.

The algorithm BUILD_ROUTING_TOPOLOGY constructs routing tree topologies for data retrieval. The algorithm implements two features: 1) it uses a special forward-link notification message, FL_Notify, to build forward links from root nodes of zones to root node of queries; 2) it changes inactive zone links to active zone links based on whether a node is in a zone that is a representing zone of a query.

Each node in the network maintains a neighbor table recording status information of its neighbors, such as id, tree level, etc. Once a node receives a new query Q that has been injected into the network, this node sets its tree level (for Q) as 1 because it is the root. Then, this root node broadcasts a Query Broadcast (QB) message, containing its own id, tree level, the query information, and the zone representation of the query. The broadcast is across one hop, so only immediate neighbors of the sender receive the QB. On receiving such a QB message from some node Z, a node A updates its own neighbor table, specifically the data about neighbor Z (including Z's tree level for Q).

Each node A periodically executes the BUILD_ROUTING_TOPOLOGY algorithm. The algorithm is executed independently for each query Q. Consider any query Q for which some query broadcasts QB have been received. Node A first tests if it has already selected a routing-tree parent node for this query. If node A has not selected such a parent node, A checks its neighbor table to find a neighbor node M with minimum tree level for query Q. A then selects M as its routing-tree parent node for query Q, and sets its own tree level for Q as M's tree level for Q plus 1. A then broadcasts its id, tree level, and Q's query information in a QB message using a 1-hop broadcast. Then, if A is a root node of a representing zone of query Q, A sets M as its forward link parent for Q, and sends a FL_Notify message for Q to M. Else, if A is in a representing zone of the query but not a root node of that zone, A sets its own inactive zone link to be an active zone link.

Consider the case where A has previously selected a parent node M for query Q, and it has also received a FL_Notify message for Q from some child node. This can happen if one of A's child nodes is a root node of a zone representation for query Q, or one of A's child nodes received a FL_Notify message for Q from one of the childs children nodes. Node A constructs a forward link for Q by setting the routing-tree parent node as the forward-link parent node for Q and sending a FL_Notify message for Q to this forward link parent. Otherwise, node A has not received the FL_Notify message for Q from any child, implying that A might not be in a tree path from root nodes of zones to the root node of query Q and A might not need to be in the active state in the data retrieval phase. Hence, A exits the algorithm.

Since the zone link of each node is unique (pre-setup by the algorithm GROUP-ING_ZONE), if two queries have a common zone in their zone sets, they both actually change the status of the same zone links in the zone from inactive to active. Hence, that sub-tree of the routing tree for both queries is common. Only the root node of that common zone may construct different forward links for different queries - the partially aggregated value of the zone will be sent separately by the root node of the zone to the different root nodes of queries along different paths.

In the BUILD_ROUTING_TOPOLOGY algorithm, for any node A, its forward-link parent is initially null. Nodes having forward links or active zone links (as established by the above algorithm) would be in the active state, meaning that these nodes can transmit data messages during query processing. Other nodes may enter a sleep state to save energy. Nodes engaged in query processing do so by executing the algorithm DATA_RETRIEVAL. Using Figure 7 as an example, node 1111 would send the tuple (1111, value) to its parent node 1110, while node 1110 would create an aggregated value for zone 111 and then send the

---

DATA_RETRIEVAL(A)   // A is the node executing the algorithm
1. If (A is a leaf node)
2.   Send zone code and sensor value to parent node
3.   Return
4. Aggregate data received from children based on zones   //merge smaller zones
5. While (there is a forward link for a query)
6.   Send along the forward link the partially aggregated value for that query
7. If (there is an active zone link)
8.   Send all partially aggregated values based on zones along the zone link
9. Return

tuple (111, aggregated-value) to its parent node 1100. For node 101, it receives (11, value) from node 1100. It cannot aggregate its zone 101 with zone 11, so it would send a two-tuple message ((101, its own value), (11, value)) to its parent node 100; and so on.

5   Experimental Evaluation

We performed simulation experiments to compare and analyze the algorithms, NMQ, OMQ, and ZMQ.

- NMQ constructs a different routing tree for each query, thereby using a different tree for each query region.
- ZMQ is implemented based on the details in Section 4.
- Based on the position data for ZMQ, we can compute the intersection regions for OMQ. Then the grouping technique is used to group nodes in the same intersection regions together in the same sub-tree to compute the results for OMQ. As OMQ assumes that the multiple query regions are known before the query processing, it forms one sub-tree for each intersection region.

Deployment: We use a 256×256 cell matrix, where a sensor can be placed at the center of a cell. The length of each side of a cell is 1. Each node, except the nodes in the border cells, can communicate directly with its eight direct neighbors in the matrix. By default, each cell has a sensor placed in it. Input parameters:

N: The number of queries (Each query defines a query region).
D: The network density, defined as the number of sensors per cell of the sensor field matrix. To be consistent with the system model, the highest value of D is 1. This is also the default value.
QR: The query region representing a query.
OP: Overlap percentage, which we define as
$\left(\frac{\sum_I sizeof(I)*(numberof(I)-1)}{\sum_Q sizeof(Q)} \right) \left(\frac{N}{N-1} \right)$, where

- I is an intersection region, defined as the largest region in which all the nodes are queried by the same set of queries.
- sizeof(I) is the number of nodes in the intersection region I,
- numberof(I) is the number of queries that each node in I receives,
- N is the total number of query regions, and
- $\frac{N}{N-1}$ is a scale factor for normalization.

Metric: Average number of messages (ANM) per node per epoch, is defined as the total number of messages used in each retrieving epoch divided by the total number of nodes in the query regions. Nodes in intersection regions are counted separately for each query, i.e., nodes in intersection regions are counted multiple times. Formally,

$$ANM = \left(\frac{total\ number\ of\ messages}{\sum_Q sizeof(Q)} \right)$$

5.1   Impact of Overlap Percentage(OP)

In this experiment, we show the impact of the overlap percentage on the performance of the three algorithms. We hold constant the input parameters QR and N, but vary the position of different query regions to change the overlap percentage for all queries. As seen in Figure 9, we can observe that ZMQ outperforms NMQ as the percentage of overlapping increases, and ZMQ has nearly the same performance as OMQ. The reason that the number of messages used by ZMQ and OMQ decreases is because these algorithms share the readings of sensor nodes in the intersection regions. As the overlapping increases, more sharing is possible, and fewer messages are needed. Notice that ZMQ uses more messages than NMQ when there is no overlap between the two query regions. This is because ZMQ slightly decreases the aggregation extent and increases the number of messages needed if a query region contains more than one pre-setup zones. However, we can also see from Figure 9 that if there is an overlap between the query regions, the number of messages reduced by sharing readings and transmissions exceeds the number of messages increased by such a decrease in the aggregation extent.

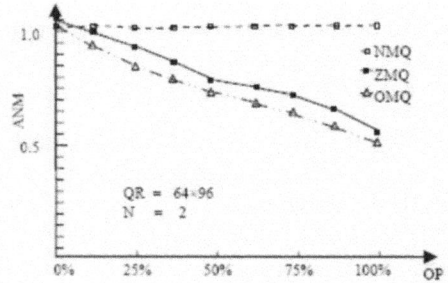

Fig. 9. Impact of OP on number of messages

The difference between ZMQ and OMQ is because each intersection region in ZMQ may consist of more than one zone, i.e., more than one sub-tree, while each intersection region in OMQ consists of just one sub-tree. The aggregation extent for intersection regions for ZMQ is slightly less than the aggregation extent for OMQ. This causes ZMQ to use more messages than OMQ.

5.2   Impact of Network Density (D)

In this experiment, we show the effect of the density of the sensor network on the performance of the three algorithms. We hold QR, N, and OP fixed, and vary the density of the sensor field. The average number of messages transmitted, as a function of density is shown in Figure 10. The density in Figure 10 is computed as the number of sensor nodes divided by the size of the sensor field. For example, for a 256×256 sensor field, the density 1/4 means that there are 128×128 sensor nodes evenly spread in the sensor field.

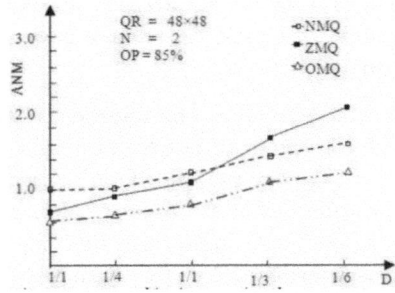

Fig. 10. Impact of density on number of messages

Observe that for all algorithms, as density decreases, the ANM increases. This is because when density decreases, the nodes become further apart, and more messages are needed to collect the readings. This can also be seen from the definition of ANM, viz., $\frac{total\ number\ of\ messages}{\sum_Q sizeof(Q)}$. As D decreases, the denominator decreases proportionately to D but the numerator does not change as rapidly. Observe that even if the query regions have significant overlap, ZMQ may perform worse than NMQ when the density is low enough. The reason is that as the density becomes lower, the number of nodes in the intersection area decreases, and the data and transmission sharing also decreases. At some threshold, the number of messages added because of the decrease in aggregation extent may exceed the number of messages decreased by data sharing.

5.3   Impact of Query Region Size (QR)

Figure 11 shows the relationship between size of query regions and the average number of messages transmitted in each epoch for the three algorithms. Observe that as QR increases, the ANM ($= \frac{total\ number\ of\ messages}{\sum_Q sizeof(Q)}$) for NMQ approaches. In addition, as QR increases, ANM of ZMQ approaches the ANM of OMQ. This implies that ZMQ performs better when the sizes of query regions are large.

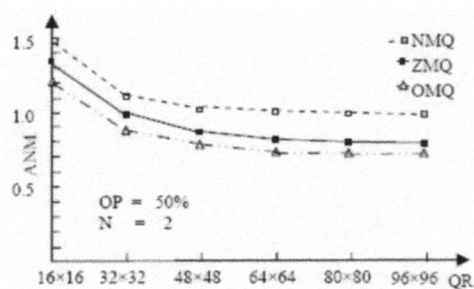

Fig. 11. Effect of QR on the algorithms

Figure 11 also shows that for all the algorithms, as the size of the query region increases, the average number of messages for each node first decreases and then becomes stable. This is due to two factors.

- Consider the ratio X:Y, where X is the number of messages used from the root nodes of each region to the root nodes of queries, and Y is the total number of sensor nodes in the query regions. This ratio represents an amortized overhead to reach the query root from the region roots. As QR increases, the numerator tends to decrease somewhat, and the denominator increases; thereby decreasing this overhead. As QR continues to increase, the value of this overhead becomes relatively small.
- With increasing QR, the predominant factor becomes the degree of node overlap and the sharing of sensor readings among regions. As QR increases, this also tends to have a saturation effect. In our example, this sets in around regions of size 48×48.

ZMQ is better than NMQ, irrespective of the query region size when the overlap percentage is not very low.

5.4 Impact of Number of Queries (N) Using Controlled Overlap Percentage

This experiment studies the effect of the number of queries on the performance of the three algorithms. We set OP and QR, and change the number of queries injected into the sensor field. Observe from Figure 12 that as the number of queries increases, the ANM for OMQ and ZMQ decreases, while it remains almost stable for NMQ. We also can find the answer from the definition of ANM as follows. Let:

- S be the number of nodes that can share reading and transmission,
- SN be the total number of sensor nodes in query regions (i.e., $\sum_Q sizeof(Q)$),
- E be the extra messages needed to send data from root nodes of sub-trees to root nodes of queries.

ANM $(= \frac{total\ number\ of\ messages}{\sum_Q sizeof(Q)}) = (SN-S+E)/SN = 1-S/SN + E/SN$. For NMQ, S is always 0, so ANM of NMQ is always more than 1 and decreases as SN increases.

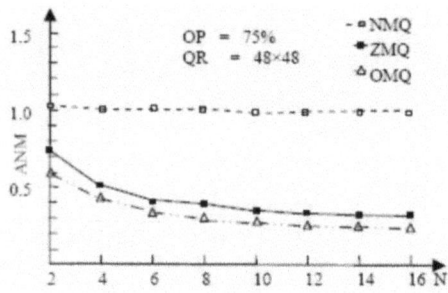

Fig. 12. Effect of N on algorithms, with fixed OP

As the number of queries increases, SN and S both increase, while E stays almost stable. Therefore, the overhead for OMQ and ZMQ approaches the value 1-S/SN as the number of queries increases.

6   Conclusion

This paper identified for the first time, multi-root, multi-query optimization for long duration aggregation queries. The paper then formulated two algorithms - an optimal algorithm (OMQ) and a heuristic algorithm (ZMQ) based on sharing the partially aggregated results of pre configured geographic regions. Simulations on OMQ and ZMQ, as well as the naive algorithm (NMQ) that does not do any sharing, indicate that the proposed algorithms provide significant energy savings under a wide range of network conditions and query region options. We found that OMQ always performs best, as expected. Furthermore, ZMQ performs generally better than NMQ when the sizes of the query regions are big, the density of the sensor field is high, and there are large overlaps among queries. ZMQ performs increasingly better than NMQ as the sizes of query regions become larger, the sensor field density increases, and overlap among query regions increases.

	Dynamic queries	Aggregation extent	Energy efficient	Time complexity	Space complexity	initialization	Latency
NMQ	Yes	Very good	Good	$O(1)$	$O(1)$	No	Small
ZMQ	Yes	Good	Very good	$O(1)$	$O(1)$	Yes, $O(n)$	Moderate
OMQ	No	Very good	Best	$O(1)$	$O(1)$	No	Small

Recommendation: The most applicable situation for ZMQ is when there are big query regions, big overlap among queries, and high network density. Otherwise, if the application requires dynamic query processing, NMQ is a good algorithm. If the queries are known a priori (before any processing), and no new queries come in during processing, OMQ is the best algorithm. ZMQ is a practical and energy efficient algorithm for multi-root, multi-query processing.

References

1. Dasgupta, K., Kalpakis, K., Namjoshi, P.: Improving the Lifetime of Sensor Networks via Intelligent Selection of Data Aggregation Trees. In: Proceedings of the Communication Networks and Distributed Systems Modeling and Simulation Conference (2003)
2. Durresi, A., Paruchuri, V., Iyengar, S.S., Kannan, R.: Optimized Broadcast Protocol for Sensor Networks. IEEE Transactions on Computers 54(8), 1013–1024 (2005)
3. Emekci, F., Yu, H., Agrawal, D., Abbadi, A.E.: Energy-Conscious Data Aggregation Over Large-Scale Sensor Networks, UCSB Technical report (2003)
4. Estrin, D., Srivastava, M.B., Sayeed, A.: Tutorial on Wireless Sensor Networks. In: ACM International Conference on Mobile Computing and Networking (MOBICOM) (2002)

5. Jia, L., Noubir, G., et al.: GIST: Group-Independent Spanning Tree for Data Aggregation in Dense Sensor Networks. In: International Conference on Distributed Computing on Sensor Systems (DCOSS) (2006)
6. Kannan, R., Sarangi, S., Iyengar, S.S.: Sensor-Centric Energy-Constrained Reliable Query Routing for Wireless Sensor Networks. Journal of Parallel and Distributed Computing 64(7), 839–852 (2004)
7. Li, X., Kim, Y.J., Govindan, R., Hong, W.: Multi-dimensional Range Queries in Sensor Networks. In: ACM Conference on Embedded Networked Sensor Systems (Sensys 2003), pp. 63–75 (2003)
8. Madden, S., Franklin, M.J., Hellerstein, J.M., Hong, W.: Tag: A Tiny Aggregation Service for Ad-hoc Sensor Networks. In: 5th Symposium on Operating Systems Design and Implementation, pp. 131–146 (2002)
9. Roumeliotis, S.I., Berkey, G.A.: Collective Localization: a Distributed Kalman Filter Approach to Localization of Groups of Mobile Robots. In: Proc. IEEE Intl Conf. on Robotics and Automation (ICRA), (2000)
10. Savvides, A., Han, C.-C., Srivastava, M.B.: Dynamic Fine-Grained Localization in Ad-Hoc Networks of Sensors. In: ACM SIGMOBILE (2001)
11. Trigoni, N., Yao, Y., Demers, A., Gehrke, J., Rajaraman, R.: Multi-Query Optimization for Sensor Networks. In: International Conference on Distributed Computing on Sensor Systems (DCOSS), pp. 307–321 (2005)
12. Wu, Q., Rao, N.S.V., Barhen, J., Iyengar, S.S., Vaishnavi, V.K., Qi, H., Chakrabarty, K.: On Computing Mobile Agent Routes for Data Fusion in Distributed Sensor Networks. IEEE Transactions on Knowledge and Data Engineering 16(6), 740–753 (2004)
13. Zhang, Z., Shatz, S.M.: A Technique for Power-Aware Query-Informed Routing in Support of Long-Duration Queries for Sensor Networks. In: International Conference on Sensing, Networking and Control (ICNSC 2006) (2006)

Snap and Spread: A Self-deployment Algorithm for Mobile Sensor Networks*

N. Bartolini, T. Calamoneri, E.G. Fusco, A. Massini, and S. Silvestri

Department of Computer Science
University of Rome "Sapienza", Italy
{bartolini,calamo,fusco,massini,simone.silvestri}@di.uniroma1.it

Abstract. The use of mobile sensors is motivated by the necessity to monitor critical areas where sensor deployment cannot be performed manually. In these working scenarios, sensors must adapt their initial position to reach a final deployment which meets some given performance objectives such as coverage extension and uniformity, total moving distance, number of message exchanges and convergence rate.

We propose an original algorithm for autonomous deployment of mobile sensors called SNAP & SPREAD. Decisions regarding the behavior of each sensor are based on locally available information and do not require any prior knowledge of the operating conditions nor any manual tuning of key parameters. We conduct extensive simulations to evaluate the performance of our algorithm. This experimental study shows that, unlike previous solutions, our algorithm reaches a final stable deployment, uniformly covering even irregular target areas. Simulations also give insights on the choice of some algorithm variants that may be used under some different operative settings.

1 Introduction

The necessity to monitor environments where critical conditions impede the manual deployment of static sensors motivates the research on mobile sensor networks. In these working scenarios, sensors are initially dropped from an aircraft or sent from a safe location, so that their initial deployment does not guarantee full coverage and uniform sensor distribution over the area of interest (AOI) as would be necessary to enhance the sensing capabilities and extend the network lifetime. Mobile sensors can dynamically adjust their position to reach a better coverage and more uniform placement. Due to the limited power availability at each sensor, energy consumption is a primary issue in the design of any self-deployment scheme for mobile sensors. Since sensor movements and, to a minor extent, message exchanges, are energy consuming activities, a deployment algorithm should minimize movements and message exchanges during deployment, while pursuing a satisfactory coverage.

The impressively growing interest in self-managing systems, starting from several industrial initiatives from IBM [2], Hewlett Packard [3] and Microsoft [4], has led to various approaches for self-deploying mobile sensors. The virtual force approach (VFA)

---

* The full version of this paper is [1].

S. Nikoletseas et al. (Eds.): DCOSS 2008, LNCS 5067, pp. 451–456, 2008.

proposed in [5,6,7], and its variants proposed in [8,9,10], model the interactions among sensors as a combination of attractive and repulsive forces. This approach requires a laborious and off-line definition of parameter thresholds, it presents oscillatory sensor behavior and does not guarantee the coverage in presence of narrows. The Voronoi approach, detailed in [11], provides that sensors calculate their Voronoi cell to detect coverage holes and adjust their position. This approach is not designed to improve the uniformity of an already complete coverage and does not support non convex AOIs.

The main contribution of this paper is the original algorithm for mobile sensor self-deployment, SNAP & SPREAD, with self-configuration and self-adaptation properties. Each sensor regulates its movements on the basis of locally available information with no need of prior knowledge of the operative scenario or manual tuning of key parameters. The proposed algorithm quickly converges to a uniform and regular sensor deployment over the AOI, independently of its shape and of the initial sensor deployment. It makes the sensors traverse small distances, avoiding useless movements, ensuring low energy consumption and stability. Furthermore, it outperforms previous approaches in terms of coverage uniformity.

2   The SNAP and SPREAD Algorithm

The deployed sensors coordinate their movements to form a hexagonal tiling, that corresponds to a triangular lattice arrangement with side R_s, where R_s is the *sensing radius*. This deployment guarantees optimal coverage (as discussed in [12]) and connectivity when $R_s \leq \sqrt{3}R_{TX}$, where R_{TX} is the *transmission radius*. To achieve this arrangement, some sensors snap to the centers of the hexagonal tiling and spread the others to uniformly cover the AOI. These snap and spread actions are performed in an interleaved manner so that the final deployment consists in having at least one sensor in each tile.

One sensor, s_{init}, is assigned the role of starter of the tiling procedure, while others may also concurrently act as starters, for fault tolerance purposes. The starter sensor gives rise to the **snap activity** selecting at most six sensors among those located in radio proximity and making them snap to the center of adjacent hexagons. Such deployed sensors, in their turn, give start to an analogous selection and snap activity thus expanding the tiling. This process goes on until no other snaps are possible, either because the AOI is completely covered, or because all the sensors that are located at boundary tiles do not have any further sensor to snap.

The **spread activity** provides that un-snapped sensors are pushed toward low density zones. Let $S(x)$ be the number of sensors located in the same hexagon as sensor x. Given two snapped sensors p and q located in radio proximity from each other, if $|S(p)| > |S(q)|$, p and q can negotiate the movement (push) of a sensor from the hexagon of p to the hexagon of q. Cyclic sensor movements are kept under control by imposing a *Moving Condition*, that we detail in [1].

The combination of these two activities expands the tiling and, at the same time, does its best to uniformly distribute redundant sensors over the tiled area.

Figure 1 shows an example of the algorithm execution. Figure (a) depicts the starting configuration, with nine randomly placed sensors and (b) highlights the role of s_{init}, which starts the hexagonal tiling. In (c) the starter sensor s_{init} snaps six sensors to the

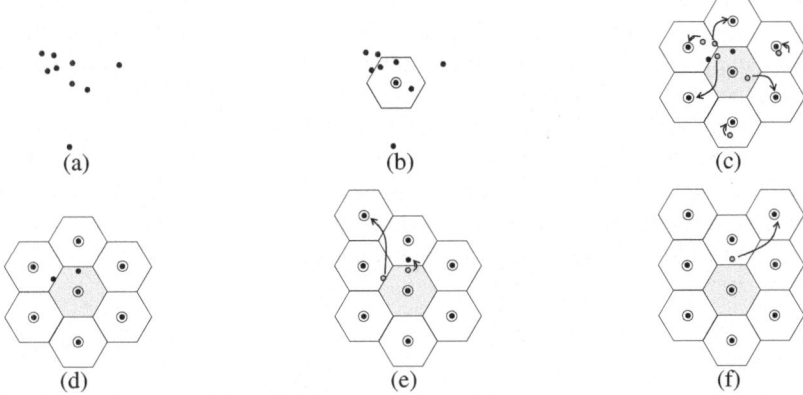

Fig. 1. Snap and spread actions: an example

center of the adjacent hexagons, according to the minimum distance criterion. Figure (d) shows the configuration after the snap action of $s_{\texttt{init}}$. In Figure (e) $s_{\texttt{init}}$ starts the spread action sending its redundant sensors to a lower density hexagon, while one of the sensors deployed in (c) starts a new snap action. Figure (f) shows the snap of the last redundant sensor just pushed by the starter, thus reaching the final configuration.

Since some performance objectives such as average traversed distance, network lifetime and coverage extension may be in contrast with each other, we introduce some algorithm variants that specifically prioritize one objective over the others.

According to the Basic Version (BV) of SNAP & SPREAD, the un-snapped sensors that are located in already tiled areas, consume more energy than snapped sensors, because they are involved in a larger number of message exchanges and movements. We introduce an algorithm variant, named Uniform Energy Consumption (UEC), to balance the energy consumption over the set of available sensors making them exchange their roles.

A second variant named Density Threshold (DT), provides that a sensor movement from the hexagon of p to the hexagon of q is allowed if, besides the Moving Condition, the constraint $|S(q)| < T_d$ is satisfied, that is the number of sensors located in the hexagon of q is lower than a *density threshold* T_d. This variant avoids unnecessary movements of sensors to already overcrowded hexagons that certainly exceed the optimal density. Notice that when $T_d \leq 1$, this variant can not be applied as it could limit the flow of redundant sensors to the AOI boundaries, thus impeding the coverage completion.

Due to space limitations we refer the reader to [1] for deeper details.

3 Simulation Results

In order to study the performance of SNAP & SPREAD and its variants, we developed a simulator on the basis of the wireless module of the OPNET modeler software [13].

In the following experiments we set $R_{\text{TX}} = 2\sqrt{3}R_S$ with $R_s = 5$ m. This setting guarantees that each snapped sensor is able to communicate with the snapped sensors located two hexagons apart. This choice allows us to show the benefits of the role exchange mechanism (UEC variant) whereas it does not imply significant changes in the qualitative analysis with respect to other settings. In all the experiments of this section we assume that the sensor speed is 1 mt/sec.

The following figures 2 e 3 show how SNAP & SPREAD performs when starting from an initial configuration where 150 sensors are sent from a high density region. In figure 2 the AOI is a square 80 m × 80 m while in figure 3 the AOI has a more complex shape in which a narrows connects two regions 40 m × 40 m. Note that previous approaches fail when applied to irregular AOIs such as the one considered in figure 3.

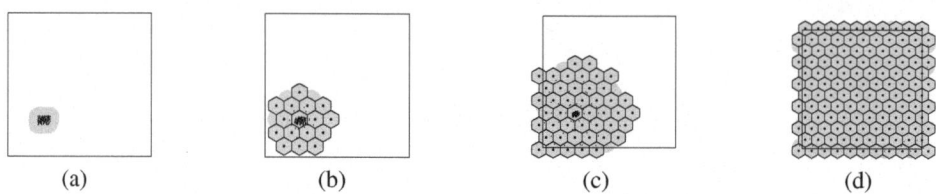

Fig. 2. Sensor deployment on a square, starting from a dense configuration

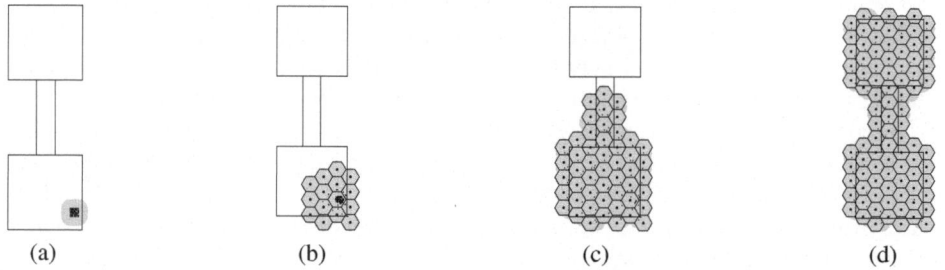

Fig. 3. Sensor deployment on an irregular AOI, starting from a dense configuration

Figure 4 shows instead how SNAP & SPREAD covers a square 80 m × 80 m starting from a random initial deployment of 150 sensors. Either starting from a high density distribution or from a random one, the algorithm SNAP & SPREAD completely covers the AOI. Of course, the coverage is much faster and consumes less energy when starting from a random configuration.

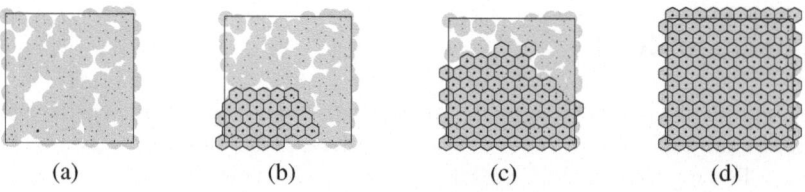

Fig. 4. Sensor deployment on a square, starting from a uniform configuration

In the figures from 5 to 8, we show some performance comparisons among the basic version BV of SNAP & SPREAD and its two variants DT and UEC. This set of simulations is conducted on the scenario described in figure 2 (a), with a high density zone in the initial sensor configuration. Notice that in all the figures, in the case of variant DT the line starts when the number of sensor is 150. This is because this variant has been designed to work when the number of sensors is sufficient to entirely cover the AOI, i.e. when the threshold T_d can be reasonably set to a value larger than 1.

Figure 5 shows the time to converge to a final deployment when varying the number of available sensors. When the number of available sensors is lower than strictly necessary to cover the area even with an optimal distribution, the time to converge to the final solution increases with the number of sensors because more sensors can cover a wider area. Instead, once the number of sensors is high enough to entirely cover the AOI, redundant sensors are helpful to complete the coverage faster. Notice that the convergence time of UEC is larger than the other ones because this variant incurs some overhead to perform role exchanges. The convergence time of DT is slightly larger than the one of BV because of the additional constraint imposed on redundant sensor movements.

Figure 6 shows the percentage of AOI being covered by SNAP & SPREAD and its variants, when increasing the number of sensors. Note that in most cases an incomplete coverage is due to the lack of the necessary number of sensors and not to a wrong behavior of the algorithm.

Since both mechanical movements and electronic communications consume energy, of which mechanical motion is the predominant part, we use the average traversed distance as a metric to highlight the energy consumption of the different algorithm

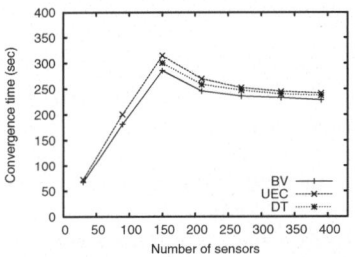

Fig. 5. Convergence Time

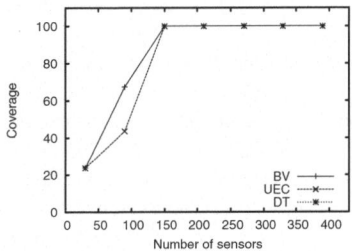

Fig. 6. Coverage

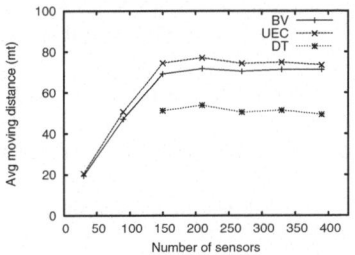

Fig. 7. Average moving distance

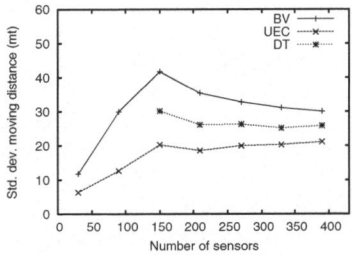

Fig. 8. Std dev of the moving distance

variants. Figure 7 shows that variant DT is highly effective in reducing the energy consumed for unnecessary movements.

Figure 8 complements the previous one by showing the effects of the two variants in terms of standard deviation of the traversed distance. Variant UEC significantly reduces the standard deviation with respect to the other variants. Indeed this variant was designed with the purpose to balance the load over all the available sensors and this obviously leads to a lower deviation. This result is important if one of the primary objectives of the deployment is the coverage endurance. The peak in the standard deviation obtained using the basic version of SNAP & SPREAD is due to the snap actions which govern the energy consumption when the number of sensors is less than 150. In the variants without role exchanges (as in BV and DT) the snap actions induce an initially high standard deviation of the traversed distance. Indeed sensors located close to the starter consume less energy than those that have to reach the farthest boundaries of the AOI. As we noticed before, the use of variant DT reduces unnecessary movements and consequently the average energy spent by each sensor. This results in a lower deviation as well.

References

1. Bartolini, N., Calamoneri, T., Fusco, E.G., Massini, A., Silvestri, S.: Snap and spread: a self-deployment algorithm for mobile sensor networks (extended version). Technical Report nr: 9/2008, University of Rome "Sapienza"
2. Ibm: the vision of autonomic computing,
   http://www.research.ibm.com/autonomic/manifesto
3. Hewlett packard: Adaptive enterprise design principles,
   http://h71028.www7.hp.com/enterprise/cache/
   80425-0-0-0-121.html
4. Microsoft: The drive to self-managing dynamic systems,
   http://www.microsoft.com/windowsserversystem/dsi/default.mspx
5. Zou, Y., Chakrabarty, K.: Sensor deployment and target localization based on virtual forces. In: Proc. IEEE INFOCOM 2003 (2003)
6. Heo, N., Varshney, P.: Energy-efficient deployment of intelligent mobile sensor networks. IEEE Transactions on Systems, Man and Cybernetics 35 (2005)
7. Chen, J., Li, S., Sun, Y.: Novel deployment schemes for mobile sensor networks. Sensors 7 (2007)
8. Poduri, S., Sukhatme, G.S.: Constrained coverage for mobile sensor networks. In: Proc. of IEEE Int'l Conf. on Robotics and Automation (ICRA 2004) (2004)
9. Pac, M.R., Erkmen, A.M., Erkmen, I.: Scalable self-deployment of mobile sensor networks; a fluid dynamics approach. In: Proc. of IEEE/RSJ Int'l Conf. on Intelligent Robots and Systems (IROS 2006) (2006)
10. Kerr, W., Spears, D., Spears, W., Thayer, D.: Two formal fluid models for multi-agent sweeping and obstacle avoidance. In: Proc. of AAMAS (2004)
11. Wang, G., Cao, G., Porta, T.L.: Movement-assisted sensor deployment. IEEE Transaction on Mobile Computing 6 (2006)
12. Brass, P.: Bounds on coverage and target detection capabilities for models of networks of mobile sensors. ACM Transactions on Sensor Networks 3 (2007)
13. Opnet technologies inc. http://www.opnet.com

An In-Field-Maintenance Framework for Wireless Sensor Networks[*]

Qiuhua Cao and John A. Stankovic

Department of Computer Science
University of Virginia
{qhua,stankovic}@cs.virginia.edu

Abstract. This paper introduces a framework for in-field-maintenance services for wireless sensor networks. The motivation of this work is driven by an observation that many applications using wireless sensor networks require one-time deployment and will be largely unattended. It is also desirable for the applications to have a long system lifetime. However, the performance of many individual protocols and the overall performance of the system deteriorate over time. The framework we present here allows the system or each individual node in the network to identify the performance degradation, and to act to bring the system back to a desirable coherent state. We implement and apply our framework to a case study for a real system, called VigilNet [5]. The performance evaluation demonstrates that our framework is effective and efficient.

1   Introduction

Many applications [5] in wireless sensor networks (WSN) typically initialize themselves by self-organizing after deployment. At the conclusion of the self-organizing stage it is common for the nodes of the WSN to know their locations, have synchronized clocks, know their neighbors, and have a coherent set of parameter settings such as consistent sleep/wake-up schedules, appropriate power levels for communication, and pair-wise security keys. However, over time these conditions can deteriorate.

The most common (and simple) example of this deterioration problem is with clock synchronization. Over time, clock drift causes nodes to have different enough times to result in application failures. While it is widely recognized that clock synchronization must re-occur, this principle is much more general. For example, even in static WSN some nodes may be physically moved unexpectedly. More and more nodes may become *out of place* over time. To make system-wide node locations coherent again, node re-localization needs to occur (albeit at a much slower rate than for clock sync).

Many WSN protocols have similar characteristics with respect to the need for re-applying computation. Reasons for this include the decentralized nature

---

[*] Supported in part by NSF grants CNS-0435060, CNS-0626616, and CNS-0720640.

of many protocols in WSN, the large scale of these systems and a high degree of loss and uncertainty experienced in these systems. WSN systems are in need of systematic mechanisms to dynamically adjust themselves at runtime based on the current performance. Moreover, many systems require mechanisms to understand which set of protocols/services are contributing to the degradation of the system performance. For example, typical tracking applications [1] might notice that the accuracy of the object classification is not acceptable, and many of these applications consider that the classification protocol is the only cause. However, the routing and localization protocols also have the impact on the classification accuracy. The unacceptable accuracy may be due to a low packet deliver ratio or incorrect location information in the detection report messages from the nodes to the base station.

In this paper we consider the entire set of services that keep or restore system coherence as *in-field-maintenance services*. We call these maintenance services since they follow the concept that systems tend to disorder unless explicit action is applied to keep them ordered. In other words, specific energy (in the form of executing specific code) has to be applied to keep or regain system-wide coherency.

Our framework allows users to (1) define the coherency requirements of the states of protocols, (2) specify the maintenance/repair policy (i.e., when to self-heal the system, and who invokes the maintenance service), and (3) define which set of protocols need maintenance services in order to satisfy a system performance requirement and the dependency constraints among the services, according to the system specification and constraints. And our framework (1) enforces the dependency requirements among maintenance services, (2) supports different memory requirements for different maintenance services, (3) implements an online monitor to measure system performances and states of protocols, and (4) provides mechanisms for on-demand real-time self-healing of the system.

However, the resource constraints in terms of memory, energy, and bandwidth of WSN systems present challenges to implement such a framework. We address the challenges in our framework by supporting different memory management strategies to satisfy different maintenance policies specified by users, by enabling both global (the whole network) and local (a specified region) maintenance services so that global maintenance services are only invoked when needed, by executing only the necessary maintenance services via the dependency enforcement instead of maintaining all the protocols in stack at each time a maintenance service required, and by implementing both global and local monitors and piggybacking report messages in the monitors.

2   Our In-Field-Maintenance Framework

Our framework works side by side with system protocols and application code as shown in Figure 1. We present the details of the two main components of our framework in subsections 2.1 and 2.2, respectively.

2.1   Maintenance Policy

The maintenance policy component takes the system specification and defines:

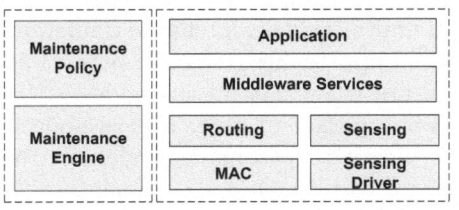

Fig. 1. System Architecture

(1) the conditions under which a maintenance service is to be invoked, (2) the dependency constraints of a particular maintenance service, (3) the region a maintenance service is to be applied, i.e., global or local, and (4) the state update policy to define when the states of a protocol are to be updated. The first 3 sub-components take energy efficiency into consideration when defining what to support in the framework. The 4th sub-component deals with the memory constraints imposed by WSN systems.

Conditions. Maintenance services might be triggered when there is (1) a system sensing coverage failure, or (2) a system communication coverage failure, or (3) a system density coverage failure or periodically.

Dependency Constraints. Dependency constraints define (1) the relationship between a system performance measurement and the protocols contributing to the measurement, and (2) the dependencies among protocols in a WSN system. Consequently, the constraints define the dependencies among the maintenance service for each system measurement and for each protocol. The dependency specifications are defined according to the system requirements and look similar to dependency graphs in the real-time scheduling research domain (but cannot be shown here due to lack of space). Briefly, the specifications define the sequence of a set of protocols to be executed.

As a result of dependency specifications, two energy inefficient situations are avoided. The first is avoiding executing unnecessary maintenance services. Two, avoiding invoking a maintenance service unnecessarily.

Regions. The region specification for a maintenance service gives the system the flexibility to accommodate different strategies for different services to minimize the energy consumption of each maintenance service. The coherent states of the protocols in WSN are organized into two categories:

Global View of Coherent States – requires the states of the protocols to be the same value within some acceptable range for every node in the system. For example, time synchronization requires every node in the system to have the consistent time view with the base station within some ϵ.

Local View of Coherent States – the local view of coherent states is defined from the node's point of view, i.e., by whether a node perceives the true state in the system. For example: node A thinks that node B is its neighbor, but actually node B is either out of its communication range due to the change of the communication quality, or unexpected movement (i.e, wind blows node B away), or node B is not functional due to running out of energy. In this case, node A does not have a coherent view of its local neighborhood states.

The two different types of coherency requirements require different capabilities to bring the system back to coherent states. Our in-field-maintenance framework is designed to provide the flexible and efficient approaches to handle both cases.

Guidelines for choosing a global service are when: (1) all the nodes in the network demonstrate similar properties over a period of time. If each node in the system has similar clock skew or drift, then it is reasonable to globally run the time synchronization maintenance service to resynchronize the clocks; (2) applications or maintenance policies require all the nodes in the network to contain the same information or to take the same action. If an application or a maintenance policy requires updating the report rate of all the nodes in the network with a new rate; (3) replacing a protocol with a new implementation. In order to defend against a newly expected security attack to the routing algorithm currently deployed in the network, it is desirable to globally disseminate a new routing algorithm to all the nodes. Otherwise, the nodes with different routing algorithms may not be able to communicate.

The recommendations to choose local services are when: (1) the states of concern only have local meanings. For example, the link quality of a node to its neighbors only makes sense to be defined as a local parameter; (2) an application or a maintenance policy only requires local repair.

Update Policies. In WSN, the memory of sensor nodes (128K ROM, 4K RAM for MICA series devices) is very limited. When applications become more complex, the memory is even a bigger concern. Moreover, some applications may require that the system does not stop from its normal processing while maintaining the system. The maintenance service requires extra available memory to be allocated. Our framework supports three update policies to give the system the capability to balance the trade-off between memory constraints and event detection delays.

Delay Update until Commit Time: This policy delays the update of the states to right before the conclusion of the maintenance process for a given protocol. It does not stop a system from its normal processing, but requires the maintenance service to allocate memory space to back up the old states.

Immediately Update, but Disable Sensor Interrupts: Here when a maintenance service is invoked, all sensor interrupts are disabled. The result is that the maintenance service does not demand substantial memory and race conditions don't exist. But the applications can not detect any new sensor events during the execution of the service.

Immediately Update, but do not Disable Interrupts: This policy delays the sensor reports until the end of the maintenance service, but allows sensor interrupts during the execution of the maintenance service. The sensor readings are stored in a buffer, so that no event is missed, but the time to report the events to the applications is delayed.

2.2   Maintenance/Repair Engine

The maintenance engine implements the policies discussed in subsection 2.1. It is composed of 4 parts: monitors, dependency checks, memory management, and in-field-maintenance services. We next discuss the functionality of each.

Monitors. Monitors collect the specified states of concern and initiate maintenance services according to the policy specification. Our framework designs a two tier monitoring architecture as shown in Figure 2. The global monitor enables the base station to collect and process the performance and/or state information from each individual node in the network via the collector. The processed information then feeds into the controller. The controller generates the list of the protocols to execute maintenance services and floods the list to the network. Each node runs a local monitor. The local monitor acquires the information on what to monitor from the base station, collects the requested information through the local collector, and reports the requested information back to the base station if required.

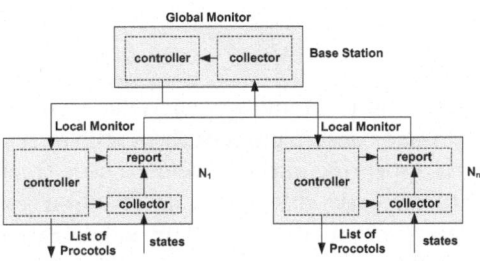

Fig. 2. Two Tire Monitor Architecture for the Global In-Field-Maintenance Policy Enforcement

The monitors are capable of collecting specified state information in 5 categories. And the 5 categories comprehensively consider different data required and system conditions so that our framework can flexibly be integrated with many existing systems. (1) States that are available via directly interfacing with the hardware layer, for instance, energy remaining, and the clock. (2) States that are obtainable through the interfaces provided in the original system without the need to interact with the nodes in the neighborhood, such as the maximum number of neighbors in a node's neighbor table or the maximum number of parents of a node. (3) States that require the cooperation among the nodes in a neighborhood with explicit message exchanges. Link quality is one example of those states. (4) States that are specified as the states to be monitored but the original system does not provide interfaces to expose those states. (5) States that are not maintained by any components in the original system.

Dependency Checks. This component checks if a maintenance service has to be executed together with some other services according to the dependency constraints specified in subsection 2.1, and generates the correct execution flow for the set of services. It also provides mechanisms to enforce the dependencies when the maintenance services are invoked.

Memory Management. To support the "Delay Update until Commit Time" policy specified in subsection 2.1, our memory management provides three primitives to manage the states of maintenance services. The primitives are:

- MaintenanceAlloc: allocates memory to backup the state information of a protocol before executing the maintenance service for the protocol.
- MaintenanceCommit: makes the new state information accessible to the applications right before the maintenance service finishes its execution.
- MaintenanceRelease: releases the allocated memory locations after the commitment.

In-Field-Maintenance Services. This component provides the mechanisms to execute the requested maintenance services.

3 A Case Study

In this section, we present one implementation of our framework for a real application, VigilNet [5] [6] [7]. We choose VigilNet because it is a typical surveillance and tracking application. Note that our framework is applicable to many other types of applications.

3.1 Brief Overview of VigilNet

VigilNet is a recent major effort to support long-term military surveillance using large-scale micro-sensor networks. The primary design goals of the VigilNet system are to detect events or moving targets appearing in the system, to keep track of the positions of moving targets, and to correctly classify the detected targets in an energy-efficient and stealthy manner. Power management and collaborative detection are two key research problems addressed in VigilNet. As discussed in [8], VigilNet provides a three level power management service, namely tripwire service, sentry service, and duty cycle scheduling.

After initialization, the tripwire service divides the system into multiple tripwire sections. A tripwire section can be either in an active or a dormant state. The uniform discharge of energy in a section is achieved through rotation strategies based on the remaining energy within individual nodes. Together with the motivation to balance the energy consumption among all the nodes in the network, at the same time to synchronize the time of all the nodes in the network with the base station, and also to heal any transient node failures, VigilNet implemented a rotation scheme to reinitialize the whole network for all the services once per day.

The timeline to control the rotation is given by the phase transition graph [5]. VigilNet starts with system initialization at phase I and follows the phase transition graph through phase VIII. The duration of each phase is a control parameter that is dynamically configurable at the base station. The initialization process from phase I to phase VII normally takes 3 minutes, but it is a tunable parameter according to the network size and system requirements. As we can see that during the rotation/reinitialization process, the system stops functioning to reinitialize itself.

3.2   One Application on VigilNet

Different from the execution sequence of the original VigilNet system, the new VigilNet system we implemented, called SelfHealingVigilNet, executes as shown in Figure 3. When the system transits into the tracking phase (phase VIII), Self-HealingVigilNet also enters to the maintenance service phase which is in parallel with the tracking phase. There is no rotation service in SelfHealingVigilNet as it would be redundant.

Maintenance Policies in SelfHealingVigilNet: We define the maintenance policy in SelfHealingVigilNet based on the system specification of VigilNet, (1) Longevity, (2) Effectiveness, (3) Adjustable Sensitivity, and (4) Stealthiness.

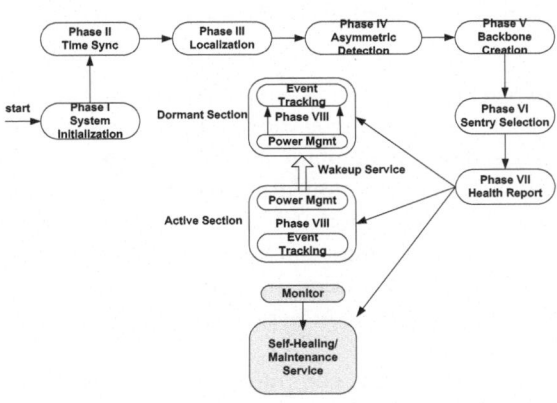

In this case study, our maintenance policy focuses on the longevity and the effectiveness of the system. More specifically, time synchronization and localization services are important for the effectiveness of the system. VigilNet requires the clock drift to be confined to the millisecond range. We use the same maintenance policy for the time synchronization protocol, once per day. Due to practical constraints, each node in VigilNet obtains its location at system deployment time.

Fig. 3. Phase Transition in SelfHealingVigilNet

Besides time synchronization (Time Syn) and localization protocols, sensing coverage (Sensing Cov) and communication coverage (Comm Cov) are also crucial to the system performance. The sensing coverage depends on the sentry selection protocol. The communication coverage involves protocols such as a asymmetric detection protocol (Asym D), and a backbone creation/robust diffusion tree protocol (R. Diff Tree). The definitions of sentry selection, asymmetric detection, neighbor discovery and backbone creation are the same as defined in [5]. For the case study of our framework, we specify the minimum sensing coverage requirement as 1 (100%), and the minimum communication coverage requirement as 1 (100%).

We also observe that time synchronization, backbone creation/robust diffusion tree creation, and self-configuration of the system need system-wide broadcast. Sentry selection (Sentry Sel) can be global or local. The region of a local maintenance service (Local R) is defined by hop counts k, meaning that a local maintenance process takes effect in its initiator's k hop neighborhood.

In summary, the maintenance policies in SelfHealingVigilNet for the case study are shown in Table 4(a). And the maintenance policies also demonstrate that our framework is capable of supporting combinations of different conditions,

	Time Syn	Sensing Cov	Comm Cov
Condition	1day	1	1
Dependency	No	Sentry Sel	Asym D
			R. Diff Tree
Region	Global	Local R	Global
Update P	Delay	Disable Int.	Immediate U

(a) Maintenance Policy.

```
BPeriod=beaconP;CollectorBTimer(CBT)=T-beaconP;
ReportTimer = T; ConnectivityTable(CT) = ø;
CBT.fired() {Broadcast(CollectorBeaconMsg (CBM));}
RecvedCollectorBeaconMsg(CBM) {Update CT;}
ReportTimer.fired() {
    signal ConnectivityReady to the report component}
```

(b) Pseudo Algorithm of the Local Collector.

different dependency check requirements, different maintenance regions, and different state update strategies to invoke a maintenance service. In Table 4(a), Delay means "Delay Update until Commit Time", Disable Int. stands for "disable sensor interrupts", Immediate U is for "Immediately Update but do not Disable Interrupts".

Monitors in SelfHealingVigilNet: The base station initiates the maintenance process on time synchronization when the invocation condition is satisfied, which is the pre-calculated period based on the one-hop clock drift/skew and the number of the hops in the network. For the communication coverage, we implement a two-tier monitor as discussed in subsection 2.2. The local collector on each individual node collects the connectivity information using the algorithm as shown in Figure 4(b) and reports the information back to the collector on the base station very period T. Period T and beacon period (BeaconPeriod) are specified by the base station and tunable.

The communication coverage of the network is computed at the base station by constructing a connectivity graph based on the connectivity information reported from the nodes. As we observe from the VigilNet system, the nodes on the communication backbone are also the sentry nodes. To be more energy efficient, we piggyback the collecting of the states of the sentry selection protocol and the connectivity.

Dependency checks in SelfHealingVigilNet: The maintenance service for the time synchronization protocol is independent from all the other protocols in VigilNet. The maintenance service for the sentry selection protocol also can be independent. However, when the base station starts the maintenance process to heal the communication coverage, it has to run the service for the asymmetric detection protocol first, then the backbone creation/robust diffusion tree protocol. These dependencies are declared and then followed by the in-field-maintenance framework.

4   Performance Evaluation

We organize our performance evaluation into two parts. First, we demonstrate that our framework works with VigilNet effectively by implementing SelfHealingVigilNet on XMS2 motes, the supported platform by VigilNet. Second, we use the overhead, energy saved, system blackout time, and event detection probability as the performance metrics to analyze our framework. We choose the latter 3 metrics because they are critical to VigilNet. Because it is not easy to deploy a large system unattended for days we use simulations to analyze SelfHealingVigilNet.

4.1 Experiments

In total, our framework uses 2,197 bytes of code memory and 115 bytes of data memory and there were some modifications to the original VigilNet mainly in the top level components (MainControlM.nc and MainControlC.nc).

In the experiments we enable the magnetic sensors. We deploy 10 XMS2 motes programmed with SelfHealingVigilNet in two lines (2 by 5) in our lab and a base station connecting to a PC. We define that maintenance period is the time between two maintenance services and maintenance phase is the time to execute a maintenance service. We set the maintenance phase for both the time synchronization and the sensing coverage to be 20 seconds and the communication coverage to be 40 seconds, to be compatible with the default parameters defined in VigilNet. However, the maintenance phase is a tunable parameter depending on the network size. In order to speed up the experiments and without effecting the correctnesses of the experiments, based on observations, we set the beacon period to 10 seconds and the beacon rate to 0.5 (one beacon message every 2 seconds) for the monitors, the maintenance period to 10 minutes, and we manually create the sensor and communication coverage failures. All experiments start the system from the initialization phase to the tracking phase as shown in Figure 3. The real testbed experiments demonstrate that our framework can (1) execute in parallel with the original system, (2) efficiently monitor specified performance measurements, (3) enforce different maintenance policy specifications, and (4) effectively manage the memory constraints. Due to lack of space the details are not presented.

4.2 Analysis

In this section, we analyze the performance of SelfHealVigilNet using the metrics defined above. The performance of VigilNet is the baseline. In our analysis, each node has a radio range of 30 meters and a sensing range of 10 meters as in VigilNet. Nodes are uniformly distributed over an area with 300 meters by 300 meters. The average density of the deployment is 10 nodes per radio range. Radio consumes $48mW$ at transmit state and $24mW$ at receive state as studied in. When a message is transmitted, the radio switches to the transmit state for 30 milliseconds, a typical time required by XSM motes to send a message under the MAC contention. The beacon period for monitors and the maintenance phases for each maintenance services are the same as in the experiments. The report period and report rate for monitors are 20 seconds and 0.2 (one report every 5 seconds), respectively. Each battery of XSM motes has an energy capacity uniformly chosen between 2,848mAh and 2,852mAh, voltage at 3V. Each analysis result is the mean of 10 runs, with a confidence interval of 95%.

Overhead: The energy consumption of the monitors is the main overhead of our framework. And the main energy consumer of the monitors is the radio,

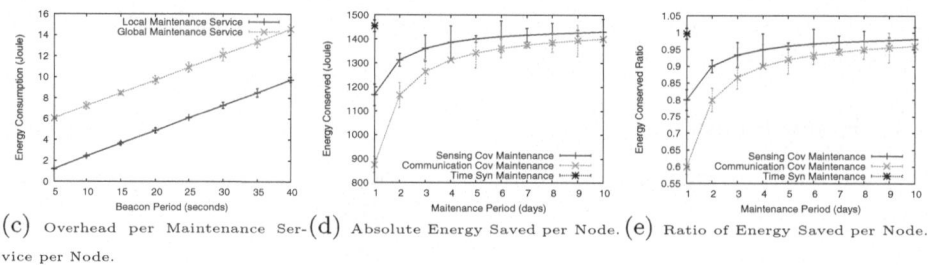

(c) Overhead per Maintenance Ser- (d) Absolute Energy Saved per Node. (e) Ratio of Energy Saved per Node.
vice per Node.

Fig. 4. Energy Analysis of the In-Field-Maintenance Framework

exchanging beacon messages around a neighborhood to collect the states information of concern and reporting the states. We do not consider the overhead to compute the connectivity graph in the communication coverage maintenance service because the computation is done at the base station, and energy is not a severe concern of the base station. Meanwhile, the computation overhead of a node to calculate the sensing coverage is minimal as compared to the energy consumed by the radio. During beacon and report periods, the radio is at either the transmit state or receive state. For local maintenance services, the energy consumed by a local collector is the overhead. For global maintenance services, the local collector and reporting cause the overhead.

Figure 4(c) shows that the overhead of a node for both a local and a global maintenance service at different beacon periods, while we fix the beacon rate at 0.5 (one beacon message every 2 seconds). We can see that the beacon period has an approximately linear impact on the overhead. But the overheads for both local and global services are minimum (less than 1.2mAh) by comparing with a battery's capacity (2,848mAh).

Energy Conserved Per Node: We analyze the impact of the maintenance period to the energy consumption of SelfHealVigilNet. Different from the original VigilNet system, our framework allows different maintenance periods for different maintenance services. We investigate different maintenance periods over a period of 60 days system lifetime. In VigilNet, all the protocols are reinitialized once per day without considering if a protocol needs the reinitialization. During these 60 days, VigilNet reinitializes the system 60 times, where each reinitialization takes approximately 24.3Joules calculated using the data available in [5]. Figure 4(d) and Figure 4(e) show that the absolute energy saved per node can be as high as 1,483.6Joules (137.4mAh), which is around 5% of a node's battery capacity. And the ratio of the energy saved is always above 60% and can save up to 98%, which is significant. The ratio of the energy conserved is the difference between the energy consumed during rotation services (eR) and the energy consumed by individual maintenance services (eE) divided by eR.

Our experiments and analysis results show that our framework can be effectively incorporated with a complex legacy system (VigilNet) to improve the performance of the system. For example, SelfHealingVigilNet saves up to 5% of a node's battery capacity, decreases the system blackout time to less than 11% of the original system, and always detects an intruder, with minimum overhead (less than 1.2mAh).

5   Related Work

SASHA [2] proposed a self-healing architecture for WSN. While SASHA is a centralized approach, our framework supports both global and local maintenance services. A timeout control mechanism with two timers was studied in [10] to enable the control center of a WSN system to be constantly aware of the existence/health of all the sensor devices in the network. The two-tier monitors in our framework can collect 5 different categories of information of concern both globally or locally.

Sympathy [13] proposed a prototype tool for detecting and debugging failures for WSN applications in both pre-deployment and post-deployment. It detects failures based on data quantity, rather than data quality at a centralized location. Different from Sympathy, our framework not only supports globally and locally collecting failure detection information, it also provides mechanisms to recover the system. LiveNet [3] provides tools to understand the behavior of or to provide a global view on the dynamics of WSNs. Passive monitoring infrastructure sniffers are implanted into the networks to collect the traffic traces. Offline trace merging and analysis tools are developed to reconstruct the network connectivity, to infer the routing path, and to identify hotspots of the system. In our system, different from LiveNet, the state information are collected, distributed or processed online by the sensor nodes themselves. Memento [14] is designed for inspecting the health state of nodes in WSNs. It provides failure detection and symptom alerts via an energy efficient protocol to deliver state summaries. Our monitored state information not only is about node failure detection, but also includes the performance and resource usage information of protocols in the stack.

[15] presented a distributed algorithm to configure the network nodes into a cellular hexagonal structure. The self-healing under various perturbations, such as node joins, leaves, deaths, movements, and corruption, is achieved by the property of the hexagonal structure. And [9] [12] [11] [4] achieve the reliability or resilience toward failures or dynamics in the system via model or analysis redundancy. However, these solutions are designed for individual protocols.

6   Conclusion

In this paper, we study an efficient and effective in-field-maintenance framework for WSN under the resource constraints. The framework is composed of two components, a maintenance policy and a maintenance engine. As far as we know, our in-field-maintenance service framework is the first effort to build an efficient and effective framework for WSN that supports the following collection of features:(1) efficiently monitor both system performance measurements and states of protocols of a system, locally and globally, (2) enforce dependency constraints, (3) provide mechanisms to recover the system, and (4) provide different strategies to manage memory. Based on the case study, both experiments and analysis, for a real application VigilNet, our framework demonstrates that it improves the performance of the original VigilNet system, such as energy conservation, system blackout time, detection probability, with minimum overhead.

References

1. Arora, A., Dutta, P., Bapat, S., Kulathumani, V., Zhang, H., Naik, V., Mittal, V., Cao, H., Demirbas, M., Gouda, M., Choi, Y., Herman, T., Kulkarni, S., Arumugam, U., Nesterenko, M., Vora, A., Miyashita, M.: A wireless sensor network for target detection, classification, and tracking. In: Computer Networks, Elsevier, Amsterdam (2004)
2. Bokareva, T., Bulusu, N., Jha, S.: Sasha:toward a self-healing hybrid sensor network architecture. In: EmNetS-II (2005)
3. Chen, B., Peterson, G., Mainland, G., Welsh, M.: Livenet: Using passive monitoring to reconstruct sensor network dynamics. Technical report, Harvard University (2007)
4. Crawford, L.S., Sharma, V., Menon, P.K.: Numerical synthesis of a failure-tolerant, nonlinear adaptive autopilot. In: CCA (2000)
5. He, T., Krishnamurthy, S., Luo, L., Yan, T., Krogh, B., Gu, L., Stoleru, R., Zhou, G., Cao, Q., Vicaire, P., Stankovic, J.A., Abdelzaher, T.F., Hui, J.: Vigilnet: An integrated sensor network system for energy-efficient surveillance. ACM Transactions on Sensor Networks 2(1), 1–38 (2006)
6. He, T., Krishnamurthy, S., Stankovic, J.A., Abdelzaher, T.F., Luo, L., Stoleru, R., Yan, T., Gu, L., Hui, J., Krogh, B.: An energy-efficient surveillance system using wireless sensor networks. In: ACM MobiSys (2004)
7. He, T., Luo, L., Yan, T., Gu, L., Cao, Q., Zhou, G., Stoleru, R., Vicaire, P., Cao, Q., Stankovic, J.A., Son, S.H., Abdelzaher, T.F.: An overview of the vigilnet architecture. In: RTCSA (2005)
8. He, T., Vicaire, P., Yan, T., Cao, Q., Zhou, G., Gu, L., Luo, L., Stoleru, R., Stankovic, J.A., Abdelzaher, T.F.: Achieving long-term surveillance in vigilnet. In: IEEE INFOCOM 2006 (2006)
9. Hoblos, G., Staroswiecki, M., Aitouche, A.: Optimal design of fault tolerant sensor networks. In: CCA (2002)
10. Hsin, C., Liu, M.: A distributed monitoring mechanism for wireless sensor networks. In: Proceedings of the 3rd ACM workshop on Wireless security (2002)
11. Marzullo, K.: Tolerating failures of continuous-valued sensors. ACM Transactions on Computer Systems (1990)
12. Provan, G., Chen, Y.: Model-based fault tolerant control reconfiguration for discrete event systems. In: CCA (2000)
13. Ramanathan, N., Chang, K., Kapur, R., Girod, L., Kohler, E., Estrin, D.: Sympathy for the sensor network debugger. In: ACM SenSys (2005)
14. Rost, S., Balakrishnan, H.: Memento: A health monitoring system for wireless sensor networks. In: SECON (2006)
15. Zhang, H., Arora, A.: Gs^3: Scalable self-configuration and self-healing in wireless networks. In: ACM PODC (2002)

Deterministic Secure Positioning
in Wireless Sensor Networks

Sylvie Delaët[1], Partha Sarathi Mandal[2,*], Mariusz A. Rokicki[3,1],
and Sébastien Tixeuil[4,5]

[1] Univ. Paris Sud, LRI - UMR 8623, France
[2] Department of Computer Science & Engineering
Indian Institute of Technology Kanpur, Kanpur - 208 016, India
[3] CNRS, France
[4] Univ. Pierre & Marie Curie - Paris 6, LIP6 - UMR 7606, France
[5] INRIA Futurs, Project-team Grand Large

Abstract. Position verification problem is an important building block
for a large subset of wireless sensor networks (WSN) applications. As a result, the performance of the WSN degrades significantly when misbehaving nodes report false location information in order to fake their actual
position. In this paper we propose the first deterministic distributed protocol for accurate identification of faking sensors in a WSN. Our scheme
does *not* rely on a subset of *trusted* nodes that cooperate and are not
allowed to misbehave. Thus, any subset of nodes is allowed to try faking
its position. As in previous approaches, our protocol is based on distance
evaluation techniques developed for WSN.

On the positive side, we show that when the received signal strength
(RSS) technique is used, our protocol handles at most $\lfloor \frac{n}{2} \rfloor - 2$ faking
sensors. When the time of flight (ToF) technique is used, our protocol
manages at most $\lfloor \frac{n}{2} \rfloor - 3$ misbehaving sensors. On the negative side, we
prove that no deterministic protocol can identify faking sensors if their
number is $\lceil \frac{n}{2} \rceil - 1$. Thus, our scheme is almost optimal with respect to
the number of faking sensors.

We discuss application of our technique in the trusted sensor model.
More specifically, our results can be used to minimize the number of
trusted sensors that are needed to defeat faking ones.

Keywords: Wireless Sensor Network, Secure Positioning, Distributed
Protocol.

1 Introduction

Position verification problem is an important building block for a large subset
of wireless sensor networks (WSN) applications. For example, environment and
habitat monitoring [15], surveillance and tracking for military [7], and geographic
routing [9], requires accurate position estimation of the network nodes.

---

* Part of this work was done while this author was a postdoctoral fellow at INRIA in
the Grand Large project-team, Univ. Paris Sud, France.

S. Nikoletseas et al. (Eds.): DCOSS 2008, LNCS 5067, pp. 469–477, 2008.
© Springer-Verlag Berlin Heidelberg 2008

In the secure positioning problem we can distinguish two sets of nodes. The set of *correct* nodes and the set of *faking* (or *cheating*) nodes. The goal of faking nodes is to mislead correct nodes about their real positions. More specifically, faking nodes cooperate and corrupt the position verification protocol to convince the correct nodes about their faked positions. This problem can be also seen as a variation on Byzantine agreement problem [10] in which correct nodes have to decide which nodes are faking their position. Initially correct nodes do not know which nodes are correct and which nodes are faking.

The faking nodes come to play in WSN in natural way due to several factors: a sensor node may malfunction, inaccurate position (coordinates) estimation [1,6]. Most of the existing position verification protocols rely on distance evaluation techniques. Received signal strength (RSS) and time of flight (ToF) techniques are relatively easy to implement and very precise. In the RSS technique, receiving sensor estimates the distance of the sender based on sending and receiving signal strengths. In the ToF technique, sensor estimates distance based on message delay and radio signal propagation time.

Related Work. Most methods [3,4,11,12] existing in the literature relies on the fixed set of trusted entities (or *verifiers*) and distance estimation techniques to filter out faking nodes. We refer to this model as the *trusted sensor* (or *TS*) model. The TS model was considered by Capkun and Hubaux [4] and Capkun et al. [3]. In [4] the authors present the protocol which relies on a distance bounding technique. This technique was proposed by Brands and Chaum [2]. Each sensor v measures its distance to the (potential) faking sensor u based on its message round-trip delay and radio signal propagation time. The protocol presented in [3] relies on set of hidden verifiers. Each verifier v measures the arrival time t_v of the (potential) faking node transmission. Verifiers exchange all such arrival times and check consistency of the declared position. The TS model presents several drawback in WSN: first the network cannot self-organize in an entirely distributed manner, and second the trusted nodes have to be checked regularly and manually to actually remain trusted.

Relaxing the assumption of trusted nodes makes the problem more challenging, and to our knowledge, has only been investigated very recently by Hwang, He, and Kim [8]. We call the setting where no trusted node preexists the *no trusted sensor* (or *NTS*) model. The approach in [8] is randomized and consists of two phases: distance measurement and filtering. The protocol may only give probabilistic guarantee for filtering faking sensors.

Our results. The main contribution of this paper is a deterministic secure positioning protocol, FINDMAP. To the best of our knowledge, this is the first fully deterministic protocol in the NTS model. The protocol guarantees that the correct nodes never accept faked position of the cheating node. The basic version of the protocol assumes that faking sensors are not able to mislead distance evaluation techniques. The protocol correctly filters out cheating sensors provided there are at most $\lceil \frac{n}{2} \rceil - 2$ faking sensors. Conversely, we show evidence that in the same setting it is impossible to deterministically solve the problem when the number of faking sensors is at least $\lceil \frac{n}{2} \rceil - 1$. We then extend the protocol to deal with faking sensors that are also allowed to corrupt the distance measure technique (RSS or ToF). In the case of RSS, our protocol tolerates at most $\lfloor \frac{n}{2} \rfloor - 2$ faking sensors (provided no four sensors are located on the same circle and no four sensors are co-linear). In the case of ToF, our protocol may handle up to $\lfloor \frac{n}{2} \rfloor - 3$ faking sensors (provided no six sensors are located on the same hyperbola and no six sensors are co-linear).

Our results have significant impact on secure positioning problem in the TS model as well. The TS protocol presented by Capkun *et al.* [3] relies on set of hidden station that detect inconsistencies between measured distance and distance computed from claimed coordinates. The authors propose the ToF-like technique to estimate the distance. Our detailed analysis shows that six hidden stations (verifiers) are sufficient to detect a faking node. The authors also conjecture that the ToF-like technique could be replaced with RSS technique. Our results answer positively to the open question of [3], improving the number of needed stations to four. Thus, in the TS model our results can be used to efficiently deploy a minimal number trusted stations.

Our FINDMAP protocol can be used to prevent Sybil attack [5]. More specifically, each message can be verified if it contains real position (id) of its sender. Each massage, which is found to contain faked position (id) can be discarded. Thus correct nodes never accept messages with faked sender position (id).

2 Technical Preliminaries

We assume NTS model unless stated otherwise. That is initially correct nodes do not have any knowledge abut the network. In particular, correct nodes do not know the position of other correct nodes. Each node is aware of its own geographic coordinates, and those coordinates are used to identify nodes. The WSN is partially synchronous: all nodes operate in rounds. In one round every node is able to transmit exactly one message which reaches all nodes in the network. The size of the WSN is n. Our protocol does not relay on n. For each transmission correct nodes uses the same transmission power.

Faking nodes are allowed to corrupt protocol's messages e.g. transmit incorrect coordinates (and thus incorrect identifier) to the other nodes. We assume that faking nodes may cooperate between themselves in an omniscient manner (*i.e.* without exchanging messages) in order to fool the correct nodes in the WSN. In the basic protocol, faking nodes cannot corrupt distance measure techniques. This assumption is relaxed in section 4 and 5 where faking sensors can corrupt the ranging technique. We assume that each faking node obeys synchrony. That is, each faking node transmits at most one message per round.

We assume that all distance ranging techniques are perfect with respect to precision. The distance computed by a node v to a node u based on a distance ranging technique is denoted by $\hat{d}(v, u)$. The distance computed by a node v to a node u using coordinates provided by u is denoted by $d(v, u)$. A particular sensor v *detects inconsistency* on distance (*i.e.* position) of sensor u if $d(v, u) \neq \hat{d}(v, u)$. Our protocols rely on detecting and reporting such inconsistency.

In the remaining of the paper, we use two distance estimation techniques:

1. In the *received signal strength* (*RSS*) technique, we assume that each node can precisely measure the distance to the transmitting node by Frii's transmission equation 1 [13]:

$$S_r = S_s \left(\frac{\lambda}{4\pi d}\right)^2 \tag{1}$$

Where S_s is the transmission power of the sender, S_r is the receive signal strength (RSS) of the wave at receiver, λ is wave length and d is distance between sender and receiver.

2. In the *time of flight* (*ToF*) technique, each sensor u transmits its message with two types of signals, that differ on propagation speed *e.g.* radio signal (RF) and ultra sound signal (US) simultaneously. Receiver v records its local arrival time t_r of RF signal and t_u of US signal from u. Then, based on the propagation speed s_r of RF and s_u of US and difference of arrival times $t = t_u - t_r$, the node v computes distance $(\hat{d})$ to u with equation:

$$t = \frac{d}{s_r} - \frac{d}{s_u} \tag{2}$$

Theoretically, we could use one signal type. However this approach requires synchronization of all the nodes in the network which is very difficult to achieve. Thus, usage of two signals allows nodes to rely on their local time.

3 Basic Protocol

In this section we present the protocol FINDMAP that essentially performs by majority voting. The protocol detects all faking sensors provided $n - 2 - f > f$. Thus, the total number of faking sensors is at most $\lceil \frac{n}{2} \rceil - 2$. In this section we consider the relatively simpler case where faking sensors cannot cheat on ranging techniques. That is faking nodes cannot change its transmission power, but they can cooperate and corrupt the protocol. In section 4 and 5 the protocol will be extended to the case where faking nodes corrupt the ranging technique. Our key assumption is that no three correct sensors are co-linear. This assumption allows to formulate the following fact.

Fact 1. *If a cheating sensor transmits a message with a faked position then at least one of three correct sensors can detect an inconsistency (see figure 1).*

Based on fact 1, we can develop FINDMAP(k), where k is the max number of correct nodes which cannot detect inconsistency. By fact 1 at most $k = 2$ correct nodes will not detect inconsistency. The protocol operates in two rounds. In *Round 1* all sensors exchange their coordinates by transmitting an initial message. Next, each node v computes the distances $\hat{d}(v, u)$ (from the ranging technique) and $d(v, u)$ (from the obtained node coordinates) of u and compare

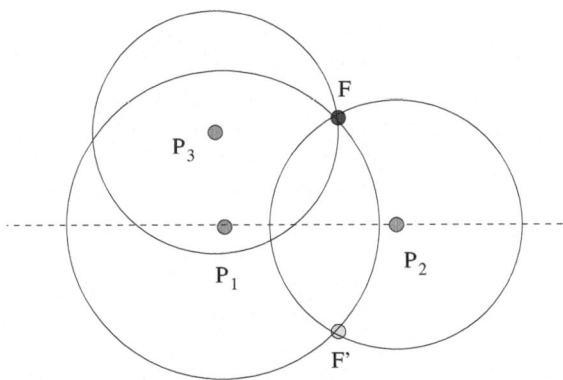

Fig. 1. An example showing F consistently fakes its location to F' against P_1 and P_2. However P_3 always detects an inconsistency since no three correct sensors are co-linear.

them. If $\hat{d}(v, u) \neq d(v, u)$ then v accuses u of faking its position. Otherwise, v approves u (v believes that u is correct). To keep record of its accusations and approvals, each node v maintains an array $accus_v$. In *Round 2* each node v transmits its array $accus_v$. Next, each node v counts accusations and approvals toward the node u including its own messages. Node v finds the node u as faking if the number of accusations is strictly larger than number of approvals minus k.

Protocol FindMap(k) executed by node v

Round 1:
1. v exchanges coordinates by transmitting $init_v$ & receiving $n-1$ $init_u$.
2. for each received message $init_u$:
3.   compute $\hat{d}(v, u)$ with ranging technique and
    $d(v, u)$ using the coordinates of u.
4.   **if** $(\hat{d}(v, u) \neq d(v, u))$ **then** $accus_v[u] \leftarrow true$
    else $accus_v[u] \leftarrow false$

Round 2:
5. v exchange accusations by transmitting $accus_v$ & receiving $n-1$ $accus_u$.
6. for each received $accus_u$:
7.   for $r = 1$ to n
8.   **if** $accus_u[r] = true$ **then** $NumAccus_r += 1$
        else $NumApprove_r += 1$
9. for each sensor u:
10.   **if** $(NumAccus_u > NumApprove_u - k)$ **then** v finds u as faking
    else v finds u as correct.

Theorem 1. *Each correct sensor, running protocol* FINDMAP($k = 2$), *never accepts position of faking sensor provided* $n - f - 2 > f$ *(at most* $\lceil \frac{n}{2} \rceil - 2$ *faking nodes) and no three sensors are co-linear.*

Proof. Let us assume $k = 2$, $n - f - k > f$ and no three sensors are co-linear. First we will show that each faking sensors will be detected by all correct sensors. By fact 1 each faking sensors v will be approved by at most $f + k$ sensors (at most f faking sensors and at most k correct sensors) and each faking sensor v will be accused by at least $n - f - k$ correct sensors. Thus the inequality $n - f - k > f + k - k$ in line 10 of the protocol will be true due to our assumption $n - f - k > f$. So faking node v will be identified by all correct nodes.

Next, we have to show that no correct sensor will be found faking. We can see that each correct sensor u can be accused by at most f faking sensors and each correct sensor u will be approved by all $n - f$ correct sensors. Thus the inequality $f > n - f - k$ in line 10 of the protocol will be false due to our assumption $f < n - f - k$. So each correct node u will be identified as a correct one by all correct nodes.   □

Next, we show that it is impossible to filter out the faking sensors when $n - 2 - f \leq f$. The assumption that faking sensors cannot corrupt the distance ranging technique makes this result even stronger. Our protocol is synchronous but this impossibility result holds for asynchronous settings also.

Theorem 2. *If $n - f - 2 \leq f$ then the faking nodes cannot be identified by a deterministic protocol.*

4   Protocol Based on RSS Ranging Technique

In this section we assume that sensors use RSS technique to measure distance. We assume that each correct sensor has a fixed common transmission signal strength of S_s. The faking sensors can change their transmission signal strength to prevent correct sensor from computing correct distance. Let F be a faking sensor that changes its signal strength S_s' and sends a suitable faked position F' to other correct sensors v. Sensor v can estimate the distance, $\hat{d}$ from the received signal strength (RSS) by Frii's transmission equation 1 assuming the common signal strength S_s has been used.

$$\hat{d}^2 = c\frac{S_s}{S_r} \implies \hat{d}^2 = \frac{S_s}{S_s'}d^2 \tag{3}$$

where $c = \left(\frac{\lambda}{4\pi}\right)^2$, $S_r = c\frac{S_s'}{d^2}$, and d is the distance from v to the actual position of F. More detailed analysis shows that correct nodes which cannot detect inconsistency on faked transmission are located on the circle (see figure 2). We can use this fact to formulate the following theorem.

Theorem 3. *If the distance evaluation is done with RSS techniques and no four sensors are located on the same circle, and no four sensors are co-linear then at least one of 4 correct sensors detects inconsistency in faked transmission.*

Theorem 3 allows to adjust the FINDMAP protocol.

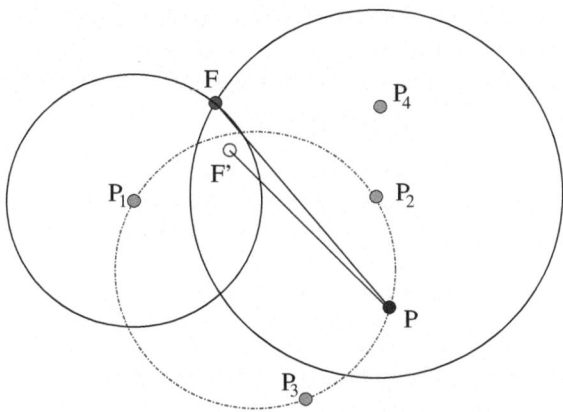

Fig. 2. An example showing a faking sensor F can lie about its position by changing signal strength to multiple number of correct sensors (P_1, P_2, P_3) which are laying on a particular circle. But the correct sensor P_4 not laying on the circle detects the inconsistency.

Corollary 1. *Each correct sensor, running the protocol* FINDMAP$(k = 3)$, *never accepts position of the faking node, in the model where faking sensors can corrupt RSS ranging technique, provided $n - f - 3 > f$ (at most $\lfloor \frac{n}{2} \rfloor - 2$ faking nodes) and no four sensors are located on the same circle and no four nodes are co-linear.*

Theorem 3 can be also applied in the protocol for the model of trusted sensors. In the protocol presented in [3], we can use theorem 3 to find deployment of the minimum number of hidden stations required to detect faking nodes.

Corollary 2. *If the four hidden stations are neither located on the same circle nor co-linear then one of the stations will always detects faked transmission.*

5   Protocol Based on ToF-Like Ranging Techniques

In this section we assume that the ToF ranging techniques is used. Each sensor u transmits each message simultaneously with two signals (*e.g.* RF and US signals). Each receiver v of the message records the difference of arrival time t between RF signal and US signal. Next, receiver v computes distance $\hat{d}(v, u)$ based on t, propagation speed s_r of RF signal and propagation speed s_u of US signal. The faking sensor u may prevent the correct sensor v from computing real distance by delaying one of the two simultaneous transmissions.

Lemma 1. *If the distance evaluation is done with ToF technique, and faking sensor F introduces shift $t' \neq 0$ between the RF and US transmissions, then all correct sensors compute the real distance to the sensor F modified by the same length b.*

Lemma 1 allows to compute the set of correct nodes which cannot detect inconsistency on the faked transmission. More detailed analysis shows that correct nodes which cannot detect inconsistency are located on the hyperbola. This fact allows to formulate the following theorem.

Theorem 4. *If the distance evaluation is done with ToF technique and no six sensors are located on the same hyperbola and no six sensors are co-linear, then at least one of six correct sensors detects inconsistency in faked transmission.*

Theorem 4 allows us to adjust the protocol FINDMAP.

Corollary 3. *Each correct sensor, running the protocol* FINDMAP$(k = 5)$, *never accepts position of the faking node, in the model where faking sensors can corrupt ToF ranging technique, provided $n - f - 5 > f$ (at most $\lfloor \frac{n}{2} \rfloor - 3$ faking sensors) and no six sensors are located on the same hyperbola and no six sensors are co-linear.*

Theorem 4 can be also applied to the TS model [3]. We can use theorem 4 to compute the deployment of the minimum number of hidden stations required to detect faking nodes.

Corollary 4. *If the six hidden stations are neither located on the same hyperbola nor co-linear then one of the stations always detects faked transmission.*

6   Concluding Remarks

We have proposed a secure deterministic position verification protocol for WSN that performs in the most general NTS model. Although the previous protocol of Hwang *et al.* [8] is randomized, it is interesting to compare number of transmitted messages. In [8], each sensor announces one distance at a time in a round robin fashion. Otherwise the faking node could hold its own announcement, collect all correct nodes information, and send a consistent faked distance claim. Thus the message complexity is $O(n^2)$. In our case, $2n$ messages are transmitted in two rounds. Thus our protocol safes on time and the power used for transmissions significantly.

Our network model assumes that correct nodes are within range of every other node. We believe that our majority voting heuristic could provide robust results for arbitrary network topology. We can observe that a correct node u identifies its faking neighbor v provided number of correct u's neighbors which report inconsistency on v is strictly larger then number of faking u's neighbors.

Extending our result to WSN with fixed ranges for every node is a challenging task, especially since previous results on networks facing intermittent failures and attacks [14] are written for rather stronger models (*i.e.* wired secure communications) than that of this paper.

References

1. Bahl, P., Padmanabhan, V.N.: Radar: An in-building rf-based user location and tracking system. In: INFOCOM, vol. 2, pp. 775–784. IEEE (2000)
2. Brands, S., Chaum, D.: Distance-bounding protocols. In: McCurley, K.S., Ziegler, C.D. (eds.) Advances in Cryptology 1981 - 1997. LNCS, vol. 1440, pp. 344–359. Springer, Heidelberg (1999)
3. Capkun, S., Cagalj, M., Srivastava, M.B.: Secure localization with hidden and mobile base stations. In: INFOCOM, IEEE (2006)
4. Capkun, S., Hubaux, J.: Secure positioning in wireless networks. IEEE Journal on Selected Areas in Communications: Special Issue on Security in Wireless Ad Hoc Networks 24(2), 221–232 (2006)
5. Douceur, J.R.: The sybil attack. In: Druschel, P., Kaashoek, M.F., Rowstron, A. (eds.) IPTPS 2002. LNCS, vol. 2429, pp. 251–260. Springer, Heidelberg (2002)
6. Fontana, R.J., Richley, E., Barney, J.: Commercialization of an ultra wideband precision asset location system. In: 2003 IEEE Conference on Ultra Wideband Systems and Technologies, pp. 369–373 (2003)
7. He, T., Krishnamurthy, S., Stankovic, J.A., Abdelzaher, T., Luo, L., Stoleru, R., Yan, T., Gu, L., Hui, J., Krogh, B.: An energy-efficient surveillance system using wireless sensor networks. In: MobiSys 2004: Proc. of the 2nd Int. Conf. on Mobile systems, applications, and services, New York, USA, pp. 270–283 (2004)
8. Hwang, J., He, T., Kim, Y.: Detecting phantom nodes in wireless sensor networks. In: INFOCOM, pp. 2391–2395. IEEE (2007)
9. Karp, B., Kung, H.T.: Gpsr: greedy perimeter stateless routing for wireless networks. In: MobiCom 2000: Proc. of the 6th Annual Int. Conf. on Mobile Computing and Networking, pp. 243–254. ACM Press, New York (2000)
10. Lamport, L., Shostak, R., Pease, M.: The byzantine generals problem. ACM Trans. on Programming Lamguages and Systems 4(3), 382–401 (1982)
11. Lazos, L., Poovendran, R.: Serloc: Robust localization for wireless sensor networks. ACM Trans. Sen. Netw. 1(1), 73–100 (2005)
12. Lazos, L., Poovendran, R., Capkun, S.: Rope: robust position estimation in wireless sensor networks. In: IPSN, pp. 324–331. IEEE (2005)
13. Liu, C.H., Fang, D.J.: Propagation. in antenna handbook: Theory, applications, and design. Van Nostrand Reinhold 29, 1–56 (1988)
14. Nesterenko, M., Tixeuil, S.: Discovering network topology in the presence of byzantine faults. In: Flocchini, P., Gąsieniec, L. (eds.) SIROCCO 2006. LNCS, vol. 4056, pp. 212–226. Springer, Heidelberg (2006)
15. Szewczyk, R., Mainwaring, A., Polastre, J., Anderson, J., Culler, D.: An analysis of a large scale habitat monitoring application. In: SenSys 2004: Proc. Int. Conf. Embedded Networked Sensor Systems, pp. 214–226. ACM Press, New York (2004)

Efficient Node Discovery in Mobile Wireless Sensor Networks

Vladimir Dyo and Cecilia Mascolo

[1] Department of Computer Science, University College London
Gower Street, London WC1E 6BT, UK
v.dyo@cs.ucl.ac.uk
[2] Computer Laboratory, University of Cambridge
15 JJ Thomson Avenue, Cambridge CB3 0FD, UK
cecilia.mascolo@cl.cam.ac.uk

Abstract. Energy is one of the most crucial aspects in real deployments of mobile sensor networks. As a result of scarce resources, the duration of most real deployments can be limited to just several days, or demands considerable maintenance efforts (e.g., in terms of battery substitution). A large portion of the energy of sensor applications is spent in node discovery as nodes need to periodically advertise their presence and be awake to discover other nodes for data exchange. The optimization of energy consumption, which is generally a hard task in fixed sensor networks, is even harder in mobile sensor networks, where the neighbouring nodes change over time.

In this paper we propose an algorithm for energy efficient node discovery in sparsely connected mobile wireless sensor networks. The work takes advantage of the fact that nodes have temporal patterns of encounters and exploits these patterns to drive the duty cycling. Duty cycling is seen as a sampling process and is formulated as an optimization problem. We have used reinforcement learning techniques to detect and dynamically change the times at which a node should be awake as it is likely to encounter other nodes. We have evaluated our work using real human mobility traces, and the paper presents the performance of the protocol in this context.

1 Introduction

Energy efficiency is a crucial aspect in wireless sensor networks. The amount of energy of a sensor network may be limited by the constrained size of devices or, for instance, by the efficiency of the source of energy, e.g., the limited size of a solar panel. In such situations, the only sensible approach to energy saving is duty cycling, i.e., the control of the awake times of sensor nodes.

Duty cycling however, limits the ability of nodes to discover each others as when nodes are sleeping they cannot detect contacts. The problem of neighbour detection is even more serious if the sensor network is mobile, as the topology in these networks changes rapidly.

S. Nikoletseas et al. (Eds.): DCOSS 2008, LNCS 5067, pp. 478–485, 2008.

Node detection is not a problem if nodes are equipped with specialized sensors, such as motion detectors or accelerometers. However, these devices increase the cost and the size of the equipment and are not always available or deployable (e.g., in some zoological applications which need cheap or very small sensors). In this paper we will assume that the detection of neighbours only happens through normal short-range radio.

The existing work on duty cycling [1] has mostly tackled static networks with fixed topologies and is not applicable to mobile scenarios, given the variability of the topology. A major challenge for some mobile networks is the uncertainty of the node arrival time. If the node arrival time is not known, the only chance a node has to discover all the nodes passing by is to be always awake, which is very energy inefficient.

MAC layer optimizations for listening times such as the ones developed in [2] offer some form of optimization of the power consumption, however not at the level of granularity which could be achieved with patterns recognition. A considerably better optimization can be achieved by using some knowledge of the encountering patterns in the network in order to decide when to switch on (and off) the radio. This, of course, can only be applied when encounter patterns exist, which however is often the case in wildlife and human applications.

In this paper we propose an energy efficient node discovery approach for mobile wireless sensor networks. The main idea of our method is the online detection of periodic patterns in node arrivals and the scheduling of wake-up activity only when contacts are expected. The approach is based on reinforcement learning techniques [3]. In this approach each node monitors the number of encounters depending on time of day and concentrates more energy budget (i.e., more awake time) into predicted busier timeslots. The approach also allows for some extra energy to monitor other timeslots in order to cope with variation in the patterns and to refine what it has learned, dynamically.

The approach can be applied to scenarios such as wildlife monitoring, as indicated above, or human-centric networks. In order to evaluate the performance of the approach we have verified it with real human mobility traces, used to drive mobile sensor movement in a synthetic way, in a simulator.

The rest of the paper is organized as follows: Section 2 contains a general overview of our approach. Section 3 contains the adaptive technique for learning arrival patterns. Section 4 describes the protocol. Section 5 present an evaluation of our approach. Section 6 discusses related work with conclusions and possible future work.

2   Overview of the Approach

The main goal of our approach is to allow nodes to detect each other's presence but, at the same time, to save energy by switching off their radio interface as much as possible. As we outlined in the introduction, the detection of neighbours allows many activities such as the relaying of the data to sinks and the logging of encounters.

Discovering nodes is expensive and requires either periodic scanning (as in Bluetooth) or periodic continuous transmission of a radio tone (if the nodes are using a Low Power Listening based protocol [2]). A high scanning rate will guarantee quick discovery but will waste energy, especially in situations, where no encounters are likely to occur. On the other hand, a low scanning rate can miss many important contacts. Specifically, the goal of the approach is to devise a simple adaptive algorithm to control the scanning rate, considering past encounter history.

We consider duty cycling as a sampling process. Intuitively, to detect more encounters, a node needs to sample more frequently when more encounters are expected. Moreover, the node should avoid to sample when no encounters are expected. Thus, the goal is to maximize:

$$R(a) = \sum_i E_i * d_i, \quad s.t. \sum_i d_i < D_{budget} \tag{1}$$

Where E_i and d_i is an expected number of encounters, a a duty cycle at timeslot i, and D_{budget} is a daily energy budget. As we see from the equation, there is a balance between number of contacts and energy consumption. Thus it seems natural to formulate the problem as a maximization of the number of successful encounters per unit of energy consumption.

As already indicated, we consider a specific class of applications, when periodic encounter patterns exists. These represent a large class, which include human and animal life.

3   Learning Arrival Patterns

In this section we describe the core ideas behind the pattern arrival mechanism we adopted. The basic behaviour of the algorithm drives each node to estimate the hourly activity of its neighbours and to progressively concentrate the discovery process only when encounters are expected. Indeed, the intuitive idea behind this behaviour is that continuously scanning for neighbours when no one will be around implies a waste of energy.

3.1   Model

We now introduce the formal model behind the approach. An agent (in the reification of our system, a node)[1] interacts with the environment through perception and action. At each step, an agent observes the state $s_t \in S$ of an environment and responds with an action $a \in A(S_t)$. The action results in a certain reward $R : S \times A \to R$. The goal of an agent is to maximize a long-term reward based on the interactions with an environment. Specifically, the goal is to learn a policy mapping from states to actions that maximizes the long-term agent reward.

---

[1] In this section we will refer to node and agents referring to the same entity: agent is the name used in the machine learning theory we adopt.

A day is modelled as N timeslots. A node has the following set of actions: i) sleep ii) wake-up ii) set duty cycle (1-100%). A node controls the duty cycle by changing the discovery beacon rate. A high duty cycling might or might not increase the chances of detecting more contacts. For example, it might be sufficient for a node to work from 11am to 12am, but with a 10% duty cycle (as opposed to 100%). A reward r is the number of successful encounters. The goal of an agent is to detect the maximum number of successful encounters within a given energy budget.

After taking each action, a node observes the outcome and updates the payoff for a given timeslot. The payoff estimation is done using an exponentially weighted moving average (EWMA) filter. The filter estimates the current payoff value by taking into account the past measurements, $r_n = r_{measured} * \alpha + r_{n-1} * (1 - \alpha)$ Where r_n and r_{n-1} are respectively the estimated and previous payoff values. $r_{measured}$ is the measured payoff over the last time slot. The weight assigned to past measurements $(1 - \alpha)$ depends on how responsive the node has to be to changing environment.

We now describe the balanced strategy which could be used to adapt the node's duty cycle and a random strategy which we will use as baseline for the evaluation.

Balanced. In a balanced strategy we propose to dynamically adjust the node's duty cycle proportionally to an expected reward. Therefore the node *does not commit* to any timeslot, but spreads its energy proportionally to the expected reward. The node sets its duty cycle according the following rule:

$$D(a) = \frac{r(a)}{\sum_{a\prime \in A} r(a)} \tag{2}$$

$D(a)$ is a duty cycle in the current timeslot, $r(a)$ is an expected reward from taking an action a. It is computed as indicated in Formula (3.1) For example, if there are several peak hours during a day, the budget will be spread evenly among all peaks. During quiet times the node continues to sample the environment but with lower intensity.

Random. In a random strategy (which we use for comparison) the node spreads its energy budget evenly throughout a day, i.e., it sends beacons with a certain fixed interval. The strategy is equivalent to normal asynchronous wake-up scheduling with fixed duty cycle, so would not require additional implementation. The obvious problem with the random strategy is that a node will waste resources when there are no nodes around. This will become evident in our evaluation.

4   Algorithm

In this section we present an algorithm for adaptive node discovery. The algorithm should allow the detection of 'quiet' periods and exclude them from the discovery process, allowing the node to sleep in that time for as much as possible.

The daily budget assignment could be performed by the application, depending on the known energy availability: for example in the scenario we envisage, it is very clear how big the batteries can be and how long the zoologists want them to last for, therefore the daily budget can be inferred.

1. The node starts by following a random strategy, i.e., spreads its duty cycle equally in each timeslot. As it discovers new nodes it dynamically reajusts its budget according to the following steps.
2. Once discovered, the nodes remain synchronized for a duration of an encounter. Short term synchronization is possible with built-in timers without the need of globally synchronized clocks. As long as there is at least one node in range, the node sends periodic keep-alive messages every $T_{keepalive}$ seconds.
3. If a node does not hear from a neighbour for T_{expire} seconds, it assumes an encounter is terminated and increments the timeslot counter. $C_t = C_t + 1$.
4. At the end of each day a node updates its timeslot counters M_t: $M_t = C_t * \alpha + M_t * (1 - \alpha), t = 0..N_{slots}$. Where M_t is an estimated encounter frequency at timeslot t and C_t is the actual number of encounters in timeslot t registered during current day. The node then resets the daily counters C_t.
5. At the beginning of the current timeslot (t), a node sets a beacon rate to be: $F_{beacon} = \frac{M_t}{\sum_i M_i} \frac{B}{E_{beacon}}$. Where B is a daily energy budget, E_{beacon} is an amount of energy required to scan the neighbourhood. The node converts the beacon frequency into interval time between beacons $T_{beacon} = 1/F_{beacon}$. If the duration of this period is longer than the timeslot duration $T_{timeslot}$, the node beacons with a probability $p = \frac{T_{beacon}}{T_{timeslot}}$. The node then schedules the next wake up by the beginning of the next timeslot. The node has preconfigured minimum and maximum beacon rates F_{min} and F_{max}. The minimum beacon rate is needed to guarantee a certain level of exploration, even when no discovery is expected. The maximum beacon rate limits the amount of energy a node spends in one timeslot.

5   Evaluation

The goal of experiments is to compare the random and adaptive algorithms presented in Section 4 on the performance of two basic applications: encounter tracking and message dissemination.

5.1   Evaluation Settings

Dataset. We evaluated our approach through simulations in TOSSIM with real human connectivity traces to emulate node mobility and dynamic encountering. We used human mobility traces from MIT reality mining bluetooth traces [4] to drive the sensor movement like they were tagged individuals. The traces were collected using 96 people carrying Bluetooth mobile phones over a duration of 292 days. The evaluation was done on 60 more active nodes over 3 months

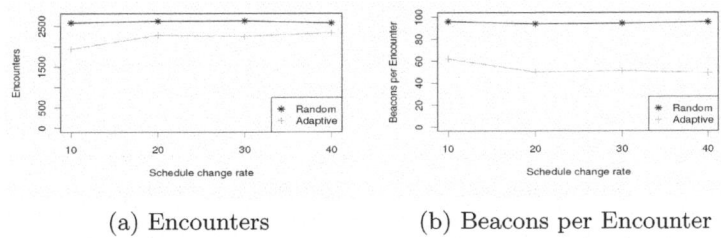

(a) Encounters   (b) Beacons per Encounter

of traces. Due to power limitations, the original traces are result of sampling every 300s, which might have missed some encounters and introduced a certain granularity of encounter duration. In this paper, however, we assume that the traces represent ground truth data about physical movement of entities and that our optimal result would be to detect all contacts. All the nodes were booted at random times between 0 and 3600 seconds. The evaluation was done over 5 runs for each algorithm x budget combination.

Impact on discovery rate. In the first experiment we measure encounters between the nodes for various wake-up algorithms and compare with a baseline random algorithm over synthetic traces in the following settings. We generated 7200 random encounters for 36 nodes for a duration of 90 days. The duration of each encounter was uniformly distributed between 300 and 900s. To model dynamic environment the network operated according to one of two schedules. In schedule A, all the links were established between times 8am and 3pm; in schedule B, all the encounters were established between times 22pm and 5am. We then generated a trace, where both schedules alternated every 10, 20, 30 and 40 days. All nodes were running encounter tracking application and were required to detect and log encounters between the nodes. The beaconing is the expensive process which requires nodes to stay up for a long period of time. In our experiments we measure the energy a node spends on beaconing (node discovery) on the performance of basic applications of encounter logging.

Figure 1a shows that the nodes running adaptive strategy managed to detect almost the same number of encounters as nodes running random strategy. At the same time, the adaptive strategy required up to 50% fewer beacons than a random one (Figure 1b). The performance degraded with more frequent schedule changes, but remained higher than random.

In the next experiment we measure encounters between the nodes over the real traces from MIT reality mining experiments. Figures 1a shows the number of detected encounters for scanning intervals from 7200 to 79200 seconds. The graph shows that the adaptive strategy detects more encounters than simple random strategy. It shows that while adaptive detected more encounters, they consumed much fewer beacons (Figure 1b). In the course of experiments, we observed that the number of detected encounters of the adaptive algorithm depends on the maximum and the minimum number of beacons in one timeslot. In the experiment we set it to maximum of 200% and 10% of average (budgeted) scanning rate. All the graphs show the percentage of encounters detected by the

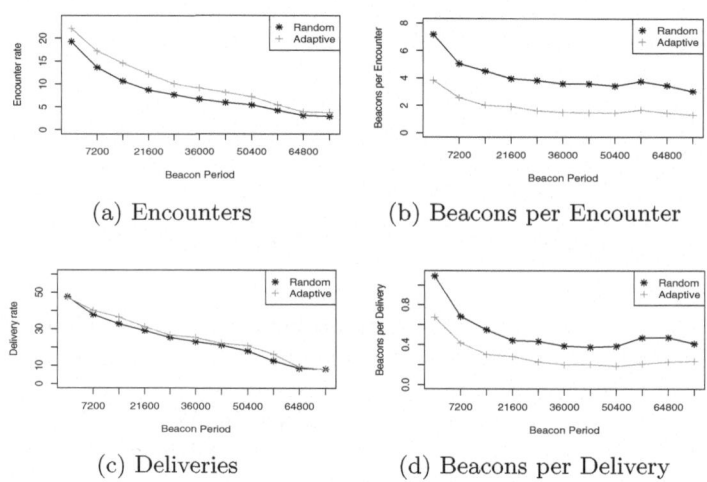

(a) Encounters

(b) Beacons per Encounter

(c) Deliveries

(d) Beacons per Delivery

Fig. 1. Impact on node discovery and message delivery rates

2 algorithms over the total number of encounters in the traces. We then tested the algorithm sensitivity for different timeslot durations and found that longer timeslots perform better for lower scanning rates. In this experiment the nodes used 3 hour timeslots.

Impact on message delivery rate. In this experiment we measure the impact of wake-up strategy on message delivery rates for a simple data collection application. The nodes are using a direct delivery algorithm, in which a sender delivers a message directly upon an encounter with a destination node (e.g., a sink in our scenario). The nodes were configured to generate one message per hour and send it towards one of six sinks. The message was considered delivered when it reached at least one of the sinks. The sinks were chosen randomly at each simulation run. All the graphs show the delivery rate in percentage from the total number of generated messages.

Figure 1c shows the number of detected message deliveries for scanning intervals from 7200 to 79200 seconds. The adaptive strategy provides better results than simple random strategy. Figure 1d shows the average number of beacon per delivery for the same experiment. It shows that while adaptive provides higher delivery rates, it consumed much fewer beacons.

It is interesting to note that at a beacon rate of one beacon per day, adaptive strategy still maintains a delivery rate of about 40%, more than twice that of random strategy while consuming twice as few discovery beacons.

6   Related Work and Conclusion

Energy efficient service discovery can be done using power efficient wake-up scheduling protocols, such as [1,2]. These protocols allow for very energy efficient communication in static wireless sensor networks. They are not, however, able to exploit the fact that contact patterns might be regular and distributed in such a way that there are periods in which nodes do not encounters other nodes. Our approach works on top of existing wake-up scheduling protocols, allowing them to make better decisions as to when to and how frequently to perform service discovery. In [5] an adaptive node discovery approach is proposed for static sink nodes to track mobile objects: some learning techniques have been used there to drive the discovery, however the network of sensors for that paper was static not allowing for the variability inherent in a mobile sensor network. A *variable inquiry rate* has been used to collect Bluetooth traces in [6]. To save power the nodes were configured to sample the neighbourhood more frequently when no nodes are detected and then reduce the sampling rate if there are nodes around. Although an approach was used to actually collect traces, there was no evaluation quantifying an impact of this technique on the number of detected encounters.

Conclusion. We have presented an approach for flexible duty cycling for mobile wireless sensor networks. We have evaluated the approach with realistic human mobility traces and have shown the performance of our proposed approach with respect to a random wake up scheduling. We are in the process of generalization of the approach to non-periodic patterns. In this case, the node needs to forecast the encounter pattern for the next N steps and then allocate energy budget accordingly.

References

1. Ye, W., Silva, F., Heidemann, J.: Ultra-low duty cycle MAC with scheduled channel polling. In: SenSys 2006: Proceedings of the 4th international conference on Embedded networked sensor systems, pp. 321–334. ACM Press, New York (2006)
2. Polastre, J., Hill, J., Culler, D.: Versatile low power media access for wireless sensor networks. In: SenSys 2004: Proceedings of the 2nd international conference on Embedded networked sensor systems, pp. 95–107. ACM Press, New York (2004)
3. Kaelbling, L.P., Littman, M.L., Moore, A.P.: Reinforcement Learning: A Survey. Journal of Artificial Intelligence Research 4, 237–285 (1996)
4. Eagle, N., Pentland, A.S.: Reality mining: sensing complex social systems. Personal Ubiquitous Comput 10(4), 255–268 (2006)
5. Dyo, V., Mascolo, C.: A Node Discovery Service for Partially Mobile Sensor Networks. In: MIDSENS 2007: Proceedings of IEEE International Workshop on Sensor Network Middleware, IEEE Computer Society, Los Alamitos (2007)
6. Su, J., Chan, K.K.W., Miklas, A.G., Po, K., Akhavan, A., Saroiu, S., de Lara, E., Goel, A.: A preliminary investigation of worm infections in a Bluetooth environment. In: WORM 2006: Proceedings of the 4th ACM workshop on Recurring malcode, pp. 9–16. ACM, New York (2006)

Decentralized Deployment of Mobile Sensors for Optimal Connected Sensing Coverage

Adriano Fagiolini[1], Lisa Tani[2], Antonio Bicchi[1], and Gianluca Dini[2]

[1] Interdep. Research Center "E. Piaggio", Faculty of Engineering,
Università di Pisa, Italy
{a.fagiolini,bicchi}@ing.unipi.it
[2] Dipartimento di Ingegneria della Informazione, Università di Pisa, Italy
gianluca.dini@ing.unipi.it

Abstract. In this paper, we address the *optimal connected sensing coverage* problem, i.e., how mobile sensors with limited sensing capabilities can cooperatively adjust their locations so as to maximize the extension of the covered area while avoiding any internal "holes", areas that are not covered by any sensor. Our solution consists in a *distributed motion algorithm* that is based on an original extension of the Voronoi tessellation.

1 Introduction

Recent technological advances in miniaturization and low–cost production of small embedded devices have greatly enhanced our capability to sense and control the physical world. Wireless Sensor Networks (WSNs) seem to represent one of the main research and application fields that will benefit from these advances, and indeed they are revolutionizing the way data is traditionally gathered from the physical world.

A crucial requirement for an efficient execution of a sensing task is an adequate sensor deployment. In this paper we consider the problem of deploying a set of mobile sensor nodes with limited sensing range in order to achieve an *optimal connected sensing coverage*. Informally, this requires to devise a deployment that maximizes the extension of the covered area while avoiding internal "holes", i.e., internal areas that are not covered by any sensor. We propose a solution to this deployment problem that consists in a *distributed motion algorithm* that makes sensor nodes cooperatively adjust their locations so as to fulfill the requirement of optimal connected sensing coverage.

The deployment of mobile sensor nodes is certainly not new [1, 2, 3]. It is particularly relevant in areas that are remote, hostile or even deadly for human operators, and its employment has been promoted by the availability of mobile sensors such as Robomote [4].

The coverage problem we solve is similar to the one studied in [2, 1]. However, they use a mix of fixed and mobile sensors and do not prove the eventual absence of coverage "holes". Our solution extends the one based on Voronoi diagrams proposed by Cortes *et al.* that assumes sensors with unlimited sensing ranges

S. Nikoletseas et al. (Eds.): DCOSS 2008, LNCS 5067, pp. 486–491, 2008.

although degrading with distance [3]. Intuitively, every sensor is mobile and moves under the effects of two contrasting forces. The one tends to keep a sensor close to its neighbors whereas the other tends to spread mobile sensors as much as possible. The former is the strongest but it is exerted only when the mutual distance between a couple of neighbors exceeds a predefined threshold. The latter is weaker but is always present. In a typical evolution of the system, mobile sensors initially assemble to fill every coverage hole and then try to cover as much area as possible without creating any hole.

The paper is organized as follows. In Section 2.1, we present the theoretical foundation of the proposed distributed motion algorithm for optimal sensor deployment. Then, in Section 2.2, we consider how to achieve a sensor deployment such that neighboring sensors are within a given desired distance and argue about the algorithm convergence toward a steady optimal solution. Then, we discuss how such a distance can be chosen so that a final optimal connected sensor deployment is eventually reached.

2   Distributed Deployment for Sensing Connectivity

2.1   Basic Framework

Consider n sensors that are to be deployed within a configuration space $\mathcal{Q}$. Let us suppose that the current configuration or location q_i of the i–th sensor is measured w.r.t. a common coordinate frame. For the sake of simplicity, we focus on a planar deployment problem, where the configuration space $\mathcal{Q}$ is a closed region $\mathcal{W} \subset \mathbb{R}^2$, and the position of the i–th sensor is $q_i = (x_i, y_i)$, although the discussion remains valid for problems in higher dimensions.

Assume that the desired *sensor distribution* is specified by a non–negative function $\phi : \mathcal{Q} \to \mathbb{R}^+$, whose value at any location q in $\mathcal{Q}$ is proportional to the need of sensing the location itself. Clearly, a null value of $\phi(q)$ means that a sensor is not required at q. Possible sensor distributions may range from a uniform $\phi(q)$, meaning that every location in $\mathcal{W}$ require the same sensing effort, to a spot–wise $\phi(q)$ representing situations where there is one or more discrete points of interest. Consider also a non–negative function $f : \mathcal{Q} \times \mathcal{Q} \to \mathbb{R}^+$ that represents a *sensing degradation*, also referred to as *sensing model*. More precisely, for a fixed position q_i of the i–th sensor, $f(q_i, q)$ is a function of q that describes how the sensor's measurement degrades as q varies from q_i. This is how we will use f in the rest of the paper.

To begin with, suppose that sensors are deployed at initial locations $q_i(0)$, for $i = 1, \ldots, n$. Then, imagine that the configuration space $\mathcal{W}$ is partitioned into n regions based on the current sensors' locations, i.e. $\mathcal{W} = \cup_{i=1}^n \mathcal{W}_i$, and $\mathcal{W}_i = \mathcal{W}_i(Q)$, where $Q = \{q_1, \ldots, q_n\}$. It is useful to assume that each sensor becomes responsible for sensing over exactly one region that will be referred to as its *dominance region*. Then, to find an optimal sensor deployment, we need to introduce a global *cost functional* $\mathcal{H}(Q, \mathcal{W})$ that measures how poor is the current sensor deployment w.r.t. the desired sensor distribution $\phi(q)$ for the given sensing model $f(q)$. One possible choice for $\mathcal{H}$ is the following cost functional [3]:

$$\mathcal{H}(Q, \mathcal{W}) = \sum_{i=1}^{n} \int_{\mathcal{W}_i} f(||q - q_i||)\, \phi(q)\, dq\,, \tag{1}$$

where $||\cdot||$ is the Euclidean norm. Let us now consider the following problem:

Problem 1 (Optimal Deployment Without Sensing Connectivity). Given a desired sensor distribution $\phi(q)$, and a sensing degradation $f(q)$, find a *distributed motion strategy* by using which n mobile sensors can iteratively adjust their locations from an initial deployment at $q_i(0)$, $i = 1, \ldots, n$, to a final optimal deployment minimizing $\mathcal{H}$.

This problem has been solved as an instance of *coverage control* by Cortes *et al.* in [3]. Therein, Corte *et al.* assumed that sensors move according to a *motion model* described by a first–order linear dynamics:

$$\dot{q}_i(t) = u_i(t), \quad \text{for } i = 1, \ldots, n\,, \tag{2}$$

where u_i are *input velocities* that can be chosen to move the sensors. This model assumes that mobile sensors can move to any location where they are asked to move. In practice, finding paths or maneuvers that take mobile sensor nodes to desired destinations is an important problem that can easily become difficult when there are obstacles in the field and kinematic constraints. This problem is studied in the area of robotics [5,6] and we do not study it any further. Moreover, Cortes *et al.* assumed an *isotropic degradation model*:

$$f(q, q_i) = ||q - q_i||^2\,, \quad \text{for all } q_i, q \in \mathcal{Q}\,, \tag{3}$$

which depends only on the Euclidean distance between the i–th sensor's position q_i and the sensed location q. They exploit the known fact that, given a current sensors' deployment $\bar{Q}$, the partitioning $W = \{\mathcal{W}_1, \ldots, \mathcal{W}_n\}$ minimizing $\mathcal{H}$ is the *Voronoi tessellation* $\mathcal{V}$ generated by $\bar{Q}$, i.e. $\mathcal{V}(\bar{Q}) = \text{argmin}_{\{\mathcal{W}_1, \ldots, \mathcal{W}_n\}} \mathcal{H}(\bar{Q}, W)$. Intuitively, given n sensors, a Voronoi tessellation is a partition of the environment into n regions, where the i–th region is formed of all locations whose distance from sensor i is less or equal than the distances from other sensors [7]. Furthermore, Cortes *et al.* consider the desired sensor distribution $\phi(q)$ as a *density function* over the domain $\mathcal{Q}$ and recall two important quantities associated with the i–th Voronoi region $\mathcal{V}_i$. In particular, they have used the generalized mass, and the centroid or center of mass of $\mathcal{V}_i$ that are respectively defined as [8]:

$$M_{\mathcal{V}_i} = \int_{\mathcal{V}_i} \phi(q)\, dq\,, \quad C_{\mathcal{V}_i} = \frac{1}{M_{\mathcal{V}_i}} \int_{\mathcal{V}_i} q\, \phi(q)\, dq\,. \tag{4}$$

Using this physical interpretation of $\phi(q)$, Cortes *et al.* proposed a *distributed gradient–based motion strategy* allowing sensors to improve their deployment and thus reduce $\mathcal{H}$. More precisely, they showed that a motion strategy where each sensor is subject to a force generated by $\phi(q)$ and pushing it toward the centroid $C_{\mathcal{V}_i}$ of its current dominance region $\mathcal{V}_i$ solves Problem 1. Note that all dominance regions $\mathcal{V}_i$s are updated at any instant by every sensor and are thus a function of time, i.e. $\mathcal{V}_i = \mathcal{V}_i(t)$. Hence we have $M_{\mathcal{V}_i} = M_{\mathcal{V}_i}(t)$, and $C_{\mathcal{V}_i} = C_{\mathcal{V}_i}(t)$.

2.2   Distance–Constrained Deployment and Sensing Connectivity

We consider a WSN where each sensor i is able to measure any quantity of interest at any location q laying within sensing range r_s from the sensor position q_i itself. This property can be modeled by the following isotropic sensing degradation with threshold r_s:

$$f_{r_s}(q, q_i) = \begin{cases} ||q - q_i||^2 & ||q - q_i|| \leq r_s\,, \\ \infty & ||q - q_i|| > r_s\,. \end{cases} \tag{5}$$

In this context, we aim at finding distributed motion strategies allowing sensors to achieve *sensing connectivity* within $\mathcal{W}$. After recalling the outward boundary ∂A of a closed set A, and its closure A^* being the set of all locations contained by ∂A, we can readily provide the following definition of connectivity:

Definition 1. *Given a closed region $A \subseteq \mathcal{W}$, a quantity of interest ξ, and a sensor deployment Q, we say that A is* sensing connected *if, and only if, for any location $q \in A^*$, there exists at least one sensor in Q that can measure $\xi(q)$.*

Reaching *sensing connectivity* means that we want sensors to be deployed in such a way that there exists a closed sub–region $\mathcal{W}_c \subseteq \mathcal{W}^*$ that contains no sensing holes and that is as large as possible compatibly with this constraint. Hence, the problem we want to solve becomes the following:

Problem 2 (Optimal Deployment With Sensing Connectivity). In addition to the requirements of Problem 1, find a motion strategy by which sensors can also establish and maintain sensing connectivity.

A distributed motion strategy allowing the sensing connectivity requirement to be fulfilled needs a *form of interaction and coordination* between *neighboring sensors*. Indeed, to solve Problem 2, it is strategically important to choose a suitable definition of neighborhood and a local form of interaction between any two neighboring sensors. Among many possible choices, we will exploit the *neighborhood relation* introduced by the Voronoi tessellation, i.e. we will require that sensors i and j coordinate their motions in order to establish and maintain sensing connectivity if, and only if, they are $\mathcal{V}$–neighbors. Furtheron, to *enforce* the sensing connectivity requirement, we introduce *virtual points of interest*, that modify the originally given density function $\phi(q)$, for any couple of sensors that are too far from each other.

To this aim, first denote with $V = \{v_{ij}\}$ the *adjacency matrix* of a Voronoi graph $G_\mathcal{V}$, where the generic element is $v_{ij} = 1$ if i is a $\mathcal{V}$–neighbor of j, or $v_{ij} = 0$ otherwise. Then, consider the *augmented sensing distribution* $\tilde{\phi}$ defined as follows:

$$\tilde{\phi}(q, d) = \phi(q) + \alpha \sum_{i=1}^{n} \sum_{j=1}^{n} v_{ij} \frac{c_d(||q_i - q_j||)}{2} \delta\left(q - \frac{q_i + q_j}{2}\right), \tag{6}$$

where $\alpha \in \mathbb{R}$ is a positive *weight* that is chosen such that $\alpha \gg \max_{q \in \mathcal{W}} |\phi(q)|$, $c_d(s)$ is a local *penalty* function, and $\delta(q)$ is Dirac's delta function. Function $c_d(s)$

should be chosen so as to penalize pair of sensors at larger relative distances then a suitable *neighborhood threshold d*. One possible choice is the following:

$$c_d(s) = \begin{cases} 0 & \text{if } 0 \leq s \leq d, \\ e^{s-d} - 1 & \text{if } s \geq d. \end{cases} \tag{7}$$

Virtual points of interest act as *contracting terms*, such as nonlinear springs, activating whenever the distance between any two $\mathcal{V}$–neighbors exceeds a certain threshold. Their action is weighted by α which is chosen to be much larger than the maximum value of $\phi(q)$. This means that the sensing connectivity requirement has higher "priority" than the WSN's deployment task. We can draw an *analogy of our strategy with a particle system* where particles are subject to an external force generated by potential field ϕ, but are also aggregated by "stronger" internal forces as for the Van der Waals phenomenon.

On the way to solve Problem 2, we will proceed by finding a distributed motion strategy by which the distance between any two $\mathcal{V}$–neighbors is constraint to be less or equal to d. On the same line of [3], we will try to minimize the following global cost functional $\tilde{\mathcal{H}}$ that measures how poor is a sensor deployment Q w.r.t. the augmented density $\tilde{\phi}$ for the sensing model f_{r_s}:

$$\tilde{\mathcal{H}}(Q, \mathcal{W}, d) = \sum_{i=1}^{n} \int_{\mathcal{W}_i} f_{r_s}(\|q - q_i\|) \, \tilde{\phi}(q, d) \, dq. \tag{8}$$

Our strategy will be to minimize $\tilde{\mathcal{H}}(Q, \mathcal{W}, d)$ by following a gradient–based motion. It is worth noting that the discontinuity of f_{r_s}, that occurs whenever the distance between q_i and q exceeds r_s, may represent a problem for the algorithm. However, it can be shown that stationary configurations of the WSN for f_{r_s} are also stationary for f, and thus we will use the continuous function f in place of f_{r_s}. Having said this, we are ready to state the following first result [9]:

Theorem 1 (Optimal Distance Constraint Deployment). *Consider a WSN where sensors move according to the linear motion model of Eq. 2. Then, given a desired neighborhood distance d, the distributed motion strategy described by:*

$$u_i(t) = -k\left(q_i(t) - \tilde{C}_{\mathcal{V}_i(t)}\right), \quad \text{for } i = 1, \ldots, n, \tag{9}$$

where k is a positive real constant, and

$$\tilde{C}_{\mathcal{V}_i}(t) = \frac{M_{\mathcal{V}_i(t)}}{\tilde{M}_{\mathcal{V}_i(t)}}\left(C_{\mathcal{V}_i(t)} + \alpha \sum_{i,j=1}^{n} v_{ij}(t) \frac{c_d(\|q_i(t)-q_j(t)\|)}{2} \frac{q_i(t)+q_j(t)}{2}\right),$$
$$\tilde{M}_{\mathcal{V}_i}(t) = M_{\mathcal{V}_i(t)}\left(1 + \alpha \sum_{i,j=1}^{n} v_{ij}(t) \frac{c_d(\|q_i(t)-q_j(t)\|)}{2}\right), \tag{10}$$

where $M_{\mathcal{V}_i(t)}$ and $C_{\mathcal{V}_i(t)}$ are computed as in Eq. 4, makes it possible to reach an optimal sensor deployment where any two $\mathcal{V}$–neighbors are within distance d from each other, and the extension of the covered region is maximized.

The proof of Theorem 1 is omitted for the sake of space and can be found in [9] along with an early implementation and a performance evaluation of a protocol realizing the specified motion strategy.

Finally, we can show that a suitable choice of d makes it possible to achieve the sensing connectivity requirement. Let us denote with I the index set of all *internal* Voronoi regions, being those regions $\mathcal{V}_i$ having all of its faces adjacent to other regions $\mathcal{V}_j$, $j \neq i$, and not to the boundaries of the considered region $\mathcal{W}$. Then, we can prove the following result [9], which solves Problem 2:

Theorem 2 (Optimal Connected Sensing Deployment). *A set of n sensors moving according to the distributed motion strategy of Equation 9, where the neighborhood threshold is chosen as $d = \sqrt{3}\, r_s$, eventually reach a final deployment such that the union of all internal Voronoi regions, $\mathcal{W}_c = \cup_{i \in I} \mathcal{V}_i$, forms an optimal sensing–connected region.*

3 Conclusion

In this paper we studied the problem of reaching optimal connected sensing coverage and proposed a fully distributed Voronoi–based motion strategy for a mobile WSN. Future work will address implementation of the motion strategy and evaluate the robustness of the obtained solution w.r.t. possible message loss.

Acknowledgment

This work has been done within the Research Project 2006 funded by Cassa di Risparmio di Livorno, Lucca e Pisa, and the NoE HYCON (IST-2004-511368).

References

1. Wang, G., Cao, G., La Porta, T.: Movement-assisted sensor deployment. IEEE Trans. on Mobile Computing (6), 640–652 (2006)
2. Wang, G., Cao, G., La Porta, T.: A bidding protocol for deploying mmobile sensors. In: IEEE Int. Conf. on Network Protocols (ICNP) (2003)
3. Cortes, J., Martinez, S., Karatas, T., Bullo, F.: Coverage control for mobile sensing networks. IEEE Trans. on Robotics and Automation, 243–255 (April 2004)
4. Sibley, G., Rahimi, M., Sukhatme, G.: Robomote: a tiny mobile robot platform for large-scale ad-hocsensor networks. In: Int. Conf. on Robotics and Automation (2002)
5. Rimon, E., Koditschek, D.: Exact robot navigation using artificial potential functions. IEEE Trans. on Robotics and Automation (1992)
6. LaValle, S.: Planning Algorithms, 1st edn. Cambridge University Press, New York (2006)
7. Vleugels, J., Overmars, M.: Approximating voronoi diagrams of convex sites in any dimension. Int. J. of Computational Geometry & Applications (IJCGA) (2), 201–221 (1998)
8. Halliday, D., Resnick, R., Krane, K.S.: Physics, 5th edn., vol. 1. John Wiley and Sons, New York (2002)
9. Fagiolini, A., Tani, L., Bicchi, A., Dini, G.: Decentralized deployment of mobile sensors for optimal connected sensing coverage. Technical report (available on demand from authors) (January 2008)

Data Collection in Wireless Sensor Networks for Noise Pollution Monitoring

Luca Filipponi[1], Silvia Santini[2,*], and Andrea Vitaletti[1,**]

[1] Dipartimento di Informatica e Sistemistica "A. Ruberti"
SAPIENZA Università di Roma, Rome, Italy
{filipponi,vitale}@dis.uniroma1.it
[2] Institute for Pervasive Computing
ETH Zurich, 8092 Zurich, Switzerland
santinis@inf.ethz.ch

Abstract. Focusing on the assessment of environmental noise pollution in urban areas, we provide qualitative considerations and experimental results to show the feasibility of wireless sensor networks to be used in this context. To select the most suitable data collection protocol for the specific noise monitoring application scenario, we evaluated the energy consumption performances of the CTP (Collection Tree Protocol) and DMAC protocols. Our results show that CTP, if used enabling the LPL (Low Power Listening) option, provides the better performances trade-off for noise monitoring applications.

1 Environmental Noise Monitoring

Conservative estimations give in about 300 millions the number of citizens within the European Community that are exposed to alarming levels of noise pollution [1]. Raising the public's awareness of this problem, the Directive 2002/49/EC of the European Parliament has made the avoidance, prevention, and reduction of environmental noise a prime issue in European policy. To better assess the extension of the problem, the European Commission required member states to regularly provide an accurate mapping of environmental noise levels for all urban areas with more than 250'000 inhabitants. While current noise maps are mostly based on sparse data and ad-hoc noise propagation models, a recent position paper by the Commission has stressed that *"every effort should be made to obtain accurate real data on noise sources,"* [2, p.6]. The demand for accurate data about noise exposure levels will increase dramatically, as this statement makes its way into mandatory regulation. Nowadays noise measurements in urban areas are mainly carried out by designated officers that collect data at a location of interest for successive analysis and storage, using a sound level meter or similar device. This manual collection method does not scale as the demand for higher

* Partially supported by the Swiss National Science Foundation (NCCR-MICS, grant number 5005-67322).
** Partially supported by the FRONTS FET Project (215270).

S. Nikoletseas et al. (Eds.): DCOSS 2008, LNCS 5067, pp. 492–497, 2008.

granularity of noise measurements in both time and space increases. Instead, a network of cheap wireless sensor nodes deployed over the area of interest could collect noise pollution data over long periods of time and autonomously report it to a central server through the sensor's on-board radio, requiring human intervention only to install and possibly subsequently remove the sensing devices. Collected noise data is typically stored in a land register and used, together with additional information about existing noise sources, to feed computational models that provide extrapolated noise exposure levels for those areas for which real data is unavailable. Even if this assessment procedure is still compliant with European regulations, today's computational models often fail to provide accurate estimations of the real noise pollution levels[1]. Indeed, while the free propagation properties of noise generated from typical noise sources are well understood [3], shadowing and reflection effects hinder accurate estimation of noise levels in complex urban settings. The accuracy of computed noise levels could be easily verified and improved by installing a wireless sensor network at those locations for which computational models are likely to provide inaccurate estimations.

2   Requirements

Before going into further details we would like to summarize the main requirements a wireless sensor network must comply with to be used for noise pollution monitoring applications.

Hardware. The high sampling rate required to properly capture acoustic signals ($\sim$ 32kHz) appears prohibitive for resource poor sensor nodes. However, commercially available platforms are able to support the required sampling rate, as long as scheduling with radio communication is properly managed. Nevertheless, to overcome this problem the most suitable solution consists in delegating sampling and signal processing to dedicated hardware and let the actual sensor node only deal with communication and possibly optimization of data collection.

Sampling. Noise levels are time-weighted averages of acoustic power, computed over variable time intervals. Sampling of noise levels may occur at sampling rates varying from fractions up to multiples of 1 Hz. For the preparation of noise maps, a spacing of about 3 meters between sampling points (i.e. sensor nodes) is recommended [2]. Observe that this results in a quite dense network deployed on a relatively small area (e.g., the internal façade of a building).

Data rate and latency. Collection of noise data does not require sensor readings to be immediately reported to the sink. Therefore, latency in packet delivery is a secondary issue for the optimization of network performances. In typical scenarios, sensor nodes generate 1 value/sec and a single packet can convey several noise samples by means of aggregation.

---

[1] The authors are indebted to Hans Huber and Fridolin Keller of the department for environmental noise protection of the city of Zurich, who pointed this out in a personal in-depth interview.

Network lifetime. To cover typical variability patterns of noise levels, measurements should ideally extend for few weeks.

Network topology. For the purpose of noise mapping, the assessment points used to measure noise levels (thus, the physical topology of the network) shouldn't change during data collection. Nevertheless, due to the spatial distribution of the nodes, reporting data to a central sink may require multi-hop communication.

Synchronization. Noise readings collected by different nodes must be ordered over a global timescale for proper processing and visualization. Since the specific network topology and data collection protocol may introduce a variable and unbounded latency on data delivery, an adequate synchronization mechanism should be adopted to allow for a correct time ordering.

3   Data Collection

To select an adequate platform to collect noise levels data, we tested the feasibility of three different hardware solutions. We considered the *Tmote Sky* prototyping platform from Moteiv Corp. equipped with either the *EasySen SBT80* multi-modality sensor board or with a custom-made noise level meter. Furthermore, we experimented with the *Tmote invent* platform, also from Moteiv Corp., which features an on-board microphone, as well as a powerful signal conditioning circuitry [4]. Taking care of sampling and processing the captured acoustic signal, the customized noise level meter behaves as an external sensor able to output noise level readings expressed in dB (with total nominal error less than 3 dB). Since the noise level is actually a time-weighted average of the captured acoustic power, its value can be sampled at much lower frequency (e.g., 1 Hz) than the acoustic signal itself. The use of a customized noise level meter allows therefore to remove the burden of computational and energy expensive operations from the sensor node itself, and represents our preferred solution at this first prototyping stage.

After being captured, noise level data needs to be transmitted to a central sink for permanent storage and further processing, imposing a typical convergecast pattern on network communication. While the design of energy-efficient medium access control received considerable attention within the wireless sensor networks research community [5], only few collection layer implementations are currently available, these including MintRoute [6] and CTP (Collection Tree Protocol) [7]. In particular, CTP is an implementation of BMAC [8] with an optional low power listening (LPL) option on trees. LPL is a power saving technique that allows to move the major costs of radio communication from receivers to transmitters by avoiding idle listening, which is known to be the main source of energy wasting in wireless sensor networks communication. To understand the performances of different data collection protocols in our specific application scenario, we evaluated both the CTP and the DMAC convergecast protocol [9]. We consider the comparison of these two protocols particularly interesting since, at the media access level, they respectively implement a contention-based and TDMA-based (Time Division Multiple Access) approach, which are the mostly exploited MAC techniques in wireless sensor networks. We embedded both implementations of the data collection layer in our software prototype and analyzed their performances in terms of energy consumption. While an implementation of CTP is available in the tinyOS-2.x repository [7], we implemented the DMAC protocol on our own.

4 Assessment and Analysis of Protocols' Performances

To analyze the energy consumption of the two data collection protocols under consideration, we connected a Tmote sensor node to the Rhode & Schwarz dual-channel analyzer/power supply NGMO2,an instrument that can accurately measure the current drain associated with all the states of a node. We then measured the node current drain during transmission and reception states, for both the CTP and DMAC protocols, using the same experimental set-up exploited in [10]. This simple setting, a single-hop network made of two nodes, allowed us to gain a clear understanding of all the major energetic aspects involved in nodes communications and to determine upper and lower bounds on nodes energy consumption. A more realistic experimental setting, considering multi-hop topologies and the effect of collisions, will be considered in further investigations. We would like to point out that measurements of power consumption in wireless sensor networks typically rely on indirect measurement methods, such as counting the number of transmitted packets or CPU duty cycles, which provide limited accuracy as pointed out in [11]. Our work aligns with few other examples in providing direct power consumption measurements [10,11,12]. Figure 1(a) shows the current consumption of a node running the CTP_{NoLPL} protocol, characterized by the transmissions spikes, occurring every second. While in idle listening, the node drains about 19mA, which represents a clear energy waste considering that the total average current drain is about 19.5 mA. Enabling the LPL option allows to dramatically reduce the power consumption of the CTP protocol, as shown in figure 1(c). The current drain while the radio is in sleep mode is negligible. Every 250ms (the value we set for nodes sleep period), the node wakes up the radio and samples the channel, as foreseen by the LPL mechanism. If it overhears a transmission on the medium, it keeps the radio active until reception is successfully completed, otherwise switches-off the radio immediately. The spikes in figure 1(c) are associated to the sampling activity, while one complete reception cycle is clearly visible in the first segment of the plot. The length of the transmission phase is considerably longer than the one observed for CTP_{NoLPL}, since LPL moves the cost of communication from the receiver to the transmitter. Nevertheless, under the same traffic load of one packet per second, the total average current drain of the CTP_{LPL} protocol is only 5mA, which represents an energy saving of 75% with respect to the 19.5mA spent on average by the CTP_{NoLPL}. The CTP_{LPL} protocol can achieve even lower average current drains using longer sleep intervals that, however, will also increase packet latency. For both CTP_{NoLPL} and CTP_{LPL}, the measured current drain values represents lower bounds on the protocol's energy consumption, since in the considered experimental setting no collisions and packet forwarding occur.

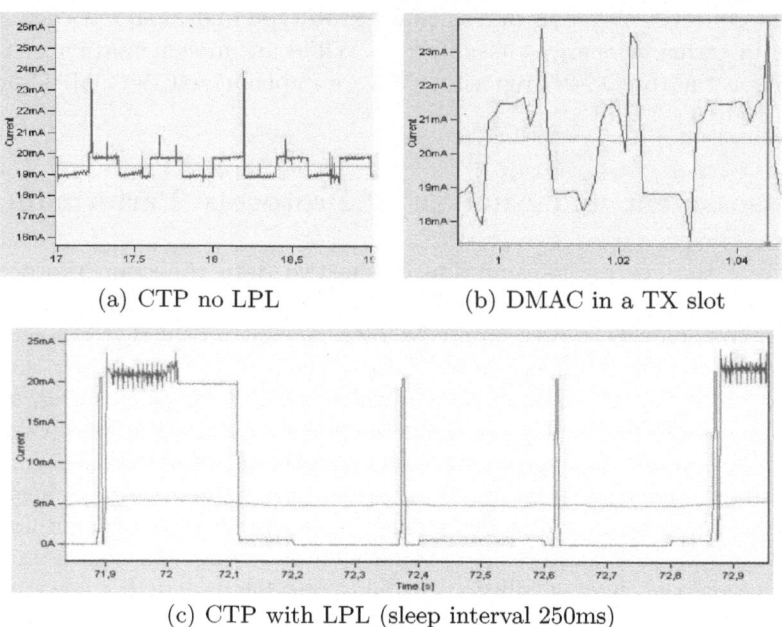

(a) CTP no LPL   (b) DMAC in a TX slot

(c) CTP with LPL (sleep interval 250ms)

Fig. 1. Current drain

For the DMAC protocol, the lower bound on energy consumption simply corresponds to the energy required to transmit (receive) a single packet per period. Determining the upper bound is less trivial and requires some additional considerations. Nodes running the DMAC protocol can, at each time instants, be in sleep, receive or transmit state. The power consumption in sleep state is negligible, so we excerpt it from the current analysis. To provide an upper bound on DMAC's power consumption, we forced nodes in transmission state to transmit for an entire period, and consequently nodes in receive state to remain active for the duration of such a period. Figure 1(b) shows the current drain of the node in transmission state, in which the average current consumption is about 20.5mA (about 21.5mA during reception). A correct estimate of the average energy consumption, requires to take into consideration the time spent in sleep state. If we consider a 10-layers network, with communication periods of 100ms, the DMAC's *time-weighted* average energy consumption is approximatively the same as for CTP_{LPL}.

5   Conclusions

Wireless sensor networks can provide a cheap and flexible infrastructure to support the collection of fine-grained noise pollution data, which is essential for the preparation of noise maps and for the validation of noise pollution models. Besides testing commercially available sensor platforms, we designed and developed a customized noise level meter that allows us to delegate costly noise levels computations to dedicated hardware. We then selected two data collection protocols, CTP (considered both with and without the LPL option) and DMAC, and performed direct measurements of their energy consumption using a simple, though representative, network topology. The CTP_{NoLPL} protocol exhibits the highest energy consumption (and lowest latency), while enabling the LPL option allows to save about 75% of the total energy (though at the cost of increased latency). DMAC's observed performances are comparable to those of CTP_{LPL}, though they depend on the number of tree levels the protocol builds up for routing. For our specific application scenario we eventually selected the simple CTP_{LPL} as the most suitable data collection protocol.

References

1. European Commission Green Paper: Future noise policy. Com (96) 540 final (November 1996)
2. European Commission Working Group Assessment of Exposure to Noise (WG-AEN): Good practice guide for strategic noise mapping and the production of associated data on noise exposure (January 2006)
3. Bies, D.A., Hansen, C.H.: Engineering Noise Control: Theory and Practice, 3rd edn. Spon Press (Taylor &Francis Group), London and New York (2003)
4. Santini, S., Ostermaier, B., Vitaletti, A.: First experiences using wireless sensor networks for noise pollution monitoring. In: Proceedings of the 3rd ACM Workshop on Real-World Wireless Sensor Networks (REALWSN 2008), Glasgow, United Kingdom, ACM Press, New York (2008)
5. Demirkol, I., Ersoy, C., Alagöz, F.: Mac protocols for wireless sensor networks: a survey. IEEE Communications Magazine 44(4), 115–121 (2006)
6. Woo, A., Tong, T., Culler, D.: Taming the underlying challenges of reliable multi-hop routing in sensor networks. In: Proceedings of the 1st international conference on Embedded networked sensor systems (SenSys 2003), ACM, New York (2003)
7. Fonseca, R., Gnawali, O., Jamieson, K., Kim, S., Levis, P., Woo, A.: Tinyos enhancement proposals (tep) 123: The collection tree protocol (ctp)
8. Polastre, J., Hill, J., Culler, D.: Versatile low power media access for wireless sensor networks. In: Proceedings of the 2nd international conference on Embedded networked sensor systems (SenSys 2004), pp. 95–107. ACM, New York (2004)
9. Lu, G., Krishnamachari, B., Raghavendra, C.S.: An adaptive energy-efficient and low-latency mac for data gathering in wireless sensor networks. In: Int. Workshop on Algorithms for Wireless, Mobile, Ad Hoc and Sensor Networks (WMAN), Santa Fe, NM, USA (April 2004)
10. van Dam, T., Langendoen, K.: An adaptive energy-efficient mac protocol for wireless sensor networks. In: Proceedings of the 1st international conference on Embedded networked sensor systems (SenSys 2003), New York, USA, pp. 171–180 (2003)
11. Landsiedel, O., Wehrle, K., Götz, S.: Accurate prediction of power consumption in sensor networks. In: Proceedings of the Second IEEE Workshop on Embedded Networked Sensors (EmNetS-II), Sidney, Australia (May 2005)
12. Hohlt, B., Doherty, L., Brewer, E.: Flexible power scheduling for sensor networks. In: Proceedings of the third international symposium on Information processing in sensor networks (IPSN 2004), pp. 205–214. ACM, New York (2004)

Energy Efficient Sleep Scheduling in Sensor Networks for Multiple Target Tracking

Bo Jiang[1], Binoy Ravindran[1], and Hyeonjoong Cho[2]

[1] ECE Dept., Virginia Tech
{bjiang,binoy}@vt.edu
[2] ETRI, Daejoen, South Korea
raycho@etri.re.kr

Abstract. This paper presents an energy-aware, sleep scheduling algorithm called SSMTT to support multiple target tracking sensor networks. SSMTT leverages the awakening result of interfering targets to save the energy consumption on proactive wake-up communication. For the alarm message-miss problem introduced by multiple target tracking, we present a solution that involves scheduling the sensor nodes' sleep pattern. We compare SSMTT against three sleep scheduling algorithms for single target tracking: the legacy circle scheme, MCTA, and TDSS. Our experimental evaluations show that SSMTT achieves better energy efficiency than handling multiple targets separately through single target tracking algorithms.

1 Introduction

Target tracking in surveillance systems is one of the most important applications of wireless sensor networks (WSNs) [1]. Some of the earlier applications in this domain have focused on rare-event tracking where targets typically enter the surveillance field one by one [2,3]. Recently, many interesting applications have emerged, which require concurrent tracking of multiple targets—e.g., search and rescue, disaster response, pursuit evasion games [4]. As one of the critical mechanism for WSNs' energy efficiency, sleep scheduling is still a challenging problem for multiple target tracking systems. This is because, it is generally difficult to optimize energy efficiency and simultaneously track all targets without missing any, when multiple targets concurrently intrude the surveillance field.

In this paper, we present an energy-aware, Sleep Scheduling algorithm for Multiple Target Tracking (or SSMTT). Our objective is to improve the energy efficiency through a sleep scheduling approach that is conscious of concurrently tracking multiple targets, in contrast to an approach which is not.

In the proactive wake-up mechanism for target tracking [5], a node that detects a target (i.e., the "root" node) broadcasts an alarm message to activate its neighbor nodes (i.e., member nodes) toward preparing them to track the approaching target. When the routes of multiple targets interfere with each other, some neighbor nodes of a root node may have already been activated by another root node's alarm broadcast. If the wake-up mechanism is carefully designed,

S. Nikoletseas et al. (Eds.): DCOSS 2008, LNCS 5067, pp. 498–509, 2008.
© Springer-Verlag Berlin Heidelberg 2008

such overlapping broadcasts can be saved, thereby saving associated energy costs on communication. SSMTT uses such a mechanism, and builds upon the tracking subarea management and sleep scheduling algorithms in [6]. A consequence of this multi-target-conscious energy efficiency mechanism is that some sensor nodes may be put into the sleep state by the alarm broadcast for a target, and thereby, they may miss the alarm broadcast for another. We present a solution to this problem by modifying the node sleep patterns.

Most of the past works on multiple target tracking aim at differentiating multiple targets from each other i.e., identifying "who is who" [7,8], or improving data fusion [9, 10]. Some of those works which also consider improving energy efficiency do not consider sleep scheduling [11]. On the other hand, most efforts on sleep scheduling do not explicitly support concurrent tracking of multiple targets [12, 13]. Liu et al. utilize multiple targets as tracking objects in their simulation studies [14]. However, this work aims at guaranteeing the quality of traffic towards the base station instead of tracking along the target's route. To the best of our knowledge, this work is the first effort to enhance energy efficiency through sleep scheduling for multiple target tracking systems.

The paper makes the following contributions: (1) We present a sleep scheduling algorithm for concurrently tracking multiple targets; (2) we further enhance energy efficiency by leveraging the awakened sensor nodes to save more energy than tracking multiple targets separately with single target tracking algorithms; and (3) we provide a simple solution for the alarm message-miss problem.

The results from our experimental evaluations show that, compared with sleep scheduling algorithms for single target tracking, the SSMTT algorithm saves alarm transmission energy by 10% ~ 15%.

The rest of the paper is organized as follows: In Section 2, we describe our assumptions and formulate the problem. Section 3 describes the rationale and design of SSMTT. In Section 4, we present detailed algorithm descriptions. Section 5 reports our evaluation results. In Section 6, we conclude and discuss future work.

2   Assumptions and Problem Formulation

2.1   Assumptions

Among the most critical assumptions are the following:

- *Node location.* Each node knows, a priori, its own position. This knowledge can be obtained during the system's initialization phase via GPS [15] or using algorithmic strategies such as [16].
- *Target's instantaneous status.* Nodes can determine a target's movement status including its position, instantaneous velocity magnitude, and direction, either by sensing or by calculating—e.g., [17, 18, 19].
- *Target identification.* Multiple targets are assumed to be distinguished from each other using a multiple target tracking algorithm such as [7].

- *Radio transmission power.* We assume that the transmission power of sensor nodes' communication radio can be adjusted to reach different distances based on popularly used radio hardware such as CC1000 [20] and CC2420 [21]. The energy consumption with variable transmission power is determined with a model developed using curve fitting based on the empirical measurements in [22].
- *Sleep level.* Nodes are assumed to operate in two states: S_0 (active) and S_1 (sleep). A node may sample and process signals, and transmit and receive packets in state S_0. In state S_1, all modules/devices of a node will be in the sleep state, except a wake-up timer that has very low energy consumption [5].
- *Sleep pattern.* We assume that the default sleep pattern is "random" i.e., before sleep scheduling is triggered by a target detection event, all the nodes switch between active and sleep modes with the same *toggling cycle* (TC) and the same *duty cycle* (DC). However, the boundaries of each node's toggling cycle are random. In each period, a node wakes up and remains active for $TC * DC$, and then sleeps for $TC * (1 - DC)$ [5].

2.2 Problem Formulation

First, we discuss the modeling of the sensor network and multiple interfering targets. We use the term *tracking subarea* to describe a sensor node set consisting of a root node and some member nodes that are awakened by the root node. Let a tracking subarea be denoted as A, a root node as γ, and member nodes as $m_0, m_1, ..., m_k$. Thus, $A = \{\gamma, m_0, m_1, ..., m_k\}$. The root node γ will broadcast an alarm message to schedule the sleep pattern of its neighbor nodes (i.e., member nodes) upon detecting a target.

Let's assume that n targets $\{T_i | i \in [0, n-1]\}$ interfere with each other. Each target will trigger an alarm message and thus form a tracking subarea. Let the tracking subareas triggered by target T_i be denoted as $\{A_{ip} | p \in N_0\}$, and their root nodes as $\{\gamma_{ip} | p \in N_0\}$, where N_0 is the non-negative integer set. Then, at the time that A_{ip} is formed, A_{ip} may overlap with $\{A_j q | j \in [0, i-1] \cup [i+1, n-1], q \in N_0\}$.

Now, we define the criteria for deciding whether or not SSMTT improves energy efficiency compared with tracking multiple targets separately with a single target tracking algorithm, and how much it improves.

Since our basic idea is to enhance energy efficiency by leveraging the targets' interference, the energy consumption is a critical aspect in making this decision. When two targets are far away from each other such that there is no node that can detect both of them at the same time, they can be handled as two single targets with single target tracking and sleep scheduling algorithms. Therefore, we only consider the difference in energy efficiency during the period when multiple targets interfere.

We define the interference period of two targets T_i and T_j as a time interval. This interval starts when the root nodes for T_i and T_j are close enough to hear each other for the first time ($|\gamma_{ip}\gamma_{jq}| \le R, (p, q \in N_0)$), and ends when the root nodes of T_i and T_j are away from each other for the first time ($|\gamma_{ip}\gamma_{jq}| \geq R, (p, q \in N_0)$).

Let the tracking energy consumed during the interference period be denoted as E_s when tracking multiple targets through single target tracking algorithms, and as E_m when tracking with SSMTT. We define the benefit that can be obtained by SSMTT as Energy Saving Ratio (or ESR), where $ESR = \frac{E_s - E_m}{E_s}$.

Detection delay is one of the most important performance feature of many surveillance sensor networks [23,24]. Since concurrently tracking multiple targets requires to guarantee the overall quality for all the targets, we use the metric, Average Detection Delay (or ADD) for measuring the average delay per target, where $ADD = \frac{1}{i} \sum_i t_i$ and t_i is the total time that target T_i is not covered by any active nodes during the interference period.

We define another metric, Tracking Degree (or TD), to evaluate tracking performance. TD is defined as the percentage of the route length of a target that is covered by successful nodes divided by the target's total route length. TD can be used to measure the probability for detecting the target (the overall detection probability is generally difficult to directly measure). TD is given by $TD = \frac{\sum_i u_i}{\sum_i L_i}$, where u_i is T_i's route length which is not covered by any active nodes, and L_i is the total route length of target T_i during the interference period.

Given these metrics, we formulate our problem as: *how to schedule the node sleep patterns and leverage the overlapping broadcasts for multiple targets, to achieve better ESR with acceptable ADD and TD loss.*

3   SSMTT Algorithm Design

The SSMTT algorithm is built upon one of our previous works, the TDSS algorithm [6]. The basic ideas of the TDSS algorithm include: (1) describe the target movement, especially its potential moving directions, with a probabilistic model; (2) manage tracking subareas to reduce the number of proactively awakened nodes; and (3) schedule the sleep patterns of the subarea member nodes to shorten their active time. Based on the TDSS framework, the SSMTT algorithm leverages the overlapping broadcasts for multiple targets to reduce the energy consumed on proactive wake-up alarm transmission.

3.1   Introduction of TDSS Algorithm

We first provide a concise introduction to make the paper self-contained.

In the real world, it is difficult to accurately predict a target's movement based on a physics-based model even for a short term. We consider probabilistic models, e.g., Linear Distribution Model (or LDM) to approximately imitate the actual target motion. Based on the target's instantaneous moving direction θ_0, we define the probability $p(\theta)$ with which the target moves along the direction $\theta_0 + \theta$ ($\theta \in (-\pi, +\pi]$) as follows.

$$p(\theta) = \begin{cases} a\theta + b, \theta \in (-\pi, 0] \\ -a\theta + b, \theta \in (0, -\pi] \end{cases}$$

Here, a and b ($a > 0, b > 0$) are constants specific for a given application.

Usually, a sensor node's communication radio range R is far longer than its sensing range r. However, only some nodes within the range R can detect the target, and others' energy consumed for being active is wasted. A more effective approach is to determine a node subset among all the neighbors and form a tracking subarea to reduce the number of awakened nodes.

Each receiver node decides a subarea's scope by determining whether or not it is in the subarea's scope. When a sensor node receives an alarm message, it computes its distance D from the root node and compares it with $d(\theta)$, where $d(\theta)$ is the subarea's radius along the direction θ. If $D < d(\theta)$, the node knows that it is a member of the subarea. The subarea radius $d(\theta)$ is decided based on both v and $p(\theta)$. For LDM, $d(\theta)$ is derived as $d(\theta) = \frac{p(\theta)}{p(\theta=0)}d(\theta = 0) = \frac{(-a|\theta|+b)}{b} \cdot (m \cdot v \cdot b + n)$.

In a tracking subarea, not all the awakened sensor nodes need to be active all the time. By scheduling their sleep pattern, we can save more energy than a single approach of reducing the number of awakened nodes. Once a node receives an alarm message and decides that it is one of the tracking subarea members, the node will enter the sleep state until the expected target arrival time. Then the node wakes up and changes its sleep pattern to the scheduled mode, in which it toggles the sleep/active states with a high duty cycle. Its sleep pattern will recover to the default mode step by step until the expected target leaving time. We define Scheduled Period (or SP) as the period that the node takes the scheduled sleep pattern instead of the default one, and Scheduled Active Period (or SAP) as the period that the node does not sleep completely within a SP.

3.2   Energy Saving for Proactive Wake-Up Alarm Transmission

Let the interfered target be denoted as T_i, and the interfering targets be denoted as $\{T_j | j \in [0, i-1] \cup [i+1, n-1]\}$. If just before a detection event, the root node γ_{ip} received alarm messages triggered by T_j, its tracking subarea A_{ip} may overlap with subareas $\{A_{jq} | j \in [0, i-1] \cup [i+1, n-1], q \in N_0\}$. If most of the member nodes in the overlapped area have already been awakened by $\{\gamma_{jq}\}$, then it will not be necessary for γ_{ip} to awaken them once again.

Now, we discuss the detailed definition and calculation of the criteria for γ_{ip} on saving this transmission energy. First, we start with the case of two targets.

Figure 1 shows the interference between two targets T_i and T_j. In the figure, the smallest circles are sensor nodes, and the dotted circles around them denote their sensing range. Assume that γ_{ip} just received an alarm message from γ_{jq} before it detects T_i, and it finds that there would be an overlapping area between A_{ip} and A_{jq} (the dashdotted line describes the alarm message from γ_{jq} to γ_{ip}). Then γ_{ip} divides A_{ip} into zones $\{Z_k | k \in N_0\}$ based on the distance from γ_{ip}, i.e., a point $P \in Z_k \Leftrightarrow |\overrightarrow{\gamma_{ip}P}| \in (kr, (k+1)r]$, where r is the sensing range of the sensor nodes. In the figure, the arcs $\{\widetilde{B_k} | k \in N\}$ are the boundaries between adjacent zones. The reason we divide the zones by the distance r is to facilitate the discussion on the criterion for saving the transmission.

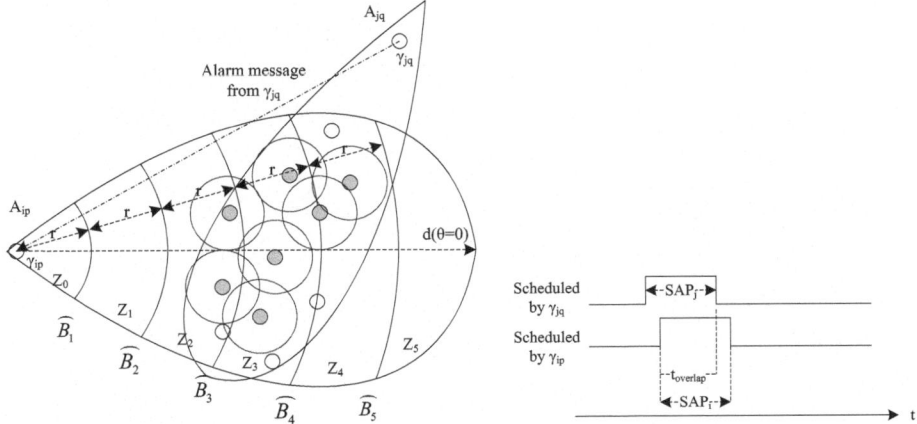

Fig. 1. Interference between Two Targets

Fig. 2. Node Overlapping Ratio

Assume that in the overlapping area of A_{ip} and A_{jq}, there are m nodes, called Overlapping Nodes (or ON) and denoted as $U_{ON} = \{ON_k | k \in [0, m-1]\}$. Now for each ON_k, we define the metric Node Overlapping Ratio (or NOR), as $NOR_k = \frac{t_{overlap}}{SAP_i}$, to describe the overlapping degree of SAP_i and SAP_j. Here SAP_i and SAP_j are the SAPs scheduled respectively for the two targets, $t_{overlap}$ is the overlapping time of SAP_i and SAP_j. Figure 2 shows their relationship.

We call those ONs whose $NOR > THS_{NOR}$ as Reusable Nodes (or RN) and denote them as $U_{RN} = \{RN_k | k \in [0, m-1]\}$, where THS_{NOR} is a threshold specific for a given implementation. In Figure 1, the RNs are shown with solid gray circles. If the sensing coverage area of RNs in a zone Z_k is large enough so as to cover most of Z_k's area, the member nodes in Z_k may be omitted in γ_{ip}'s alarm broadcasting. For each zone Z_k, we define the metric Zone Overlapping Ratio (or ZOR) to describe the overlapping degree of two targets in this zone. The overlapping ratio of Z_k, denoted ZOR_k, is calculated as:

$$ZOR_k = \frac{\bigcup S_{RN}}{S_{all}} \qquad (1)$$

Here S_{RN} is the covered area in Z_k of all the Z_k's RNs, and S_{all} is the total area of Z_k. Figure 3 shows the definition of ZOR_k, in which we use Z_3 in Figure 1 as the example. In Figure 3, $\bigcup S_{RN}$ is shown as the dotted area, and S_{all} equals to the area of $\overline{ABCD}$. We call those zones whose $ZOR_k > THS_{ZOR}$ as Reusable Zones (or RZ) and denote them as $U_{RZ} = \{Z_k | k \in N_0\}$, where THS_{ZOR} is a threshold specific for a given implementation.

The core idea of the energy saving effort for proactive wake-up alarm transmission is (1) to cancel the alarm broadcast completely if a zone that is close to the root node is reusable, and (2) to reduce the transmission power of the alarm broadcast if a zone that is far from the root node is reusable and all the zones that are closer to the root node than it are not reusable.

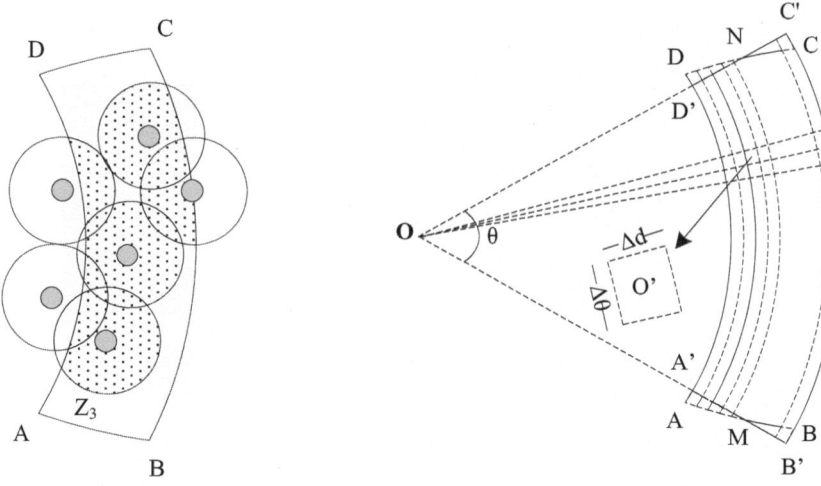

Fig. 3. Definition of ZOR   **Fig. 4.** Calculation of Covered Area

Based on the case of two targets interference, we calculate ZOR_k using Equation 1 for the multiple target case, too. However, the difference is that the RNs for calculating $\bigcup S_{RN}$ includes the reusable nodes in all of the overlapping areas of A_{ip} and each interfering target.

Next, we present the calculation of $\bigcup S_{RN}$. To reduce the computational complexity, we adopt an approximate approach and again discuss with Z_3 in Figure 1 as the example. In Figure 4, $Z_3 = \overline{ABCD}$ is determined by the tracking subarea A_{ip}'s edges, and the boundaries $\widetilde{B_3}$ and $\widetilde{B_4}$. O is the position of γ_{ip}. M and N are points on the intersection of A_{ip}'s edges and a circle with center O and radius $3.5r$ (i.e., $|OM| = |ON| = 3.5r$). θ is the central angle corresponding to the arc $\widetilde{MN}$. Now, A', B', C', and D' are points on the intersection of line OM, line ON, and the boundaries $\widetilde{B_3}$ and $\widetilde{B_4}$.

We use the area $\overline{A'B'C'D'}$ as the approximation of $\overline{ABCD}$, and divide $\overline{A'B'C'D'}$ evenly into small "sectors" with concentric circles centered at O and lines through O. Here, sectors are not mathematical sectors but more like disk sectors in the context of computer disk storage. Each sector has a segment with the radius $\Delta d = \frac{r}{a}$ and corresponds to a central angle $\Delta\theta = \frac{\theta}{b}$, where a and b are constants specific for a given implementation. Let a sector be denoted as $C(d, \theta)$. In polar coordinates with the radial coordinate ρ and the polar angle α, $C(d, \theta)$ is determined by the circle $\rho = d$, circle $\rho = d + \Delta d$, line $\alpha = \theta$, and line $\alpha = \theta + \Delta\theta$. Sector $C(d, \theta)$'s central point O' (i.e., $(d + \Delta d/2, \theta + \Delta\theta/2)$ in polar coordinates) is used as the representative of the sector. If O' is covered by a sensor node's sensing range, we consider that $C(d, \theta)$ is covered by this node.

Now, we approximate the calculation of ZOR_k in Equation 1 as $ZOR_k = \frac{SN_{RN}}{SN_{total}}$, where SN_{RN} is the number of sectors that are covered by reusable nodes, and SN_{total} is the total number of sectors in a zone.

Algorithm 1: Calculation of Reusable Zones

```
 1  for each entry in AMD do
 2  │   Decide U_ON;                        /* Calculate the set of overlapping nodes
    │   */
 3  U_RN = φ;                               /* Initialize the set of reusable nodes */
 4  for each ON_k in U_ON do
 5  │   Calculate NOR_k;
 6  │   if (NOR_k > THS_NOR) then
 7  │   │   U_RN = U_RN + ON_k;             /* Calculate the set of reusable nodes */
 8  U_RZ = φ;                               /* Initialize the set of reusable zones */
 9  for each zone Z_k in A_ip do
10  │   SN_RN = 0;
11  │   SN_total = 0;
12  │   for each sector C(λ, α) in Z_k do
13  │   │   SN_total + +;                   /* Summarize all the sectors in a zone */
14  │   │   if central point O' of C(λ, α) is covered by a RN_k (RN_k ∈ U_RN) then
15  │   │   │   SN_RN + +;                  /* Account for the sectors that are
    │   │   │   covered by reusable nodes */
16  │   ZOR_k = SN_RN / SN_total;
17  │   if (ZOR_k > THS_ZOR) then
18  │   │   U_RZ = U_RZ + Z_k;              /* Calculate the set of reusable zones */
19  return U_RZ;
```

3.3 Preventing Alarm Messages from Being Missed

The consequence of the SSMTT mechanism is that, when a sensor node is scheduled to sleep by an alarm message, it may miss the alarm message broadcast for other approaching targets. Our solution to this problem is to force the member node, which has been scheduled to sleep until the expected target arrival time, to wake up with the default toggling cycle and an extremely short duty cycle. The only purpose of this is to check alarm messages from other approaching targets.

4 SSMTT Algorithm Description

To record each approaching target, a sensor node needs to manage an Alarm Message Database (or AMD). Each entry in the AMD records all the information transmitted by an alarm message including target ID, subarea ID, γ's position, the target's instantaneous movement status, and TTL (i.e., time to live) et al. The AMD is updated whenever the node receives an alarm message, irrespective of whether it will become a member of the tracking subarea.

When a node wakes up, it changes its sleep pattern according to the scheduled result and sets the wake-up timer for the subsequent wake-up. During this active period, it may detect a target or receive an alarm message, and corresponding interrupt handlers for them will be released for execution. The main function of the SSMTT algorithm is implemented in Algorithm 1.

5   Performance Evaluation

A. Simulation Environment. We evaluated SSMTT algorithm against three sleep scheduling algorithms for single target tracking: the legacy circle-based proactive wake-up scheme (or CIRCLE) [5], MCTA awakened node reducing algorithm [25], and TDSS. As discussed in Section 1, the reason that we did not compare SSMTT with other sleep scheduling algorithms with multiple target tracking support is because, we are not aware of any previous works in this area.

In the evaluation, 400 nodes were deployed in a 20×20 grid structure in a $100m \times 100m$ area. We tracked $2 \sim 10$ targets concurrently, which move with the speeds $\{3, 6, 9, 12, 15, 18, 21, 24, 27\}$ in Uniform Rectilinear Motion (URM) mode. For each combination of algorithm, number of target, target speed, and interfering angle, we simulated 50 cases and reported the average results. The energy consumption data comes from the actual Mica2 platform [22, 26]. As discussed previously, the transmission power is developed from the curve fitting based on the empirical measurements.

B. Simulation Results. Figure 5 shows ESR on alarm communication of SSMTT over the other three reference single target tracking algorithms under different numbers of targets. We can observe that the energy saved increases as the number of interfering targets increases. This is because that the more targets interfere, the more overlapping broadcasts can be saved. Although TDSS algorithm's energy consumption on alarm transmission is the most, its overall energy consumption is still better than CIRCLE and MCTA [6]. Figure 6 shows ADD of the four simulated algorithms under different numbers of targets. Both TDSS and SSMTT introduce an increasing detection delay. However, this performance loss is acceptable, since in most of the cases the increased delay is within 0.1 second for each target per interference. Figure 7 shows TD of the four simulated algorithms under different numbers of targets. SSMTT introduces a little decrease on the tracking coverage. Similar to ADD, this performance loss is negligible compared with the ESR enhancement.

We also studied the correlation between ESR and the target speed. In the two targets case, the correlation is shown in Figure 8. Basically ESR decreases as the target speed increases. This is because that the randomness of targets will increase significantly as the speed increases, therefore the overlap among interfering targets' scheduling will decrease. Figure 9 shows the correlation between ESR and the interfering angle for two targets case. We can observe that little interfering angle for two targets moving on the same direction presents the best energy saving ratio, and the worst case occurs when the interfering angle is close to $\frac{2\pi}{3}$.

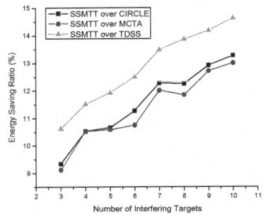

Fig. 5. ESR vs. Number of Targets

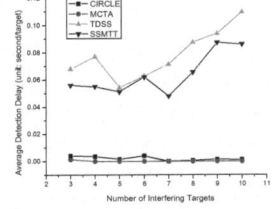

Fig. 6. ADD vs. Number of Targets

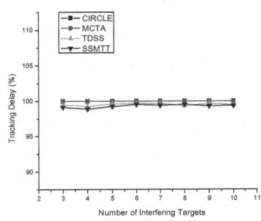

Fig. 7. TD vs. Number of Targets

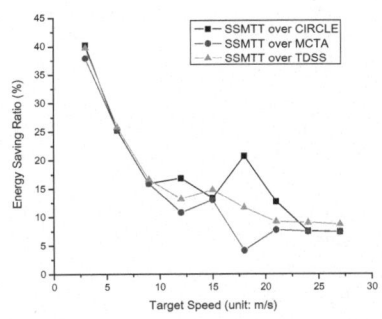

Fig. 8. ESR vs. Target Speed

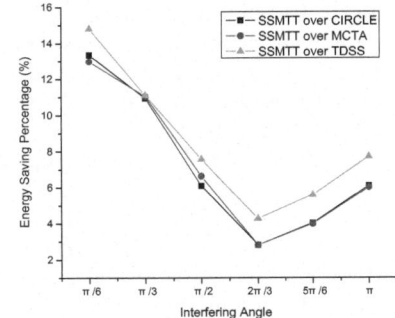

Fig. 9. ESR vs. Interfering Angle

6   Conclusions

In this paper, we present an energy-aware, Sleep Scheduling algorithm for Multiple Target Tracking (or SSMTT). Our objective is to improve the energy efficiency through a sleep scheduling approach that is conscious of concurrently tracking multiple targets, in contrast to an approach which is not. We introduced a linear target movement model as the foundation for energy efficiency optimization. Based on the movement model, we presented a tracking subarea management mechanism and sleep scheduling for nodes. We introduced an energy saving approach to reduce the transmission energy for alarm broadcasts. Our experimental evaluation shows that SSMTT can achieve better energy efficiency and suffer less performance loss than single target tracking algorithms.

Directions for future work include: (1) Further enhance energy efficiency on the alarm message transmission with collaboration among the subareas of multiple targets; and (2) discuss the energy efficiency given specific tracking performance requirements.

Acknowledgment. This work was supported by the IT R&D program of MIC/IITA, Korea.

References

1. Akyildiz, I.F., Su, W., Sankarasubramaniam, Y., Cayirci, E.: Wireless sensor networks: a survey. Computer Networks 38(4), 393–422 (2002)
2. Cao, Q., Abdelzaher, T., He, T., Stankovic, J.: Towards optimal sleep scheduling in sensor networks for rare event detection. In: IPSN, vol. 4 (2005)
3. He, T., Vicaire, P., Yan, T., Cao, Q., Zhou, G., et al.: Achieving long-term surveillance in vigilnet. In: INFOCOM (2006)
4. Oh, S., Schenato, L., Chen, P., Sastry, S.: A scalable real-time multiple-target tracking algorithm for sensor networks. Memorandum (2005)
5. Gui, C., Mohapatra, P.: Power conservation and quality of surveillance in target tracking sensor networks. In: MOBICOM, pp. 129–143 (2004)
6. Jiang, B., Han, K., Ravindran, B., Cho, H.: Energy efficient sleep scheduling based on moving directions in target tracking sensor network. IEEE IPDPS (2008)
7. Liu, J., Chu, M., Reich, J.: Multitarget tracking in distributed sensor networks. Signal Processing Magazine 24(3), 36–46 (2007)
8. Oh, S., Schenato, L., Sastry, S.: A hierarchical multiple-target tracking algorithm for sensor networks. In: International Conference on Robotics and Automation (2005)
9. Fan, L., Wang, H., Wang, H.: A solution of multi-target tracking based on fcm algorithm in wsn. In: IEEE International Conference on Pervasive Computing and Communications Workshops, p. 290 (2006)
10. Chen, L., Cetin, M., Willsky, A.: Distributed data association for multi-target tracking in sensor networks. In: International Conference on Information Fusion (2005)
11. Shin, J., Guibas, L., Zhao, F.: A distributed algorithm for managing multi-target identities in wireless ad-hoc sensor networks. IPSN (2003)
12. Chen, Y., Fleury, E.: A distributed policy scheduling for wireless sensor networks. In: INFOCOM (2007)
13. Denga, J., Hanb, Y.S., Heinzelmanc, W.B., Varshney, P.K.: Balanced-energy sleep scheduling scheme for high density cluster-based sensor networks. In: Computer Communications: special issue on ASWN 2004, vol. 28, pp. 1631–1642 (2005)
14. Liu, S., Fan, K.W., Sinha, P.: Dynamic sleep scheduling using online experimentation for wireless sensor networks. In: SenMetrics (2005)
15. Hightower, J., Borriello, G.: Location systems for ubiquitous computing. IEEE Computer 34(8), 57–66 (2001)
16. Stoleru, R., Stankovic, J.A., Son, S.: Robust node localization for wireless sensor networks. In: EmNets (2007)
17. Arora, A., Dutta, P., Bapat, S., Kulathumani, V., Zhang, H., et al.: A line in the sand: A wireless sensor network for target detection, classification, and tracking. Computer Networks 46(5), 605–634 (2004)
18. Wang, X., Ma, J.J., Wang, S., Bi, D.W.: Cluster-based dynamic energy management for collaborative target tracking in wireless sensor networks. Sensors 7, 1193–1215 (2007)
19. Yang, L., Feng, C., Rozenblit, J.W., Qiao, H.: Adaptive tracking in distributed wireless sensor networks. In: Engineering of Computer Based Systems, IEEE International Symposium and Workshop, p. 9 (2006)
20. Chipcon: Cc1000 a unique uhf rf transceiver, http://www.chipcon.com
21. Chipcon: Cc2420 2.4 ghz ieee 802.15.4 / zigbee-ready rf transceiver, http://www.chipcon.com

22. Xing, G., Lu, C., Zhang, Y., Huang, Q., Pless, R.: Minimum power configuration in wireless sensor networks. In: MobiHoc, pp. 390–401 (2005)
23. He, T., Vicaire, P., Yan, T., Luo, L., et al.: Achieving real-time target tracking using wireless sensor networks. ACM TECS (2007)
24. Lu, G., Sadagopan, N., Krishnamachari, B., Goel, A.: Delay efficient sleep scheduling in wireless sensor networks. In: INFOCOM (2005)
25. Jeong, J., Hwang, T., He, T., Du, D.: Mcta: Target tracking algorithm based on minimal contour in wireless sensor networks. In: INFOCOM, pp. 2371–2375 (2007)
26. CrossBow: Mica data sheet, http://www.xbow.com

Optimal Rate Allocation for Rate-Constrained Applications in Wireless Sensor Networks*

Chun Lung Lin, Hai Fu Wang, Sheng Kai Chang, and Jia Shung Wang

Department of Computer Science, National Tsing Hua
Univeristy, Hsinchu, Taiwan 30043
{cllin,hfwang,skchang}@vc.cs.nthu.edu.tw
jswang@cs.nthu.edu.tw

Abstract. This paper addresses the optimal rate allocation (ORA) problem as follows: given a target bit rate constraint, determine an optimal rate allocation among sensors such that the overall distortion of the reproduction data is minimized. Optimal rate allocation algorithms are proposed to determine the coding bit rate of each sensor in single hop and multi-hop sensor networks, given a target rate constraint. Extensive simulations are conducted by using temperature readings of the real world dataset. The results show that at low bit rates the optimal rate allocation improves about 2.745 dB on the uniform rate allocation in terms of SNR, and improves nearly 7.602 in terms of MSE. Spatial-temporal range queries are also evaluated to confirm that our approach is often sufficient to provide approximate statistics for range queries.

Keywords: Wireless sensor networks, Optimal rate allocation, Range query, Linear regression, Embedded Zerotree Wavelet (EZW) coding.

1 Introduction

Wireless sensor networks (WSNs) have many current and future envisioned applications, including climate monitoring, environment monitoring, health care and so on [1]. Consider the severe resource constraints on sensor nodes. Designing energy-efficient transmission algorithms is important because that as compared with sensing and computation, communication is the most energy-consuming operation in WSNs [1], [2], [3], [4], [5].

Consider that a lossy transform coder in wavelet or block cosine bases is installed at each sensor node. In this paper, we address the *Optimal Rate Allocation* (ORA) problem in sensor networks as follows: Given a target bit rate and a distortion measure, find an optimal bit rate allocation among sensors such that the overall distortion of the reproduction data at the base station is minimized. Such an optimization model can be useful for collecting minimum amount of sensor data that is often sufficient to provide approximate answers of user queries, e.g., spatial-temporal range queries: "what was the average temperature value

---

* This work was supported by 96-EC-17-A-04-S1-044.

S. Nikoletseas et al. (Eds.): DCOSS 2008, LNCS 5067, pp. 510–515, 2008.

in sub-region Y of region X in the last 30 minutes?". In many realistic applications, according to these approximate answers, users can efficiently and quickly go through a large number of sensor nodes, thereby choosing the small subsets of nodes of interest. Subsequently, higher overhead queries can be described, if necessary, to obtain the chosen data with a much better quality level. In other words, the coarse view of sensor data can provide users a priori knowledge about what to look for or where is event of interest in an energy-efficient way.

2   Problem Statement and Our Solutions

Consider the *Mean-Squared Error* (MSE) as the distortion measure $D_i\left(R_i\right)$ with respect to the coding bit rate R_i. In this paper, rate allocation means the process by which coding bit rates are assigned to the various sensors inside the network. For a given desired overall bit rate, we may try to minimize MSE, while achieving the target rate. Given a target bit rate R, our goal is to determine the bit rate R_i for each sensor i, $1 \leq i \leq n$, such that the following objective is optimized:

$$\text{Minimize} \quad \sum_{i=1}^{n} D_i\left(R_i\right) \quad \text{subject to} \quad \sum_{i=1}^{n} R_i \leq R \tag{1}$$

Or, equivalently, for a given desired distortion D, we may try to minimize the overall bit rate while achieving the target distortion:

$$\text{Minimize} \sum_{i=1}^{n} R_i\left(D_i\right) \quad \text{subject to} \quad \sum_{i=1}^{n} D_i \leq D \tag{2}$$

2.1   Optimal Rate Allocation in Single Hop Sensor Networks

We begin by ignoring network topology and consider a single hop networks which is rooted at the base station with full computing capabilities. Later on, we will extend our solutions to multihop sensor networks. Let $\bar{R}_i$ denote the averaged coding bit rate per reading for sensor i. Suppose that the buffer of each sensor is of size m, i.e. $\bar{R}_i = \frac{R_i}{m}$. Mallat and Falzon [6] had shown that at low bit rates, i.e., $\bar{R}_i \leq 1$, the distortion rate of transform coders in wavelet and block cosine bases can be approximately depicted by

$$D_i\left(\bar{R}_i\right) = C_i \bar{R}_i^{(1-2\gamma_i)} \tag{3}$$

where both $C_i > 0$ and $\gamma_i > \frac{1}{2}$ are constants. According to Eq. (3), the optimal rate allocation problem in Eq. (1) can be reformulated as follows.

$$\text{Minimize} \quad \sum_{i=1}^{n} C_i \bar{R}_i^{(1-2\gamma_i)} \quad \text{subject to} \quad \sum_{i=1}^{n} \bar{R}_i \leq \frac{R}{m} \tag{4}$$

This optimization problem can be solved by the Largrange multiplier as follows. Its Lagrangian L is given as

$$L\left(R_1, ..., R_n, \lambda\right) = \sum_{i=1}^{n} C_i \bar{R}_i^{(1-2\gamma_i)} + \lambda \left(\sum_{i=1}^{n} \bar{R}_i - \frac{R}{m}\right) \tag{5}$$

which leads to the following optimization conditions:

$$\frac{\partial L}{\partial \bar{R}_i} = C_i \cdot (1 - 2\gamma_i) \cdot \bar{R}_i^{-2\gamma_i} + \lambda = 0, i = 1, 2, ..., n \tag{6}$$

$$\sum_{i=1}^{n} \bar{R}_i = \frac{R}{m} \tag{7}$$

Since $C_i > 0$ and $\gamma_i > \frac{1}{2}$, Eq. (6) implies that $\lambda > 0$. Substitute Eq. (6) into Eq. (7) yields the necessary condition of Eq. (4)

$$\sum_{i=1}^{n} \left(\alpha_i^{\beta_i} \cdot \lambda^{-\beta_i} \right) = \frac{R}{m} \tag{8}$$

where $\alpha_i = C_i \cdot (2\gamma_i - 1) > 0$ and $\beta_i = \frac{1}{2\gamma_i} > 0$. We shall show in Theorem 1 that Eq. (8) are both necessary and sufficient. The proof of Theorem 1 can be found in [10].

Theorem 1. *Eq. (8) is both necessary and sufficient condition for the optimization problem in Eq. (4); hence, if λ^* is a zero root of Eq. (8), then the optimal bit rates are*

$$\bar{R}_i = \alpha_i^{\beta_i} \cdot (\lambda^*)^{-\beta_i} , i = 1, 2, ..., n \tag{9}$$

2.2   Estimation of the R-D Functions

In this paper, we propose a linear regression-based R-D estimation method. By taking logarithm on both sides of Eq. (3), a linear equation in the two variables $\ln \bar{R}_i(D_i)$ and $\ln D_i$ is given by

$$\ln D_i(\bar{R}_i) = (1 - 2\gamma_i) \cdot \ln R_i + \ln C_i \tag{10}$$

At sensor i, our goal is to compute a best-fit line $y = \hat{a}_i x + \hat{b}_i$ by using a sequence of rate-distortion pairs $(d_1, \bar{r}_1), (d_2, \bar{r}_2), ..., (d_u, \bar{r}_u)$ which can be obtained by running the coding system u times. Then, according to Eq. (10), C_i and γ_i can be estimated as follows.

$$\gamma_i = \frac{1 - \hat{a}}{2} \quad \text{and} \quad C_i = 2^{\hat{b}} \tag{11}$$

2.3   Optimal Rate Allocation in Multihop Sensor Networks

Consider tiered architecture sensor networks in which the root node is the *base station* (BS), leaf nodes are *sensing* nodes and internal nodes act as *forwarding* nodes that are responsible for receiving and forwarding messages coming from their child nodes.

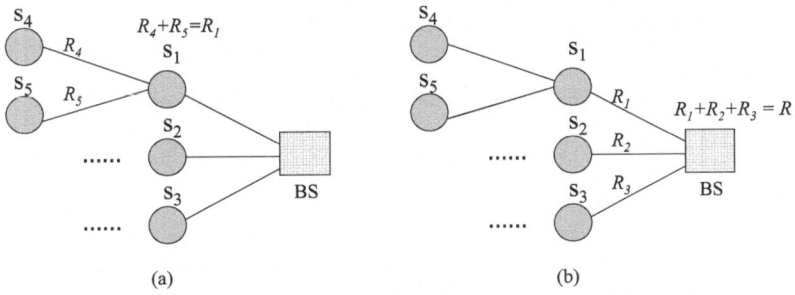

Fig. 1. An illustrative example of the distributed optimal rate allocation

Distributed Approach. Fig. 1 gives an illustrative example of the distributed optimal rate allocation. Consider the forwarding node s_1 in Fig. 1(a). By solving Eqs. (8) and (9), s_1 is able to determine the optimal bit rates R_4 and R_5 if C_i and γ_i, where $i = 3, 4$, and a rate constraint R_1 are given to s_1. The optimal rates R_4 and R_5 guarantee $R_4 + R_5 \leq R_1$ and $D_4(R_4) + D_5(R_5)$ is minimized. At this point, let us first assume that s_1 knows the rate-distortion function $D_1(R)$. (Note that the proposed R-D estimation in section 2.2 can not be applied to a forwarding node directly.) Similarly, the rate R_1 can be computed by the BS if all the parameters C_j and γ_j, where $j = 1, 2, 3$, and a rate constraint R are given, as illustrated in Fig. 1(b). The optimal rates R_1, R_2 and R_3 again guarantee the target rate R is complied with and the overall distortion of the collected data is minimized. Our distributed optimal rate allocation is the reverse of this procedure. Given a target rate, the optimal bit rates of nodes are calculated level by level in the routing tree, beginning from the BS until leaf nodes are reached. In the following, we propose a distributed estimation approach to estimate C_z and γ_z for a forwarding node s_z by using the R-D parameters of its child nodes. Let $child(s_z)$ denote the set of child nodes of s_z and $|child(s_z)| = p$. According to Eq. (10), the set of parameters C_i and γ_i, $i \in child(s_z)$, can be transformed into p lines. The basis of our estimation is to compute a best-fit line $y(x) = \hat{a}x + \hat{b}$ for the constructed lines, and then C_z and γ_z are given by Eq. (11). Theorem 2 states that C_z and γ_z can be estimated by the geometric mean and the arithmetic mean of the R-D parameters of child nodes respectively if the best-fit line is computed in least-square sense. The proof of Theorem 2 can be found in [10].

Theorem 2. *Let $a_i = (1 - 2\gamma_i)$ and $b_i = \ln C_i$, for all $i \in child(s_z)$. Let $y(x) = \hat{a}x + \hat{b}$ be the best-fit line in least-square sense for lines $y_i(x) = a_i x + b_i$, $i \in child(s_z)$. Then C_z and γ_z are*

$$\gamma_z = \frac{1 - \hat{a}}{2} = \frac{1}{p} \cdot \sum_{i \in child(s_z)} \gamma_i \tag{12}$$

$$C_z = 2^{\hat{b}} = \sqrt[p]{\prod_{i \in child(s_z)} C_i} \tag{13}$$

Table 1. Mean, Relative Error(RE), SNR and MSE

	Raw data	$RD(0.05)$	$U(0.05)$	$RD(0.03)$	$U(0.03)$	$RD(0.01)$	$U(0.01)$
Mean	20.24	20.23	21.12	20.22	21.12	20.16	21.95
RE		0.05%	4.35%	0.10%	4.35%	0.40%	8.45%
SNR(dB)		21.162	17.274	18.333	16.152	16.868	14.123
MSE		3.209	7.857	6.156	10.172	8.628	16.230

3   Simulation Results

3.1   Data Sets and Experimental Settings

In our simulations, we used the real-world sensor data collected from 14 Crossbow MicaZ [8] sensors deployed in our lab and that collected by the Intel Berkeley Research lab [9]. The temperature readings are extracted as test data. We simulated a two-level tree network with $n = 64$ sensing nodes, four forwarding nodes and the BS. We let each of the four forwarding nodes connect 16 sensing nodes on which the embedded zerotree wavelet (EZW) coding were implemented [7]. More details of the simulations can be found in [10].

3.2   Rate-Distortion Evaluation

Table 1 shows the simulated results of the optimal rate allocation (ORA) and the uniform rate allocation (URA) respectively in terms of mean, relative error (RE), MSE and SNR. $RD(x)$ and $U(x)$ in Table 1 denote the simulated results of the ORA and the URA respectively with respect to data compression ratio x. Obviously, the ORA outperforms the URA in all the three bit rates. The results suggest that in rate-constrained applications deliberately allocating bit rates for sensors indeed improve the overall data quality.

3.3   Spatial-Temporal Range Query

The collected data streams are stored as a two-dimensional array, where x-axis corresponds to time and y-axis corresponds to the ordered set of sensors. Fig. 2 shows the frequency distribution, i.e., occurrence count, of the range query $T_{16}S_{16}$ with respect to 5% compression ratio ($T_{16}S_{16}$ indicates the result was evaluated from the aggregated results obtained by first averaging along with x-axis per 16 readings and then averaging along with y-axis per 16 sensors). Clearly, the frequency distribution of the ORA nearly matches with that of the raw data; however, the URA tends to congregate in certain of temperature values. This outcome becomes more clear at lower bit rates, which confirms that the ORA can be utilized to collect a small amount of sensor data that is often sufficient to provide approximate statistics for spatial-temporal range queries.

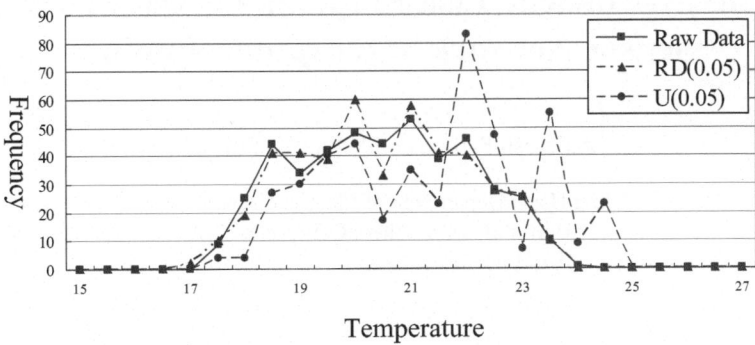

Fig. 2. The frequency distribution of the range query $T_{16}S_{16}$

4   Conclusion

In this paper, we focused on transform coders and proposed optimal rate allocation algorithms in single hop and multihop sensor networks. In our model, sensors are assumed to have identical distribution of data source over a long period of time, but differ in a short period of time. The basis of our approach is to capture this variation in a short period of time and optimally allocate bit rates for sensors so as to prolong the network lifetime.

References

1. Akyildiz, I.F., Melodia, T., Chowdury, K.: Wireless multimedia sensor networks: A survey. Wireless Communications, IEEE 14(6), 32–39 (2007)
2. Ganesan, D., Estrin, D., Heidemann, J.S.: Dimensions: why do we need a new data handling architecture for sensor networks? Computer Communication Review 33(1), 143–148 (2003)
3. Krishnamachari, B., Estrin, D., Wicker, S.: The Impact of Data Aggregation in Wireless Sensor Networks. In: 22nd International Conference on Distributed Computing Systems Workshops, pp. 575–578 (2002)
4. Yoon, S., Shahabi, C.: The Clustered AGgregation (CAG) technique leveraging spatial and temporal correlations in wireless sensor networks. ACM Trans. Sen. Netw. 3(1), 3 (2007)
5. Madden, S., Franklin, M.J., Hellerstein, J.M., Hong, W.: TAG: a Tiny AGgregation service for ad-hoc sensor networks. In: 5th ACM Symposium on Operating System Design and Implementation, vol. 36, pp. 131–146. ACM, Boston (2002)
6. Mallat, S., Falzon, F.: Analysis of low bit rate image transform coding. Signal Processing, IEEE Transactions on 46(4), 1027–1042 (1998)
7. Shapiro, J.M.: Embedded image coding using zerotrees of wavelet coefficients. Signal Processing, IEEE Transactions on 41(12), 3445–3462 (1993)
8. Crossbow Technology, http://www.xbow.com/
9. Intel Berkeley Research Lab, http://berkeley.intel-research.net/labdata
10. Visual and Communication Lab, http://vc.cs.nthu.edu.tw/~cllin/dcoss.pdf

Energy-Efficient Task Mapping for Data-Driven Sensor Network Macroprogramming*

Animesh Pathak and Viktor K. Prasanna

Ming Hsieh Department of Electrical Engineering,
University of Southern California, USA
{animesh,prasanna}@usc.edu

Abstract. Data-driven macroprogramming of wireless sensor networks (WSNs) provides an easy to use high-level task graph representation to the application developer. However, determining an energy-efficient initial placement of these tasks onto the nodes of the target network poses a set of interesting problems. We present a framework to model this task-mapping problem arising in WSN macroprogramming. Our model can capture task placement constraints, and supports easy specification of energy-based optimization goals. Using our framework, we provide mathematical formulations for the task-mapping problem for two different metrics — energy balance and total energy spent. Due to the complex nature of the problems, these formulations are not linear. We provide linearization heuristics for the same, resulting in mixed-integer programming (MIP) formulations. We also provide efficient heuristics for the above. Our experiments show that the our heuristics give the same results as the MIP for real-world sensor network macroprograms, and show a speedup of up to several orders of magnitude.

1 Introduction

Various high-level programming abstractions have been proposed recently to assist in application development for Wireless Sensor Networks (WSNs). Specifically, *Data-driven macroprogramming* [1] refers to the general technique of specifying the WSN application from the point of view of data-flow. In sense-and-respond applications such as traffic management [2], building environment management [3], target tracking etc., the system can be represented as a set of tasks running on the system's nodes – producing, processing and acting on data items or streams to achieve the system's goals. The mapping of these tasks onto the nodes of the underlying system (details of which are known at compile time) is an important part of the compilation of the macroprogram, and optimizations can be performed at this stage for energy-efficiency.

Although the initial information (positions, energy levels) about the target nodes is known, during the lifetime of the WSN, changing conditions, either external or internal may alter the circumstances. We do not address these unpredictable situations, and instead aim to provide a "good" initial mapping of tasks. We assume that during the lifetime of the system, remapping of tasks will occur to face these circumstances, for

---

* This work is partially supported by the National Science Foundation, USA, under grant number CCF-0430061 and CNS-0627028.

S. Nikoletseas et al. (Eds.): DCOSS 2008, LNCS 5067, pp. 516–524, 2008.

example, a distributed task-remapping algorithm can be triggered when the energy at any node goes below a certain fraction of its initial energy level. Our work attempts to utilize the global knowledge available at compile-time to obtain efficient results.

In this paper, we make three **contributions**. In Sect. 2 we provide a **modeling framework** for the problem of task-mapping for data-driven sensor network applications. In Sect. 3 we propose a **mixed integer programming (MIP) formulation** to obtain task mappings in order to optimize for the energy balance and total-energy minimization goals. Since the formulation is non-linear, we provide substitution-based techniques to linearize the MIPs. Although the MIP formulations give optimal results, they may take inordinately large times to terminate for large real-world scenarios. In Sect. 4, we provide **greedy heuristics** for the two problem instances.

Our experimental results, discussed in Sect. 5, show the performance comparison between the techniques, using realistic applications and deployment scenarios. Our heuristics are shown to obtain the optimal solution for these scenarios, while gaining significant speedups over the MIP technique. Section 6 discusses the differences of our work from other closely related work in parallel and distributed systems and sensor networks. Section 7 concludes.

2   Problem Formulation

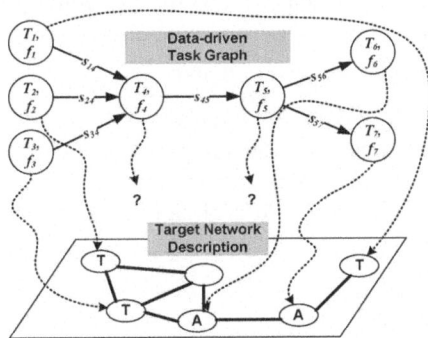

Fig. 1. Temperature mgmt. application

As an example of data-driven macroprogramming representation, consider the following (simple) application – A room is instrumented with six wireless nodes, with three nodes equipped with temperature sensors, and two nodes connected to actuators for controlling the temperature of the room. We need to periodically determine the average temperature in the room, compare it with a threshold, and produce the corresponding actuation. One way of designing such an application at a high-level using a data-driven approach is shown in the top part of Fig. 1. Tasks T_1, T_2 and T_3 are temperature sampling tasks, which fire at a rates of f_1, f_2, f_3 and generate ambient temperature readings of size s_{14}, s_{24}, s_{34}. Task T_4 calculates the average of these readings and feeds it to T_5, which determines the action to be taken. Tasks T_6 and T_7 act upon the data generated by T_5, and control the actuators. The system for which this application is being designed is shown in the lower part of the same figure. The nodes equipped with temperature sensors and actuators are marked with a **T** and **A** respectively.

The mapping of tasks T_1 through T_7 onto the nodes of the target network is an instance of the problem faced while compiling data-driven macroprograms for WSNs. The placement of the sensing tasks (T_1, T_2, T_3) and the actuating tasks(T_6 and T_7) are pre-determined to the nodes with the relevant capabilities. This fact is shown using curved broken lines in the figure. However, tasks T_4 and T_5 can be placed on any of the nodes in the floor, thus allowing for optimizations in this process.

Our aim is to capture the various aspects of systems like the one above, including the placement constraints and firing rates of the tasks, the data-flow between them, the heterogeneity in the nodes and links of the system, and the energy spent in sensing, computation and communication.

Application and System Model

A **Network Description** N represents the target system of nodes where the WSN application is to be deployed. Each node k $(k = 1, \ldots, n)$ has an initial energy reserve e_k^0. We assume that the system operates in *rounds*, and denote the energy remaining at node k after t rounds by e_k^t. A **round** is defined as the least time-period after which the system behavior repeats itself.

A **Data-driven Task** i represents the sensing, processing or actuation activity in a WSN. Its firing rate f_i, denotes the number of times it is invoked in one round.

A **Data-driven Task Graph** $D = (DT, DE)$ is a directed acyclic graph (DAG) consisting **a)** A set $DT = \{1, \ldots, i, \ldots, m\}$ of data-driven tasks, and **b)** A set $DE \subseteq DT \times DT$ of edges. Each edge (i, j) is labeled with the size s_{ij} of the data that task i produces for task j upon each invocation.

The **Task Execution Energy Matrix** $\mathcal{T}$ is an $m \times n$ matrix, where $\mathcal{T}_{ik}$ denotes the energy spent by node k per invocation of task i, if i is mapped onto node k.

The **Routing Energy Cost Matrix** $\mathcal{R}$ for N is a $n \times n \times n$ matrix, with $\mathcal{R}_{\beta\gamma k}$ denoting the energy consumed at node k while routing one unit of data from node β to γ.

The **Task Mapping** is a function $M : DT \to N$, placing task i on node $M(i)$.

Energy Costs: In a sensor network, the *cost* that developers are largely concerned with is the *energy spent* by the nodes as the system operates. We therefore use the terms *cost* to mean the energy spent at a node throughout this paper, unless otherwise stated. Using the model defined above, we compute the following costs[1]. At each node $k \in N$, the **computation cost** in each round is given by

$$C_{\text{comp}}^k = \sum_{i:M(i)=k} f_i \cdot \mathcal{T}_{ik} \tag{1}$$

and the **energy cost of communicating messages** in each round is given by

$$C_{\text{comm}}^k = \sum_{(i,j)\in DE} f_i \cdot s_{ij} \cdot \mathcal{R}_{M(i)M(j)k} \tag{2}$$

Performance Metrics: The above modeling framework can be used to easily model two common optimization goals. The first is *energy balance*, which we consider to be achieved when the maximum fraction of energy spent by any node in the system is minimized.

$$\text{OPT}_1 = \min_{\text{all Mappings } M} \max_{k \in N} \frac{1}{e_o^k} \cdot (C_{\text{comp}}^k + C_{\text{comm}}^k) \tag{3}$$

The second performance goal we model using our framework is the more commonly used *total energy spent* in the entire system. Although we believe that energy balance is

---

[1] Note that the cost of sensing is included in the $\mathcal{T}_{ik}$ of the sensing tasks.

a better metric to measure the quality of task placement, we use the goal of minimizing the total energy spent in the system to illustrate the modeling power of our framework.

$$\text{OPT}_2 = \min_{\text{all Mappings } M} \sum_{k \in N} (C^k_{\text{comp}} + C^k_{\text{comm}}) \tag{4}$$

For each of the two metrics, a *feasible* solution is possible only when all nodes have non-zero energy left at the end of one round. If there are no mappings possible for which this holds, the task-mapping algorithms should report failure. In addition to the above, our framework can be used to model other application scenarios also, e.g. when multiple paths between two nodes are possible.

3   Mathematical Formulations for Task Mapping on WSNs

3.1   Mixed Integer Programming Formulation for OPT$_1$

To formulate the problem as a mixed integer programming (MIP) problem, we represent task mapping M by an $m \times n$ *assignment matrix* X, where x_{ik} is 1 if task i is assigned to node k, and 0 otherwise. The problem can then be defined as:

Inputs:

- $D = (DT, DE)$: Data-driven Task Graph, f_i: Firing rate for task i, and s_{ij}: Size of data transferred from task i to j on each invocation of i
- N: Network description, $\mathcal{T}$: Task execution energy matrix, $\mathcal{R}$: Routing energy cost matrix

Output: X: Assignment Matrix. x_{ik} is binary.
Optimization Goal:

$$\text{minimize } c$$

Constraints:

$$\sum_{k=1}^{n} x_{ik} = 1 \text{ for } i = 1, 2, \ldots, m \tag{5}$$

$$\frac{1}{e^0_k}\left(\sum_{i=1}^{m} f_i \cdot \mathcal{T}_{ik} \cdot x_{ik} + \sum_{(i,j) \in DE} \sum_{\beta=1}^{n} \sum_{\gamma=1}^{n} f_i \cdot s_{ij} \cdot x_{i\beta} \cdot x_{j\gamma} \cdot \mathcal{R}_{\beta\gamma k}\right) \leq c, \forall k \in \{1, \ldots, n\} \tag{6}$$

$$x_{ik} \in \{0, 1\} \text{ for } (i, k) = (1, 1), \ldots, (m, n) \tag{7}$$

$$0 \leq c < 1 \tag{8}$$

The summation terms in (6) denote C^k_{comp} and C^k_{comm} respectively. The final constraint ensures that the MIP fails if no feasible solution exists. Note that the above is an MIP since c is real whereas x_{ik} are binary integers. Also, it is not a linear program since product terms $x_{i\beta} \cdot x_{j\gamma}$ appear in the constraints.

The above problem can be converted to a linear MIP by replacing each $x_{i\beta} \cdot x_{j\gamma}$ term with a binary variable $y_{i\beta j\gamma}$, and adding the following constraints:

$$y_{i\beta j\gamma} - x_{i\beta} \leq 0 \tag{9}$$

$$y_{i\beta j\gamma} - x_{j\gamma} \leq 0 \tag{10}$$

$$x_{i\beta} + x_{j\gamma} - y_{i\beta j\gamma} \leq 1 \tag{11}$$

Using techniques similar to the ones above, we also designed a linear MIP to solve the task-mapping problem for OPT_2, i.e., minimizing the *total energy* spent by the system. Owing to the space limitations, it is not discussed in detail here.

4 Heuristic for Task Mapping

4.1 Greedy Algorithms for Task Mapping

Although the MIP formulation leads to optimal results, solving an MIP can be quite time consuming in practice. Our greedy heuristic for the goal of minimizing the maximum fraction of energy spent at a node (OPT_1) is detailed in Algorithm 1. The main intuition is that the algorithm sorts the edges in the task graph in non-increasing order of the traffic going on them, and then tries to map the still unmapped endpoints of each edge (i, j) so as to achieve the minimum increase in the objective function. We also appropriately modified this algorithm to obtain *GreedyMinTotal()* for OPT_2.

Algorithm 1. *GreedyMinMax*: for OPT_1

Input: $D(= DT, DE), N, \mathcal{T}[m][n], \mathcal{R}[n][n][n], f[m], s[m][m], e_o[n]$
Output: $M[m]$: Task Assignment
 1: Initialize $M[i] = -1$ for $i = \{1, \ldots, m\}$
 2: Sort $(i, j) \in DE$ in non-increasing order of $f[i] \cdot s[i][j]$
 3: **for all** (sorted) (i, j) in DE **do**
 4:     $minmaxCost = \infty$; $minPath = (-1, -1)$ // Initialize $minmaxCost$ and $minPath$
        for this iteration
 5:     **for all** (α, β) such that (i, j) can be assigned to them **do**
 6:         $M[i] = \alpha$, $M[j] = \beta$ // Temporarily assign (i, j) to $(\alpha \rightarrow \beta)$
 7:         $maxCost = maxCost(D, N, \mathcal{T}, \mathcal{R}, f, s, e_0, M)$
 8:         **if** $maxCost < minmaxCost$ **then**
 9:             $minmaxCost = maxCost$; $minPath = (\alpha, \beta)$ // Update $minmaxCost$ and
                $minPath$
10:     **if** $minmaxCost > 1$ **then**
11:         **declare failure. stop.** // Checking for feasibility
12:     $M[i] = minPath.\alpha$; $M[j] = minPath.\beta$
13: **return** M

Computational Complexity: Each invocation of *maxCost* takes $\theta(n(m+|DE|))$ time. During Algorithm 1, the sorting takes $O(|DE| \log(|DE|))$ time, and the main loops invokes Algorithm 2 for evaluating the *maxCost* $O(|DE|n^2)$ times. The total time complexity of the algorithm is $O(|DE|(\log(|DE|) + n^3(m + |DE|)))$.

Algorithm 2. *maxCost*: for determining the maximum fraction of energy spent at a node

Input: $D(= DT, DE), N, \mathcal{T}[m][n], \mathcal{R}[n][n][n], f[m], s[m][m], e_o[n], M[m]$
Output: *maxCost*: Maximum fraction of energy spent at any node
1: $maxCost = 0$ // Initialize max cost
2: **for all** $k \in N$ **do**
3:    $cost = 0$ // Initialize node cost
4:    **for all** $i \in DT$ **do**
5:      **if** $M[i] == k$ **then**
6:        $cost = cost + f[i] \cdot \mathcal{T}[i][k]$ // Increment computation cost
7:    **for all** $(i, j) \in DE$ such that $M[i] \neq 1$ AND $M[j] \neq 1$ **do**
8:      $cost = cost + f[i] \cdot s[i][j] \cdot \mathcal{R}[M[i]][M[j]][k]$ // Increment communication cost
9:    **if** $cost/e_0[k] > maxCost$ **then**
10:      $maxCost = cost/e_0[k]$
11: **return** $maxCost$

4.2   Worst-Case Analysis

Since both *GreedyMinMax* and *GreedyMinTotal* are heuristics, we explored the situations when they can give sub-optimal results. We introduce the notion of the *cost of an algorithm* for this purpose – the cost of *GreedyMinMax* is defined a the maximum fraction of energy spent in one round at any node in N, while the cost of *GreedyMinTotal* is the total energy spent by all the nodes in N in one round.

Theorem 1. *For any integer $\upsilon \geq 1$, there are problem instances for which the cost of* GreedyMinMax *(*GreedyMinTotal*) is arbitrarily close to $\upsilon \times OPT_1$ ($\upsilon \times OPT_2$).*

Proof. Consider a situation as illustrated in Fig. 2. $\mathcal{T}_{ax} = \mathcal{T}_{ay} = 0$, and the other tasks can only be placed on the nodes indicated by the arrows. Let us also assume that $f_a = 1$, $e_0^x = e_0^y = e_0$, and both nodes in N spend one unit of energy per unit of data transmitted on the link between them. Finally, $e_0 \gg \sigma \gg \varepsilon > 0$. The optimal solution, both for OPT_1 and OPT_2, is to place a on node x, thereby causing only the data on the (a, b_0) edge in DE to go on the network, costing σ units of energy to be spent by node x (and the entire system) in each round. The greedy algorithms, however, start with placing the costliest edge (a, b_0) in the best possible manner, co-locating a and b_0 on node y. This leads to $\upsilon \times (\sigma - \varepsilon)$ traffic to go over the $y \to x$ link. We thus get:

$$\text{OPT}_1 = \frac{1}{e_0}\sigma \tag{12}$$

$$\Rightarrow \text{cost}(GreedyMinMax) = \frac{1}{e_0}\upsilon \times (\sigma - \varepsilon) \approx \upsilon \times \text{OPT}_1 \tag{13}$$

$$\text{Similarly, OPT}_2 = 2\sigma \tag{14}$$

$$\Rightarrow \text{cost}(GreedyMinTotal) = 2\upsilon \times (\sigma - \varepsilon) \approx \upsilon \times \text{OPT}_2 \tag{15}$$

hence proving the theorem   □

Theorem 2. *There are problem instances for which* GreedyMinMax *and* GreedyMinTotal *will terminate in failure although a feasible solution exists.*

Proof. Consider the situation as illustrated in Fig. 2. However, in this case, assume that $e_0 = \sigma \gg \varepsilon > 0$. The optimal solution (given by the MIP formulation) will still place task a on node y, while the greedy algorithms will try to place it on node y. Note that for $v \geq 2$, this will lead to an infeasible solution, as the nodes end up spending $> e_0$ energy and the heuristics will declare failure.   □

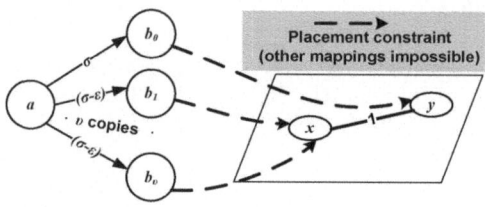

Fig. 2. Scenario for worst case performance of *GreedyMinMax* and *GreedyMinTotal*

5   Evaluation

The worst-case performance bounds discussed above apply to cases where *arbitrary* task graphs and constraints are permitted. However, for realistic sensor network applications, the relationships between the tasks are not completely arbitrary. For evaluating the relative performance of our heuristics in realistic applications, we applied them on the task graphs of the building environment management (HVAC) and traffic management applications discussed in [4]. We used our algorithms to map their tasks onto various simulated target deployments based on real-world scenarios.

In our experiments, we assumed that all nodes started with a sufficiently high initial energy level e_0. The routing energy cost matrix $\mathcal{R}$ was obtained by using a shortest path algorithm on the network, assuming equal energy spent by all nodes on a route, and all data items were assumed to be of unit size ($s_{ij} = 1$). The task execution energy matrix

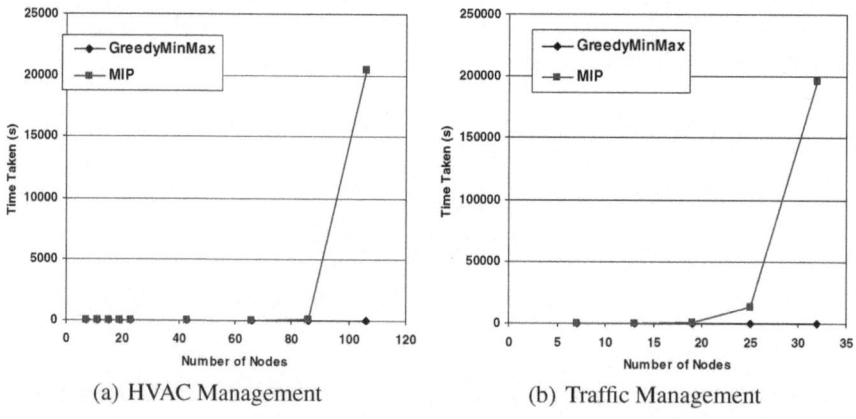

Fig. 3. Time taken to compute task-mapping for minimizing maximum energy spent

$\mathcal{T}$ was set up to represent placement constraints: $\mathcal{T}_{ik} = 0$ when task i could be placed on node k, ∞ when it could not. The tasks which performed sensing and actuating were tied to a node with the relevant capabilities. Finally, the f_i for each task was computed as follows: For sensing tasks, f_i was set to 10, and for all other tasks j, f_j was set to the sum of the firing rates of tasks on the other ends of the incoming edges. This represented the fact that task j fires whenever there is data available for it.

We ran our experiments on a PC with dual quad-core Xeon processors running at 2GHz, with 16GB of RAM. We implemented our greedy algorithm in Java, and solved the MIPS using the `lp_solve` linear programming toolkit. We recorded the average time taken for task placement over 500 runs for each data point. The time taken for computing task placements for both the applications for OPT_1 is shown in Fig. 3. For all instances of the traffic application, and the for *all* experiments with OPT_1, the solution given by the greedy algorithm was the same as the one given by the MIP. Our experiments clearly show that the greedy algorithms take much less time that the MIP formulation to determine the mappings. The speedups were similar with OPT_2. This showcases the efficacy of the algorithms in solving the task-mapping problem for complex real-world WSN applications.

6 Related Work

Parallel and Distributed Computing: The task mapping problem [5] is a well studied problem is parallel and distributed computing. In [6], the authors have covered a wide range of mapping problems in such systems and approaches to solve them to minimize system latency in cases where tasks do not have placement constraints. In [7], the authors present a genetic algorithm for placing tasks on a parallel processor, with an extension for the case where not all tasks can be run on all nodes, by way of assigning each node to a class, and associating a class number with each task. Their algorithm is designed to work for a range of metrics, and they focus on the *minimize total execution time* metric in the paper. However, unlike us, they assume full control over routing.

Wireless Sensor Networks: Task placement on sensor networks has also been addressed recently. Efforts such as [8] approach the task-mapping problem for WSNs from a protocol-centric point of view, whereas we take a high-level perspective of the problem. [9] proposes a greedy solution to the *service placement problem*, which is applicable to our context of compiling macroprograms. Similar to our case, their application also has task placement constraints, where certain tasks can be placed only on certain nodes. However, they focus only on the specific goal of minimizing the total energy spent for *trees*. The work in [10] solves the generic role assignment problem, where task placements are specified using *roles*. Their algorithm allows ILP solutions of role assignment onto the nodes of the target system, based on a global optimization criteria represented in terms of the number of nodes with a particular role. Unlike their case, our heuristics are meant for solving an offline version of the problem, and the optimization goals more tied to the energy-consumption at the nodes.

7 Concluding Remarks

In this paper, we formalized the task-mapping problem as it arises in the context of designing applications for wireless sensor networks using data-driven macroprogramming. We provided mathematical formulations for two energy-related optimization goals – minimizing the maximum fraction of energy consumed in a node and minimizing the total energy consumed in the sensor network. We used our modeling framework to provide mathematical formulations to solve these two problem instances, and demonstrated linearization techniques to convert them into mixed-integer programs (MIP). We also provided greedy heuristics for the above problem scenarios, and provided worst-case performance bounds for the same. In spite of the worst-case performance possible for specially crafted problem instances, our heuristics were shown to out-perform the MIP formulation by several orders of magnitudes of time for real-world WSN applications, while not compromising in the quality of the solutions. We acknowledge that later in the life of the WSN applications, distributed protocols will be needed to re-assign the tasks in view of changing operating circumstances. However, our techniques (and other technique based on our models) will provide good initial task placements. Our immediate future work is to reduce the complexity of the greedy approaches, as well as to explore better polynomial time approximation algorithms. Additionally, we are working on integrating our algorithms into the compiler [4] of a pre-existing data-driven macroprogramming framework.

References

1. Bakshi, A., Prasanna, V.K., Reich, J., Larner, D.: The Abstract Task Graph: A methodology for architecture-independent programming of networked sensor systems. In: Workshop on End-to-end Sense-and-respond Systems (EESR) (2005)
2. Hsieh, T.T.: Using sensor networks for highway and traffic applications. IEEE Potentials 23(2) (2004)
3. Dermibas, M.: Wireless sensor networks for monitoring of large public buildings. Technical report, University at Buffalo (2005)
4. Pathak, A., Mottola, L., Bakshi, A., Picco, G.P., Prasanna, V.K.: A compilation framework for macroprogramming networked sensors. In: Int. Conf. on Distributed Computing on Sensor Systems (DCOSS) (2007)
5. Bokhari, S.H.: On the mapping problem. IEEE Transactions on Computers (March 1981)
6. El-Rewini, H., Lewis, T.G., Ali, H.H.: Task scheduling in parallel and distributed systems. Prentice-Hall, Inc., Upper Saddle River (1994)
7. Ravikumar, C., Gupta, A.: Genetic algorithm for mapping tasks onto a reconfigurable parallel processor. In: IEE Proceedings on Computers and Digital Techniques (March 1995)
8. Low, K.H., Leow, W.K., Ang Jr., M.H.: Autonomic mobile sensor network with self-coordinated task allocation and execution. IEEE Transactions on Systems, Man and Cybernetics, Part C: Applications and Reviews 36(3), 315–327 (2006)
9. Abrams, Z., Liu, J.: Greedy is good: On service tree placement for in-network stream processing. In: ICDCS 2006: Proceedings of the 26th IEEE International Conference on Distributed Computing Systems, Washington, DC, USA, p. 72. IEEE Computer Society, Los Alamitos (2006)
10. Frank, C., Römer, K.: Solving generic role assignment exactly. In: IPDPS (2006)

Robust Dynamic Human Activity Recognition Based on Relative Energy Allocation*

Nam Pham[1] and Tarek Abdelzaher[2]

[1] Department of Electrical and Computer Engineering
[2] Department of Computer Science
University of Illinois at Urbana Champaign
nampham2@uiuc.edu, zaher@cs.uiuc.edu

Abstract. This paper develops an algorithm for *robust* human activity recognition in the face of imprecise sensor placement. It is motivated by the emerging *body sensor networks* that monitor human activities (as opposed to environmental phenomena) for medical, entertainment, health-and-wellness, training, assisted-living, or entertainment reasons. Activities such as sitting, writing, and walking have been successfully inferred from data provided by body-worn accelerometers. A common concern with previous approaches is their sensitivity with respect to sensor placement. This paper makes two contributions. First, we explicitly address robustness of human activity recognition with respect to changes in accelerometer orientation. We develop a novel set of features based on relative activity-specific body-energy allocation and successfully apply them to recognize human activities in the presence of imprecise sensor placement. Second, we evaluate the accuracy of the approach using empirical data from body-worn sensors.

1   Introduction

User activities inferred from data collected from wearable devices are useful in many medical and social applications. For example, user activity patterns along with biometrics measured over a long period of time can be helpful to physicians for diagnostic purposes or for monitoring the progress of recovery from injury. Real-time activity recognition applications can serve personal safety purposes by sending alarms if unusual activity is detected (e.g., a fall detector for assisted living). Several physical activity recognition algorithms have been developed using infrastructure sensors such as video [1], as well as personal sensors such as accelerometers [2,3], gyroscopes, and magnetometers [4]. In this paper, we use body-worn accelerometers to estimate the distribution of body energy.

Since the frequency of most human activities ranges from 0.3 to 3.5 Hz [5,6], the low-frequency components of the acceleration signal play an important role in recognizing human activities. A low-frequency model and a Kalman filter

---

* This work is supported in part by NSF grant CNS 06-15318, CNS 05-5759 and the Vietnam Education Foundation (VEF).

S. Nikoletseas et al. (Eds.): DCOSS 2008, LNCS 5067, pp. 525–530, 2008.

are develpoped to track the low-frequency components of activities. To gain immunity with respect to orientation, instead of using functions of absolute acceleration data from each axis, we use values calculated as the sum squared of all projected accelerations. This measure does not change with the change in the orientation of sensors. We call this metric *energy*. Finally, although the energy of an activity may change from time to time and from person to person, the relative distribution of energy across different parts of the body is more uniform across people and better characterizes the activity. We call features based on this metric *relative energy* features. Section 2 describes the low frequency model along with the Kalman filter and the details on the calculation of these features.

We evaluate our classification algorithm based on the above features in Section 3. The results show that by using both the filtered signal and error to compute features, the algorithm achieves a very high accuracy (often more than 97%) which is approximately 10% higher than the case where features from either the filtered signal or error are used alone. Furthermore, unlike the case with prior approaches, the accuracy of the classifier does not appreciably change with changes in the orientation of sensors. These results demonstrate the robustness of the new activity recognition approach.

2   Design

We first develop a continuous model to track the low-frequency components of human activities. Each sensor can be seen as an object moving in a three-dimensional space, namely the xyz space, with acceleration measured on at least two axes, x and y. Since our physical sensors use biaxial accelerometers, in the model below we assume that two acceleration components are measured. For biaxial accelerometers, in the continuous case, we can choose $\mathbf{x} = (x, y, \dot{x}, \dot{y}, \ddot{x}, \ddot{y})$ to be the state vector for the model. Elements of the state vector are the components of sensor position, velocity, and acceleration. The *Wiener-process acceleration model* [7, Section 6.2] is used to model the movement of the sensor in two dimensions:

$$\frac{d\mathbf{x}(t)}{dt} = \underbrace{\begin{pmatrix} 0\,0\,1\,0\,0\,0 \\ 0\,0\,0\,1\,0\,0 \\ 0\,0\,0\,0\,1\,0 \\ 0\,0\,0\,0\,0\,1 \\ 0\,0\,0\,0\,0\,0 \\ 0\,0\,0\,0\,0\,0 \end{pmatrix}}_{F} \mathbf{x}(t) + \underbrace{\begin{pmatrix} 0\,0 \\ 0\,0 \\ 0\,0 \\ 0\,0 \\ 1\,0 \\ 0\,1 \end{pmatrix}}_{L} w(t) \tag{1}$$

where, $w(t)$ is a two dimensional independent white noise process. The model above is then discretized and is used by the Kalman filter to seperate the noise from te low-frequency component of activities. Readers are referred to [7, Section 5.2] for more details on the Kalman filter.

Two sets of energy features are obtained; one for the low-frequency (filtered) signal and one for the error (due to high-frequency components and noise). For

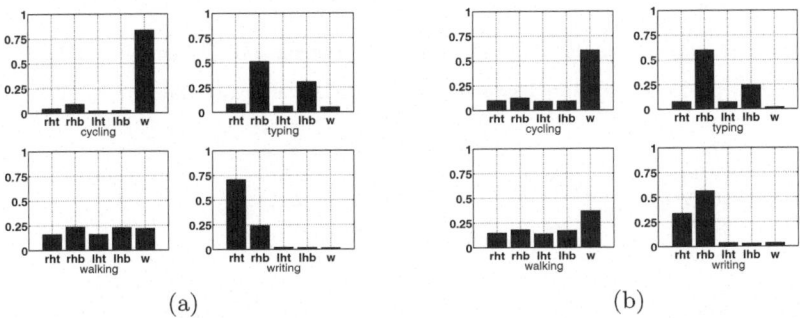

Fig. 1. Energy allocation of (a) the low-frequency acceleration component, and (b) the high-frequency component

low-frequency components, we compute the energy of the filtered acceleration signal for each axis. The total acceleration energy of the ith sensor is calculated as $E^i = \sqrt{(E_x^i)^2 + (E_y^i)^2}$. The relative acceleration energy, E_r^i, is the fraction of energy for each sensor i to the total energy of all sensors, $E_r^i = \frac{E^i}{\sum_i E^i}$. It measures the average fraction of energy exerted by each part of the body in each activity. In a similar manner, we define *error energy* to be $(x_k - \hat{x}_k)^2$ and compute a triaxial (or biaxial) energy then the relative error energy for each sensor.

Typical acceleration energy allocation and error energy allocation (Figure 1(b)) for four activities including cycling, typing, walking and writing are shown in Figure 1(a) and Figure 1(b) respectively. The five sensors are placed at right-hand top (*rht*), right-hand bottom (*rhb*), left-hand top (*lht*), left-hand bottom (*lhb*), and waist (*w*) to monitor acceleration. Each bar in a graph represents the fraction of energy for one part of the body to the total energy.

3   Experimental Evaluation

For our experiments, we used the dataset recorded by Ganti et al. [8]. Subjects were required to wear a jacket with five biaxial accelerometers attached on it. Two sensors were mounted on the each arm (one below and one above the elbow) and one sensor was placed close the waist, with all the y-axes pointing downwards when in standing position (with the arms pointing down). Two people were involved in the data collection process. Each person was asked to perform four activities: cycling, typing, walking, and writing.

For comparison purposes, three recognition algorithms were implemented and tested. *Recognition based on relative energy allocation*: features were computed from 128-sample windows with 64 samples overlapping between consecutive windows. Activity recognition based on these features was performed using a naive Bayes learner [9, Chapter 8]. *Recognition based on absolute features*: The same

Table 1. Comparison of recognition rates for three recognition algorithms

Activity	Training and Evaluating on the same person			Training and Evaluating on different people		
	Relative Energy	Absolute Features	HMM	Relative Energy	Absolute Features	HMM
Cycling	94.20	78.78	76.63	93.15	69.91	73.51
Typing	97.55	99.32	69.07	96.33	94.32	67.75
Walking	99.92	96.76	74.85	97.52	82.88	72.38
Writing	96.33	99.91	79.48	93.94	93.51	78.52
Overall	97.00	93.86	75.01	95.23	85.15	73.04

Bayesian learner is used for activity recognition based on the absolute features discussed above. *Recognition based on HMMs* comparison was also conducted with the activity recognition algorithm proposed in [8] that uses HMMs.

In the first experiment, the orientation of the motes was fixed. The average F_1 statistics [10] for each activity in both settings are presented in Table 1. In both cases, the classifier that uses relative energy shows the highest recognition rate on all activities. Although the HMM algorithm used more features than the Bayesian classifier with absolute features (some of those are the same), the recognition rate of the HMM is the worst of the three. Furthermore, the accuracy of the classifier using relative energy is almost the same for all activities. These results suggest that features based on relative energy allocation are not only a good characterization of dynamic activities but also possibly more stable across subjects.

Next, we analyze the effect of rotation on the performance of the recognition algorithms. Figure 2 shows the recognition rates for four activities under variation of the orientation of the motes. Variation in the orientation has virtually

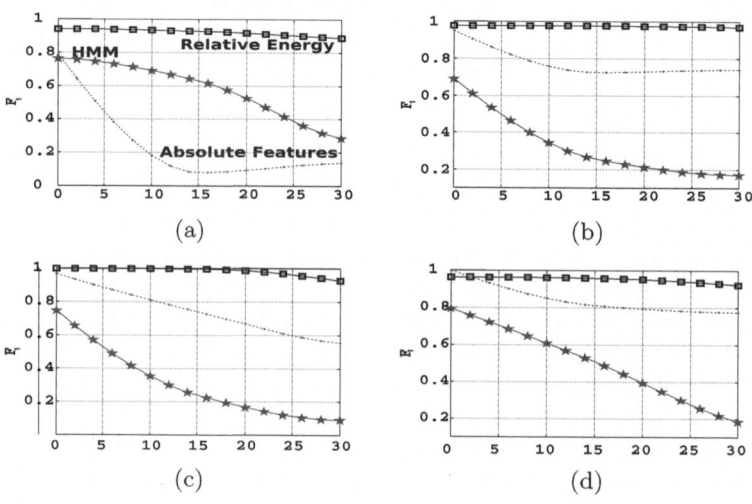

Fig. 2. The recognition rates of activities under the variation of the orientation of the motes. (a) Cycling, (b) typing, (c) walking, and (d) writing.

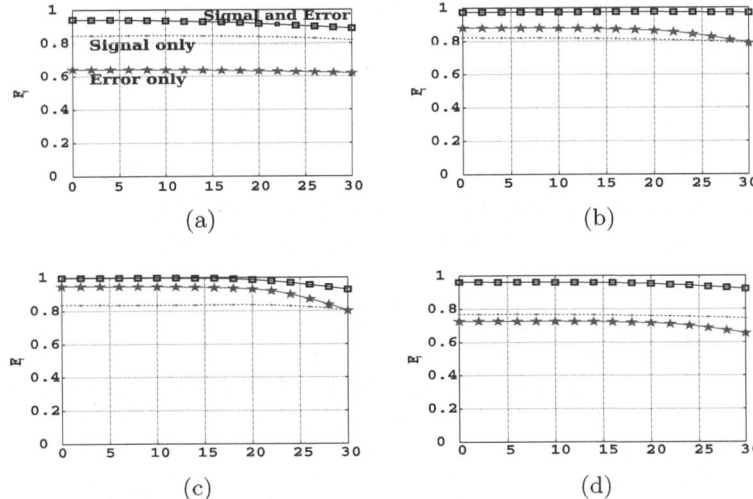

Fig. 3. The recognition rates of the classifiers with relative features in three settings: signal only, noise only, and both. (a) Cycling, (b) typing, (c) walking, and (d) writing.

no effects on our proposed algorithm over all activities. On the other hand, the rate of recognition of the algorithm based on absolute features decreases from 93.86% at $\psi = 0$ to 66.67% at $\psi = 30$. The F_1 of the algorithm based on HMM decreases from 75.01% at $\psi = 0$ to 15.23% at $\psi = 30$, which is the worst of the three. This result suggests that the HMM-based algorithm may not be a good choice for human activity recognition in practice when inaccuracies in sensor orientation may occur.

In the last experiment, we compared the performance of the classifier with relative energy features in three cases: using relative filtered signal energy alone, using relative error energy alone and using both relative feature sets. Figure 3 compares the performance of the three cases. The accuracy of the algorithms using only one feature set ranges from 72% to 90% while the algorithm using both features achieves a 97% accuracy on average. Furthermore, the low-frequency energy features are better at recognizing cycling and writing, while the relative error energy features are better at recognizing typing and walking. This result confirms that both feature sets are complementary and are of equal importance in improving the accuracy of the classifier.

4   Conclusion

This work presented an algorithm for dynamic activity recognition which is more robust than prior art with respect to sensor orientation inaccuracies. The algorithm uses a new feature set for activity recognition based on relative body-energy distribution characteristic to different activities. It considers separately both low-frequency acceleration components and the resulting modeling error.

The algorithm maintained its high accuracy despite changes in sensor orientation while the accuracy of other recognition algorithms was shown to decrease substantially.

References

1. Xie, L., Chang, S.F., Divakaran, A., Sun, H.: Unsupervised Mining of Statistical Temporal Structures in Video, pp. 280–307. Kluwer Academic Publishers, New York (2003)
2. Bao, L., Intille, S.S.: Activity recognition from user-annotated acceleration data. In: Ferscha, A., Mattern, F. (eds.) PERVASIVE 2004. LNCS, vol. 3001, pp. 1–17. Springer, Heidelberg (2004)
3. Ravi, N., Dandekar, N., Mysore, P., Littman, M.L.: Activity recognition from accelerometer data. In: AAAI, California, pp. 1541–1546. AAAI Press / The MIT Press (2005)
4. Bachmann, E.R., Yun, X., McGhee, R.B.: Sourceless tracking of human posture using small inertial/magnetic sensors. In: Proceedings of IEEE International Symposium on Computational Intelligence in Robotics and Automation, July 2003, vol. 2, pp. 822–829 (2003)
5. Antonsson, E.K., Mann, R.W.: The frequency of gait. Journal of Biomechanics 18(1), 39–47 (1985)
6. Sun, M.S., Hill, J.O.: A method for measuring mechanical work. Journal of Biomechanics 26, 229–241 (1993)
7. Bar-Shalom, Y., Li, X.R.: Estimation with Applications to Tracking and Navigation. John Wiley & Sons, Inc., New York (2001)
8. Ganti, R.K., Jayachandran, P., Abdelzaher, T.F., Stankovic, J.A.: Satire: A software architecture for smart attire. In: MobiSys 2006: Proceedings of the 4th International Conference on Mobile Systems, Applications and Services, pp. 110–123 (2006)
9. Bishop, C.M.: Pattern Recognition and Machine Learning. Springer, New York (2006)
10. Van Rijsbergen, C.J.: Information Retrieval. 2 edn. Dept. of Computer Science, University of Glasgow, London (1979)

SenQ: An Embedded Query System for Streaming Data in Heterogeneous Interactive Wireless Sensor Networks*

Anthony D. Wood, Leo Selavo, and John A. Stankovic

Department of Computer Science
University of Virginia
{wood,selavo,stankovic}@cs.virginia.edu

Abstract. Interactive wireless sensor networks (IWSNs) manifest diverse application architectures, hardware capabilities, and user interactions that challenge existing centralized [1], or VM-based [2] query system designs. To support in-network processing of streaming sensor data in such heterogeneous environments, we created SenQ, a multi-layer embedded query system. SenQ enables user-driven and peer-to-peer in-network query issue by wearable interfaces and other resource-constrained devices. Complex virtual sensors and user-created streams can be dynamically discovered and shared, and SenQ is extensible to new sensors and processing algorithms. We evaluated SenQ's efficiency and performance in a testbed for assisted-living, and show that on-demand buffering, query caching, efficient restart and other optimizations reduce network overhead and minimize data latency.

1 Introduction

Wireless sensor networks enable fine-grained collection of sensor data about the real world. Applications are growing in military, environmental, health-care, structural monitoring, and other areas and many sensing modalities are now available. Integrating them on embedded platforms and efficiently managing their data remains a challenge due to application constraints on form-factor and high cost sensitivity.

One growth area for wireless sensor networks is the health-care domain, which already uses a wide variety of sensors, and could benefit from their dynamic deployment. For example, based on an assisted-living resident's health, a doctor may give a box of sensors to place in the apartment or to be worn by the resident. An emplaced sensor network provides a rich context for residents' environmental conditions—but it must integrate with body area networks, embedded user interfaces, and back-end control and storage. The result is a heterogeneous and highly *interactive* wireless sensor network (IWSN) that presents new challenges for query systems.

This work describes SenQ, an embedded query system for IWSNs that makes several contributions to the state of the art:

---

* Supported by NSF grants CNS–0435060, CNS–0626616, and CNS–0720640.

S. Nikoletseas et al. (Eds.): DCOSS 2008, LNCS 5067, pp. 531–543, 2008.

- We identify core requirements for emerging interactive wireless sensor networks (IWSNs) and present the design of SenQ, a query system for in-network monitoring that addresses challenges from heterogeneity of: application architectures, user interfaces, device capabilities, and information flows.
- SenQ flexibly supports both hierarchical application architectures and ad hoc decentralized ones, by providing a stack with loosely coupled layers that may be placed independently on devices according to their capabilities, and by enabling in-network peer-to-peer query issue for streaming data. Its standardized software interfaces and protocol mechanisms support extensibility for new sensors, processing algorithms, and context models.
- A novel shared stream abstraction uses virtual sensors to encapsulate complex data processing and to discover and re-use dynamic user-created streams.
- Evaluation on a real implementation shows good performance of the software stack: query caching with restart halves the internal first-data latency to $670\,\mu s$, and we show sampling rates up to $1\,KHz$ with low jitter without ADC DMA or other specialized hardware.

The features of SenQ enable embedded control loops, user reminders, real-time delivery of body network data and other rich interactions within the system that are not supported in any integrated way by existing query systems. We evaluate SenQ in the context of AlarmNet [3], a testbed for assisted-living.

2   Research Challenges in Interactive WSNs

Emerging IWSN systems present research challenges and constraints that are not satisfied by existing query system designs in an integrated way. Here we identify the key challenges that SenQ addresses.

Heterogeneity. Diversity in IWSNs spans device types, capabilities, interface modalities, user-created data flows, and system architectures.

Deployment dynamics. Because IWSNs are human-centric and interactive, network membership and data flow patterns may change continuously.

In-network monitoring. Point-to-point data streams that remain entirely within the network are needed for several situations. 1) *Decentralized control loops* require fast access to sensor data to provide predictable performance. 2) *Embedded user interfaces* allow the ad hoc creation of data streams for personal consumption, and are a vital interaction method for applications.

Localized aggregation. Due to the heterogeneity of sensor types and information flows in IWSNs, most spatial aggregation is within small areas, such as a body area network that combines wearable accelerometer data to detect falls.

Resource constraints. Processor, energy, and memory capabilities of embedded sensor and interface devices are limited, especially for unobtrusive wearable devices with small form factors and low cost.

Data querying and management has received considerable attention in WSN literature, but we are aware of no systems that comprehensively address the additional requirements and challenges of interactive WSNs identified here. After briefly reviewing related work, we present the design of SenQ in Section 3.

2.1 Brief Examples of Related Work

Emerging IWSN systems must support *distributed data access* not just by back-end servers, but by users possessing a wide range of expertise. For traditional back-end interfaces, we want to retain the benefits of declarative query languages like that provided by TinyDB [1]. But they are unsuitable for most *embedded interfaces* with limited capabilities. Virtual machines like Maté [2] and SwissQM [4] provide a flexible programmatic approach useful for sophisticated in-network processing, but their low-level abstractions and compilation and interpretation overhead are not well suited for in-network query issue.

In TAG [5], sampling periods or epochs are sub-divided among nodes in a path from the source to the sink. Data flows up the tree in a synchronized fashion to ensure parents can receive and process the data before relaying it themselves. TinyDB [1] distributes queries by flooding, and uses semantic routing trees to prune the sensors involved in query execution. For relatively high-rate streams, however, the delays involved in these approaches may be prohibitive.

TinyDB and Cougar [6] provide declarative query languages similar to SQL that hide many of the details of network operation from the user and ease construction of queries. However, textual query languages are less useful for embedded user interfaces or sensor devices themselves, and efficiency suffers if the queries must be relayed to a server for parsing and execution. SenQ provides a uniform programmatic abstraction and network protocol that can be used directly by embedded applications.

In the TENET architecture [7], only resourceful nodes are allowed to perform data fusion in a strictly tiered network. VanGo [8] similarly requires the use of micro-servers for adaptive processing. They take advantage of an ADC DMA capability to provide high rate sampling, and have a static processing chain compiled into the motes.

3 SenQ Query System Architecture

The requirement for decentralized, in-network monitoring had a large impact on the design of SenQ, particularly its layered structure shown in Figure 1(b). In a full system deployment, the lowest layers 1 and 2 (sensor sampling and query processing, respectively) reside on embedded sensor devices. Since these are heavily resource-constrained, the software must be efficient with small memory footprint. Layer 3 (query management and storage) resides on micro-server gateways with more abundant resources, such as a connected database and back-end systems. Layer 4 is a high-level declarative language, SenQL, similar to SQL and TinyDB [1] for external user-issued queries. Due to space constraints, we only present the design details of SenQ's bottom two layers.

Arrows in Figure 1(b) show the nominal data flow: from query language to micro-server for authorization, binding, and translation, into the WSN to the

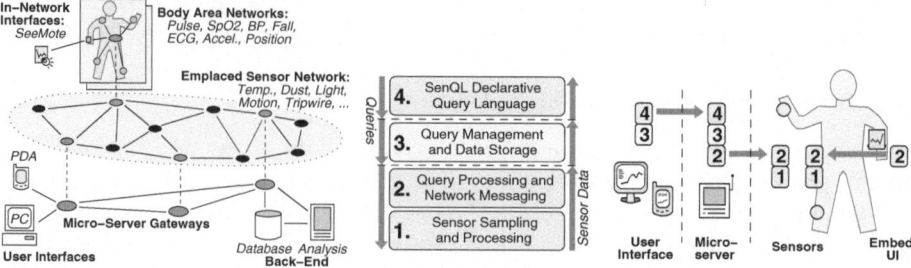

(a) Topology and components of AlarmNet, a prototypical IWSN for assisted-living.

(b) SenQ query system stack. In ad hoc mode the upper two layers are optional.

(c) Loosely coupled layers may be separated and placed on heterogeneous devices as resource constraints allow.

Fig. 1. SenQ supports both hierarchical and ad hoc architectures by maximizing layer independence

sensor device where it is parsed and activated; then streams of data flow back through the micro-server to the user's interface.

However, embedded user interfaces and sensor-initiated queries characteristic of IWSNs are more efficiently supported when the upper layers are bypassed. And not every system will have an architecture with centralized control, even if only temporarily. For example, an elder wearing a body area network disconnects from AlarmNet's infrastructure (shown in Figure 1(a)) while visiting the doctor. It is desirable for the body network to continue to monitor health status and allow the doctor to query vitals in real-time. Other systems may use only sensors and embedded UIs all the time in a low-cost ad hoc network topology.

To satisfy diverse application needs, we designed the layers of SenQ's stack to be *loosely coupled* with well-defined interfaces throughout.

3.1   Query and Data Model

To solve the research challenge of supporting significantly different information flows simultaneously with low delay, SenQ provides snapshots and streams.

Streaming queries specify the sensor, sampling rate, processing chain, and whether to perform local-area spatial aggregation. As data is collected, reports are streamed back to the requester until a *Stop* command is received or an optional maximum duration is reached. *Snapshot* queries provide efficient point-in-time samples of raw sensor values that bypass the entire processing chain for minimal response times.

For both types of queries, the returned sensor data is composed of *<timestamp, value>* tuples. Query caching allows them to be efficiently *stopped and restarted* later with a short command from the originator that minimizes communication, parsing, and startup overhead.

Queries are uniquely identified in the network by *<source address, ID>* tuples to allow multiple concurrent queries on sinks and sources—a requirement for the interactive, peer-to-peer traffic flows in IWSNs. The 4 KB of data memory on the MicaZ mote limits SenQ to 21 concurrent queries on each sensor.

3.2 Sensor Sampling and Processing Layer

In contrast to many environmental monitoring networks, IWSNs support a wide variety of sensor types—there are currently twenty in AlarmNet. Sensor data is accessed using internal components, external ADC channels, UART serial links, or interrupts. This heterogeneity complicates the addition of new sensors.

The Sampling and Processing layer (shown in Figure 2) encapsulates access to onboard resources to insulate applications from the complexity. Standard interfaces for the sensor drivers and processing blocks allow them to be easily incorporated and plugged-in (wired) at compile time, and enables the virtual sensor feature described later in Section 3.3.

SenQ treats the sensor and processing types opaquely, so that old devices ignore unknown types. This lets new sensor types be deployed dynamically into the network without having to reloading code on existing devices, and maintains continuity of operation for environments—such as health-care—where it is not practical or safe to download new code and reset the system.

Sensor Drivers. Sensor drivers are categorized by the timing and regularity properties of their access, since these determine the most efficient way for SenQ to sample them. *EventSensors* generate data sporadically as it becomes available, such as from an interrupt, and so need not be sampled periodically. *SplitPhaseSensor* represents sensors which must read and convert data upon request, such as from analog sensors connected to an ADC. Data is provided asynchronously to SenQ. Data that may be quickly read synchronously uses the *PollableSensor* interface. Drivers also provide a *SensorInfo* interface to aid runtime discovery of nodes' capabilities and types.

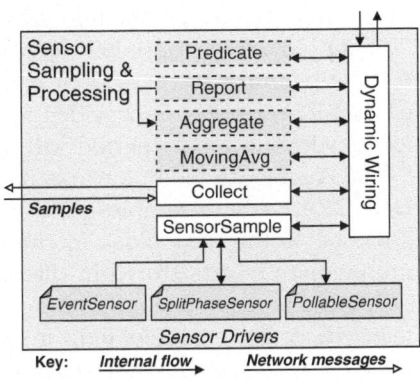

Fig. 2. Layer one in SenQ samples and processes sensor data

Physiological sensors in AlarmNet include pulse oximetry (Harvard [9] and UVA designs), wearable two-lead electrocardiography (Harvard [9]), and body weight and blood pressure devices by A&D Medical. Long-running background streams monitor residents' environmental conditions, such as air quality, light, temperature, and humidity. Sensors detect motion and activity to inform context-aware back-end algorithms using PIR motion, optical tripwires, magnetic reed switches, and wearable accelerometers for classifying movement-based activities.

Processing Chain. Some sensors require little in-network processing, but for those that provide a high-volume of data, it is essential to reduce both the energy cost and network congestion from sensor streams by filtering at the source.

Above the hardware and sensor drivers is a group of modules comprising a scheduler and data processing chain, collectively called the Sampler. They act in concert to manage sensor sampling for multiple concurrent queries, and filter generated data according to the application query.

Processing modules provide a *ProcessControl* interface with standard control and configuration methods. These allocate structures, pass query configuration to the process, and supervise sampling. As with sensor drivers, the interface eases the extension of SenQ to new, application-specific processing algorithms.

DataProducer interfaces on each block provide incoming and outgoing paths for sensor data. Instead of a static sequence (as in VanGo [8]), blocks are wired to a dynamic data flow coordinator. This provides more flexibility to the application, since each query has its own ordering of the processing chain.

The *SensorSample* module maintains a schedule for sensor sampling to satisfy multiple ongoing queries. A single timer tracks the next sampling operation. Upon expiry, data is requested from the driver according to its category, *Event*, *SplitPhase*, or *Pollable*. When data is available it is propagated up the processing chain to the next consumer, as determined by the query's dynamic wiring.

All queries that have concurrently requested the same type of sensor data are notified upon its availability. This sample caching is necessary to promptly service queries despite the limited bandwidth of the ADC.

SenQ provides an optional *Aggregate* module that is optimized for functions or descriptive statistics that can be calculated while keeping little state. In addition, a "latch" aggregator is provided for polled binary EventSensors. An event that occurs within a report period will be remembered until the next report.

The Aggregate module works in conjunction with the *Report* module, which drops intermediate samples until a specified period has passed. Then the data is passed to the next block in the chain, and the module flushes or resets the intermediate results stored in the associated Aggregate module.

Specifying a report period longer than the sample period provides *temporal aggregation*. For example, light may be sampled every second but only reported every 4s, with intermediate results averaged by the Aggregate module.

An optional *Moving Average* module provides a windowed average, moving average, or exponentially-weighted moving average (EWMA) with specified window or α parameters.

Finally, queries may cull irrelevant or redundant data by specifying relational *Predicate* filters. The query specifies the argument values for comparison.

Figure 3 shows the memory tradeoff between the Aggregate and Moving Av-

Software Configuration	Code	Data
Base (4 query, 70B payload)	19010	1751
Process: Aggregate	+ 1016	+ 36
Process: Aggregate + Report	1620	98
Process: Moving Average	968	240
Process: Predicate	602	44

Fig. 3. Memory consumption for processing modules in bytes on MicaZ

erage modules. With four concurrent queries and a window of ten 32-bit elements, the increase in data memory usage over the Aggregate + Report configuration is 142 bytes, but with 652 fewer bytes of code memory. The variety of such tradeoffs among applications and hardware platforms is the reason SenQ preserves modularity of processing algorithms in the design.

Spatial Aggregation. IWSNs may not need to sample and aggregate sensor data from the entire network, due to their heterogeneity. For example, although aggregate environmental data is useful in AlarmNet, physiological and activity data must not be mixed among residents.

Spatial aggregation is needed more often for collecting data from other nodes in the local area, such as in a body area network. A flag in the stream query specifies that the recipient is to act as the coordinator of a spatial query. This coordinator node sends the query to its immediate neighbors, including its own network address and sample ID.

Neighboring devices possessing the requested sensor type then process the delegated query. Sensor data is sampled and flows up the processing chain, eventually passing through the *Collect* module. The samples are redirected over the network to the query coordinator's Collect module in a *Sample* message (shown in Figure 2), where they are combined with local samples and inserted into the coordinator's processing chain.

Overall, the Sensor Sampling layer provides flexible mechanisms for access to heterogeneous sensor types and extension to new ones. However, so far the data is available only locally. For application and external access we need a Query Processing layer, which is now described.

3.3   Query Processing and Network Messaging Layer

A Query Processing layer (Figure 4) provides stream and snapshot abstractions to local applications via a software API, and remote ones via a network protocol. Queries for local sensors allocate resources in the Sampling and Processing Layer below, configure the processing chain, and start collection of data. Data is buffered (if the query allows) and reported to the originator, whether local or remote.

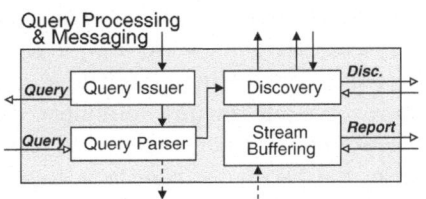

Fig. 4. Layer two in SenQ may stand alone or above the Sampler

SenQ provides *location transparency* with respect to embedded applications, since they use the same interface to issue queries for local or remote sensors, and the networking aspects are hidden behind a QueryProcessor software interface. Queries for remote sensors are marshalled and routed to the destination, while local queries are managed by the resident Query Processing and Sampling layers.

Embedded query issue capabilities distinguish SenQ from most existing solutions and open many possibilities for smart in-network processing, human interactivity, and embedded control loops. This accrues particular benefits in decentralized, large-scale networks where many data flows exist. Centralized approaches like TinyDB [1] and SwissQM [4] do not cater well to networks in which many point-to-point streams are dynamically created between embedded query issuers and sensors.

Control loops can be embedded in the network so, for example, a light controller sends long-lived queries to a room's door or motion sensors and acts directly and locally on the lights. Control input, decision, and actuation are all close to the data source and suffer lower delays than a centralized approach.

SenQ maximizes *layer independence* to provide flexibility for heterogeneous platforms and architectures. An embedded device with no sensors includes only a Query Processing layer, so it can issue queries. In AlarmNet, the MicaZ-based SeeMote [10] (shown at right) includes SenQ, a graphics library, and a real-time data visualization application in only 4 *KB* of SRAM.

A *Discovery* component allows an embedded application to locate nearby devices, sensor types, and processing modules. Rather than relay queries through the gateway, which incurs extra delay and energy costs, smart sensors can discover each other and issue queries directly.

Virtual Sensors. User-driven creation of data streams in an open environment is a central characteristic of IWSNs, and a query system should support three classes of users: developers, application domain experts, and the ordinary user. An "ordinary user" of an IWSN could be an assisted-living resident, an elder or soldier recovering at home from an injury, or a student immersed in a campus-wide sensor network.

High-level declarative languages are good for application experts knowledgable about relational database abstractions and the capabilities of the system. However, these languages are a poor choice for system developers who must create specialized processing algorithms, and as a basis for network protocols they are too verbose and require complicated parsing.

SenQ enables a developer to use its embedded query issue capabilities to collect data streams from both local and remote sensors for custom processing, and then export the results as a *virtual sensor* at the bottom of the stack that conforms to the sensor interfaces described in Section 3.2. This encapsulates the complex, hierarchical stream processing as a low-level sensor type that can be discovered, queried, and viewed as any other.

Shared Streams. Declarative and programmatic access methods support domain experts and developers—but they do not consider the ordinary user. This user may face challenges of: 1) unfamiliarity with relational databases and programming, 2) embedded interfaces with poor input capabilities, and 3) uncertainty of domain parameters (e.g., age-appropriate "normal" heart rates).

Shared streams build on the virtual sensor capability to address these challenges. A domain expert crafts a custom stream Q at runtime using an appropriate interface, and enables sharing of the query. The Query Processing layer dynamically allocates a virtual EventSensor VS in the Data Sampling layer, and then the Discovery component advertises VS as a primitive sensor type. When a new query Q' is received for *VS*, the Sampling layer recursively activates the Query Processing layer for query Q.

Together, virtual sensors and stream sharing enable novel ad hoc user-to-user interactions in the IWSN that are usually outside the scope of other query approaches. Systems using TinyDB or Cougar for declarative data access, or Maté or SwissQM for virtual machine-based access would have to develop additional protocols or user interfaces to provide this capability. By supporting re-use of custom-crafted sensor streams, SenQ helps to address the challenges of providing open access to ordinary users.

Network Efficiency. The query processing layer uses several techniques to maximize performance for streams in resource-constrained embedded systems.

Combining multiple samples received by the Query Processor into a single report message saves overhead and reduces radio traffic—at the expense of latency. Query originators may specify full buffering or on-demand buffering.

On-demand buffering is used when a sample has been received in the Query Processor and is ready to be transmitted, but the outgoing message buffer is busy due to channel congestion. This incurs less average latency than full-buffering (though has more overhead) and avoids dropping high-rate samples. It represents a tradeoff between latency and loss.

Reports with data are timestamped to allow the receiver to properly sequence the data and detect drop-outs, in case the underlying routing provides out-of-order delivery or messages are lost in-transit. Reports also bear status changes, such as positive and negative acknowledgements with cause codes, which are important for meaningful user feedback on embedded UIs.

Compacting reports reduces energy wasted in the transmission of redundant data. Other compression schemes, such as run-length encoding, can be added as processing plug-ins if an application warrants the additional computation.

Memory constraints of WSN devices limit the number of queries that may be simultaneously serviced or stored. Inactive queries are replaced using a least-recently used policy to maximize the ability of applications to restart them later. Restarting a *cached query* that has already been parsed and configured is twice as fast as issuing a new one.

3.4   Query Management and SenQL Layers

Upper layers in Figure 1(b) are described here only briefly due to lack of space.

In hierarchical networks, a *Query Management* layer provides services that are important for usability, context-awareness, connectivity, and data analysis. It manages device registration and client connections, and maps queries for abstract entities (e.g., people and places) onto particular devices.

The *SenQL* layer provides a declarative query language to users, allowing them to specify what data is desired independent of how it will be collected. It uses a constrained subset of SQL–99 with extensions for SenQ functionality.

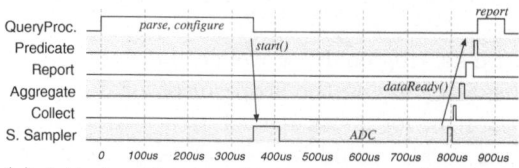

Query Operation		mean	stddev
Stream start (initial)		654.18	176.352
	(subseq.)	217.62	0.003
Stream stop		36.93	2.639
Stream restart		267.10	10.904
Snapshot		601.05	1.745
Snapshot restart		301.28	1.684

(a) A timing diagram showing receipt of a stream query, message parsing, Sampler configuration, and propagation of first datum up the processing chain until a report is ready to send.

(b) Time in μs from start of operation until first report is ready (if applicable). Does not include 366.42 μs ADC time. 100 trials.

Fig. 5. SenQ timing measurements on the MicaZ mote

4   Evaluation

We present SenQ's consumption of memory and CPU resources and show the performance limitations of the sampling and processing chain to demonstrate runtime efficiency.

4.1   Resource Consumption and Efficiency

The program and data memory required for SenQ depends on three application-specific parameters: the size of TinyOS messages, the sensor drivers linked in, and the number of maximum concurrent queries supported. Table 1 shows the size in bytes of TinyOS applications with different parameters. In the minimum configuration, all non-query related modules (such as localization, configuration, etc) are removed from the mote application that runs in AlarmNet. No sensor drivers are included, and the default TinyOS message length is used. The code takes about 16KB, with 711B data memory. A configuration more typical for use in AlarmNet is also shown: "Base" provides access to the internal mote voltage, supports four concurrent queries (per-node), and uses 70 byte payloads. Each included sensor driver requires code and data memory in addition to the Base.

We measured SenQ's load on the sensor device using an Intronix LogicPort logic analyzer. This gave accurate profiling of processing times with very little measurement overhead.

A timing trace from one experiment shows the relative magnitudes of processing times (shown in Figure 5(a)). A stream query samples the node's internal battery voltage every 73 ms, with no processing or buffering in the QueryProcessor. The timeline starts when the query is received from the network, and shows the time to parse it, allocate data structures, and configure and start the Sampler (358 μs).

SensorSampler started the sampling timer and requested data from the Voltage driver (50 μs). ADC conversion takes 25 ADC clock cycles, or 366.42 μs from the request until the data is available. Then the SensorSampler propagates it up the processing chain. Since no processing was specified, it reaches the QueryProcessor 68 μs later.

QueryProcessor generates a report immediately since buffering is not enabled. The total time spent on the first sample is $918\,\mu s$, of which $366\,\mu s$ is waiting for the ADC. Subsequent samples begin at the SensorSampler module when the sample timer fires.

The mean and standard deviation of worst-case execution times from 100 trials are shown in Figure 5(b) for each query operation. Queries sampled battery voltage, used *mean* aggregation, a *range* predicate filter, no buffering, and four-bytes of reported data. These parameters together give the largest possible execution overhead of a non-coordinated (spatially distributed) query.

Table 1. Memory consumption for sensor drivers in bytes. Includes required TinyOS components (radio, timer, ADC, etc). Values for sensors are relative to the Base configuration.

Software Configuration	Code	Data
Minimum (no sensors, 1 query, 29B payload)	16756	711
Base (voltage, 4 queries, 70B payload)	18830	1576
Sensor: Blood Pressure	+ 698	+ 29
Sensor: Pulse, SpO2, Heartbeat	1140	32
Sensor: ECG (Tmote Sky)	138	4
Sensor: Scale	1366	27
Sensor: Dust	414	25
Sensor: Motion, Light	594	0
Sensor: MTS300 (temp, photo)	1524	33
Sensor: MTS310 (+accel, mag)	2032	45
Sensor: Switch	502	4
Sensor: Tripwire	2834	76
Sensor: Fall	4654	105

Even on the $8\,MHz$ MicaZ, these worst-case execution times leave most CPU resources for application demands. At a sampling rate of $100\,Hz$, the steady-state overhead of SenQ is only 2.18%.

4.2 Sampling Performance

The *maximum effective sampling* rate of SenQ depends on the execution overhead (described above) and the query's sampling rate. As the sampling rate increases beyond a point, we expect to see worse performance in sampling jitter and dropped messages or samples. This is especially true due to the limited processing capability of the MicaZ and the non-real-time design of TinyOS.

To find SenQ's limits on sampling, we use a stream query with relatively costly parameters: mean aggregate, range predicate filter, four-byte data size, and no buffering so a message is transmitted for *every* sample. The timing was captured precisely by the logic analyzer.

Figure 6(a) shows sampling jitter (difference from the requested rate) as the sampling rate varies from $4\,KHz$ to $32\,Hz$. We focus on small sampling periods here, where difficulty is more likely to occur. The actual sampling rate tracks the specified rate closely down to $2\,ms$, with low variance and only occasional aberrations as shown by the plotted maximum. Below $1\,ms$, the microprocessor fails to service the timer as fast as requested, due to frequent radio and ADC interrupts and high CPU utilization from SenQ and other tasks. Without DMA to lighten the load on the microprocessor, it saturates.

Congested conditions benefit from SenQ's on-demand data buffering, shown in Figure 6(b). When message transmission became a bottleneck at around a $4\,ms$ sampling period, reports included more samples. At $0.25\,ms$, half of the reports included nine or more samples. A maximum sample loss of 11% was

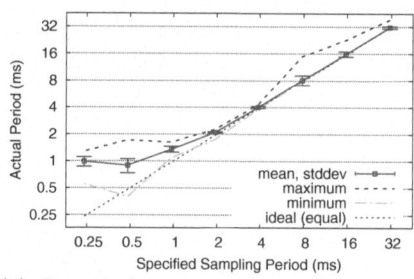

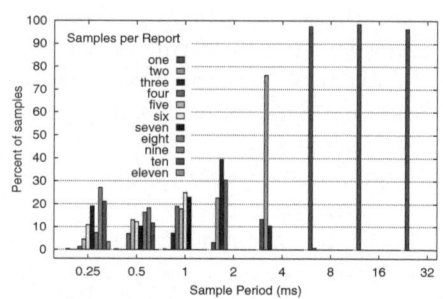

(a) Specified versus actual (measured) sample period for over 300 trials. The ideal linear plot, when specified equals actual, is also shown. Timer performance degrades for $S <= 1\,ms$.

(b) Distribution of report sizes resulting from on-demand buffering during congestion. The percentage of variable-sized reports bearing 1–11 samples is shown.

Fig. 6. SenQ sampling jitter for a processing- and transmit-intensive query, and impact on on-demand buffering

recorded at the sender for these trials. By applying this aggregation at the data source, network overhead is reduced while low latency and loss are preserved.

5   Conclusion

SenQ supports heterogeneous device types, user interfaces, data flows, processing algorithms, and application architectures. Network load and energy consumption is reduced by using temporal and spatial aggregation, filtering at data sources, data compaction, and on-demand report buffering. Virtual sensors and stream sharing enable rich user interactions, and embedded interfaces and sensor devices can issue queries without the aid of powerful back-end servers. SenQ enables distributed smart networking for the kind of interactive systems we expect to see in the near future.

References

1. Madden, S., Franklin, M.J., Hellerstein, J.M., Hong, W.: TinyDB: An acqusitional query processing system for sensor networks. ACM TODS 30(1), 122–173 (2005)
2. Levis, P., Culler, D.: Maté: a tiny virtual machine for sensor networks. In: Proc. of ASPLOS, pp. 85–95 (2002)
3. Wood, A., Virone, G., Doan, T., Cao, Q., Selavo, L., Wu, Y., Fang, L., He, Z., Lin, S., Stankovic, J.: ALARM-NET: Wireless sensor networks for assisted-living and residential monitoring. Technical Report CS-2006-11, Department of Computer Science, University of Virginia (2006)
4. Müller, R., Alonso, G., Kossmann, D.: A virtual machine for sensor networks. In: Proc. of EuroSys, pp. 145–158 (2007)
5. Madden, S.R., Franklin, M.J., Hellerstein, J.M., Hong, W.: TAG: a Tiny AGgregation service for ad-hoc sensor networks. In: Proc. of OSDI, pp. 131–146 (2002)

6. Yao, Y., Gehrke, J.E.: The Cougar approach to in-network query processing in sensor networks. SIGMOD Record 31(3), 9–18 (2002)
7. Gnawali, O., Greenstein, B., Jang, K.Y., Joki, A., Paek, J., Vieira, M., Estrin, D., Govindan, R., Kohler, E.: The TENET architecture for tiered sensor networks. In: Proc. of SenSys, pp. 153–166 (2006)
8. Greenstein, B., Mar, C., Pesterev, A., Farshchi, S., Kohler, E., Judy, J., Estrin, D.: Capturing high-frequency phenomena using a bandwidth-limited sensor network. In: Proc. of SenSys, pp. 279–292 (2006)
9. Malan, D., Fulford-Jones, T., Welsh, M., Moulton, S.: Codeblue: An ad hoc sensor network infrastructure for emergency medical care. In: Proc. of BSN (2004)
10. Selavo, L., Zhou, G., Stankovic, J.A.: SeeMote: In-situ visualization and logging device for wireless sensor networks. In: Proc. of BASENETS, pp. 1–9 (2006)

SESAME-P: Memory Pool-Based Dynamic Stack Management for Sensor Operating Systems[*]

Sangho Yi[1], Seungwoo Lee[2], Yookun Cho[1], and Jiman Hong[3],[**]

[1] School of Computer Science and Engineering, Seoul National University
{shyi,cho}@os.snu.ac.kr
[2] Flash Development Division, Hynix Cooperation
seungwoo1.lee@hynix.com
[3] School of Computing, Soongsil University
jiman@ssu.ac.kr

Abstract. In wireless sensor networks, each sensor node has very small memory space compared with any other embedded computing systems. For this reason, operating systems running on the sensor nodes cannot allocate sufficient fixed-size stack space for all threads. In the previous work, SESAME was proposed to allocate stack space more space-efficiently, but there is a problem of time overhead. In this paper, we present SESAME-P, which is a dynamic stack allocation scheme based on memory pool. The size of memory pool is predetermined by using static analysis of each function's stack usage information. Using the determined memory pool, SESAME-P reduces the dynamic stack allocation cost. Our experimental results show that SESAME-P significantly reduces the time overhead compared with the existing SESAME.

1 Introduction

In these days, many kinds of operating systems[1,2,3,4,5,6] have been proposed for wireless sensor networks. They can be categorized into two classes according to how they operate and design their kernel architecture. The two kinds of classes show significant difference in their performance and efficiency of the systems and applications.

TinyOS[1], SOS[2], and Contiki[5,7] are the event-driven sensor operating systems. The internal structure of event-driven system gives us small context switching latency and less memory usage based on a single stack management. However, response time and preemptivity are much poorer than that of the multi-threaded systems[8]. MANTIS OS[4], Nano-Qplus[3], and RETOS[6] are the multi-threaded sensor operating systems. The multi-threaded design enables full preemption, but necessary memory space for the threads' stack space is much greater than that of the event-driven[9]. Each sensor nodes typically 2~10kB

---

[*] This Research was supported by the Soongsil University Research Fund.
[**] Corresponding author.

S. Nikoletseas et al. (Eds.): DCOSS 2008, LNCS 5067, pp. 544–549, 2008.

RAM, and therefore, it is very hard to allocate sufficient stack space for all threads. If the stack space is insufficient, stack overflow may occur on the sensor nodes. To remove stack overflow problem, we proposed SESAME[10] in the previous work. SESAME adaptively adjust stack size by allocating or releasing additional stack frame based on the amount of each function's stack usage. In this way, SESAME can eliminate stack overflow problem while improving space-efficiency. However, the dynamic stack management cost significantly increases the thread's execution time because stack allocation and release operations are required for each function call instruction.

In this paper, we present SESAME-P, which is a memory pool-based dynamic stack allocation scheme for sensor operating systems. The size of memory pool is determined based on the static analysis of each function's stack usage information. Using the memory pool, SESAME-P reduces the dynamic stack allocation cost. Our experimental results show that SESAME-P significantly reduces the time overhead compared with the existing SESAME.

2   Related Works

In [11], Torgerson proposed an automatic thread stack management scheme on MANTIS sensor operating system based on compile-time calculation of the upper-bound of the thread stack size. This scheme mitigates trade-off relation between stack overflow and stack space efficiency on the multi-threaded task model. However, it cannot remove the space inefficiency which caused by static allocation of each thread stack area.

In [12], Gustafsson briefly introduced concepts of the stackless implementation of C/C++ programs to minimize memory usage. For example, C compiler can predict maximum required stack space, and the compiler would allocate space for function arguments and local variables on the heap instead of a stack.

In [10], Yi et al. proposed a dynamic threads' stack allocation scheme for resource-constrained sensor nodes, called SESAME. In order to efficiently utilize memory usage, SESAME provides adaptive and on-demand stack allocation for each thread based on the amount of stack used for each function call. The information of each function's stack usage is pre-calculated using source codes at compile-time, and dynamic threads' stack allocation is performed at run-time. However, there is a significant overhead in terms of execution time, because SESAME allocates and releases stack frame at every function call instruction.

3   Design and Implementation of SESAME-P

The original SESAME[10] was proposed to minimize stack usage on multi-threaded embedded systems. In order to efficiently utilize memory usage, it provides dynamic stack allocation for each thread based on the calculation method of the amount of stack used for each function call. In SESAME, the space efficiency will be significantly improved, but the several times of stack allocation may increase total execution time. Such multiple memory allocation operations

Table 1. List of notations used in this paper

Notation	Description
$M_i(j)$	stack space requirement for a jth interrupt service routine
n_{isr}	number of interrupt service routines
$M_f(k)$	stack space requirement for a kth function
n_{fnc}	number of functions
$P(x,y)$	a memory pool for functions from x^{th} to y^{th}
$S(x,y)$	set of memory pools $P(1, x-1)$, $P(x, y-1)$, and $P(y, n_{fnc})$
$D(x,y)$	discriminant value of a set $S(x,y)$
$M_{max-isr}$	maximum stack space requirement for interrupt service routines

will increase by the number of function call instructions, and it may consume significant time. This is a big problem in the energy-constrained embedded systems such as wireless sensor nodes.

The major objective of SESAME-P is minimizing time overhead of the existing SESAME based on predetermined memory pool. SESAME-P calculates the amount of memory pool and the size of each memory chunk by using each function's stack usage. Table 1 shows several notations used in this paper.

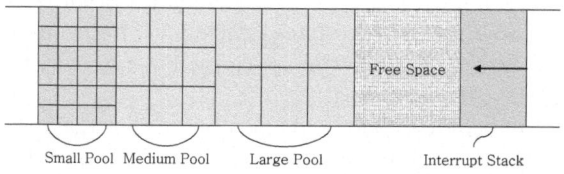

Fig. 1. Internal Memory Structure of SESAME-P

Figure 1 briefly shows a memory structure used in SESAME-P. There are three kinds of memory pools and a stack space for interrupt service routines. The size of each memory pool is determined by using the functions' stack usage information.

Determination of Interrupt Stack Space. In Table 1, we defined $M_i(j)$ and $M_f(k)$ as stack space requirement for an interrupt handler and a function, respectively. We assume that the number of interrupt service routines and functions are n_{isr} and n_{fnc}, respectively. Using the notations, we can derive $M_{max-isr}$, the maximum requirement for interrupt service rouitines, as follows.

$$M_{max-isr} = \sum_{j=1}^{n_{isr}} M_i(j)$$

The size of each $M_i(j)$ can be calculated by analyzing the assembly-level source codes. By summing the stack requirement of all interrupt service routines, we can determine the stack space for the interrupt handlers.

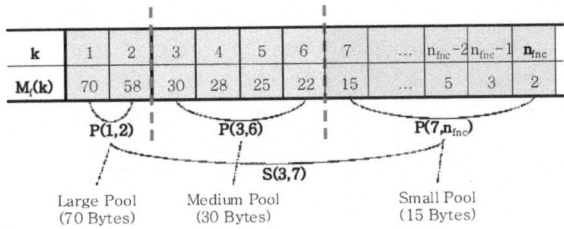

Fig. 2. An Example of Memory Pools and Related Notations

Determination of Memory Pool Size. In terms of space-efficiency, determining the size of memory pools is the most important because each function's stack requirement is different each other. In SESAME-P, each thread's stack space is allocated from an appropriate memory pool between small, medium, and large.

Figure 2 presents an example of three kinds of memory pools and the related notations used in this paper. In SESAME-P, the set of memory pools is determined by using the information of stack space requirement for each functions, $M_f(k)$. First of all, SESAME-P sorts all $M_f(k)$ by stack space requirement in decreasing order. Then, SESAME-P need to find an appropriate size of the memory pools. When using a pool-based memory allocator, it is important to minimize the amount of the internal fragmentation. The following $D(x, y)$ represents sum of the amount of internal fragmentation of every function.

$$D(x,y) = \sum_{k=1}^{x-1}(M_f(1) - M_f(k)) + \sum_{k=x}^{y-1}(M_f(x) - M_f(k)) + \sum_{k=y}^{n_{fnc}}(M_f(y) - M_f(k))$$

By calculating $D(x, y)$ for all cases, SESAME-P finds a set which has the lowest $D(x, y)$ value. Then, SESAME selects $S(x, y)$ as the set of memory pools. In this case, the size of the large, medium, and small pools becomes $M_f(1)$, $M_f(x)$, and $M_f(y)$, respectively.

4 Performance Evaluation

In our performance evaluation, we used Octacomm's Nano-24 wireless sensor platform[13]. To compare performance of SESAME-P with the original SESAME, we implemented those schemes on the Nano-Qplus[3]. Nano-Qplus is a multi-threaded sensor operating system, and it statically allocates 200 bytes of stack for each thread. We used several kinds of sensor applications, but the page limit is 6. Thus, we can show the results using only one application as follows.

  - *Tx_app* application senses and transmits information to a receiver node via wireless channel. It has 6 threads. Those are the determined size of the

Table 2. Total Execution Time and Time Overhead of *Tx_app*

Round	Total Execution Time			Time Overhead	
	Fixed Stack	SESAME	SESAME-P	SESAME	SESAME-P
100	3,120 ms	3,284 ms	3,148 ms	164 ms	28 ms
300	9,540 ms	9,997 ms	9,628 ms	457 ms	88 ms
500	15,960 ms	16,710 ms	16,108 ms	750 ms	148 ms

interrupt stack and the memory pools; interrupt stack: 115 bytes, large pool: 98 bytes, medium pool: 32 bytes, small pool: 24 bytes.

In Table 3, SESAME-P has 0.9 percent of time overhead while the original SESAME has 5 percent compared with the fixed stack management scheme. In other words, SESAME-P reduces the time overhead of the original SESAME by 82 percent.

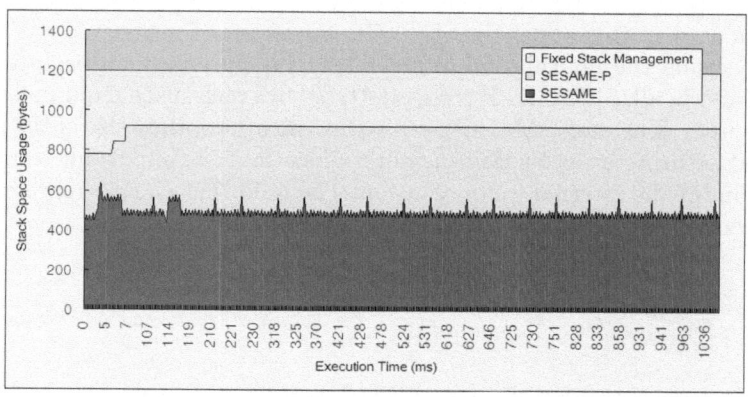

Fig. 3. Run-time stack space usage of *Tx_app*

In Fig. 3, SESAME requires about 610 bytes, and SESAME-P requires about 920 bytes of stack space. In SESAME-P, the amount of internal fragmentation may increase stack usage, but SESAME-P is still space-efficient compared with the fixed stack management scheme.

5   Conclusions

In the previous work, SESAME was proposed to allocate stack space more space efficiently, but there is a problem of significant time overhead. In this paper, we present SESAME-P, which is a dynamic stack management scheme based on memory pool. SESAME-P stands for a memory Pool-based SESAME, and it uses memory pool optimization to minimize time overhead. The size of memory pool is determined by using static analysis of each function's stack usage information.

Using the determined memory pool, SESAME-P reduces the dynamic stack allocation cost. Our experimental results show that SESAME-P significantly reduces the time overhead compared with the existing SESAME.

References

1. Levis, P., Madden, S., Gay, D., Polastre, J., Szewczyk, R., Woo, A., Brewer, E., Culler, D.: The emergence of networking abstractions and techniques in tinyos. In: First USENIX/ACM Symposium on Networked Systems Design and Implementation (NSDI 2004) (2004)
2. Han, C.C., Kumar, R., Shea, R., Kohler, E., Srivastava, M.B.: A dynamic operating system for sensor nodes. In: MobiSys, pp. 163–176 (2005)
3. Lee, K., Shin, Y., Choi, H., Park, S.: A design of sensor network system based on scalable and reconfigurable nano-os platform. In: IT-Soc International Conference (2004)
4. Bhatti, S., Carlson, J., Dai, H., Deng, J., Rose, J., Sheth, A., Shucker, B., Gruenwald, C., Torgerson, A., Han, R.: Mantis os: An embedded multithreaded operating system for wireless micro sensor platforms. ACM Kluwer Mobile Networks and Applications (MONET) Journal, Special Issue on Wireless Sensor Networks (2005)
5. Dunkels, A., Gronvall, B., Voigt, T.: Contiki - a lightweight and flexible operating system for tiny networked sensors. In: First IEEE Workshop on Embedded Networked Sensors (2004)
6. Kim, H., Cha, H.: Towards a resilient operating system for wireless sensor networks. In: The 2006 USENIX Annual Technical Conference (USENIX 2006) (2006)
7. Dunkels, A., Finne, N., Eriksson, J., Voigt, T.: Run-time dynamic linking for reprogramming wireless sensor networks. In: The 4th ACM Conference on Embedded Networked Sensor Systems(SENSYS 2006) (2006)
8. Ousterhout, J.K.: Why threads are a bad idea (for most purposes). In: Presentation given at the 1996 Usenix Annual Technical Conference (1996)
9. von Behren, R., Condit, J., Brewer, E.: Why events are a bad idea (for high-concurrency servers). In: HotOS IX: The 9th Workshop on Hot Topic in Operating Systems, pp. 19–24 (2003)
10. Yi, S., Min, H., Lee, S., Kim, Y., Jeong, I., Shin, S.Y.: Sesame: Space-efficient stack allocation mechanism for multi-threaded sensor operating systems. In: 2007 ACM Symposium on Applied Computing (ACM SAC 2007), pp. 1201–1202 (2007)
11. Torgerson, A.: Automatic thread stack management for resource-constrained sensor operating systems (2005)
12. Gustafsson, A.: Threads without the pain. ACM Queue 3, 34–41 (2005)
13. Octacomm: http://www.octacomm.net/

Author Index

Lecture Notes in Computer Science

Sublibrary 5: Computer Communication Networks and Telecommunications

Vol. 4320: R. Gotzhein, R. Reed (Eds.), System Analysis and Modeling: Language Profiles. X, 229 pages. 2006.

Vol. 4311: K. Cho, P. Jacquet (Eds.), Technologies for Advanced Heterogeneous Networks II. XI, 253 pages. 2006.

Vol. 4272: P. Havinga, M. Lijding, N. Meratnia, M. Wegdam (Eds.), Smart Sensing and Context. XI, 267 pages. 2006.

Vol. 4269: R. State, S. van der Meer, D. O'Sullivan, T. Pfeifer (Eds.), Large Scale Management of Distributed Systems. XIII, 282 pages. 2006.

Vol. 4268: G. Parr, D. Malone, M. Ó Foghlú (Eds.), Autonomic Principles of IP Operations and Management. XIII, 237 pages. 2006.

Vol. 4267: A. Helmy, B. Jennings, L. Murphy, T. Pfeifer (Eds.), Autonomic Management of Mobile Multimedia Services. XIII, 257 pages. 2006.

Vol. 4240: S.E. Nikoletseas, J. Rolim (Eds.), Algorithmic Aspects of Wireless Sensor Networks. X, 217 pages. 2006.

Vol. 4238: Y.-T. Kim, M. Takano (Eds.), Management of Convergence Networks and Services. XVIII, 605 pages. 2006.

Vol. 4235: T. Erlebach (Ed.), Combinatorial and Algorithmic Aspects of Networking. VIII, 135 pages. 2006.

Vol. 4217: P. Cuenca, L. Orozco-Barbosa (Eds.), Personal Wireless Communications. XV, 532 pages. 2006.

Vol. 4195: D. Gaiti, G. Pujolle, E.S. Al-Shaer, K.L. Calvert, S. Dobson, G. Leduc, O. Martikainen (Eds.), Autonomic Networking. IX, 316 pages. 2006.

Vol. 4124: H. de Meer, J.P.G. Sterbenz (Eds.), Self-Organizing Systems. XIV, 261 pages. 2006.

Vol. 4104: T. Kunz, S.S. Ravi (Eds.), Ad-Hoc, Mobile, and Wireless Networks. XII, 474 pages. 2006.

Vol. 4074: M. Burmester, A. Yasinsac (Eds.), Secure Mobile Ad-hoc Networks and Sensors. X, 193 pages. 2006.

Vol. 4033: B. Stiller, P. Reichl, B. Tuffin (Eds.), Performability Has its Price. X, 103 pages. 2006.

Vol. 4026: P.B. Gibbons, T. Abdelzaher, J. Aspnes, R. Rao (Eds.), Distributed Computing in Sensor Systems. XIV, 566 pages. 2006.

Vol. 4003: Y. Koucheryavy, J. Harju, V.B. Iversen (Eds.), Next Generation Teletraffic and Wired/Wireless Advanced Networking. XVI, 582 pages. 2006.

Vol. 3996: A. Keller, J.-P. Martin-Flatin (Eds.), Self-Managed Networks, Systems, and Services. X, 185 pages. 2006.

Vol. 3976: F. Boavida, T. Plagemann, B. Stiller, C. Westphal, E. Monteiro (Eds.), NETWORKING 2006. Networking Technologies, Services, and Protocols; Performance of Computer and Communication Networks; Mobile and Wireless Communications Systems. XXVI, 1276 pages. 2006.

Vol. 3970: T. Braun, G. Carle, S. Fahmy, Y. Koucheryavy (Eds.), Wired/Wireless Internet Communications. XIV, 350 pages. 2006.

Vol. 3964: M.Ü. Uyar, A.Y. Duale, M.A. Fecko (Eds.), Testing of Communicating Systems. XI, 373 pages. 2006.

Vol. 3961: I. Chong, K. Kawahara (Eds.), Information Networking. XV, 998 pages. 2006.

Vol. 3912: G.J. Minden, K.L. Calvert, M. Solarski, M. Yamamoto (Eds.), Active Networks. VIII, 217 pages. 2007.

Vol. 3883: M. Cesana, L. Fratta (Eds.), Wireless Systems and Network Architectures in Next Generation Internet. IX, 281 pages. 2006.

Vol. 3868: K. Römer, H. Karl, F. Mattern (Eds.), Wireless Sensor Networks. XI, 342 pages. 2006.

Vol. 3854: I. Stavrakakis, M. Smirnov (Eds.), Autonomic Communication. XIII, 303 pages. 2006.

Vol. 3813: R. Molva, G. Tsudik, D. Westhoff (Eds.), Security and Privacy in Ad-hoc and Sensor Networks. VIII, 219 pages. 2005.

Vol. 3462: R. Boutaba, K.C. Almeroth, R. Puigjaner, S. Shen, J.P. Black (Eds.), NETWORKING 2005. XXX, 1483 pages. 2005.